Topics in
Current Physics

21

Topics in Current Physics Founded by Helmut K. V. Lotsch

1 **Beam-Foil Spectroscopy**
Editor: S. Bashkin

2 **Modern Three-Hadron Physics**
Editor: A. W. Thomas

3 **Dynamics of Solids and Liquids by Neutron Scattering**
Editors: S. W. Lovesey and T. Springer

4 **Electron Spectroscopy for Surface Analysis**
Editor: H. Ibach

5 **Structure and Collisions of Ions and Atoms**
Editor: I. A. Sellin

6 **Neutron Diffraction**
Editor: H. Dachs

7 **Monte Carlo Methods** in Statistical Physics
Editor: K. Binder

8 **Ocean Acoustics**
Editor: J. A. DeSanto

9 **Inverse Source Problems** in Optics
Editor: H. P. Baltes

10 **Synchrotron Radiation** Techniques and Applications
Editor: C. Kunz

11 **Raman Spectroscopy** of Gases and Liquids
Editor: A. Weber

12 **Positrons in Solids**
Editor: P. Hautojärvi

13 **Computer Processing of Electron Microscope Images**
Editor: P. W. Hawkes

14 **Excitons** Editor: K. Cho

15 Physics of **Superionic Conductors**
Editor: M. B. Salamon

16 **Aerosol Microphysics I** Particle Interactions
Editor: W. H. Marlow

17 **Solitons**
Editors: R. Bullough, P. Caudrey

18 **Magnetic Electron Lens Properties**
Editor: P. W. Hawkes

19 Theory of **Chemisorption**
Editor: J. R. Smith

20 **Inverse Scattering Problems** in Optics
Editor: H. P. Baltes

21 **Coherent Nonlinear Optics** Recent Advances
Editors: M. S. Feld and V. S. Letokhov

22 **Electromagnetic Theory of Gratings**
Editor: R. Petit

23 **Structural Phase Transitions I**
Editors: K. A. Müller and H. Thomas

24 **Amorphous Solids** Low-Temperature Properties
Editor: W. A. Phillips

Coherent Nonlinear Optics

Recent Advances

Edited by M. S. Feld and V. S. Letokhov

With Contributions by
F. Biraben B. Cagnac C. D. Cantrell
V. P. Chebotayev M. S. Feld G. Grynberg
W. Kaiser A. L. Laubereau V. S. Letokhov
M. D. Levenson J. C. MacGillivray A. A. Makarov
J. J. Song

With 134 Figures

Springer-Verlag Berlin Heidelberg New York 1980

Professor Michael S. Feld, Phd

Department of Physics and Spectroscopy Laboratory, Massachusetts Institute of Technology, Cambridge, Massachusetts 02139, USA

Professor Vladilen S. Letokhov

Institute of Spectroscopy, Academy of Sciences USSR, Podol'skii Rayon Troitzk, Moscow 142092, USSR

ISBN-13: 978-3-642-81497-6 e-ISBN-13: 978-3-642-81495-2
DOI: 10.1007/ 978-3-642-81495-2

Library of Congress Cataloging in Publication Data. Main entry under title: Coherent nonlinear optics. (Topics in current physics ; v. 21). Includes bibliographies and index. 1. Nonlinear optics. 2. Coherence (Optics). 3. Multiphoton processes. 4. Laser spectroscopy. I. Feld, Michael S., 1940–. II. Letokhov, V. S. III. Biraben, F. IV. Series. QC446.2.C63 535'.2 80-36704

Softcover reprint of the hardcover 1st edition 1980

2153/3130-543210

Rem Viktorovich Khokhlov

Sergio Pereira Da Silva Porto

Dedication

In Memory of

Rem Viktorovich Khokhlov and **Sergio Pereira Da Silva Porto**

The international scientific community has recently lost two distinguished scholars, Rem V. Khokhlov, Rector of the Moscow State University, USSR, and Sergio P.S. Porto, Vice President of the University of Campinas, Brazil. Both were pioneers in quantum electronics, Professor Khokhlov in the field of nonlinear optics and Professor Porto in the field of laser Raman spectroscopy. In addition to their outstanding scientific achievements, both were prominent organizers and participants in activities of the international science community. The loss of these men to us is particularly great because their work combined scientific excellence with the persuit of international cooperation and understanding. The contributions made by each of us to mutual scientific and humanistic endeavors will be the best memorial to our departed colleagues.

Cambridge, Massachusetts, USA
Troitzk, Moscow, USSR

Professor M.S. Feld
Professor V.S. Letokhov

Contents

List of Contributors

Biraben, François

Laboratoire de Spectroscopie Hertzienne de l'ENS, 4, place Jussieu, F-75230 Paris, Cedex 05, France

Cagnac, Bernard

Laboratoire de Spectroscopie Hertzienne de l'ENS, 4, place Jussieu, F-75230 Paris, Cedex 05, France

Cantrell, Cyrus D.

Center for Quantum Electronics and Applications, Department of Physics, The University of Texas at Dallas, Richardson, TX 75080, USA

Chebotayev, Veniamin P.

Institute of Thermophysics, Siberian Branch of the USSR, Academy of Sciences, Novosibirsk-90, USSR

Feld, Michael S.

Department of Physics and Spectroscopy Laboratory, Massachusetts Institute of Technology, Cambridge, MA 02139, USA

Grynberg, Gilbert

Laboratoire d'Optique Quantique, Ecole Polytechnique, F-91128 Palaiseau, France

Kaiser, Wolfgang

Physik Department der Technischen Universität München, D-8000 München, Fed. Rep. of Germany

Laubereau, Alfred, L.

Physikalisches Institut, Universität Bayreuth, D-8580 Bayreuth, Fed. Rep. of Germany

Letokhov, Vladilen S.

Institute of Spectroscopy, Academy of Sciences USSR, Podol'skii Rayon Troitzk, Moscow 142092, USSR

Levenson, Mark D.

IBM Research Laboratory, 5600 Cottle Road, San Jose, CA 95193, USA

MacGillivray, Jeffrey C.

Optical Systems Division, Itek Corporation, 10 Maguire Road, Lexington, MA 02173, USA

Makarov, Alexander A.

Institute of Spectroscopy, Academy of Sciences, USSR, Podol'skii Rayon Troitzk, Moscow 142092, USSR

Song, J.J.
Physics Department, University of Southern California, Los Angeles, CA 90007, USA

1. Coherent Nonlinear Optics

M. S. Feld and V. S. Letokhov

The gold rush phenomenon — an intense period of rapid discovery and exploitation — is a well-known phase experienced by many fields of the natural sciences at certain points in their evolution. Optical physics, stimulated by major advances in laser research and technology, is currently in the midst of such a period. Everywhere new veins of gold are being uncovered and mined by thousands of prospectors, most of whom have crossed over into this exciting research area from a diversity of other disciplines. Their principal tool is the laser and its associated analytical, spectroscopic and dynamical techniques. The purpose of this volume is to make available to our co-workers in optical physics in-depth reports on the current status of a set of important topics in this rapidly changing field.

The laws of optical physics were formulted 50 to 100 years ago. They remain true for laser light and, in fact, form the basis of operation for the laser itself. However, *coherence* and *nonlinearity* are relatively new concepts which have become central to describing the interactions of laser light with matter. These ideas are fundamental to the many new techniques becoming available for studying atoms and molecules. Coherent nonlinear optics is the theme of this book, and the editors have brought together a selected set of specifically prepared reviews of current topics of active interest. Three major areas — coherent resonance effects, multiphoton resonant processes and coherent Raman processes — are covered.

1.1 Introductory Comments

The coherent nature of the interaction of laser light with matter manifests itself in different ways, with interesting applications. About twenty-five years ago DICKE [1.1] pointed out that the spontaneous emission from an ensemble of excited quantum systems can occur at a greatly accelerated rate, via a mechanism he termed *superradiant emission*. In this process the particles respond cooperatively because of their mutual interaction with the common radiation field, giving rise to coherent radiation. DICKE's original discussion considered both optical and microwave emission, but it emphasized the latter regime, where the sample is small compared to the wavelength λ of the emitted radiation. Four years later the laser was proposed [1.2] as a means of extending the maser principle of generating coherent microwave emission into the optical spectral range. The production of coherent radiation by lasers and masers is based on the principle of *stimulated emission* from excited particles into a small number of modes of an optical resonator (one mode in the ideal case), a process which is distinct from superradiant emission. The laser principle is, of course, now widely used to generate coherent light. Nevertheless, the production of coherent optical radiation via cooperative spontaneous emission from an ensemble of excited particles outside the resonator is of great physical interest. However, the theoretical results developed for long wavelength superradiance are not directly applicable to the optical regime, since the sample volume is much larger than λ^3. This was recognized in the first demonstration of superradiant emission [1.3], and much recent attention, both experimental and theoretical, has been given to the exploration of this interesting effect. The present state of our understanding of superradiance is reviewed in Chap.2 by Feld and MacGillivray.

High-resolution laser spectroscopy is based on the ability of laser radiation to induced nonlinear behavior and phase coherence in atomic and molecular systems. These principles have led to a set of new methods for producing extremely narrow "Doppler-free" spectral resonances in two-level and multilevel systems. The concept of saturating an atomic transition by means of monochromatic laser radiation was first worked out by LAMB [1.4] in his 1974 analysis of a gas laser, and narrow laser saturation resonances were observed shortly thereafter [1.5]. The extension of these ideas to three-level systems soon followed [1.6]. These initial developments have led to a series of techniques which are now standard tools for precision studies of atoms and molecules [1.7,8]. One of the most important recent developments in this area is the extension of the *method of separated fields* to the optical regime [1.7,8]. The separated fields technique is based on the fact that when a beam of atoms traverses a region of electromagnetic field, phase memory of the resonant interaction is preserved. Constructive interference can then occur as the atoms pass through a second field. This property is fundamental for producing very narrow absorption resonances in separated microwave fields [1.9]. Unfortunately, it cannot be directly applied in the optical range because the separation between the fields is large compared to λ. Indeed, because of the divergence of the atomic beam, the atoms traversing the first light beam at a given point, and hence acquiring an optical polarization of a given phase, intersect the second light beam at a range of points, thus giving rise to destructive interference. CHEBOTAYEV and co-workers [1.10] showed that this obstacle can be overcome by using two-photon transitions or, in the case of single-photon transitions, by using three optical fields. These techniques, which are closely connected with the photon echo effect [1.11], are now among the most powerful and elegant tools in the field of high-resolution laser spectroscopy. Coherence in high-resolution laser spectroscopy is reviewed in Chap. 3 by Chebotayev.

Multiphoton processes are one of the main sources of nonlinearity in the interaction of intense laser fields with atoms and molecules. Resonant multiphoton processes are of special interest: First, multiphoton transition probabilities are enhanced under resonance conditions, and can be observed in fields of moderate intensity. Furthermore, such processes find various applications in laser spectroscopy. Two-quantum resonant transitions in a standing wave field is an important method for eliminating Doppler broadening [1.12]. One of the techniques of high-resolution laser spectroscopy is based on this approach [1,7,8]. Resonant multi-step processes permit selective photoionization of atoms. This approach is fundamental for laser methods of single-atom detection [1.13], which have recently been demonstrated experimentally [1.14]. The field of multiphoton resonant processes in atoms is reviewed in Chap.4 by Biraben, Cagnac, and Grynberg.

Over the past few years impressive progress has been made in studying multi-photon vibrational transitions in polyatomic molecules induced by intense infrared e.m. fields. The first successful demonstration of isotopically selective multi-photon excitation and dissociation by intense CO_2 laser pulses [1.15], in BCl_3, followed by similar experiments in other molecules, has triggered a torrent of experimental and theoretical activity. This wide research area, now called multi-photon (or multiple photon) infrared photochemistry, has already become the subject of special reviews [1.16,17]. Among these works are numerous papers devoted to the coherent interaction of a multilevel quantum system, the levels of which are almost equidistant, with powerful quasi-resonant radiation. Although models of this type are only simple approximations to a real polyatomic molecule, they do provide a physical basis for describing a variety of the features of such multi-quantum processes, and a qualitative picture for their interpretation. The major results of this field to date are summarized in Chap.5 by Cantrell, Letokhov, and Makarov.

In a condensed medium the relaxation time of the phase memory (i.e., phase coherence) induced by an intense optical field is extremely short, in the picosecond range. Nevertheless, progress in generating intense ultrashort laser pulses [1.18] has made possible systematic studies of coherent interactions between picosecond laser pulses and molecular vibrations. The method first used for this purpose [1.19] has proved to be the most productive. In this approach the sample is simultaneously irradiated by two coherent collimated ultrashort light pulses, their frequency difference being exactly equal to the molecular vibrational frequency. This induces Raman-type excitation of the N molecules contained in a coherent interaction volume. An ultrashort pulse of variable delay then probes the state of the system as it decays. Both the intensity and direction of the scattered probe pulse can be studied. Because of the coherent nature of the interaction between the excitation and probe pulses, the interaction efficiency for short delay ($\lesssim$ phase memory time) is proportional to N^2, and depends on the relative orientation of the wave vectors of the exciting and probe fields. As the molecular vibrations dephase, however, the interaction becomes incoherent, leading to isotropic efficiency proportional only to N. These features make it possible to separate coherent and incoherent processes occurring on a picosecond time scale. Furthermore, since the details of the dephasing process depend on the extent of inhomogeneous broadening of the vibrational transition and its internal structure, picosecond pulse techniques can be used to study these features. Recent progress in the field of coherent picosecond interactions is reviewed in Chap.6 by Laubereau and Kaiser.

The interaction of an ensemble of molecules with two laser fields offset in frequency can give rise to two-quantum Raman transitions. This nonlinear process had led to an important spectroscopic technique, coherent Raman spectroscopy. Although a comprehensive review of this field appeared in 1977 [1.20], many important advances in the basic method, coherent antistokes Raman spectroscopy (CARS), have since occurred. These new developments are presented in Chap.7 by Levenson and Song. In contrast to [1.20], Chap.7 describes the steady-state aspects of the Raman scattering process, rather than the transient ones. In this regime the pulse duration of both exciting and probe fields is long compared to the phase memory relaxation time, T_2, (or even cw), hence excitation and interrogation processes occur simultaneously. Thus, coherence must be maintained at all times by the applied fields. Taken together, Chaps.6 and 7 form a comprehensive review of coherent Raman processes under transient and steady-state conditions.

In summary, the contributions to this volume cover recent advances in three major areas of coherent nonlinear optics, coherent resonance effects (Chaps.2 and 3), multiphoton resonant processes (Chaps.4 and 5), and coherent Raman processes (Chaps.6 and 7). It is hoped that these reviews will serve as useful summaries of recent developments, and perhaps stimulate new advances in the field.

References

1.1 R.H. Dicke: Phys. Rev. *93*, 99 (1954)

1.2 A.L. Schawlow, C.H. Townes: Phys. Rev. *112*, 1940 (1958)

1.3 N. Skribanowitz, I.P. Herman, J.C. MacGillivray, M.S. Feld: Phys. Rev. Lett. *30*, 309 (1973)

1.4 W.E. Lamb, Jr.: Phys. Rev. *134*, A1429 (1964)

1.5 A. Szöke, A. Javan: Phys. Rev. Lett. *10*, 521 (1963)
R.A. MacFarlane, W.R. Bennett, Jr., W.E. Lamb, Jr.: Appl. Phys. Lett. *2*, 189 (1963)

1.6 A. Javan: In *Fundamental and Applied Laser Physics*, ed. by M.S. Feld, A. Javan, N.A. Kurnit (Wiley-Interscience, New York 1973) p.295
See also M.S. Feld, J.H. Parks, H.R. Schlossberg, A. Javan: In *Physics of Quantum Electronics*, ed. by P.L. Kelley, B. Lax, P.E. Tannenwald (McGraw-Hill, New York 1966) p.567

1.7 K. Shimoda (ed.): *High-Resolution Laser Spectroscopy* (Springer, Berlin, Heidelberg, New York 1976)

1.8 V.S. Letokhov, V.P. Chebotayev: *Nonlinear Laser Spectroscopy* (Springer, Berlin, Heidelberg, New York 1977)

1.9 N.F. Ramsey: *Molecular Beams* (Clarendon Press, Oxford 1956)
1.10 E.V. Baklanov, B.Yu. Dubetzkii, V.P. Chebotayev: Appl. Phys. *9*, 171 (1976)
E.V. Baklanov, V.P. Chebotayev, B.Yu. Dubetzkii: Appl. Phys. *11*, 201 (1976)
1.11 N.A. Kurnit, I.D. Abella, S.R. Hartmann: Phys. Rev. Lett. *13*, 367 (1964)
1.12 L.S. Vasilenko, V.P. Chebotayev, A.V. Schishaev: ZhETF Pis'ma Red. *12*, 161 (1970) [JETP Lett. *12*, 113 (1970)]
1.13 V.S. Letokhov: In *Tunable Lasers and Applications*, ed. by A. Mooradian, T. Jaeger, P. Stokseth (Springer, Berlin, Heidelberg, New York 1976) p.122
1.14 G.S. Hurst, M.H. Nayfeh, J.P. Young: Appl. Phys. Lett. *30*, 229 (1977)
G.I. Bekov, V.S. Letokhov, V.I. Mishin: ZhETF Pis'ma Red. *27*, 52 (1978)
1.15 R.V. Ambartzumian, V.S. Letokhov, E.A. Ryabov, N.V. Chekalin: ZhETF Pis'ma Red. *20*, 597 (1974) [JETP Lett. *20*, 273 (1974)]
1.16 R.V. Ambartzumian, V.S. Letokhov: In *Chemical and Biochemical Applications of Lasers*, Vol.3, ed. by C.B. Moore (Academic Press, New York 1977) p.166
1.17 C.D. Cantrell, S.M. Freund, J.L. Lyman: In *Laser Handbook*, Vol.3 (North-Holland, Amsterdam 1979)
1.18 S.L. Shapiro (ed.): *Ultrashort Laser Pulses* (Springer, Berlin, Heidelberg, New York 1977)
1.19 D. Von der Linde, A. Laubereau, W. Kaiser: Phys. Rev. Lett. *26*, 954 (1971)
R.R. Alfano, S.L. Shapiro: Phys. Rev. Lett. *26*, 1247 (1971)
1.20 S.A. Akhmanov, N.I. Koroteev: Adv. Phys. Sci. (Sov.) *123*, 405 (1977)

2. Superradiance *

M. S. Feld and J. C. MacGillivray

With 14 Figures

Superradiance, the idea that the spontaneous emission rate of an assembly of atoms (or molecules) can be much greater than that of the same number of isolated atoms, has been the subject of much interest since it was orginally proposed by DICKE [2.1] in 1954. In the process of superradiant emission the atoms are coupled together by their common radiation field, and so decay cooperatively. The intensity emitted by N atoms in this case is proportional to N^2 instead of N, as in ordinary emission. Thus, superradiance is a fundamental effect.

2.1 Background Material

The reason why the cooperative emission rate differs from the single-atom emission rate can be understood by considering the radiative decay of two-level atoms. The emission rate of an isolated atom is $1/T_{sp}$, where T_{sp} is the single-atom radiative lifetime. However, if this excited atom is brought close to a second excited atom it radiates in a much different manner. A system of two two-level atoms separated by a distance much smaller than the wavelength λ of the emitted radiation must be properly symmetrized. Thus, the emission process takes the system from the fully excited state ($\uparrow\uparrow$) to the triplet intermediate state $(\uparrow\downarrow + \downarrow\uparrow)/\sqrt{2}$, and then to the ground state ($\downarrow\downarrow$). The radiation rate for each of these steps is $2/T_{sp}$, twice that of isolated atoms. The emitted intensity is thus $I = 2\hbar\omega/(T_{sp}/2) = 4\hbar\omega/T_{sp}$, 4 times as great as in the single-atom case.

Similarly, in considering the radiation emitted by N closely spaced atoms the total wave function must be symmetrized. The state associated with the maximum emission rate, analogous to the triplet intermediate two-atom state above, is the totally symmetric state which has zero population difference. The radiation rate of such a state is N/4 times greater than that of N independent atoms.

Symmetrizing the wave function is equivalent to stating that the atoms interact with a common radiation field, and thus emit cooperatively rather than independently. The maximum emission state then corresponds to the arrangement of atoms in which the net dipole moment is maximized.

*Work supported in part by the U.S. National Science Foundation.

In the optical region the sample is usually much larger than the wavelength λ. DICKE showed that such an "extended" sample of two-level atoms can also radiate collectively [2.1,2]. Maximum emission from such a system occurs when a phased array of dipoles has developed. The ensuing radiation has all the characteristics expected from a classical dipole array. The emitted intensity is large only over a small solid angle along which the dipoles are phased. In the case of a disc-shaped sample[1] this solid angle $\sim \lambda^2/A$, where A is the cross-sectional area of the sample. Thus, the emitted intensity is enhanced by a factor $\sim N\lambda^2/A$ compared to incoherent spontaneous emission.

Superradiance has been observed in the far infrared (in HF [2.3-6] and CH_3F [2.7-9]); in the near infrared (in Na [2.10,11], Tl [2.11,12], Cs [2.11,13-18], Li [2.19], and Rb [2.20,21]); and recently in the visible (in Sr and Eu [2.21]) and submillimeter (in Cs and Na [2.22]). In all of the experiments a long (compared to λ), optically thick sample of two-level atoms is prepared in the excited state, inverted indirectly so as to create a complete inversion between the two levels of interest. Feedback is absent, and there are no mirrors. After a sizeable delay during which the system evolves into a superradiant state, it emits a burst of coherent radiation (Fig.2.1), a sequence of events sometimes referred to as "superfluorescence" [2.23]. In contrast to ordinary spontaneous emission, the radiation is emitted as a highly directional pulse of peak output power $\propto N^2$, often accompanied by ringing. The radiation rate is orders of magnitude faster than that of ordinary spontaneous emission.

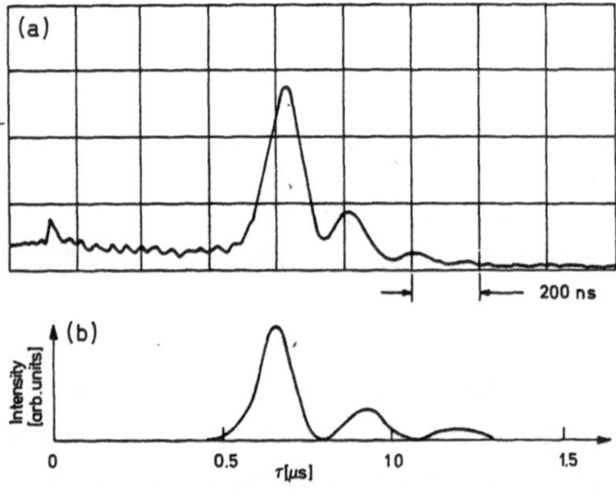

Fig.2.1a,b. Oscilloscope trace of a superradiant pulse. The data shown is for the $\lambda = 84$ µm transition in the first excited vibrational state of HF. The small peak on the scope trace at $T = 0$ is the 2.5 µm pump pulse, highly attenuated. (b) Theoretical fit to (a). The parameters are $I = 1$ kW/cm^2, $p = 1.3$ mTorr and $\kappa L = 2.5$ for $L = 100$ cm. The corresponding values of T_2^* and T_R are 330 and 6.1 ns, respectively. Note that $\alpha L = T_2^*/T_R = 54$. [2.3]

This radiation process can be viewed as the collective radiative damping of a completely inverted medium. Although the initial decay of such a system is intrinsically quantum mechanical, its subsequent evolution to a superradiant state, i.e., a state of maximum polarization, and all the features of the ensuing radiation burst can be described semiclassically (quantized atoms, classical fields).

[1]A disc is a cylinder of length L with $A \gg \lambda L$ [2.39].

Since virtually all of the stored energy is released very rapidly, superradiant emission of this type is the optimal process for extracting coherent energy from an inverted system [2.24].

Superradiance differs greatly from other cooperative emission effects such as free-induction decay [2.25,26] and echoes [2.27], in which only a small fraction of the stored energy is emitted cooperatively. In those phenomena the collective behavior of the sample does not significantly perturb its state, and so the decay is determined primarily by incoherent processes. Thus, these effects can be termed "limited superradiance" [2.24].

2.2 Physical Principles

Consider an extended, optically thick sample of two-level atoms interacting with a common radiation field. In the Bloch formalism [2.28] the atoms comprising this system correspond to a spatial distribution of Bloch vectors which are coupled together via the field. Due to propagation effects, however, the individual Bloch vectors of the extended medium may evolve differently.

An initially inverted system of this type is analogous to an array of rigid pendula balanced exactly on end. Just as the pendula are unstable to small fluctuations, so the excited atomic system is unstable to a small perturbing field, initiated by spontaneous emission from one of the excited atoms (more important for $\lambda < 50$ μm) or by background thermal radiation ($\lambda > 50$ μm). This weak propagating electric field induces a small macroscopic polarization in the medium, which acts as a source to create additional electric field in the medium which, in turn, produces more polarization. This regenerative process gives rise to a growing electric field and an increasing polarization throughout the medium (Fig.2.2a). In this way a superradiant state slowly evolves over a sizable portion of the sample, at which time radiation is emitted at a greatly enhanced rate. This process leads to a rapid deexcitation of that region of the medium, after which essentially all of the population is in the lower level. Deexcited regions can then be reexcited by radiation from other parts of the sample, which gives rise to the "ringing" often observed in the output radiation.

Figure 2.2 plots the polarization envelope P and the population inversion density n, respectively, throughout the sample at several instants of time. These values have been calculated from the semiclassical model of MacGillivray and Feld (Sect.2.3.2), which is based on the intuitive picture presented above. Notice that P and n vary slowly in space and time throughout the medium, giving rise to several regions of locally uniform polarization. These spatial variations in P and n are due to propagation effects in a high gain medium, a crucial point which was not appreciated in some earlier work. The ringing in the output radiation (Fig.2.1) is a direct consequence of these spatial variations.

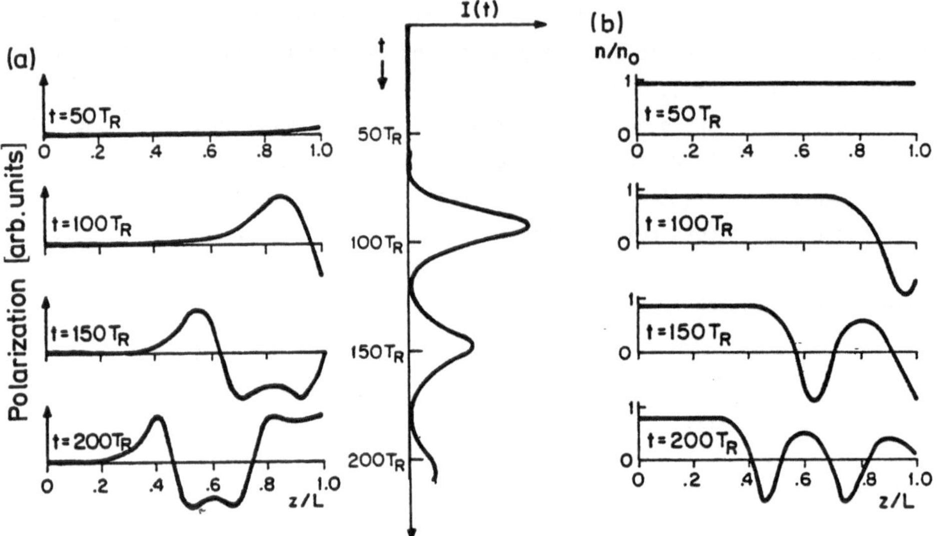

Fig.2.2a,b. Sketch of (a) the polarization envelope P and (b) the population inversion density n in the medium as functions of x at $T = 50T_R$, $100T_R$, $150T_R$, and $200T_R$. The corresponding output intensity pattern is shown at the right of (a). As explained in [2.24], double peaks in $P(x)$ occur whenever the ringing is sufficiently large. [2.24]

The time evolution of the radiation emitted by an initially inverted extended sample of two-level atoms depends on many factors, including broadening, diffraction loss, and level degeneracy. However, in a high gain system the major features of the output radiation are determined by two parameters: I) T_R, the characteristic radiation damping time of the collective system, and II) θ_0, a measure of the conditions which initiate the superradiant process.

I) The quantity T_R is defined as [2.29]

$$T_R \equiv T_{sp}(8\pi/n_0\lambda^2 L) \quad , \tag{2.1}$$

where $n_0 = N/AL$ is the initial inversion density, T_{sp} is the lifetime of an isolated atom with dipole moment μ,

$$\frac{1}{T_{sp}} = \frac{4}{3}\frac{\mu^2\omega^3}{\hbar c^3} \quad , \tag{2.2}$$

and L is the length of the cylindrical sample. The quantity T_R may be interpreted in several ways. First, the radiated power, I, of N cooperatively emitting atoms is proportional to N^2. However, as mentioned above, the atoms in a disc-shaped array can only radiate coherently into a solid angle $\lambda^2/4\pi A$. Hence,

---

$\hbar = h/2\pi$ (normalized Planck's constant)

$$I = N^2 \frac{\hbar\omega}{T_{sp}} \frac{\lambda^2}{4\pi A} \simeq N \frac{\hbar\omega}{T_R} \quad . \tag{2.3}$$

Thus, when the sample radiates as a collective system, the radiative enhancement factor changes in lifetime from T_{sp} to $\sim T_R$. This justifies the interpretation of T_R as the characteristic collective radiation damping time.

Essentially the same result may be obtained by considering the radiation emitted by a linearly polarized array of atomic dipoles whose macroscopic polarization is in the form of a plane wave

$$\underline{P} = n\mu\hat{z}\, e^{i(\omega t - kx)} \quad . \tag{2.4}$$

This polarization gives rise to an electric field $\underline{E} = E_0\hat{z}\, e^{[i(\omega t-kx)]}$ which can be computed from Maxwell's equations, hence a radiated power $I = (cE_0^2/8\pi)A$. One obtains

$$I = \frac{\pi A}{2c} (n\mu\omega L)^2 \quad , \tag{2.5}$$

which, using (2.2), is the same as (2.3) except for a numerical factor of order unity.

In order for the collective mode to dominate it is necessary that $T_R \ll T_{sp}$. This requires that $n\lambda^2 L \gg 1$, i.e., there must be a large number of molecules in a "diffraction volume" $\lambda^2 L = (\lambda^2/A)AL$. This requirement also insures that there are many molecules in a cylinder of cross section λ^2 and length L, so that a coherent wavefront can be properly reconstructed as the wave propagates down the sample.

II) The quantity θ_0 is the effective initial tipping angle of the array of Bloch vectors (or pendula) described previously. It is a parameter which characterizes the effectiveness of spontaneous emission and background thermal radiation in initiating the buildup of superradiance in a medium which is initially totally inverted. Expressions for observed quantities (Sect.2.4) depend logarithmically on Θ_0, through the parameter

$$\Phi = \ln(2\pi/\Theta_0) \quad . \tag{2.6}$$

Recent semiclassical [2.24] and quantum mechanical [2.30-34] derivations of Θ_0 find that

$$\Theta_0 = N^{-1/2} \text{ (slowly varying function)} \quad , \tag{2.7}$$

hence

$$\Phi \simeq \frac{1}{2} \ln N \quad . \tag{2.8}$$

Since $N \gg 1$ it is difficult to experimentally confirm the exact theoretical form of Θ_0. The current experimental evidence is described in Sect.2.6.

The major features of the superradiant emission process are completely characterized by T_R and θ_0 provided that the small-signal field gain αL is sufficiently large. Here

$$\alpha = 2\pi\mu_z^2\omega n_0 T_2'/\hbar c \tag{2.9}$$

is the usual expression for the gain coefficient, with T_2' the inverse linewidth of the transition and $\mu_z^2 = \mu^2/3$ the average z component of the dipole-moment squared, averaged over M levels. Although the growth of the electric field in the sample does not follow simple exponential gain (e.g., Beer's Law), the time integral of the electric field at any point in the sample obeys an exponential law, the area theorem of McCALL and HAHN [2.35]. This gives rise to the high gain condition for strong superradiance (as opposed to limited superradiance [2.36])

$$\alpha L \gtrsim \Phi \quad . \tag{2.10}$$

It should be noted that αL, T_2' and T_R are related by the equation [2.37]

$$\alpha L = T_2'/T_R \quad . \tag{2.11}$$

Thus, the high gain requirement insures that coherent decay processes dominate over incoherent processes ($T_R \ll T_2'$), as is necessary for superradiance to occur.

For sufficiently large gain a normalized curve [2.24] can be drawn for any given value of Φ, which gives the output intensity $I(T)$ multiplied by T_R^2 as a function of time in units of T_R (T is the retarded time $T = t - x/c$ at the end of the sample, $x = L$). This curve (Fig.2.3) depicts a burst of radiation with ringing, preceded by a long delay. The scaling of the curve is such that when T_R is halved, the system radiates twice as fast and the peak intensity is quadrupled. This curve exhibits the N^2 intensity dependence which is characteristic of superradiant emission, since the peak intensity is proportional to T_R^{-2} and T_R is proportional to N^{-1}. Experimental data exhibiting this scaling are given in Fig.2.4.

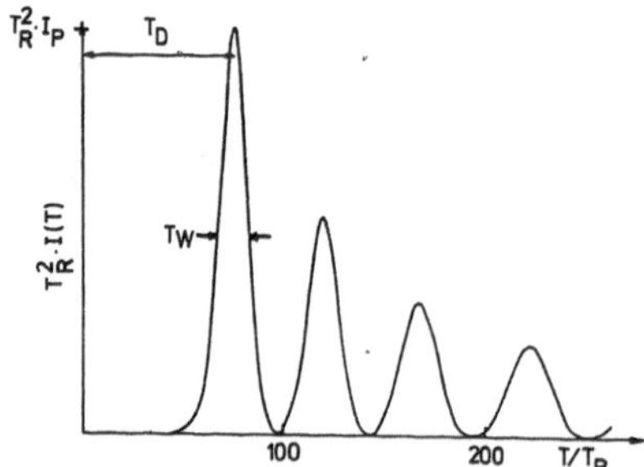

Fig.2.3. Normalized output curve. The time scales as T_R and the intensity scales as T_R^{-2}. Note that the shape of the normalized output curve depends on Φ. I_P, T_D, and T_W can all be expressed in terms of T_R and Φ (Sect.2.4.1). [2.5]

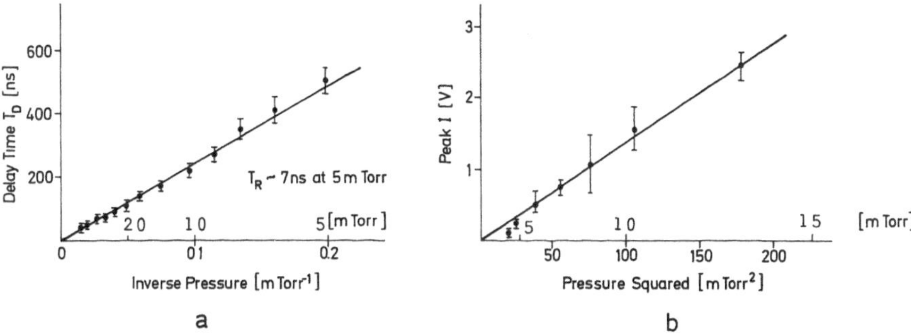

Fig.2.4. (a) Time delay from pump pulse to intensity peak and (b) peak intensity of superradiant output as functions of inverse pressure and pressure squared, respectively, in sample cell. The data shown are for the same HF transition as in Fig.2.1.[2.5]

2.3 Theoretical Treatments

Since DICKE's original proposal many theoretical descriptions of superradiance have been presented. Most of these treatments quantize the electromagnetic field. A large group of these studies considers cases in which the emitting radiators are confined to a point sample, are permitted to couple with only one radiation mode, or are very few in number (e.g., 2 to 8) [2.38]. Although these conditions permit important mathematical simplifications which allow "exact" solutions for the time evolution of the system, the results are not directly applicable to the problem of superradiance in an extended medium. A description of many of these treatments can be found in the introduction to [2.39a] and in a review article by STENHOLM [2.40].

Another set of quantized field theories [2.23,39,41] treats an extended sample in which all modes of the radiation field interact with the atomic medium, but fails to include spatial variation in the field amplitudes. As discussed in Sect. 2.2, this assumption introduces difficulties, since superradiance takes place only in optically thick samples, where spatial variations do occur. Critical discussion of some of these treatments is given in [2.24].

Recently POLDER and co-workers [2.30,31], and GLAUBER and co-workers [2.32-34] have developed quantized field treatments which analyze the initial stage of the superradiant emission process. These theories allow for spatial variations of the field amplitudes, and thus fully include propagation effects. They show that beginning at the early phases of the evolution process, the e.m. field behaves classically and the system can be described by the coupled Maxwell-Schrödinger equations (quantized atoms, classical field). The initial fluctuations which trigger the superradiant pulse can be treated stochastically, using either a fluctuating field source or a polarization source, or both.

These results provide an important link to semiclassical treatments of superradiance and confirm the validity of a semiclassical description of the pulse initiation process. Important contributions toward developing the semiclassical description have been made by ARECCHI and BONIFACIO [2.42], HOPF and SCULLY [2.43], and ICSEVGI and LAMB [2.44], whose pulse propagation studies include methods for solving the coupled Maxwell-Schrödinger equations of the semiclassical theory in several limits. FRIEDBERG and HARTMANN [2.37] have shown that high gain is necessary for superradiance by pointing out the interdependence of αL, T_2' and T_R (2.11). ARECCHI and COURTENS [2.45] have described a maximum cooperation length which can limit the number of atoms that contribute to a single superradiant emission pulse. For certain inversion conditions (described in Sect.2.4.3) a sample of excessive length for a given density would divide into regions which superradiate independently.

BURNHAM and CHIAO [2.46] have treated pulse propagation under conditions very similar to the ideal limit of Sect.2.4.1. They pointed out the significance of T_R, and showed that the delay time and the amount of ringing increase as the initial state of the system becomes more nearly inverted (i.e., as θ_0 becomes smaller).

MACGILLIVRAY and FELD [2.24] have used the coupled Maxwell-Schrödinger equations to develop a theory of superradiance for realistic systems having decay, Doppler dephasing, level degeneracy and intensity loss. In their model the spontaneous emission which initiates the pulse evolution process is simulated by a distributed polarization source, an approach subsequently validated from quantized field considerations by GLAUBER and HAAKE [2.32]. This formalism is used to derive simple expressions for observable quantities [2.24], and to explore the factors which limit ideal superradiant behavior and the resulting modifications which can occur in the emitted radiation [2.36,47,48].

2.3.1 Initiation of Superradiance: Quantized Field Treatment

As mentioned above, POLDER and co-workers [2.30,31] and GLAUBER and co-workers [2.32-34] have independently developed theories which describe the quantum fluctuations that initiate the evolution of a superradiant pulse in an initially inverted, high gain sample. These treatments take advantage of the simplification pointed out in [2.24], that in the initial phases of the emission process the inversion density is essentially constant, and so the equations of motion can be linearized. To further simplify the analyses, level degeneracy, Doppler dephasing, loss, and decay (other than spontaneous emission) are all neglected. (The conditions for the validity of these assumptions are discussed in Sect.2.4.) The results show that such a system behaves semiclassically from the early phases of the pulse evolution process.

This section outlines these two treatments and summarizes their major conclusions. In the two approaches, which are quite similar, the atom-field system is described by Heisenberg atomic and electric field operators, from which Heisenberg equations of motion are derived. The atomic operators at a given point $\underline{R}$ in the medium are collective operators, composed of the individual atomic operators in a small region of space around $\underline{R}$. The operators are assumed to vary continuously along the length of the sample ("smooth fields" approximation). Furthermore, interest is restricted to samples having Fresnel number $A/\lambda L \sim 1$, so that transverse field variations can be neglected (plane wave approximation). It will be convenient (although it is not necessary) [2.30,34], to restrict attention to fields propagating in the +x direction, thus neglecting the interaction between oppositely travelling waves. The validity of the plane wave approximation and of the neglect of forward-backward wave interaction is discussed in Sect.2.4.4.

Making the rotating-wave approximation and assuming the envelopes of the field operators to be slowly varying, the following equations of motion are obtained [2.30,34]

$$\partial \underset{\sim}{E}/\partial x = 2\pi k \underset{\sim}{P} \quad , \tag{2.12a}$$

$$\partial \underset{\sim}{P}/\partial T = \mu_z^2 \underset{\sim}{n}\underset{\sim}{E}/\hbar \quad , \tag{2.12b}$$

$$\partial \underset{\sim}{n}/\partial T = -\frac{1}{2\hbar}(\underset{\sim}{E}\underset{\sim}{P}^+ + \underset{\sim}{E}^+\underset{\sim}{P}) \quad . \tag{2.12c}$$

Here $\underset{\sim}{P}(x,T)$, $\underset{\sim}{n}(x,T)$ and $\underset{\sim}{E}(x,T)$ are the slowly varying envelopes (denoted by $\sim$) of the Heisenberg polarization, inversion density, and electric field operators, respectively, at position x and retarded time $T = t - x/c$. They obey the commutation relations [2.34][2]

$$[\underset{\sim}{P}^+(x,T),\underset{\sim}{P}(x',T)] = \frac{4}{A}\mu_z^2\delta(x-x')\underset{\sim}{n}(x,T) \quad , \tag{2.13a}$$

$$[\underset{\sim}{n}(x,T),\underset{\sim}{P}(x',T)] = -\frac{2}{A}\delta(x-x')\underset{\sim}{P}(x,T) \quad , \tag{2.13b}$$

$$\frac{1}{8\pi}[\underset{\sim}{E}(x,T),\underset{\sim}{E}^+(x',T)] = \frac{\hbar\omega}{A}\delta(x-x') \quad . \tag{2.13c}$$

The Heisenberg operator equations of motion are formally identical to the coupled Maxwell-Schrödinger semiclassical equations in the slowly varying envelope approximation. This is not unexpected, since quantum mechanical equations of motion can be

---

[2] The set of commutation relations derived by Polder, Schuurmans and Vrehen includes an additional relationship, valid for early times (such that $n \approx n_0$),

$$\frac{cA}{8\pi}[\underset{\sim}{E}(x,T),\ \underset{\sim}{E}^+(x,T')] = \hbar\omega\delta(T-T') \quad . \tag{2.13d}$$

See [2.31] for details.

obtained from the corresponding classical equations by replacing the c-number dynamical variables by Heisenberg operators subject to the appropriate commutation relations.

For N large the commutators all become small. This is most easily seen by re-writing (2.13) in terms of normalized dimensionless variables: $\hat{n} = \underset{\sim}{n}/n_0$, $\hat{P} = \underset{\sim}{P}/\mu_z n_0$, $\hat{E} = \mu_z \underset{\sim}{E} T_R/\hbar$, $x \to x/L$, $T \to T/T_R$. This gives

$$[\hat{P}^+(x,T), \hat{P}(x',T)] = \frac{4}{N}\,\delta(x - x')\hat{n}(x,T) \quad , \tag{2.14a}$$

$$[\hat{n}(x,T), \hat{P}(x',T)] = -\frac{2}{N}\,\delta(x - x')\hat{P}(x,T) \quad , \tag{2.14b}$$

$$[\hat{E}(x,T), \hat{E}^+(x',T)] = \frac{4}{N}\frac{cT_R}{L}\,\delta(x - x') \quad . \tag{2.14c}$$

These commutators all approach zero for $N \gg 1$. Therefore, the dynamical variables n, P and E will tend to behave classically for large N. In other words, in the limit of large N the semiclassical Maxwell-Schrödinger equations correctly describe the time development of the system.

In the quantum state from which the pulse evolution process begins (at retarded time T = 0) the atoms are fully inverted and the e.m. field is a vacuum. In this state expectation values of field and polarization envelopes vanish, and the inversion density operator has eigenvalue n_0

$$\langle \underset{\sim}{E}(x,0) \rangle = 0 \quad , \tag{2.15a}$$

$$\langle \underset{\sim}{P}(x,0) \rangle = 0 \quad , \tag{2.15b}$$

$$\langle \underset{\sim}{n}(x,0) \rangle = n_0 \quad . \tag{2.15c}$$

The expectation values of the second-order correlation functions follow from (2.13). For the normally ordered pairs,

$$\langle \underset{\sim}{E}^+(x,0)\underset{\sim}{E}(x',0) \rangle = 0 \quad , \tag{2.16a}$$

$$\langle \underset{\sim}{P}^+(x,0)\underset{\sim}{P}(x',0) \rangle = \frac{4}{A}\,\mu_z^2\delta(x - x')n_0 \quad . \tag{2.16b}$$

For the anti-normally ordered pairs,

$$\langle \underset{\sim}{E}(x,0)\underset{\sim}{E}^+(x',0) \rangle = \frac{\hbar\omega}{A}\,\delta(x-x') \quad , \tag{2.17a}$$

$$\langle \underset{\sim}{P}(x,0)\underset{\sim}{P}^+(x',0) \rangle = 0 \quad . \tag{2.17b}$$

In addition, for large N even moments of higher order can be expressed as sums of products of the corresponding second-order moments in a way characteristic of Gaussian statistics. Thus, each pulse behaves as if it were initiated by a random classical fluctuating source which follows Gaussian statistics.

Either choice of ordering may, of course, be used. When normal ordering is chosen, as in the Glauber-Haake treatment, pulse evolution appears to be triggered by the initial uncertainty in the atomic polarization. In anti-normal ordering, as in the Polder, Schuurmans, Vrehen treatment, initiation of the emission process is ascribed to fluctuations in the vacuum field.

In the early stages of the pulse evolution process the inversion density remains essentially constant. Approximating

$$\underset{\sim}{n}(x,T) \simeq n_0 \underset{\sim}{I} \quad , \tag{2.18}$$

with $\underset{\sim}{I}$ the identity operator, the operator equations of motion become linear. Their solution is then formally identical to that of the linearized semiclassical Maxwell-Schrödinger equations. Furthermore, once the atomic polarization becomes large compared to its initial fluctuations, the behavior of the system becomes completely semiclassical. As discussed in Sect.2.4.3, this occurs after a few T_R, and long before nonlinear behavior due to decreasing inversion density sets in.

Based on these considerations, Polder, Schuurmans and Vrehen developed a stochastic variable description to model the pulse evolution process, considering the initiating fluctuations as arising from a classical fluctuating field source. Similarly, Glauber and Haake concluded that a valid quantum description of the emission process can be obtained by solving the semiclassical Maxwell-Schrödinger equations subject to a random classical initial polarization which follows Gaussian statistics. Numerical solutions are then used to predict the form of the output radiation and the shot-to-shot variations in pulse delay and intensity.

To summarize, the quantum treatments of [2.30-34] show that the superradiant emission process can be properly described by solving the coupled Maxwell-Schrödinger equations subject to a fluctuating source. This confirms the validity of the semiclassical treatment presented in the following sections. Quantum mechanical derivations of the explicit form of the fluctuating source also make it possible to derive formulas for the initial tipping angle and the statistical fluctuations in the intensity and delay time of superradiant output pulses. These results are discussed in Sect.2.4.2.

2.3.2 Semiclassical Theory

A detailed theory of superradiance in an extended sample has been developed by MACGILLIVRAY and FELD [2.24]. The semiclassical formalism was chosen since it implicitly includes propagation effects and is well suited to incorporating level degeneracy, decay, intensity loss and other factors which can be important in actual systems. In this model spontaneous emission is simulated by a randomly phased polarization source distributed throughout the medium, an approach which has been validated by the recent quantized field treatments discussed in the last section.

The coupled Maxwell-Schrödinger equations in the slowly varying envelope approximation, written in complex form, are

$$\partial E/\partial x = -\kappa E + 2\pi k \sum_{v,M} P , \tag{2.19a}$$

$$\partial P/\partial T = -\left(\frac{1}{T_2} - ikv\right)P + \frac{\left(\mu_z^2\right)_M}{\hbar} En + \Lambda_p , \tag{2.19b}$$

$$\partial n/\partial T = \Lambda - \frac{n}{T_1} - \frac{1}{\hbar}\,\mathrm{Re}\{EP^*\} . \tag{2.19c}$$

Here $E(x,T)$ and $P(x,T,v,M)$ are the complex, slowly varying envelopes of the electric field and the polarization density per velocity interval dv of a particular transition between degenerate M states, respectively, at position x and retarded time $T = t - x/c$; $n(x,T,v,M)$ is the inversion density; κE is a term which accounts for diffraction and other loss; T_1 and T_2 are the time constants for population decay and polarization decay, respectively; Λ is a source term describing the rate of production of n; $(\mu_z)_M$ is the dipole moment component parallel to the direction of polarization; and $\sum_{v,M}$ denotes a velocity integral and a sum over degenerate M states. The remaining notation is the same as in [2.24].

Spontaneous emission from the excited state is represented in this model by a polarization source term Λ_p in (2.19b), which describes the rate of production of P. To properly simulate the effects of spontaneous emission Λ_p should be randomly phased and distributed throughout the medium. For computer evaluation the medium is divided into small regions of volume $A\Delta x (\Delta x \ll L)$, and time into intervals $\Delta t \lesssim T_R$. A convenient form for Λ_p is then

$$\Lambda_p = \xi_0 \frac{(\mu_z)_M}{A} \sqrt{\frac{\Delta T}{T_{sp}} \frac{\Delta x}{L} (n_2 AL)} \sum_{j=1}^{j'} \delta(x - x_j)$$

$$\cdot \sum_{\ell=-\infty}^{\infty} \delta(T - T_\ell) \sum_{k=1}^{k'} \delta(v - v_k) \exp(i\Phi_{j\ell kM}) , \tag{2.20}$$

where $x_j = (j - 1/2)\Delta x$, $\Delta x = L/j'$, $T_\ell = \ell\Delta T$, v_k are k' discrete velocities, $\Phi_{j\ell kM}$ are independent random phases, and $n_2(x,T,v,M)$ is the population density of the upper level of the superradiant transition. The dimensionless constant ξ_0 is determined by the initial polarization fluctuations. Note that Λ_p is proportional to $(\mu_z)_M$ and depends on the total number of excited molecules $n_2 A\Delta x \Delta T$ associated with one point in the space-time grid; the squareroot dependence occurs because the radiation from independent spontaneous emission events adds incoherently.

The pulse evolution process can also be initiated by background thermal radiation, the influence of which becomes more important with increasing wavelength. It predominates at $\lambda > 50\ \mu m$, where $\hbar\omega/kT < 1$. In the computer model background thermal radiation is simulated by means of a randomly phased input electric field, the intensity of which depends on the bandwidth and solid angle of the input radiation which interacts with the system. The electric field amplitude $E(x=0,T)$ is given by

$$\frac{|E|^2}{8\pi} = \frac{\hbar\omega}{e^{\hbar\omega/kT}-1} \cdot \frac{2\pi(\omega/2\pi)^2}{c^3} \cdot \frac{\lambda^2}{4\pi A} \cdot \frac{\Delta\omega_{eff}}{2\pi} \quad , \tag{2.21a}$$

or

$$E(0,T) = \left(\frac{2}{cA}\,\frac{\hbar\omega_0\Delta\omega_{eff}}{e^{\hbar\omega/kT}-1}\right)^{\frac{1}{2}} \exp(i\varphi_{random}) \quad . \tag{2.21b}$$

The factors on the right-hand side of (2.21a) are, respectively, the background thermal energy per mode, the number of modes of one polarization per unit volume per frequency interval, the solid angle factor, and the effective bandwidth of the system.

With this value of $E(x=0,T)$ as an input boundary condition, and using (2.20) for Λ_p, (2.19) can be numerically integrated to obtain the output radiation intensity. Examples of these computer results are given in Figs.2.1b, 2 and 5. Expressions for ξ_0 and $\Delta\omega_{eff}$ are given in Sect.2.4.2.

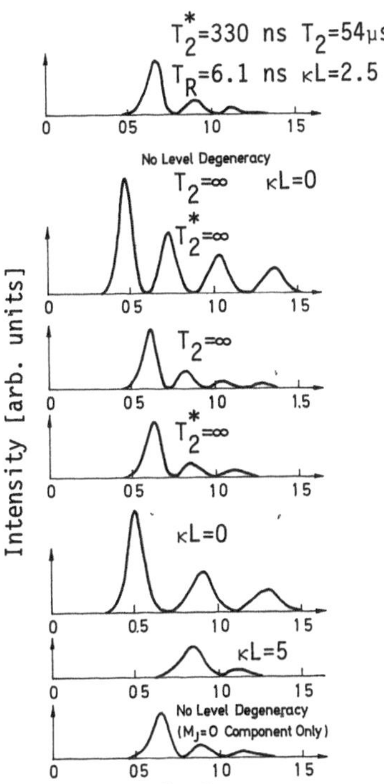

Fig.2.5. Computer results showing the influence of parameters on pulse evolution. The uppermost curve is the theoretical fit shown in Fig.2.1b. All parameters have the same values as in this curve except when stated otherwise. The values of the modified parameters are indicated in the figure. The same intensity scale is used throughout. [2.24]

Three basic Assumptions are incorporated in (2.19).

1) *The semiclassical model with a polarization source term to simulate spontaneous emission can be used instead of a quantized field model.* The justification of this Assumption from quantized field considerations has been discussed in Sect.2.3.1. The validity of the semiclassical approach for $T \gg T_R$ also follows from general grounds, since after an interval $\sim T_R$ after the sample is inverted there are many photons in the radiation field mode. To see why this is so consider a long sample of cross section A which is completely inverted at T = 0. The rate ρ at which photons are incoherently emitted into a (single) diffraction mode along the axis of the sample is

$$\rho = \frac{n_2 AL}{T_{sp}} \cdot \frac{\lambda^2}{4\pi A} \quad . \tag{2.22}$$

Hence, the first photon will be emitted in a time $\rho^{-1} \simeq T_R$, at which time the coherent emission process begins. Hence for $T \gg T_R$ the field behavior becomes classical.

Although, strictly speaking, the semiclassical description breaks down for $T < T_R$, we are only interested in the output radiation when the field has grown large, which occurs after a delay which is typically 25-100 T_R. Fluctuations in the fields during the first T_R will have little effect on the output at that time due to the logarithmic dependence of the output pulse characteristics on the initial tipping angle (Sect.2.4.1). Thus, a polarization source term of the type described above should give correct results for $T \gg T_R$.

2) *The interaction of forward and backward travelling waves can be ignored.* Computer analysis of the interaction between forward and backward travelling waves shows that this effect is virtually negligible in all swept inversion systems, and it is also negligible in uniformly inverted systems for which $L/c \lesssim T_D$ [i.e., $L \lesssim L_c$, (2.57)]. This is so because the forward and backward waves only become sizable in the same region after much of the stored energy has been radiated. Further discussion is given in Sect.2.4.4.

3) *The plane wave approximation can be utilized.* Thus, effects associated with finite beam diameter are neglected. Discussion of this Assumption is deferred to Sect.2.4.4.

Although (2.19) can be used directly to obtain computer simulations of superradiant behavior for precise comparison with experimental data, computer simulations show that two additional simplifying Assumptions can almost always be made.

4. *Effects of level degeneracy may be neglected in (2.19) by summing n over degenerate M states and averaging $(\mu_z)^2_M$ over orientations.* Computer simulations show that this simplifying procedure has little effect on the output radiation. Therefore, the influence of level degeneracy is insignificant. This is different from an absorber, where level degeneracy can inhibit coherent pulse propagation [2.49].

5) *The distributed polarization source* Λ_p *and the randomly phased blackbody input field may both be replaced by an equivalent input electric field,* $E_{eq}(x = 0,T)$, *of constant amplitude.* Computer simulations show that the influence of both Λ_p and $E(0,T)$ is limited to the first few T_R's. In addition, output pulses of the same intensity, shape and delay are obtained when these source terms are replaced by an input field of the appropriate amplitude. The output is insensitive to the exact shape and phase of this input field. Delta function, step function, Gaussian pulses, and pulse trains of varying phase all give output pulses of about the same shape and size, as long as their partial areas, $(\mu_z/\hbar)\int E_{eq}dT$, integrated over the first few T_R's, are equal. However, changes in input area do cause corresponding variations in the delay and peak intensity of the output pulses. The equivalent input field approximation works because in a high gain system the spontaneous emission which occurs near the input face is most important in initiating the superradiant evolution process.

Since the distributed source can be replaced by an equivalent input field and phase fluctuations in the input field do not significantly affect the computer curves, the initiation of the superradiant process in an initially inverted medium can be simulated by a step function pulse of the form

$$E_{eq}(0,T) = E_0 u(T) \quad , \tag{2.23}$$

with u(T) the unit step function. As discussed in Sect.2.4.1, an initiating field of this form is equivalent to considering the atomic dipoles throughout the medium to be initially tipped, on the average, by a small angle

$$\theta_0 = (\mu_z E_0/\hbar)T_R \quad . \tag{2.24}$$

This effective tipping angle is a key parameter which characterizes the behavior of the system.

An alternate interpretation of (2.24) is that the equivalent input field acts as a short pulse of partial area [see (2.35)]

$$\psi(x = 0,T_R) = \int_0^{T_R} \frac{\mu_z E_{eq}(0,T')}{\hbar} dT' = \theta_0 \quad . \tag{2.25}$$

This implies that the input field is effective only for a short time T_R after the medium is inverted. The situation is analogous to the falling of a rigid pendulum initially balanced on one end: A small perturbation can initiate the motion, but similar perturbations have almost no effect once the motion is under way. Under the influence of E_0 a macroscopic polarization builds up in the medium. As the field produced by this polarization grows, the importance of E_0 diminishes. The output field more than doubles in a time T_R after excitation (see, for example, [Ref.2.5b, Sect.VI]) and E_0 has no further influence after this time.

This discussion also gives insight into the implicit assumption of (2.19) of a particular axis along which the emitted radiation is polarized. The polarization of the E field is determined by, at most, the first few incoherently emitted photons.[3] Subsequent incoherent photons have little effect on the rapidly evolving pulse.

The picture of pulse initiation being due to an input pulse of effective area Θ_0 makes it possible to use the area theorem [2.35], which states that for a non-degenerate Doppler-broadened collisionless system subjected to an incident field $E(x=0,T)$ of constant phase, the pulse area $\theta(x)$ evolves according to

$$\tan\frac{1}{2}\theta(x) = \left(\tan\frac{1}{2}\theta_0\right)e^{\alpha x} \quad , \tag{2.26}$$

where

$$\theta(x) = \psi(x,T\to\infty) = \frac{\mu_z^2}{\hbar}\int_{-\infty}^{\infty} E(x,T)dT \quad , \tag{2.27}$$

and $\theta_0 = \theta(x=0)$. Equation (2.26) shows that in a high gain medium a pulse of small initial area begins to grow exponentially,

$$\theta(x) = \theta_0\, e^{\alpha x} \quad , \tag{2.28}$$

and then evolves towards π. In this process the field envelope may develop positive and negative lobes (ringing) whose contributions to the area substantially cancel one another. Accordingly, in a high gain system the pulse energy continues to grow even though the area remains constant. Notice that the requirement that the area of the superradiant pulse be fully developed $[\Theta(L)\simeq\pi]$ places a lower limit on the gain

$$\theta_0\, e^{\alpha L} \gtrsim 2\pi \quad , \tag{2.29}$$

hence

$$\alpha L \gtrsim \ln(2\pi/\theta_0) = \Phi \quad , \tag{2.30}$$

as in (2.10). Typically, gains of the order of 20 are needed for strong superradiant emission to occur.

2.4 Results of the Theory

2.4.1 Superradiance in the Ideal Limit

Although computer solutions of (2.19) should be used for precise comparisons with experimental data, reliable analytical approximate solutions can be obtained in certain limiting cases. These results are useful in estimating relevant parameters and as an aid to understanding the underlying physical processes. One particularly

[3]The polarization can also be affected by level degeneracy effects. See the discussion of the experiments of Crubellier et al. in Sect.2.5.2.

useful approximation is the "ideal limit", in which loss is unimportant and cooperative decay is so rapid that influence of relaxation processes is negligible.

To motivate this approximation consider the computer plots of Fig.2.5, generated using (2.19), which show the changes in pulse shape and delay in a realistic high gain system caused by eliminating the effects of homogeneous and inhomogeneous broadening, level degeneracy and loss (κL), one at a time. These results show that over a wide range of experimental conditions the influence of these parameters on the superradiant emission pulse is relatively small, so that a fairly good description of the major features may be obtained by neglecting their effects. An exact solution of the resulting equations can then be obtained, with simple expressions for experimentally observable quantities such as output intensity, pulse width, and delay time. These equations also lead to the normalized output curve in Fig.2.3.

In order to obtain the ideal limit, six additional Assumptions are added to Assumptions 1-5:

6) The medium is inverted instantaneously.
7) The system is inverted by a pulse travelling longitudinally through the medium at the speed of light ("swept inversion").
8) Loss is negligible ($\kappa = 0$).
9) Incoherent relaxation and dephasing processes are negligible ($1/T_1 = 1/T_2 = 1/T_2^* = 0$, where T_2^* is the Doppler dephasing time). [This implies very large gain, (2.11)].
10) Feedback is absent.
11) No polarization is present at the superradiant transition when the medium is initially inverted.

The remainder of this section shall consider the behavior of a superradiant system when these six additional Assumptions are valid. Section 2.4.3 discusses the extent to which each of these Assumptions can be relaxed without departing significantly from ideal behavior, and the modifications in behavior when they are relaxed further.

Given Assumptions 4-11, (2.19) become

$$\partial E/\partial x = 2\pi k P \quad , \tag{2.31a}$$

$$\partial P/\partial T = \mu_z^2 n E/\hbar \quad , \tag{2.31b}$$

$$\partial n/\partial T = - EP/\hbar \quad , \tag{2.31c}$$

and n, E and P are all real. The solution of these equations is

$$n = n_0 \cos\psi \tag{2.32}$$

and

$$P = \mu_z n_0 \sin\psi \tag{2.33}$$

with $n_0 = n(x,0)$, and

$$d\psi/dT = \mu_z E/\hbar \quad , \tag{2.34}$$

where

$$\psi(x,T) = \int_{-\infty}^{T} (\mu_z/\hbar)E(x,T')dT' \tag{2.35}$$

is the partial area of the pulse. [The total area $\theta(x) \equiv \psi(x,\infty)$, (2.27)]. Applying the transformation $w = 2\sqrt{xT}$ to (2.31a) and (2.34) gives the pendulum equation

$$\psi'' + (1/w)\psi' = \sin\psi/(T_R L) \quad , \tag{2.36}$$

where $\psi = \psi(w)$.

Equations (2.31b,c) give rise to the familiar Bloch vector picture. As can be seen from the xT dependence of (2.36), this system is analogous to a spatial array of coupled pendula, initially tipped at a uniform small angle $\psi(w = 0) = \theta(x = 0)$, which fall as a phased array.

For a step function initial field of the form (2.23), the solution of (2.36) gives the normalized output curve of Fig.2.3, which relates $T_R^2 I_p$ to T/T_R and is completely determined by Θ_0, as defined by (2.24). Analytical solutions to (2.36) in the small θ limit can be obtained, from which approximate expressions in terms of $\Phi = \ln(2\pi/\theta_0)$ can be derived for the peak output power

$$I_p \approx 4N\hbar\omega/T_R\Phi^2 \propto N^2 \quad , \tag{2.37}$$

the width of the output pulse

$$T_w \approx T_R\Phi \propto N^{-1} \quad , \tag{2.38}$$

and the energy contained in the first lobe of emitted radiation

$$E_p \approx 4N\hbar\omega/\Phi \propto N \quad . \tag{2.39}$$

The delay time from the inversion to I_p is

$$T_D \approx T_R\Phi^2/4 \propto N^{-1} \quad , \tag{2.40}$$

so that $T_D \approx T_w\Phi/4$. These results summarize the major features of the normalized output curve of Fig.2.3.

Equations (2.37-40) describe the basic characteristics of superradiant emission: The duration of the radiation pulse varies inversely with both the inversion density of the sample and its length, and the peak output power increases as the square of each of these parameters. Note that the time delay scales as the pulse width. This has important experimental consequences, since the delay time can usually be measured more accurately than the width. Data confirming the dependence of I_p and T_D on N are given in Fig.2.4.

The expression for the delay time, (2.40), can also be used to check the accuracy of the step function equivalent input field approximation, Assumption 5 and (2.24). For $10^{-12} < \Theta_0 < 10^{-1}$, agreement of better than 1% is obtained between (2.40) and the values of T_D obtained from computer solutions of both the original model with polarization source term[4] *and* the simplified model with a step function input electric field replacing the source term. Equation (2.40) is also valid for a delta function input field or, equivalently, input conditions of the form

$$n(x,0) = n_0 \cos\Theta_0 \quad , \tag{2.41a}$$

$$P(x,0) = \mu n_0 \sin\Theta_0 \quad . \tag{2.41b}$$

Accurate agreement with the polarization source model is then obtained for $\Theta_0 < 10^{-3}$ by using $\Phi = |\ln\theta_0|$ in place of $\Phi = \ln(2\pi/\theta_0)$.

The delay time expression, (2.40), was originally derived [2.3,5] in the semiclassical framework by linearizing (2.36). The accuracy of this approximation has recently been verified from an analysis based on a series expansion solution to the sine-Gordon equation [2.50]. This expression has also been derived from the quantized field treatment of [2.30], with $\Phi = \ln\sqrt{2\pi N}$.

2.4.2 Influence of Quantum Fluctuations

As explained in Sect.2.3.1, the dynamical evolution of a superradiant pulse can be considered as "classical", with the initial conditions determined by quantum fluctuations. Hence, the quantum initiation process determines the value of the initial tipping angle (or the effective area of the equivalent input pulse), Θ_0. As seen in Sect.2.4.1, the delay time, peak intensity, and other quantities which characterize the output pulse all depend logarithmically on this parameter through Φ. In addition, quantum fluctuations should produce small shot-to-shot variations in the observed pulse characteristics.

Expressions for Θ_0 and Φ have been derived from both quantum and semiclassical considerations. In the Glauber-Haake quantized field treatment, where the initiating disturbance is interpreted as arising from polarization fluctuations, the average initial tipping angle, defined by (2.41b)

$$\langle P^2(x,0)\rangle = (\mu_z n_0 \sin\Theta_0)^2 \quad , \tag{2.42}$$

may be obtained from (2.166)

$$\frac{1}{L^2} \iint \langle \underset{\sim}{P}^+(x,0)\underset{\sim}{P}(x',0)\rangle dx'dx = \frac{4}{AL}\,\mu_z^2 n_0 \quad . \tag{2.43}$$

This gives, since $\Theta_0 \ll 1$,

$$\Theta_0 = \frac{2}{\sqrt{N}} \quad . \tag{2.44}$$

BONIFACIO and LUGIATO [2.23] have also obtained this result.

---

[4]Equation (2.20), with ξ_0 evaluated as in (2.49,51), below.

The Polder, Schuurmans, Vrehen treatment considers the initiating quantum noise as arising from fluctuations in the vacuum field. As explained in Sect.2.3.1, the field operators $\underset{\sim}{E}$ and $\underset{\sim}{E}^{+}$ of the quantized field treatment of [2.30] may be replaced by complex c-numbers E and E^{*} and, for early times, $E(x,T)$ may be considered as a classical Gaussian fluctuating field source. Equation (2.13d) then yields, for the second-order correlation function,

$$\frac{cA}{8\pi}\langle E(0,T)E^{+}(0,T')\rangle = \hbar\omega\delta(T - T') \quad . \tag{2.45}$$

As in the discussion of Sect.2.3.2, this gives an equivalent input pulse of effective area

$$\Theta_0 = \left[\frac{\mu_z^2}{\hbar^2}\int_0^{T_R}\int_{-\infty}^{\infty}\langle E(0,T)E^{*}(0,T')\rangle dTdT'\right]^{1/2} \tag{2.46a}$$

$$= \frac{2}{\sqrt{N}} \quad . \tag{2.46b}$$

A more detailed calculation [2.30] gives an effective tipping angle of

$$\frac{2}{\sqrt{N}}\left[\ln(2\pi N)^{1/8}\right]^{1/2} \quad , \tag{2.47}$$

and a corresponding value of

$$\Phi = \ln\sqrt{2\pi N} \quad . \tag{2.48}$$

A somewhat different expression for Φ has been given by MACGILLIVRAY and FELD [2.24], using the semiclassical description of Sect.2.3.2. This approach evaluates ξ_0, the amplitude of the polarization source of (2.30), from detailed balance considerations similar to those [2.51] used to derive the ratio of the Einstein A and B coefficients

$$\xi_0 = \sqrt{\frac{4\Delta\omega_{eff}T_{sp}}{\pi}} \quad . \tag{2.49}$$

This leads to an initial tipping angle

$$\Theta_0 = \left[\frac{\Delta\omega_{eff}T_R}{\pi N_2}\left(\frac{n_2}{n_2 - n_1}\right)\right]^{1/2} \quad , \tag{2.50}$$

with n_1 and n_2 the initial population densities of ground and excited states of the superradiating transition, respectively, and $N_2 = n_2AL$. $\Delta\omega_{eff}$, the effective bandwidth of the transition near the steady state, is given by

$$\Delta\omega_{eff} = \frac{1/T_2'}{\sqrt{2\alpha L/\pi}} \quad . \tag{2.51}$$

It is smaller than the transition linewidth $1/T_2'$ because of the "narrowing" that occurs in a high gain system ($\alpha L \gg 1$) [2.52]. For large inversion density ($n_2 \gg n_1$) this gives

$$\Phi = |\ln(\Theta_0/2\pi)| = \ln\left[\sqrt{2\pi N_2}(2\pi\alpha L)^{3/4}\right] \quad . \qquad (2.52)$$

For $\lambda \gtrsim 50$ µm, where blackbody radiation can also initiate the pulse evolution process, N_2 must be replaced by $N_2\left[1 + (e^{\hbar\omega/kT} - 1)^{-1}\right]$, as in (2.21).

For typical values of N, (2.48,52) only differ by ~20%. Hence, it is difficult to decide experimentally between the two. A recent experiment [2.16] favoring (2.48) is described in Sect.2.6. Other expressions for Θ_0 and T_D are reviewed in [2.30].

References [2.30,34] include analyses of the pulse-to-pulse variations in the time delays of the output radiation caused by quantum fluctuations. Reference [2.30] gives an expression for the relative standard deviation in delay time,

$$\sigma_D = \left[\frac{\langle T_D^2\rangle}{\langle T_D\rangle^2} - 1\right]^{1/2} = \frac{2.3}{\ln N} \quad . \qquad (2.53)$$

For typical experimental conditions this gives standard deviations of ~10% or less, indicating rather small shot-to-shot variations in delay time. A similar expression has been obtained by DEGIORGIO [2.67]. A numerically calculated plot of the distribution of delay times of first intensity maxima, generated in [2.34] using a model similar to that of Sect.2.3.2, is shown in Fig.2.6.

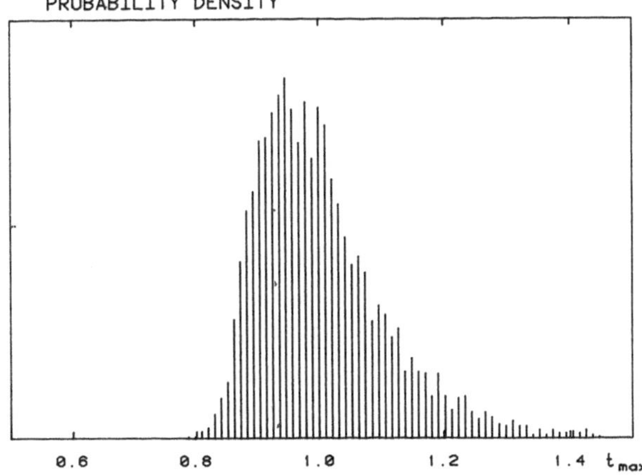

Fig.2.6. Distribution of delay times of the first intensity maxima of superradiant pulses, as calculated numerically in [2.34], for a short ($L \ll L_c$) sample with initial population inversion $N = 4 \times 10^9$

2.4.3 Deviations from Ideal Behavior

From (2.37-40) it would appear that increasing the length or inversion density of a superradiating system indefinitely will result in extremely short pulses of arbitrarily high intensity. Obviously, in any physically realizable system Assumptions 6-11 of the ideal limit will not be satisfied. While small deviations from ideality are of little importance, large deviations can cause significant changes in the form of the output radiation. It is of considerable practical interest to establish the extent to which the various Assumptions can be relaxed without modifying the output characteristics, and the consequences of changing them further. This is particularly important in considering achieving superradiant emission in new systems and regimes, such as in X-ray lasers and ultrashort pulse generation schemes.

This section examines the effects of relaxing Assumptions 6-11. We shall first ascertain the extent to which a given Assumption can be relaxed without causing a significant departure from ideal behavior, and then explore the consequences of changing it further. In many cases, as an aid to understanding, simple analytical expressions similar to those derived in the ideal limit are given. All of the predictions and formulas discussed below have been verified by numerically integrating (2.19). More detailed discussions will be found in [2.36].

Finite Inversion Time (Assumption 6)

In any actual system the process which populates the upper level of the superradiant transition will occur over a finite time τ, rather than instantaneously. For simplicity, the present discussion will consider the particular case of a constant pump rate [i.e., Λ of (2.19c) is a constant]. More general formulas can be found in [2.36,47].

Consider a system which obeys all the Assumptions of the ideal limit except that $\tau \neq 0$. As long as the inversion process is completed before the superradiant pulse is emitted, the effect on the output is small, other than to increase the observed delay time from T_D to $T_D + \tau/2$ (Fig.2.7a,b). This formula applies to the case

$$\tau \lesssim 2T_D \quad . \tag{2.54}$$

For $\tau \gtrsim 2T_D$ superradiant output occurs before the inversion process terminates, so that only the early part of the population inversion can contribute to the first burst of radiation (Fig.2.7c). Such a system will radiate before it has been fully inverted and will no longer be able to emit all of its stored energy in one coherent series of bursts of radiation. In this case (2.37-40) are no longer valid. Calculations [2.36] show that in this limit the observed delay time T'_D is given by

$$T'_D = \sqrt{4\pi T_{sp}\Phi^2/\lambda^2 L\Lambda} \quad , \tag{2.55a}$$

the peak intensity is

$$I_p = 2\hbar\omega A L\Lambda \quad , \tag{2.55b}$$

and the width of the emitted radiation pulse is

$$T_W = 8\sqrt{\pi T_{sp}/\lambda^2 L\Lambda} \quad . \tag{2.55c}$$

The maximum length of an efficient system, L_1, occurs when $\tau = 2T_D$

$$L_1 = \frac{4\pi T_{sp}\Phi^2}{\lambda^2 \Lambda \tau^2} \quad . \tag{2.56}$$

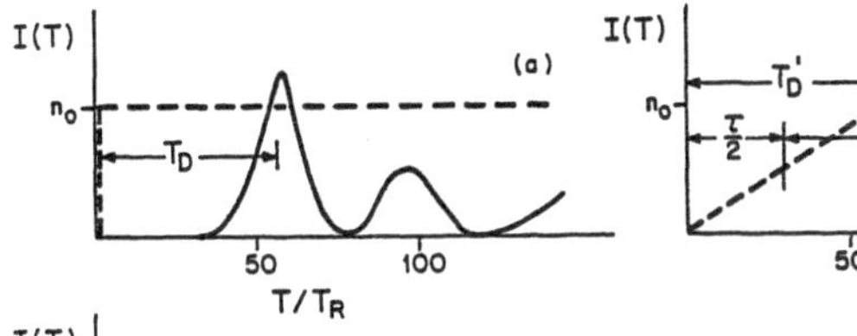

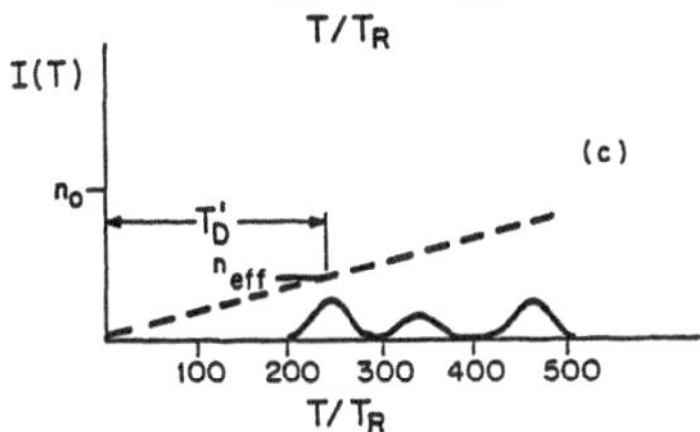

Fig.2.7a-c. Effect of changes in τ on I(T). In Figs.2.7-9 a system with parameters similar to those of [2.13-15] is used: Unless otherwise indicated $n_0 = 5\times 10^{10}$ cm^{-3}, L = 2 cm, T_R = 0.5 ns, and $\Phi \approx 15$. The solid lines in Fig.2.7 indicate output intensity in arbitrary units as a function of retarded time (same scale throughout). The dashed lines indicate the total inversion density created up to that time. (a) Instantaneous inversion (τ = 0); (b) $\tau \approx T_D$ (τ = 60 ns, $T_D \approx$ 57 ns). Note that $T_D' = T_D + \tau/2$; (c) $\tau \gg T_D$ (τ = 600 ns). Note the change in horizontal scale. Only a fraction $n_{eff}/n_0 \approx 0.4$ of the atoms can contribute to the first lobe of radiation. [2.36]

In experiments up to now L_1 has typically been 10-100 times L (Table 2.1).

Table 2.1. Typical experimental parameters for various observations of superradiance, and calculated values of L_1 (2.56) and L_c (2.57). [2.36,48]

	CH_3F	HF	Na	Cs	Tl
λ [µm]	496	84	3.4	2.9	1.3
L [cm]	600	100	14	2	15
T_2^* [ns]	(T_1 = 60)	220	1.7	32	1
n_0 [cm^{-3}]	3×10^{12}	10^{12}	2×10^{10}	4×10^{10}	2×10^{15}
T_R [ns]	0.3	5	0.2	0.7	0.05
αL	200	45	9	50	20
T_D [ns]	100	400	7	12	12
ϕ	25	17	13	14	24
τ [ns]	65	100	2	2	5
L_1 (Calculated) [cm]	900	700	120	70	40
L_c (Calculated) [cm]	1300	1500	120	60	80
Reference	[2.7-9]	[2.3-6]	[2.10,11]	[2.11, 2.13-18]	[2.11,12]

Note that for samples with $L > L_1$ $I_p \propto \Lambda$, not Λ^2, but N^2 radiation still occurs because the effective number of radiators $\propto \Lambda^{1/2}$. However, although I_p increases with increasing Λ, the fraction of the stored energy which is radiated in the first lobe of superradiant emission *decreases*.

When the pumping process is very long the system will only emit a few bursts of radiation of this type. As the system continues to radiate, the inversion density is replenished as fast as it is depleted, leading to a quasi-steady state [2.24,53] with $I \approx \hbar\omega AL\Lambda$ (Sect.2.5.3). Some evidence for the transition of such a system from its transient, N^2-dependent superradiant behavior to its steady state has been observed in Na by GROSS et al. [2.10].

Uniform Inversion: Cooperation Length (Assumption 7)

The ideal limit assumes the medium to be prepared by swept inversion, in which the system is inverted by a pulse travelling longitudinally through the medium at the speed of light. There is then no upper limit on the length L of the sample. This can be contrasted with uniform inversion, in which the medium is inverted simultaneously along its entire length. In this case long samples superradiate differently from short ones.

For samples in which $L \lesssim L_c$, with [2.36,47]

$$L_c = \Phi\sqrt{4\pi c T_{sp}/n_0\lambda^2} \quad , \tag{2.57}$$

the method of preparing the sample is irrelevant: In either case computer simulations, extended from the model of (2.19) to allow for the possibility of E fields propagating in both directions along the length of the sample, predict the evolution of two symmetrical oppositely propagating waves.[5] The interaction between these waves is negligible for cases of experimental interest (Sect.2.4.4). The only effect of uniform inversion is then to increase the observed delay by L/2c, the average transit time.

For sufficiently long samples, however, a uniformly inverted medium cannot radiate as a whole, because atoms near $x = L$ (the output end for the "forward" travelling wave) start to superradiate before radiation near $x = 0$ can reach them. The changeover occurs at $L \simeq L_c$, the length where the transit time $T_{tr} = L/c$ equals the observed delay, $T_D + T_{tr}/2$. Longer systems break up into independently radiating segments [2.36] in a manner described by ARECCHI and COURTENS [2.45].

For a uniformly inverted medium of $L > L_c$ which otherwise obeys the assumptions of the ideal limit, the percentage of input energy radiated in the first output lobe drops markedly, the ratio T_W/T_R increases, and I_p is no longer proportional to L^2. For $L \gtrsim L_c$ the peak power of the first lobe is

$$I_p \approx \frac{4N\hbar\omega}{T_R\Phi^2} = \frac{4n_0AL_c\hbar\omega}{T_{sp}(8\pi/n\lambda^2L_c)\Phi^2} = 2n_0A\hbar\omega c \quad , \tag{2.58}$$

---

[5]When simulating a long uniformly inverted sample a distributed source *must* be used, as opposed to an initial boundary condition, since polarization and the electric field are building up simultaneously in various regions of the medium.

since only the length L_c can contribute to the first lobe of the output radiation. However, the rest of the medium contributes to later output lobes, which as a result can be more intense than the first lobe (Fig.2.8).

ARECCHI and COURTENS [2.45] have derived an expression for L_c in the case of a system inverted uniformly with a large-area ($\Theta_0 \sim 1$) pulse, where the delay time is $\sim T_R$. Their expression,

$$L_c' = \sqrt{4cT_{sp}/n_0\lambda^2} \quad , \tag{2.59}$$

is the large Θ_0 limit of (2.57).

The above considerations do not apply for samples prepared via swept inversion, in which the inversion pulse reaches all parts of the medium at the same retarded time. Hence, transit time effects are irrelevant. However, in long ($L \gtrsim L_c$) samples oppositely propagating waves are no longer symmetrical. For the wave propagating in the direction of the inversion pulse all atoms can contribute coherently to the output, giving rise to essentially ideal behavior. The oppositely propagating wave, however, can be affected by cooperation length effects [2.36]. This has been observed by EHRLICH et al. [2.8] in the form of an increased ratio of forward wave intensity to backward wave intensity as length increases.

Loss (Assumption 8)

The growing electric field may be attenuated by diffraction and other losses, parametrized by κ in (2.19a). Effects of loss can be conveniently illustrated in the case of constant κ. In a medium with high gain the effects of loss are negligible for $\kappa L \ll \phi/4$. However, for $\kappa L \gtrsim \phi/4$ significant deviations from ideal behavior can occur even if the other conditions for ideality are satisfied, even for $\alpha L \ggg 1$ (Fig.2.9) [2.24,54]. For such large values of κL, E approaches the "steady-state" limit [2.42,44,55]

$$\mu_z E(T)/\hbar = (1/\kappa L T_R)\operatorname{sech}[(T - T_D)/\kappa L T_R] \quad , \quad [T \gtrsim 0] \tag{2.60}$$

with

$$T_D = \kappa L T_R \operatorname{sech}^{-1}(\kappa L \Theta_0) \tag{2.61a}$$

$$\simeq \kappa L T_R \tag{2.61b}$$

for $\kappa L \Theta_0 \ll 1$, as in many experiments.

Equation (2.60) describes a pulse of area π, without ringing, the peak of which is delayed from the inversion pulse (which occurs at retarded time $T = 0$) by T_D. The peak intensity is

$$I_p = (cA/8\pi)E^2 = (cA\hbar^2/8\pi\mu_z^2)(1/\kappa L T_R)^2 \tag{2.62a}$$

$$= N\hbar\omega/4T_R(\kappa L)^2 \quad , \tag{2.62b}$$

and its (full) width is

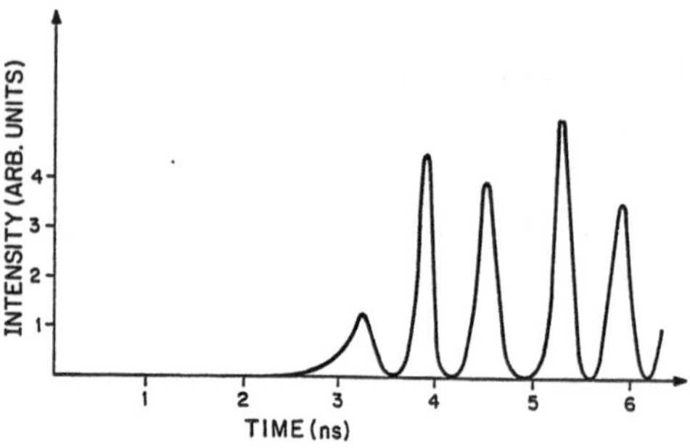

Fig.2.8. Output intensity I(T) of a uniformly inverted sample of length $L = 200$ cm $\simeq 3.5\ L_c$ and otherwise the same as in Fig. 2.7. The transit time is 6.67 ns. The corresponding I(T) for a much shorter system ($L \ll L_c$) is given in Fig.2.1b. [2.36]

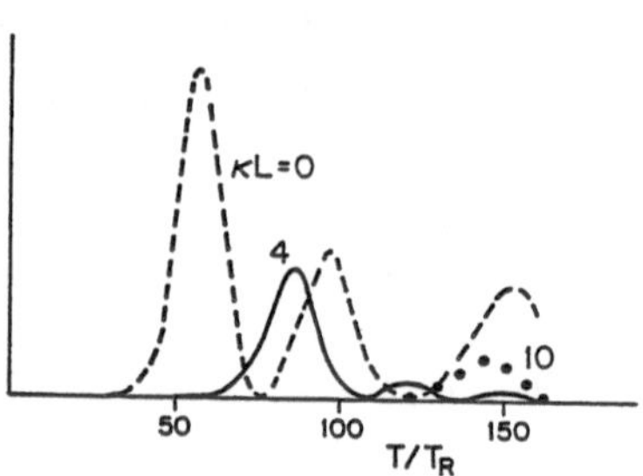

Fig.2.9. Effect of changes in κL on output intensity. All parameters of the dashed curve are as in Fig.2.7a. In the solid curve $\kappa L = 4$, and in the dotted curve $\kappa L = 10$. [2.36]

$$T_W \simeq 1.8\kappa L T_R \quad . \tag{2.63}$$

Equations (2.60-63) are the limiting expressions for $\kappa L \gtrsim \phi/4$, and (2.37-40) are appropriate for limit $\kappa L \ll \phi/4$. This suggests a characteristic length [2.36]

$$L_2 = \Phi/4\kappa \quad , \tag{2.64}$$

which demarks the boundary between ideal and steady-state behavior. However, the changeover is a gradual one, and computer solutions should be used for intermediate values.

For systems in which T_1 is the dominant decay process ($T_1 = T_2$) BONIFACIO et al. [2.55] have derived an additional condition, $\alpha/\kappa \gtrsim \phi$, in the steady-state limit in order for superradiance to occur. It should be noted that combining this condition with $L > L_2$ gives $\alpha L \gtrsim \phi^2/4$, a condition which is not only applicable to the case of large κL, but is also a general condition on T_1/T_R, as shown below (2.70b).

Another condition on the length follows from the area theorem, generalized to include loss [2.44]

$$d\Theta/_{dx} = -\kappa\Theta + \alpha\sin\Theta \quad . \tag{2.65}$$

This equation implies that in order for a pulse of area $\Theta_0 \ll 1$ to grow to $\sim\pi$,

$$(\alpha - \kappa)L > \Phi \quad . \tag{2.66}$$

Equation (2.66), the generalization of (2.10) when loss is present, provides a necessary condition on the gain αL in order for ideal behavior to occur. It is not sufficient, however, since in most cases (2.70) is far more stringent. The behavior of systems in which (2.66) does not hold is discussed in Sect.2.5.1.

Equation (2.66) implies that the duration of a superradiant pulse, $T_W \approx T_R\Phi$, must always be shorter than the inverse bandwidth of the transition, $T_2' = T_R\,\alpha L$: Since (2.66) requires that $\alpha L > \Phi$ for full superradiance, $T_W < T_2'$. This relationship is important in understanding the transient nature of the superradiant process (Sect.2.5.3).

When loss is due to diffraction κ is a function of x. For a Gaussian beam of waist cross-sectional area A, where

$$\kappa(x) = \frac{x}{x^2 + (A/\lambda)^2} \quad , \tag{2.67}$$

all the proceeding formulas involving κ apply, with κL replaced by $\int_0^L \kappa\, dx = \frac{1}{2}\ln[1 + (\lambda L/A)^2]$, which is always negligible for Fresnel number $2A/\lambda L \gtrsim 0.1$. For smaller Fresnel numbers other considerations discussed in Sect.2.4.4 become important.

Decay and Dephasing Times (Assumption 9)

As mentioned above, (2.66) is a necessary condition for ideal behavior. This places a restriction on T_2', hence on T_1 and T_2^*, since $\alpha L = T_2'/T_R$

$$T_2'/T_R > \Phi + \kappa L \quad . \tag{2.68}$$

The form the emission takes when (2.68) is not satisfied is discussed in Sect.2.5.1.

A second condition on relaxation processes can often be more restrictive. The homogeneous decay time T_1, due to spontaneous emission, collisions, and other mechanisms, must not be so short that incoherent decay becomes the dominant mode of deexcitation, which would prevent superradiance. In order for a system with finite T_1 to reach full superradiance [2.36],

$$T_1 \gtrsim eT_R\Phi^2/4 \tag{2.69a}$$

$$= eT_{sp}(8\pi/n_0\lambda^2L)\Phi^2/4 \quad , \tag{2.69b}$$

or, equivalently,

$$\alpha L \gtrsim e(\Phi^2/4)T_2'/T_1 \quad . \tag{2.70a}$$

For a system where T_1 is the dominant mode of incoherent decay, (2.70a) becomes

$$\alpha L \gtrsim e\Phi^2/4 \quad , \tag{2.70b}$$

a far more stringent requirement than (2.66).

The effects of finite dephasing time T_2 and inhomogeneous broadening time T_2^* are different from the effect of finite T_1, in that high gain increases the effective T_2 and T_2^* to $T_2\alpha L$ and $T_2^*\alpha L$, respectively [2.5]. Since whenever $\alpha L > \Phi$ both $T_2\alpha L$ and $T_2^*\alpha L$ are greater than T_D, there is no new condition on T_2 and T_2^* similar to that of (2.69).

Feedback (Assumption 10)

The basic features of superradiance are not changed by the presence of feedback. However, the details of the radiation process may be modified. In particular, the delay time decreases and the amount of ringing increases. This is because feedback tends to increase the effective length of the sample or, alternatively, to increase the initial tipping angle θ_0, a situation analogous to continuing to push an initially inverted pendulum after it has started to fall. A similar problem, the radiative damping of an inverted NMR system in a resonant circuit, has been studied previously [2.57].

In the HF experiments [2.3,4], when feedback was deliberately introduced using mirrors, superradiant emission was still observed, but with drastically reduced delay times. The pulse shapes and delays with feedback were similar to those observed in much higher density systems without feedback.

The effect of feedback on the output will be negligible as long as the output field due to the initializing spontaneous emission is significantly greater than the additional field which results from the feedback process. This leads to a condition on the maximum fraction F of the output radiation which can be added to the input without significantly altering ideal behavior [2.36]. For a short system, where transit time is negligible,

$$F \ll e^{-0.35\Phi}\Phi^{-1/2} \quad . \tag{2.71a}$$

For systems in which it takes appreciable time (compared to T_D) for the emerging radiation to return to the input face via the feedback process, the influence of feedback decreases. Thus, the value of F for which feedback becomes important will be larger

$$F \ll \frac{(1 - T_X/T_D)^{1/2}}{e^{0.35\Phi}\Phi^{1/2}} \exp\left\{-1.35\Phi\left[(1 - T_X/T_D)^{1/2} - 1\right]\right\} \quad , \tag{2.71b}$$

where T_X is the round-trip transit time. For $\Phi = 15$ and $T_X = 0$ feedback up to $\sim 3\times10^{-3}$ is negligible. For $T_X/T_D = 0.1$ this value increases to $\sim 10^{-2}$.

As F gradually increases beyond the value allowed by (2.71) the delay time decreases and the peak power increases. For F close to unity the feedback process in some ways resembles that of a laser amplifier whose effective length, and consequently gain, is increased by a factor $\sim 1/(1 - F)$. Thus, a single pass system with $\alpha L = 1$ and $F = 0.95$ is to some extent similar to a system with $F = 0$ and $\alpha L = 20$. However, there are important differences. Due to the small transmission $(1 - F)$ associated with near-unity feedback, the system's energy recirculates and is released slowly, rather than in a few bursts. Furthermore, unlike a long single pass system, the growth of the E field is influenced by the polarization which has built up in the medium during previous passes.

Initial Polarization (Assumption 11)

The presence of initial ($T = 0$) polarization at the superradiant transition will not significantly affect the output radiation provided that the associated tipping angle $\theta' = P(z,T=0)/\mu_z n_0$, is small compared to θ_0 due to spontaneous emission. This gives

$$P_0 \lesssim \mu_z n_0 \theta_0 \quad (2.72)$$

Larger values of P_0 are equivalent to increasing the initial tipping angle of the Bloch vector, which shortens the delay time and reduces the ringing. This increases the difficulty of completing the inversion process before coherent emission begins. In principle, a pulse of area exactly π could completely invert an initially absorbing medium without residual polarization. An energy conservation argument (Sect.2.5.2) shows that such a pulse would have to be shorter than T_R in order not to lose area as it traverses the medium; for longer pulses the effects of self-induced transparency [2.35] and pulse propagation [2.42-44] become relevant. As a practical matter, schemes to directly invert two-level systems in this manner are probably not feasible due to problems associated with loss, level degeneracy, transverse variations in the electric field associated with beam profile, and the difficulty of generating a pulse of exactly area π.

These problems can be circumvented by using indirect excitation methods such as three-level pumping and two-photon excitation with a nonresonant intermediate state. All observations of superradiance up to now have employed such schemes. However, the problem remains that when the pump radiation is turned off, a large residual polarization could be left at the pump transition. This can result in superradiance at *this* transition, which could deplete the population available for superradiance at the desired wavelength. This problem can be overcome by using an incoherent pump pulse, or by choosing a much shorter wavelength for the pump transition (to increase its T_R).

One should also note that in indirect excitation schemes the background emission which initiates superradiance can be modified by the presence of the pump field through Raman-type transitions [2.58]. In this case background spontaneous emission is effectively increased by the presence of the intense pump field. Although these processes tend to increase the effective initial tipping angle, estimates show that this effect is small in all experiments performed to date.

2.4.4 Further Discussion of the Basic Assumptions

This section further explores the validity of Assumptions 2 and 3 of the semiclassical model of Sect.2.3.2.

Neglect of Interaction of Forward and Backward Waves (Assumption 2)

As explained in Sect.2.3.2, interaction between oppositely travelling waves is negligible, because forward and backward waves only become sizable in the same region after much of the stored energy has been radiated. This statement is true only for sufficiently small values of θ_0. As θ_0 becomes larger the fraction of stored energy which the system radiates in the first lobe increases, and the two waves begin to interact. SAUNDERS et al. [2.59,60] have studied this interaction in the $\theta_0 \sim 0.1$ regime, where the forward-backward interaction tends to significantly increase delay times and decrease ringing.

To study this θ_0 dependence, the forward-backward wave interaction was added to (2.19) by writing separate $\partial E/\partial z$ and $\partial P/\partial T$ equations for the two waves and using a single $\partial n/\partial T$ equation combining EP^* from both waves. Separate solutions for the forward and backward waves were then iterated until covergence was reached. Computer results in short samples [$L \lesssim L_c$ (2.57)] show that the interaction starts to become significant for the second lobe near $\theta_0 \sim 10^{-3}$, and for the first lobe near $\theta_0 \sim 10^{-2}$. Since estimates of θ_0 for all superradiance experiments reported to date give $\theta_0 \lesssim 10^{-2}$, the forward-backward wave interaction should not be significant.

Longer systems could not be explored by computer due to the failure of this algorithm to converge. There is no reason to expect the forward-backward wave interaction to become significant in longer systems for small experimentally relevant values of θ_0. However, if θ_0 were much larger than indicated by current theories (Sect.2.4.2) and by experiments to date, interaction between oppositely propagating waves could become particularly important in long ($L > L_c$) uniformly inverted samples [2.36].

Unwanted feedback between forward and backward waves can also influence the output. The effect of this feedback is similar to that occurring in the absence of the forward-backward wave interaction (Sect.2.4.3), except that the feedback is less important in the early stages of the evolution process, when the system is linear and the phases and polarization directions of the two waves are unrelated. Since sufficiently large feedback between forward and backward waves will cause their phases to become correlated, a lower limit on the acceptable amount of feedback can be estimated by using the one-way transit time in place of T_x in (2.71).

Limitations of the Plane Wave Approximation (Assumption 3)

The semiclassical model of Sect.2.3.2 assumes that the E field is a uniform one-dimensional plane wave. This model is a reasonable approximation as long as (a) the entire cross section radiates as one coherent plane wave; (b) transverse field components are small; and (c) in the coupled Maxwell-Schrödinger equations the E field envelope can be approximated by its average value over the cross section. These assumptions are not always valid, particularly in systems with small or large Fresnel number $F = 2A/\lambda L$.

The limitations of the plane wave approximation can be avoided by including transverse variations and transverse fields in the wave equation. Such an approach has been used by MATTAR and NEWSTEIN [2.61] and others [2.62] to study the effects of non-plane wave behavior on the propagation of coherent pulses in absorbers. The following discussions explore experimental circumstances for which non-plane wave effects may be important.

I) Small Fresnel Number. In samples with small Fresnel number, where the radiating volume is long and thin ("pencil"), the dependence of T_R on the system parameters differs from that of (2.1), which is valid for large Fresnel number ("disc"). This can be visualized by noting that for a long thin array of radiators, the solid angle over which phases sum constructively is very different from that of a short flat array. REHLER and EBERLY [2.39a] found that a sharp change in the shape factor μ used in the formula for T_R occurs near $F \sim 0.1$ [Ref.2.39, Fig.5]. For $F \lesssim 0.1$ T_R is no longer given by (2.1), but instead becomes [2.37,39] approximately

$$T_R = T_{sp}(8\pi/n_0\lambda A) = T_{sp}(8\pi/n_0\lambda^2 L)(\lambda L/2A) \quad . \tag{2.73}$$

The maximum length L_3 of a system which can be described by (2.1) and (2.37-40) is therefore given by $2A/\lambda L_3 \sim 0.1$, or

$$L_3 \sim 20A/\lambda \quad . \tag{2.74}$$

For $L > L_3$, T_R remains constant as the length increases. Since T_R determines the time scale of the entire superradiant process, output pulses can no longer be shortened by increasing the length of the sample. However, the output intensity should still be proportional to the square of the inversion density.

The expressions for diffraction loss and θ_0 are also modified for samples with small Fresnel number. The diffraction loss $\kappa L = 1/2 \ln[1 + 4/F^2]$ (Sect.2.4.3) is no longer negligible (as it is for large F) and must be taken into account. Also, the derivations of the θ_0 expressions of Sect.2.4.2 all employ solid angle factors which (implicitly) depend on the shape factor μ as does T_R; for $F \lesssim 0.1$, this factor changes from $\lambda^2/4\pi A$ to $\lambda/4L$. The effect of this change is small, however, due to the logarithmic dependence of the expressions for the output parameters on θ_0.

These changes in T_R and θ_0 and this choice of κL may be adequate to allow the use of the plane wave solutions for small Fresnel numbers.

II) Large Fresnel Number. A large Fresnel number system ($F \gtrsim 1$) may be unable to evolve to the coherent plane wave assumed in our model. In an initially inverted system spontaneous emission occurs independently in different sections of the cross section, and coupling between sections due to diffraction is needed to produce a single coherent phase front. In the $F \lesssim 1$ case the output face of the medium ($x = L$) is in the far field zone of $x = 0$, where spontaneous emission is most important. Therefore, coupling between transverse sections should be sufficiently large for a single coherent wave to evolve. For large F, however, independent regions each with $F \gtrsim 1$ should evolve independently, since $x = L$ is in the near field zone of $x = 0$. Therefore, diffraction coupling may not be strong enough to create a single coherent wavefront. Thus, different sections may have difficult phases which would not add constructively.

In addition, in high gain media nonlinear effects such as self-focusing, self-defocusing, and beam trapping may place a limit on the largest coherent phase front which can be sustained. Similar limitations occur in pulse propagation in absorbers. For example, GIBBS et al. [2.63] have shown that self-induced transparency can be inhibited by transverse effects of this type.

III) Initial Nonuniform Cross Section. A related experimental difficulty which can lead to non-plane wave behavior is an initial inversion density which is nonuniform over the cross section of the system (e.g., near the perimeter of the cross section of the inverted region). Such nonuniformity could lead to independent radiation in distinct sections of the cross section, with different T_R's and different delays, and could also cause the effects of transverse fields to be significant.

As an example, consider the case of a *coherent* input pulse of Gaussian cross section which is used to indirectly populate the upper level of an initially unpopulated two-level system. If θ_0 at the center of the beam is several times π, the inversion density will vary greatly over the cross section, since variations in E will cause some sections to be completely populated (where θ_0, a linear function of E, equals an odd multiple of π) and others to be completely unpopulated, forming a pattern of concentric rings. As a result, different sections of the cross section with different T_R's begin to evolve independently. This evolution may be further complicated by diffraction coupling between these rings. For large Fresnel number small diffraction coupling could result in independent sections of output radiation with different phase fronts and delays. A detector incapable of resolving these sections would then average ringing from different sections into a single asymmetric radiation lobe with a long tail. Such an effect may contribute to the absence of ringing in some of the Cs experiments of [2.14,15].

The behavior of transverse variations and transverse fields during the evolution of a superradiant system merits further attention.

2.4.5 Point Sample Superradiance

Early discussions of superradiance emphasized observation of the effect in the long wavelength regime, where the sample size is small compared to the radiation wavelength (point sample: $\lambda \ll L$). Derivations of point sample superradiance, which are usually quantum mechanical [2.38,39], predict a sech^2 radiation pulse. It is interesting to consider a semiclassical derivation of this result, which points out the different ways the induced polarization couples to the radiation field in point and extended samples. The Schrödinger (Bloch) equations, (2.19b,c), are the same in the two cases (except that $T \to t$ and $\mu_z \to \mu$), but in the point sample case the reduced Maxwell equation differs from (2.19), reflecting the different relationship of $E(P)$ when effects of propagation and retardation are absent. The appropriate E in this case is the radiation reaction field. This can be easily seen from the energy balance relationship,

$$\frac{\hbar\omega}{2}\frac{\partial N}{\partial t} = -\frac{4}{3}\frac{|\ddot{\underline{P}}|^2}{c^3} \quad , \tag{2.75}$$

where $N = nAL$ and the right-hand side is the familiar expression for the power radiated by a point dipole with polarization P. For a point sample the complex envelope function P can be defined by $P = \mathrm{Re}\{P\, e^{i\omega t}\}$. Using (2.31c), the slowly varying envelope approximation gives

$$E = \frac{4}{3}\frac{\omega^3}{c^3} P \quad , \tag{2.76}$$

which is just the expression for the radiation reaction field, usually derived in other ways [2.64]. For a point sample (2.76) replaces (2.19). Equations (2.32-35) then give

$$\frac{d\varphi}{dt} = \left(\frac{4\mu^2\omega^3}{3\hbar c^3}\right) N_0 \sin\varphi \quad , \tag{2.77}$$

which has the solution

$$\sin\varphi = \mathrm{sech}\; t/T_R^0 \quad , \tag{2.78a}$$

with

$$\frac{1}{T_R^0} \equiv \frac{4\mu^2\omega^3}{3\hbar c^3} N_0 \quad . \tag{2.78b}$$

The radiated power is then

$$I = \frac{2}{3}\frac{\mu^2\omega^4}{c^3} N_0^2\, \mathrm{sech}^2(t/T_R^0) \quad , \tag{2.79}$$

the well-known expression for the superradiant emission power in a point sample.

No experiment to date has demonstrated superradiance in a point sample. FRIEDBERG et al. [2.65] have pointed out that in samples small compared to a wavelength superradiant emission is severely limited by the dephasing due to dipole-dipole interactions among the atoms. [Such terms are omitted in the derivation of (2.79)]. These lead to frequency shifts which vary spatially throughout the medium, hence a dephasing of the macroscopic polarization. This dephasing will occur in a time of the order of [2.65]

$$T_\Delta = \frac{L}{\lambda} T_R^0 \ll T_R^0 \quad . \tag{2.80}$$

Since the dephasing rate greatly exceeds the superradiant emission rate, strong cooperative emission is inhibited.[6]

Dephasing due to dipole-dipole interactions becomes important when the interatomic distance becomes comparable to λ. This gives a condition for superradiant emission

$$n\lambda^3 \ll 1. \tag{2.81}$$

---

[6]Pathological behavior can occur in a small spherical sample. See [2.66] for details.

On the other hand, superradiance in an extended sample requires (Sect.2.2) that

$$n\lambda^2 L \gg 1 \quad . \tag{2.82}$$

These two conditions place bounds on the range of inversion densities for which superradiant emission can occur

$$\frac{1}{\lambda^2 L} \ll n \ll \frac{1}{\lambda^3} \quad . \tag{2.83}$$

This condition cannot be satisfied in a point sample, where the characteristic sample dimension $L \ll \lambda$.

2.5 Relation to Other Coherent Phenomena

Superradiance can be compared with other processes in which atoms respond cooperatively and emit coherent radiation. These include 1) coherent transient phenomena in optically thin samples; 2) transient phenomena in optically thick media, such as self-induced transparancy; and 3) "ordinary" stimulated emission. This section explores the relationships between superradiance and these various phenomena.

2.5.1 Limited Superradiance

As mentioned earlier, cooperative emission can occur even if (2.66) is not satisfied, i.e., when

$$\alpha L < \kappa L + \Phi \quad . \tag{2.84}$$

All that is required is that $T_R < T_{sp}$ (i.e., $n_0\lambda^2 L/8\pi > 1$), so that collective decay dominates over incoherent spontaneous emission. Historically, observations of cooperative emission of this type go back at least as far as HAHN's spin echo experiment of 1950 [2.68]. In the ensuing years this technique and related ones such as free induction decay and optical nutation have been developed for studying incoherent relaxation processes, both at long wavelengths and in the optical regime [2.25-27]. Such phenomena can be classified as "limited" superradiance in that, although the emitted radiation is due to the cooperative behavior of the medium, only a small fraction of the energy stored in the sample is emitted collectively, the decay of the sample being due primarily to incoherent relaxation processes. This situation can occur when

$$T_2' \ll T_R \ll T_{sp} \quad , \tag{2.85}$$

so that cooperative radiative decay is more rapid than incoherent spontaneous emission, yet slow compared to nonradiative relaxation. Note that the left hand inequality implies that $\alpha L \ll 1$, i.e., that the sample is optically thin. In this case the emitted radiation does not significantly react back on the sample, in contrast to the behavior of an optically thick sample, and so the coherent emission does not affect the decay rate. Thus, the coherent emission acts as a probe of the system without significantly affecting its state. This is precisely why such phenomena are useful for studies of incoherent decay processes. Examples of calculations in the limited superradiance regime are given in [Ref.2.56, Sect.5].

In the intermediate regime $1 < (\alpha - \kappa)L < \Phi$, the peak intensity will be much less than that of (2.37), and analytical results can be obtained from the linear theory of CRISP [2.69]. The output from an initially inverted system will become more directional as the gain is increased beyond one. The onset of directionality may prove to be an important indicator of superradiance in systems such as X-ray lasers, where the gain and other parameters may be difficult to measure directly.

2.5.2 Transient Phenomena in Optically Thick Media

Coherent phenomena in optically thick absorbers are a second class of processes closely related to superradiance. These include self-induced transparancy [2.35] and propagation of short pulses in a resonant absorber [2.42-44]. Such processes are characterized by an exchange of the energy of the incident pulse with the atoms as the pulse propagates through the sample, leading to pulse delay and reshaping. They differ from superradiance in that they are not primarily processes for releasing stored energy. However, all these effects are closely related, and all can be described semiclassically.

An important distinction between self-induced transparancy and superradiance can be drawn by considering inducing superradiance in an ideal two-level medium by *directly* inverting the sample with a pulse of area π (not the usual way of achieving superradiance). To attain complete inversion along the entire length of the sample the area of the inversion pulse must not change as it travels through and interacts with the medium. To avoid this the energy of the pulse must be much larger than that needed to completely invert the medium. For a square pulse of amplitude E and duration τ this requires that

$$\frac{E^2}{8\pi} cA\tau \gg \hbar\omega n_0 AL \quad . \tag{2.86}$$

For a π pulse ($\mu_z E\tau/\hbar = \pi$), this gives $\pi^2 T_2' \gg 2\tau\alpha L$, i.e.,

$$\tau \ll T_R \quad . \tag{2.87}$$

This is a far more stringent requirement on τ than (2.54).

It should be noted that the other extreme, $\tau \gg T_R$, insures complete absorption of the pulse by the medium, as occurs in self-induced transparency [2.35]. Thus, for direct inversion of a two-level system $\tau \ll T_R$ and $\tau \gg T_R$ serve to distinguish conditions under which superradiance and self-induced transparency, respectively, may be observed.

2.5.3 Stimulated and Superradiant Emission

Stimulated emission of the type which occurs in a laser amplifier can be viewed as a cooperative emission process, although it is not usually considered as such. Although this effect and superradiance appear to be quite different, they are actually closely related. Both can be described by the coupled Maxwell-Schrödinger equations (2.19). They are, in fact, two distinct limiting cases [2.56].

Superradiance is the transient limit, in which the polarization envelope and inversion density change rapidly on the time scale of incoherent decay processes

$$\frac{\partial P}{\partial T} \gg \frac{P}{T_2} \quad , \quad \frac{\partial n}{\partial T} \gg \frac{n}{T_1} \quad . \tag{2.88}$$

This requires, (a) high gain (2.66)

$$(\alpha - \kappa)L \gg 1 \quad ; \tag{2.89a}$$

and, (b) rapid inversion of the sample (2.54)

$$\tau \lesssim 2T_D \quad . \tag{2.89b}$$

As discussed in Sect.2.4.1, it then follows that the peak intensity of the emitted radiation pulse is proportional to N^2.

In the stimulated emission limit, on the other hand, P and n change slowly compared to incoherent decay rates, leading to a quasi-steady state. This can occur in high gain samples, as well as those with small gain, as long as (2.89b) is not satisfied. The emitted radiation intensity (for a high gain sample) is then proportional to N.

To see how stimulated emission fits into the framework of the Maxwell-Schrödinger equations, consider (2.19b,c) in the limit in which

$$\frac{\partial P}{\partial T} \ll \frac{P}{T_2} \quad , \quad \frac{\partial n}{\partial T} \ll \frac{n}{T_1} \quad . \tag{2.90}$$

As in Sect.2.4.1, we simplify to the homogeneously broadened limit (kv = 0) and neglect loss (κ = 0), level degeneracy (Assumption 4), and the polarization source term (Λ_p = 0). This gives[7]

$$P = \left(\frac{\mu^2 T_2}{\hbar}\right) nE \quad , \tag{2.91}$$

and

$$n = \frac{n_0}{1 + \left|\frac{\mu E}{\hbar}\right|^2 T_1 T_2} \quad , \tag{2.92}$$

---

[7]In some cases solutions (2.91,92) may not be stable, and spiking may occur in the output radiation [2.71].

where $n_0 = \Lambda T_1$ is the steady state inversion density when $E = 0$. Equation (2.92) is often written in the form

$$n = \frac{n_0}{1 + I/I_s} \quad , \tag{2.93}$$

where

$$I(x) = (cA/8\pi)EE^* \tag{2.94}$$

is the power emitted at position x, and I_s is the saturation power

$$I_s = \frac{cA}{8\pi}\frac{\hbar^2}{\mu^2 T_1 T_2} \quad . \tag{2.95}$$

The growth of I follows from (2.19a) and (2.94)

$$\frac{dI}{dx} = \frac{4\pi k\mu^2 T_2}{\hbar} nI \quad . \tag{2.96}$$

For weak fields ($I \ll I_s$) (2.96) gives the stimulated emission intensity gain equation,

$$\frac{dI}{dx} = 2\alpha I \quad , \tag{2.97a}$$

with α as defined in (2.9), leading to exponential growth

$$I(L) = I(0)e^{2\alpha L} \quad . \tag{2.97b}$$

For high intensities ($I \gg I_S$), however, $n \propto I^{-1}$, and so dI/dx becomes independent of I,

$$\frac{dI}{dx} = 2\alpha I_S \quad , \tag{2.98a}$$

leading to linear intensity growth

$$I(L) = I(0) + \frac{1}{2}\frac{\hbar\omega}{T_1} n_0 AL \quad . \tag{2.98b}$$

If I(0) is small, as when the intensity builds up from noise, this term may be dropped, giving

$$I(L) = \frac{1}{2}\hbar\omega\Lambda AL \tag{2.99a}$$

$$= \frac{1}{2} N \frac{\hbar\omega}{T_1} \quad , \tag{2.99b}$$

with $N = n_0 AL$. This equation can also be written as

$$\left|\frac{\mu E}{\hbar}\right|^2 T_1 T_2 = 2\alpha L \gg 1 \quad , \tag{2.100}$$

which shows that high gain is a necessary condition for linear buildup of coherent emission from noise.

When $I >> I_s$, so that (2.100) holds, (2.91,92) can be written as

$$n = n_0/2\alpha L << n_0 \quad , \tag{2.101a}$$

$$P = \mu n_0 \sqrt{\frac{T_2/T_1}{2\alpha L}} \quad . \tag{2.101b}$$

Thus, the effective number of radiators is severely reduced. In this limit both P and E are proportional to $\sqrt{N}$ (rather than N), and so $I \propto N$. Hence in the high intensity limit stimulated emission is proportional to the square of the effective number of radiators, just as is superradiant emission.

Equations (2.98-101) describe stimulated emission in a high gain, single-pass saturated amplifier (with no feedback). It is no less "coherent" than superradiant emission, being produced by a phase-coherent macroscopic polarization of the form (2.91). Coherent emission of this type also occurs in high gain "mirrorless" lasers of various kinds—nitrogen, dye, metal vapor, etc.[8] In exciting such systems, which are essentially single-pass saturated noise amplifiers, it is usually exceedingly difficult to satisfy (2.89b), because T_{sp} is typically 10^{-9} - 10^{-10} s and, since $\alpha L >> 1$, T_D is typically several orders of magnitude shorter. Hence such systems are generally prepared defacto in the quasi-steady state, even with ns or sub-ns inversion processes, giving rise to linear intensity growth, not superradiance. Intensity dependence of this type is sometimes called amplified spontaneous emission [2.52,53,70].

The high gain condition (2.100) necessary to achieve efficient single-pass emission—either superradiant or stimulated—is related to the threshold condition in a conventional laser, i.e., a low gain amplifying medium with regenerative feedback. For an amplifier of length L in a high-Q resonator of mirror reflectivity R, the single pass gain must exceed the loss. This gives the threshold condition $2\alpha L > 1 - R$, or

$$2\alpha L_{eff} > 1 \quad , \tag{2.102}$$

where $L_{eff} = L/(1 - R)$ is the effective length of the medium when placed in a resonator which can support $1/(1 - R)$ passes. This equation can be compared to the condition for efficient single-pass emission, $2\alpha L >> 1$, (2.100).

In summary, superradiant emission and stimulated emission in a high gain medium are the transient and steady-state forms of phase-coherent amplification, respectively. The transient nature of the superradiant process is further illustrated by the fact that observed superradiant pulses are shorter than the inverse bandwidth

[8]Unfortunately, these lasers are sometimes referred to as "superradiant" or "superfluorescent".

of the transition [2.5], a direct result of the requirement of high gain for achieving superradiance

$$\frac{\text{pulse duration}}{\text{inverse bandwidth}} = \frac{T_W}{T_2^1} \approx \frac{\Phi}{\alpha L} < 1 \quad . \tag{2.103}$$

In the HF experiments, for example, pulses as short as 10 ns were observed, which is several times shorter than the inverse bandwidth of the HF transition.

2.6 Experiments

2.6.1 Experimental Observation of Superradiance

Beginning in 1973, superradiance has been observed in a variety of atomic and moleculear transitions in the visible, infrared and microwave spectral regions [2.3-22]. The experiments have been made possible by the realization that an optical transition can be prepared in a state of complete inversion with no residual polarization by means of indirect pumping. In the initial experiments, [2.3-6] for example, a low pressure cell of HF gas was optically pumped by means of a short pulse from an HF infrared laser, thereby populating a particular rotational level, J, of the excited vibrational state. This three-level pumping scheme produced a population inversion in the excited state $J \rightarrow J-1$ rotational transition, which then emitted a superradiant burst of far infrared radiation, delayed by several hundred ns (Fig.2.1a). All subsequent superradiance experiments have utilized the principle of indirect inversion. Other inversion schemes have included three-level pumping [2.7-9,13-18,20,21], stimulated Raman pumping [2.11,12], and two-photon excitation, both direct [2.10,22] and indirect [2.19,21].

In indirect pumping the pump field is decoupled from the superradiant transition, thus avoiding the complications inherent in direct pumping schemes (Sect. 2.4.3, Assumption 11). Furthermore, indirect pumping enables one to invert a sample without leaving residual polarization. This can give rise to relatively long delays between pump pulse and superradiant pulse, which make it possible to study a "fast" process slowly. The absence of residual polarization also makes it possible to explore the quantum fluctuations which initiate coherent emission in a medium prepared in a completely inverted state (Sect.2.3.1).

The experimental arrangement for producing superradiance is thus relatively simple. A radiation pulse from the pump laser(s) is incident on the sample cell. After an appropriate delay the superradiant pulse is emitted along the propagation axis of the pump pulse, at the wavelength of the coupled transition. The pump pulse is separated from the superradiant pulse in time, and can be further discriminated against by means of filters or a monochromator, since the two pulses are at different wavelengths. The pump pulse must be short compared to the delay time of the emitted pulse. Otherwise, the sample will start to superradiate before complete inversion is achieved (Sect.2.4.3) and, in addition, unwanted coupling between radiation at the pump and superradiant transitions will occur. Furthermore, feedback must be kept to a minimum to avoid shortening the pulse delays and distorting the pulse shapes (Sect.2.4.3).

The parameter most easily measured is the delay time which, according to (2.40), should scale as N^{-1} (inversely with pressure for a gas sample). Another key parameter is the peak intensity which should, of course, scale as N^2 (pressure squared). These dependences have been verified in several of the experiments. Data for HF are given in Fig.2.4.

Up to now all experiments have been performed in gases at low pressures. Both sample cell and atomic beam (to minimize T_2^*) configurations have been used. A summary of relevant experimental parameters is given in Table 2.1.

2.6.2 Recent Experimental Results

This section summarizes some of the interesting phenomena which have been observed in superradiance experiments.

Several types of beats have been observed in superradiant output pulses (Fig. 2.10). Superradiant beats—i.e., oscillatory modulation superimposed on the superradiant pulse envelope—can occur when the inverted medium is able to simultaneously amplify two (or more) closely spaced frequencies. VREHEN et al. [2.13] have observed hyperfine-structure beats in the 7P-7S transitions of atomic cesium (λ = 2.9 and 3.1 μm). For example, by pumping to the $7P_{1/2}$ level with a broadband laser they induce modulation beats at 400 MHz., corresponding to the $7P_{1/2}F = 3 \rightarrow F = 4$ hyperfine splitting. GROSS et al. [2.18] have observed *Doppler* beats at this same transition, by inverting the system by means of a *narrowband* (100 MHz) pump laser. In this case a specific velocity group is excited in each hyperfine level, giving rise to beat frequencies offset from the $F = 4 \rightarrow F = 3$ hyperfine splitting by 59 MHz, the Doppler shift between the two velocity groups at the superradiating transition.

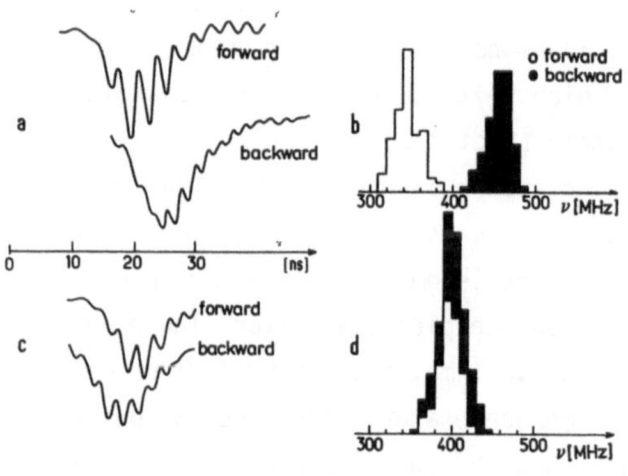

Fig.2.10a-d. Superradiant beats observed in cesium. (a,b): Doppler beats observed in forward and backward directions induced by narrowband excitation; (c,d): Hyperfine beats induced by broadband excitation. Histograms (b) and (d) give the beat frequencies. As can be seen, the Doppler beats are shifted in frequency from the hyperfine beats by ± 59 MHz. [2.18]

Superradiant beats are inherently different from the quantum beats observed in ordinary spontaneous emission [2.72], in that the latter are due to quantum interference between coherently prepared states of a single atom. In contrast, superradiant beats are caused by interference between (classical) superradiant e.m. fields, each generated by coherent emission from many atoms. Hence, superradiant beats can be observed in cases in which quantum beats are forbidden, and even between physically different atoms. As an extreme example, Vrehen, Andersen and der Weduwe (reported in [2.17]) have observed the beatnote between superradiant pulses emitted from two independently radiating samples.

The ringing (Fig.2.1) predicted by the semiclassical model of Sect.2.3.2 can be viewed as a beatnote of a different type occurring in a nondegenerate two-level transition—a Rabi modulation of the superradiant emission field due to its interaction with the collective atomic system. The appropriate Rabi frequency is $\sim\mu E_p/h$, with E_p the peak value of the superradiant emission field. Equations (2.37, 38) give

$$\frac{\mu E_p}{4\hbar} T_w \approx 1 \quad , \tag{2.104}$$

which supports the above interpretation.

Cascade superradiance, i.e., superradiant emission between sequential cascade transitions, has been observed in several systems [2.4,10,20,21]. Cascading can occur when the population buildup in the lower level of the initial superradiating transition is sufficient to produce high gain at the coupled sequential transition. A detailed study of this effect has been carried out by CRUBELLIER et al. [2.20,21] who studied the polarization characteristics of the light emitted in the $6p \rightarrow 6s \rightarrow 5p$ and $6p \rightarrow 4d \rightarrow 5p$ cascade transitions of rubidium (Fig.2.11). They found that the polarization states of sequential pulses depend on the polarization of the pump pulse and the angular momenta of the energy levels which form the cascade (Table 2.2). The results are explained using a semiclassical model in which level degeneracy is taken into account. In certain transitions, where no preferential polarization direction is expected, linearly polarized emission is observed, the orientation of which changes randomly from pulse to pulse. This is an interesting manifestation of the initiating quantum fluctuations.

GROSS et al. [2.10] have reported evidence of the transition from superradiant emission to stimulated emission, discussed in Sect.2.5.3, in the $5S \rightarrow 4P$ transition of sodium (λ = 3.4 μm). They observed that at densities which are sufficiently high so that the superradiant pulse is emitted before the inversion pulse terminates, the emitted intensity becomes proportional to N rather than N^2. This illustrates the discussion of Sect.(2.5.3): When the upper level continues to be pumped while the superradiant pulse is being emitted, the transient condition (2.54) cannot be sustained and the response of the system shifts to a quasi-steady state. "Pre-superradiant" behavior, which occurs when $\alpha L < \Phi$, has also been studied, in the $3S \rightarrow 2P$ transition (λ = 0.81 μm) in Li [2.19].

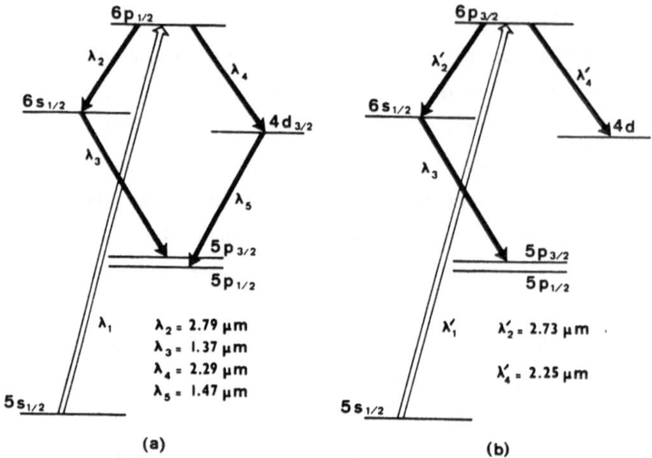

Fig.2.11. Energy level diagrams for studying cascade superradiance in Rb. The pump wavelengths are $\lambda_1 = 421.5$ nm, $\lambda_1' = 420.2$ nm. [2.21]

Table 2.2. Polarization states observed in cascade superradiance in Rb (Fig.2.11). For the linearly polarized cases (↑) indicates that the polarization direction is the same as for the exciting light, and (R) indicates that it varies randomly from shot to shot. [2.21]

Pump transition	$5s_{1/2} \to 6p_{1/2}$		$5s_{1/2} \to 6p_{3/2}$	
Pump polarization	circ(σ^+)	linear(↑)	circ(σ^+)	linear(↑)
Superradiant Transition				
6p → 6s	circ(σ^-)	linear(R)	circ(σ^-)	linear(↑)
6s → 5p	circ(σ^-)	linear(R)	circ(σ^+)	linear(R)
6p → 4d	circ(σ^+)	linear(R)	circ(σ^+)	linear(↑)
4d → 5p	circ(σ^-)	linear(R)	circ(σ^-)	linear(↑)

BOWDEN and co-workers [2.8,9] have studied far infrared ($\lambda = 496$ μm) superradiant emission in a long, homogeneously broadened sample of CH_3F prepared by means of swept inversion produced by a truncated CO_2 laser pulse. For short samples pulses of similar delays and intensities are observed in both forward and backward directions. (The "forward" pulse is emitted in the propagation direction of the pump radiation). However, for longer samples ($L \gtrsim cT_2/2$ [2.8,9]) the forward pulses become narrower and more intense; at sufficiently high pressures the forward-backward intensity ratio is reported to exceed 100. This type of behavior is expected, since the forward emission should continue to grow as predicted by (2.19), whereas the backward wave should become negligible because of retardation effects of the type referred to in the discussion of Assumption 7 (Sect.2.4.3).

Recent experiments have extended the wavelength range over which superradiance has been observed at both ends of the spectrum. Production of visible superradiance has been reported by CRUBELLIER et al. [2.21] in three strontium transitions, λ = 647,655 and 461 nm, the latter being a ground state transition. Visible superradiance in Eu has also been reported (see [Ref.2.21, Refs.12,13]). Superradiant emission at submillimeter wavelengths has been observed by HAROCHE et al. [2.23] from nS and nD Rydberg states ($20 < n < 50$) in cesium and sodium in an atomic beam. (See also T.F. Gallagher and W.E. Cooke, [Ref.2.22, Ref.5]). In most of the transitions studied so far the sample length is longer than the wavelength. Quantitative studies of long wavelength Rydberg superradiance may serve to test theories of point sample superradiance (Sect.2.4.5).

2.6.3 Comparison with Theory

This section reviews the current state of understanding of superradiance, as indicated by the extent of agreement of the experimental results with theory. There is now general agreement about the principal features: The N^2 dependence of the peak intensity (2.37) and the N^{-1} dependence of the pulse delay and width (2.38, 40) has been confirmed (as in Fig.2.4) in several different systems. However, there is not yet a complete understanding of the origin and extent of the ringing observed under various experimental conditions, nor agreement on the correct expression for the effective tipping angle. The experimental study of shot-to-shot variations in the features of the emitted radiation caused by quantum fluctuations has just begun.

GIBBS, VREHEN and their collaborators have performed a series of careful experiments [2.13-17] on the $7P_{3/2}\rightarrow 7S_{1/2}$ transition of atomic cesium (λ = 2.93 μm) to explore the detailed behavior of superradiant emission. Pumped by a circularly polarized 2 ns dye laser pulse on the $6S_{1/2}\rightarrow 7P_{3/2}$ transition (λ = 455 nm) in a ~2.8 kG magnetic field, a nondegenerate two-level system is achieved. The superradiant emission occurs along the propagation axis of the pump pulse, hence the inversion is swept (Sect.2.4.3, Assumption 7). Both cell and atomic beam samples are used, both with a Fresnel number $F \sim 1$; and care is taken to minimize feedback, achieve a pump pulse with a Gaussian profile, and carefully measure experimental parameters such as the density of the sample gas.

Examples of superradiant output pulses observed in cesium are shown in Fig. 2.12 [2.14a]. A major finding of these experiments is that the extent of ringing is considerably smaller than that predicted by the semiclassical equations (2.19). Ringing is always less pronounced than expected and it is completely absent for delays longer than about 7 ns [2.14b]. The changeover to the "single-pulse" regime is rather abrupt, and is observed to occur at approximately $T_R = 2L/c$ for samples

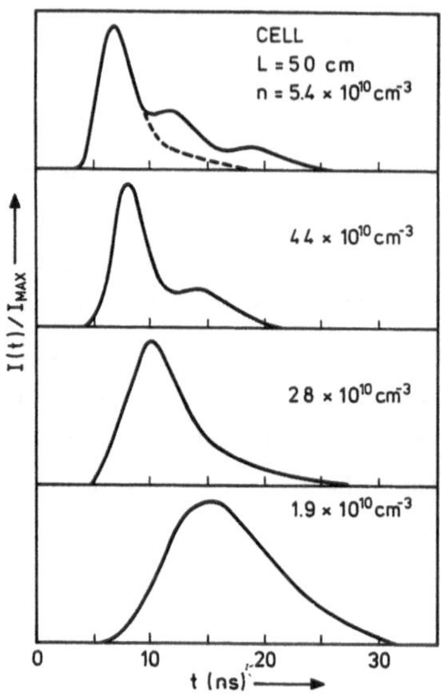

Fig.2.12. Superradiant emission in cesium, normalized pulse shapes. The data is for a 5 cm cell in which T_{sp} = 551 ns, T_1 = 70 ns, T_2 = 80 ns, T_2^* = 5 ns. The inversion densities are accurate to (+60, -40)%. [2.14b]

of several different lengths [Ref.2.14a, Table II]. More recent experiments [2.15] using samples of various Fresnel numbers indicate that the observed ringing in the cesium experiments is better described as multiple pulse emission caused by transverse field variations: In samples of large cross section, with $F \gg 1$, the medium breaks into regions which tend to superradiate independently (see the discussion of Sect.2.4.4).

Pulses without ringing are predicted by the semiclassical model under some conditions—for sufficiently large loss, or when the delay time is sufficiently long that effects of decay and Doppler dephasing set in (Fig.2.5). The predicted changeover, however, is more gradual than that observed in cesium. The HF experiments also observed a regime of no ringing for sufficiently long delays (see, for example, [Ref.2.3, Fig.2b]) but this behavior was in agreement with the semiclassical predictions, based on the measured values of the experimental parameters. These parameters, however, were not known as precisely as in the cesium experiments, and could have been in error. Ringing has also been observed in other systems, for example, in the sodium experiments of [2.10].

Earlier, it was suggested [2.14] that the lack of ringing observed in cesium for $T_R \gtrsim 2L/c$ might be evidence for the mean field theory of BONIFACIO and LUGIATO [2.23], which predicts a regime of "pure superfluorescence" with sech^2 pulse shapes when T_R exceeds the "escape time" $\tau_E = L/c$. In fact, some of the cesium pulses fit well to a sech^2 pulse shape [2.14a]. However, the recent quantized field treatments discussed in Sect.2.3.1 confirm the validity of the semiclassical model, the predictions of which do not agree with mean field theories, at least for cases of experimental interest. Lack of ringing is now generally attributed [2.31,34,36] to non-plane wave effects of the type discussed in Sect.2.4.4.

The details of superradiant pulse shapes and the importance of ringing under various conditions is not yet well understood, and requires further exploration, both experimental and theoretical.

VREHEN and SCHUURMANS [2.16] have measured the effective tipping angle, Θ_0, in an inverted sample of cesium (Sect.2.4.2). The experimental setup (Fig.2.13a) consists of two cesium cells inverted in tandem by a pulse from the pump laser. By means of an infrared attenuator placed between the cells, the superradiant emission from the first cell serves as a coherent pulse of (small) variable area, Θ, resonant with the superradiating transition. This pulse is injected into the second cell (which is filled to a lower pressure and, therefore, emits pulses having much longer delays), and the delay time of its superradiant output is studied as a function of Θ (Fig.2.13b). For sufficiently small values of Θ there is no change in delay, but for $\Theta > \Theta_0$ the delay time decreases. The crossover occurs at $\Theta = \Theta_0$.

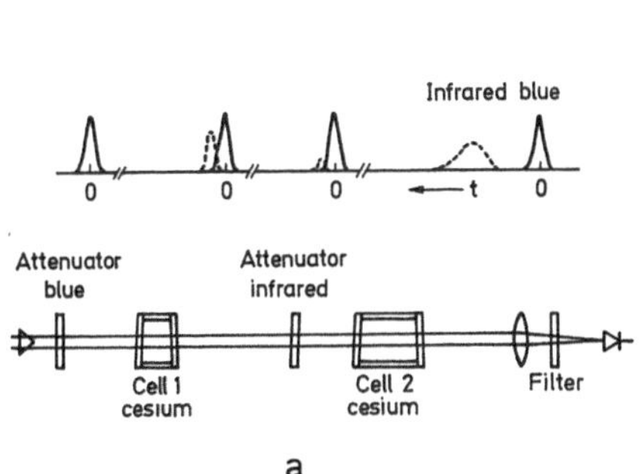

a

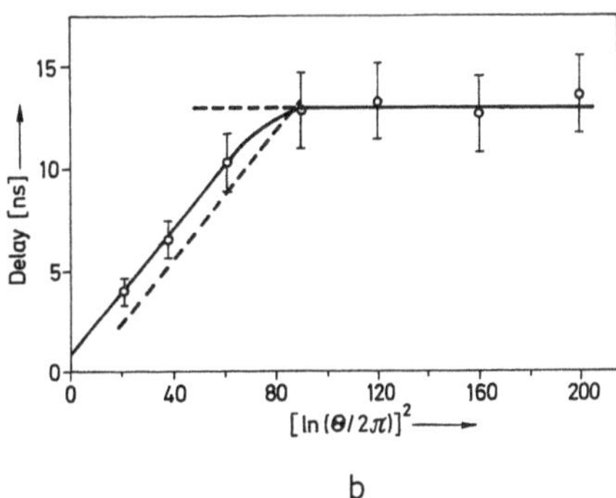

b

Fig.2.13a,b. Measurement of Θ_0 in cesium. (a) Experimental setup; (b) Delay time of output pulse vs $[\ln(\Theta/2\pi)]^2$. The dashed line is used to correct for the delay of the injected pulse with respect to the pump pulse. [2.16]

The data of Fig.2.13b, obtained by adjusting the density of the second cell to give a delay of 13 ns, give a value of Θ_0 in the range $10^{-4} < \Theta_0 < 2.5 \times 10^{-3}$. Under these conditions the number of inverted cesium atoms is measured to be $N \simeq 2 \times 10^8$. Using this number, it is found that the expressions for Θ_0 given by (2.44,47) are included in this range, while the Θ_0 of (2.52) differs by at least one order of magnitude. This experiment thus supports the Θ_0 expressions of GLAUBER-HAAKE [2.32] and POLDER et al. [2.30].

This experiment has the advantage that it measures Θ_0 directly, and thus, avoids the logarithmic dependence entailed in measuring the pulse delays or other associated output parameters. It is not, however, unaffected by measurement errors in N (as the authors claim), since N must be known for comparison with theory.

In a recent experiment in cesium VREHEN [2.17] has studied the shot-to-shot variations in superradiant pulse delay due to quantum fluctuations (Sect.2.4.2).

These measurements are difficult because pump pulse variations can also cause fluctuations in the delay time which are difficult to differentiate.[9] In the cesium experiments this problem is circumvented by measuring differences in delays from pairs of pulses emitted from two similar sample cells inverted simultaneously by the same pump pulse. The experimental setup (Fig.2.14a) consists of two cells of the same length filled to the same cesium pressure. The detected signal from one of the cells is given an additional delay so that both pulses can be displayed on the same trace. Figure 2.14b plots the observed distribution of delay time differences. These data yield a relative standard deviation of $\sigma_D = (13 \pm 3)\%$. This result agrees with (2.53), derived by POLDER et al. [2.30], which predicts $\sigma_D = 12.5\%$ for the conditions of the experiment, and also with the numerical prediction of $\sigma_D = 12\%$ obtained by HAAKE et al. [2.34] (Fig.2.6).

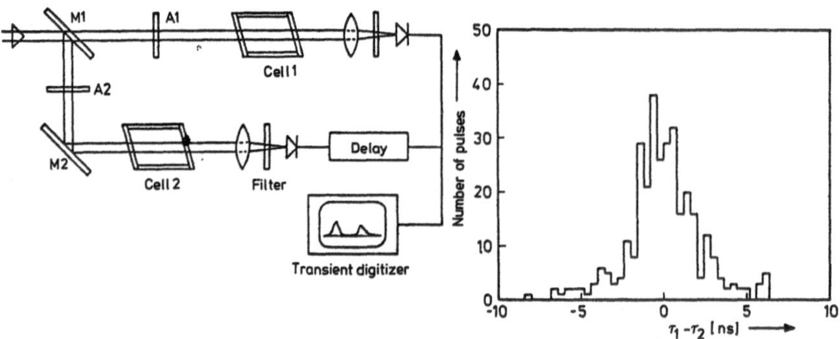

Fig.2.14a,b. Measurement of delay time fluctuations in cesium superradiant emission pulses. (a) Experimental setup; (b) Distribution of delay time differences for a set of 323 pulses (compare Fig.2.6). The average delay time is 10.2 ns. [2.17]

2.7 Concluding Remarks

2.7.1 Applications

I) *Applications to Basic Research.* Basic applications of superradiance include its use in the submillimeter as a detector-amplifier and its application to the study of paramagnetic systems (spin-phonon superradiance). HAROCHE and co-workers have emphasized the use of superradiance in Rydberg atoms as amplifiers of submillimeter and microwave radiation.[10] By cooling the sample cell (to reduce blackbody radiation, which triggers the emission pulse for $\lambda > 50$ µm) these systems should be

[9] Fluctuations in pulse delay have been observed in earlier experiments [2.3,4,12], but their origin was unclear; those observed in the HF studies, e.g. [Ref.2.3, Fig.2b], could have been caused by pump pulse variations.

[10] Ref.2.22. See also Ref.2.4.

capable of amplifying very weak signals. Tunability could be achieved by Stark-shifting the highly polarizable Rydberg levels. Rydberg microwave experiments may be useful for fundamental studies of matter-radiation interactions, as well as for technological applications.

Spin-phonon superradiance is another potential area of interest. It may be possible to observe this acoustical analog of superradiance in the spin-phonon interaction process in paramagnetic crystals [2.73]. In such a system the paramagnetic spins are coupled to the lattice vibrations. As shown in [2.74], in the slowly varying envelope approximation the coupled spin-phonon equations become almost identical in form to (2.19).[11] Acoustical gain can be suitably defined, and so in a high gain medium it should be possible for an initially inverted ensemble of spins, perturbed by kT fluctuations, to rapidly transfer its stored energy to the lattice. The ensuing acoustic waves should have all the properties of the coherent emission observed in optical superradiance.

Recently, HAHN and WILSON [2.73] proposed a related experiment to observed superradiant emission in a spin-phonon system by preparing the spins in a phased array. The phonon avalanche experiment of BRYA and WAGNER [2.75], although probably not a true coherent effect, is an interesting advance along these lines.

II) *Practical Applications.* The requirements for efficient superradiant emission should also be of interest to designers of X-ray laser systems. Due to the short lifetimes of the transitions and the lack of suitable mirrors in this regime, most proposed schemes use a single pass high gain swept-excitation system. Thus, X-ray lasers will superradiate. Consequently, the conventional rate equation analysis is not applicable, and the above considerations can be useful to estimate the output behavior. Some of the discussions in Sect.2.4.3 are especially relevant to the X-ray regime; in particular, T_1 is usually so short that the inversion process will not be completed by the time superradiance occurs (Assumption 6).

As mentioned above, the rate equation analysis gives incorrect results. For example, in the Na scheme of DUGUAY and RENTZEPIS [2.77], rate equation analysis predicts (at the threshold value) I_p about 10 times smaller and T_W about 10 times larger than the semiclassical predictions [2.76]. In addition, the threshold inversion density is a factor of 10 smaller than the corresponding rate equation threshold.

Specific applications of these requirements to X-ray laser schemes are discussed further in [2.76].

It is also interesting to consider superradiance as a method for generating ultrashort pulses, since superradiance is the optimum method for extracting coherent energy from an inverted medium. Although in the ideal case T_W decreases

[11] The similarity between the Maxwell-Schrödinger and spin-phonon equations has also been exploited by SHIREN [2.78] to observe an acoustical analogue of self-induced transparency.

with increasing N, many of the conditions listed above restrict the shortness of output pulses one can hope to achieve. Combining (2.38,40 and 54) shows that the inversion time τ places a particularly restrictive limit on the minimum T_W which can be generated via superradiance

$$T_W \gtrsim 2\tau/\Phi \quad . \qquad (2.105)$$

Therefore, ultrashort pulse generation by this method requires swept excitation, small κ, and as short an inversion time as possible. Values of T_W/τ less than 1/10 appear possible.

Note that superradiance is a transient process, and so the generation of ultrashort pulses by this method is inherently different from the mode locking approach, where short pulses are generated by mixing a set of equally spaced phase correlated modes to synthesize a Fourier spectrum.

2.7.2 Summary

It is now firmly established that superradiance is a simple, fundamental effect. It can be viewed as the transient form of stimulated emission and can be described semiclassically, in terms of the coupled Maxwell-Schrödinger equations subject to appropriate initial conditions. It has been observed in a variety of systems, both atomic and molecular, spanning a wide range of the electromagnetic spectrum.

The conditions required to achieve superradiant emission are now well established. Also well established are its inherent features, $I \propto N^2$, and T_W and $T_D \propto N^{-1}$. However, the understanding of the shape of the emitted pulses, the occurrence of ringing, and the exact details of the quantum processes which determine the pulse delay and other parameters are still under debate.

Theories describing the quantum initiation process have been developed, and some initial studies have been performed to explore effects due to quantum fluctuations. A set of conditions under which ideal superradiance can be observed has been derived and also expressions for the modifications which occur when these conditions are not satisfied, although most of the latter predictions have not been tested. Several potentially important applications of superradiance based on this understanding have been proposed.

The time is right to use the detailed knowledge obtained from theory and the recent experience of the experiments to branch out into new regimes and applications.

Acknowledgement. The authors thank Richard Forber for helpful comments.

References

2.1 R.H. Dicke: Phys. Rev. *93*, 99 (1954)
2.2 R.H. Dicke: In *Proceedings of the Third International Conference on Quantum Electronics*, Paris, 1963, ed. by P. Grivet, N. Bloembergen (Columbia University Press, New York 1964) p.35
2.3 N. Skribanowitz, I.P. Herman, J.C. MacGillivray, M.S. Feld: Phys. Rev. Lett. *30*, 309 (1973)
2.4 N. Skribanowitz: Ph.D. Thesis, Massachusetts Institute of Technology (1973) unpublished
2.5a R.G. Brewer, A. Mooradian (eds.): *Laser Spectroscopy* (Plenum Press, New York 1974)
2.5b I.P. Herman, J.C. MacGillivray, N. Skribanowitz, M.S. Feld: In Ref.2.5a
2.6 P.T. Ho: S.M. Thesis, Massachusetts Institute of Technology (1975) unpublished
2.7a C.M. Bowden, D.W. Howgate, H.R. Robl (eds.): *Cooperative Effects in Matter and Radiation* (Plenum Press, New York 1977)
2.7b A.T. Rosenberger, S.J. Petuchowski, T.A. DeTemple: In Ref.2.7a
2.8a L. Mandel, E. Wolf (eds.): *Coherence and Quantum Optics IV* (Plenum Press, New York 1978)
2.8b J.J. Ehrlich, C.M. Bowden, D.W. Howgate, S.H. Lehnigk, A.T. Rosenberger, T.A. DeTemple: In Ref. 2.8a
2.9 A.T. Rosenberger, T.A. DeTemple, C.M. Bowden, C.C. Sung: J. Opt. Soc. Am. *68*, 700 (1978)
2.10 M. Gross, C. Fabre, P. Pillet, S. Haroche: Phys. Rev. Lett. *36*, 1035 (1976)
2.11 A. Flusberg, T. Mossberg, S.R. Hartmann: In Ref.2.7a, p.37
2.12 A. Flusberg, T. Mossberg, S.R. Hartmann: Phys. Lett. *58*A, 373 (1976); Phys. Rev. Lett. *38*, 59 (1977)
2.13 Q.H.F. Vrehen, H.M.J. Hikspoors, H.M. Gibbs: Phys. Rev. Lett. *38*, 764 (1977)
H.M. Gibbs: In Ref.2.7a, p.61
2.14a Q.H.F. Vrehen: In Ref.2.7a
2.14b H.M. Gibbs, Q.H.F. Vrehen, H.M.J. Hikspoors: Phys. Rev. Lett. *39*, 547 (1977)
2.14c J.L. Hall, J.L. Carlsten (eds.): *Laser Spectroscopy III*, Springer Series in Optical Sciences, Vol.7 (Springer, Berlin, Heidelberg, New York 1977)
2.14d H.M. Gibbs, Q.H.F. Vrehen, H.M.J. Hikspoors: In Ref.2.14c, p.213
2.15 Q.H.F. Vrehen: In Ref.2.8a
2.16 Q.H.F. Vrehen, M.F.H. Schuurmans: Phys. Rev. Lett. *42*, 224 (1979)
2.17a H. Walther, K.W. Rothe (eds.): *Laser Spectroscopy IV*, Springer Series in Optical Sciences, Vol.21 (Springer, Berlin, Heidelberg, New York 1979)
2.17b Q.H.F. Vrehen: In Ref.2.17a, p.471
2.18 M. Gross, J.M. Raimond, S. Haroche: Phys. Rev. Lett. *40*, 1711 (1978)
2.19 J. Okada, K. Ikeda, M. Matsuoka: Opt. Commun. *27*, 321 (1978)
2.20 A. Crubellier, S. Liberman, P. Pillet: Phys. Rev. Lett. *41*, 1237 (1978)
2.21 A. Crubellier, C. Brechignac, P. Cahuzac, P. Pillet: In Ref.2.17a, p.480
2.22 M. Gross, P. Goy, C. Fabre, S. Haroche, J.M. Raimond: Phys. Rev. Lett. *43*, 343 (1979)
S. Haroche, C. Fabre, P. Goy, M. Gross, J.M. Raimond: In Ref.2.17a, p.244
2.23 R. Bonifacio, L.A. Lugiato: Phys. Rev. A*11*, 1507 (1975); *12*, 587 (1975) see also panel discussion in Ref.2.7a
2.24 J.C. MacGillivray, M.S. Feld: Phys. Rev. A*14*, 1169 (1976)
2.25 R.G. Brewer, R.L. Shoemaker: Phys. Rev. A*6*, 2001 (1972)
R.L. Shoemaker, R.G. Brewer: Bull. Amer. Phys. Soc. *17*, 66 (1972)
P.F. Liao, S.R. Hartmann: Phys. Lett. *44*A, 361 (1973)
C.L. Tang, B.D. Silvermann: In *Physics of Quantum Electronics*, ed. by P.L. Kelley, B. Lax, P. Tannenwald (McGraw-Hill, New York 1966)
2.26 R.G. Brewer, A.Z. Genack, S.B. Grossman: In Ref.2.14c
2.27 N.A. Kurnit, I.D. Abella, S.R. Hartmann: Phys. Rev. Lett. *13*, 567 (1964); I.D. Abella, N.A. Kurnit, S.R. Hartmann: Phys. Rev. *141*, 391 (1966); C.K.N. Patel, R.E. Slusher: Phys. Rev. Lett. *20*, 1087 (1968); J.P. Gordon, C.H. Wang, C.K.N. Patel, R.E. Slusher, W.J. Tomlinson: Phys. Rev. *179*, 294 (1969); B. Bölger, J.C. Diels: Phys. Lett. *28*A, 401 (1968); R.G. Brewer, R.L. Shoemaker: Phys. Rev. Lett. *27*, 631 (1971)

2.28 F. Bloch: Phys. Rev. *70*, 460 (1946)
2.29 Various definitions of collective damping times are discussed by R. Friedberg, S.R. Hartmann: Phys. Rev. A*13*, 495 (1976)
2.30 D. Polder, M.F.H. Schuurmans, Q.H.F. Vrehen: Phys. Rev. A*19*, 1192 (1979)
2.31 M.F.H. Schuurmans, D. Polder: In Ref.2.17a, p.459
2.32 R. Glauber, F. Haake: Phys. Lett. *68*A, 29 (1978)
2.33 F. Haake, H. King, G. Schröder, J. Haus, R. Glauber, F. Hopf: Phys. Rev. Lett. *42*, 1740 (1979)
2.34 F. Haake, H. King, G. Schröder, J. Haus, R. Glauber: Phys. Rev. A*20*, 2047 (1979)
F. Haake: In Ref.2.17a, p.451
2.35 S.L. McCall, E.L. Hahn: Phys. Rev. *183*, 457 (1969)
2.36 J.C. MacGillivray, M.S. Feld: Phys. Rev. A (to be published)
2.37 R. Friedberg, S.R. Hartmann: Phys. Lett. *38*A, 227 (1972); *37*A, 285 (1971)
2.38 A. Gamba: Phys. Rev. *110*, 601 (1958); A.D. Gazazyan: Zh. Eksp. Teor. Fiz. *51*, 1863 (1966) [Sov. Phys. JETP *24*, 1254 (1967)]; M. Tavis, F.W. Cummings: Phys. Rev. *170*, 379 (1968); R.H. Lehmberg: Phys. Rev. *181*, 32 (1969); R. Bonifacio, G. Preparata: Nuovo Cimento Lett. *1*, 887 (1969); V.S. Letokhov: Zh. Eksp. Teor. Fiz. *53*, 2210 (1967) [Sov. Phys. JETP *26*, 1246 (1968)]; L.A. Shelepin: Zh. Eksp. Teor. Fiz. *54*, 1463 (1968) [Sov. Phys. JETP *27*, 784 (1968)]; W.R. Mallory: Phys. Rev. *188*, 1976 (1969); C.R. Stroud, Jr.: In *Proceedings of 1970 Rochester Symp.*, University of Rochester (1970), ed. by J.H. Eberly, p.77 (unpublished); J.H. Eberly, N.E. Rehler: Phys. Rev. A*2*, 1607 (1970); F.T. Arecchi, D.M. Kim, I.W. Smith: Nuovo Cimento Lett. *3*, 598 (1970), R. Bonifacio, G. Preparata: Phys. Rev. A*2*, 336 (1970); A.M. Ponte Goncalves et al.: Phys. Rev. *188*, 576 (1969); Phys. Rev. A*1*, 1472 (1970); E.T. Jaynes, F.W. Cummings: Proc. IEEE *51*, 89 (1963); M.D. Crisp, E.T. Jaynes: Phys. Rev. *179*, 1253 (1969); C.R. Stroud, Jr., E.T. Jaynes: Phys. Rev. A*1*, 106 (1970); M.J. Stephen: J. Chem. Phys. *40*, 669 (1964); M. Dillard, H.R. Robl: Phys. Rev. *184*, 312 (1969); R. Bonifacio, P. Schwendimann, F. Haake: Phys. Rev. A*4*, 302, 854 (1971); R. Bonifacio, P. Schwendimann: Lett. Nuovo Cimento *3*, 509, 512 (1970)
2.39a N.E. Rehler, J.H. Eberly: Phys. Rev. A*3*, 1735 (1971)
2.39b J.H. Eberly: Am. J. Phys. *40*, 1374 (1972)
2.40 S. Stenholm: Phys. Rep. *6*C, 1 (1973) pp.88-109
2.41 V. Ernst, P. Stehle: Phys. Rev. *176*, 1456 (1968)
G.S. Agarwal: Phys. Rev. A*2*, 2038 (1970)
R.H. Lehmberg: Phys. Rev. A*2*, 883, 889 (1970)
D. Dialetis: Phys. Rev. A*2*, 599 (1970)
C.M. Bowden, C.C. Sung: Phys. Rev. A*20*, 2033 (1979) and references therein
2.42 F.T. Arecchi, R. Bonifacio: IEEE QE-*1*, 169 (1965)
2.43 F.A. Hopf, M.O. Scully: Phys. Rev. *179*, 399 (1969)
2.44 A. Icsevgi, W.E. Lamb, Jr.: Phys. Rev. *185*, 517 (1969)
2.45 F.T. Arecchi, E. Courtens: Phys. Rev. A*2*, 1730 (1970)
2.46 D.C. Burnham, R.Y. Chiao: Phys. Rev. *188*, 667 (1969)
2.47 J.C. MacGillivray, M.S. Feld: In Ref.2.7a, p.1
2.48 J.C. MacGillivray: Sc.D. Thesis, Massachusetts Institute of Technology (1978) unpublished
2.49 C.K. Rhodes, A. Szöke, A. Javan: Phys. Rev. Lett. *21*, 1151 (1968)
2.50 J.A. Hermann: Phys. Lett. *69*A, 316 (1979)
2.51 A. Einstein: Phys. Z. *18*, 121 (1917); Verh. Dtsch. Phys. Ges. No. 13/14 (1916)
M.W.P. Strandberg: Phys. Rev. *106*, 617 (1957)
2.52a M.S. Feld, N.A. Kurnit, A. Javan (eds.): *Fundamental and Applied Laser Physics* (Wiley, New York 1973)
2.52b See, for example, J.H. Parks: In Ref. 2.52a and references therein
2.53 M.F.H. Schuurmans, D. Polder: Phys. Lett. (to be published)
2.54 J.P. Wittke, P.J. Warter: J. Appl. Phys. *35*, 1668 (1964)
2.55 R. Bonifacio, F.A. Hopf, P. Meystre, M.O. Scully: Phys. Rev. A*12*, 2568 (1975)
2.56 M.S. Feld: In *Frontiers of Laser Spectroscopy*, ed. by R. Balian, S. Haroche, S. Liberman (North-Holland, Amsterdam 1977) Sect.5.3
2.57 N. Bloembergen, R.V. Pound: Phys. Rev. *95*, 8 (1954)

2.58 See, for example, M.S. Feld: In Ref.2.52a, p.369
2.59 R. Saunders, S.S. Hassan, R.K. Bullough: J. Phys. A*9*, 1725 (1976)
R. Saunders, R.K. Bullough: In Ref.2.7a, p.209
2.60 R.K. Bullough, R. Saunders, C. Feuillade: In Ref.2.8a
2.61 F.P. Mattar, M.C. Newstein: IEEE QE-*13*, 507 (1977); in Ref.2.7a; see also panel discussion therein
2.62 N. Wright, M.C. Newstein: Opt. Commun. *9*, 8 (1973)
B.R. Sudyam: NBS special publication 387, p.42-48 (1973); IEEE QE-*10*, 837 (1974); IEEE QE-*11*, 225 (1975)
J.A. Fleck, Jr., R.L. Carman: Appl. Phys. Lett. *22*, 546 (1973)
2.63 H.M. Gibbs, B. Bölger, F.P. Mattar, M.C. Newstein, G. Forster, P.E. Toschek: Phys. Rev. Lett. *37*, 1743 (1976)
2.64 J.D. Jackson: *Classical Electrodynamics* (Wiley, New York 1962)
2.65aR. Friedberg, S.R. Hartmann, J.T. Manassah: Phys. Lett. *40*A, 365 (1972)
2.65bL. Mandel, E. Wolf: *Coherence and Quantum Optics* (Plenum Press, New York 1973)
2.65cR. Friedberg, S.R. Hartmann, J.T. Manassah: In Ref.2.65b
2.65dR. Friedberg, S.R. Hartmann: Opt. Commun. *10*, 298 (1974)
2.66 R. Friedberg, S.R. Hartmann: Phys. Rev. A*10*, 1728 (1974)
2.67 V. Degiorgio: Opt. Commun. *2*, 362 (1971)
2.68 E.L. Hahn: Phys. Rev. *80*, 580 (1950)
2.69 M.D. Crisp: Phys. Rev. A*1*, 1604 (1970)
2.70 L.W. Casperson, A. Yariv: IEEE J. QE-*8*, 80 (1972)
G.I. Peters, L. Allen: J. Phys. A*4*, 238 (1971)
L. Allen, G.I. Peters: J. Phys. A*4*, 564 (1971)
2.71 M.M. Miller, A. Szöke: In Ref.2.65b, p.885
2.72 Single atom quantum beats are reviewed by I.D. Abella, A. Compaan, L.Q. Lambert: In Ref.2.5a
2.73 See references contained in E.L. Hahn and R. Wilson: In Proc. XIXth Congress Ampère, Heidelberg (1976)
2.74 C. Leonardi, J.C. MacGillivray, S. Liberman, M.S. Feld: Phys. Rev. B*11*, 3298 (1975)
2.75 W.J. Brya, P.E. Wagner: Phys. Rev. *157*, 400 (1967)
2.76 J.C. MacGillivray, M.S. Feld: Appl. Phys. Lett. *31*, 74 (1977)
2.77 M.A. Duguay, P.M. Rentzepis: Appl. Phys. Lett. *10*, 350 (1967)
2.78 N. Shiren: Phys. Rev. B*2*, 2471 (1970)

3. Coherence in High Resolution Spectroscopy

V. P. Chebotayev

With 30 Figures

Such well-known techniques as radiation echo, the method of quantum beats, and superradiance are widely used in studying relaxation processes both in gases and in condensed media. These techniques are based on coherent processes. The advent of lasers extended extremely the potentialities of these methods in the optical band. Recent advances in this field are surveyed in [3.1]. Over the last years new spectroscopic methods were developed in optics on the basis of narrow nonlinear optical resonances. The methods proved to be very useful in examining low-pressure gases as they permitted obtaining Doppler free resonances with a homogeneous width. This increased to a great extent a spectroscopic resolution in the optical band and simultaneously permitted studies into relaxation phenomena. A lot of methods to obtain Doppler free narrow resonances have been developed in optics by now[1] [3.2].

The following methods are widely used in nonlinear laser spectroscopy:

1) Method of saturated absorption.
2) Two-photon resonances.
3) Saturation resonances of stimulated Raman scattering.
4) Method of separated optical fields.

The first method is closely related to the processes of level population saturation in the nonlinear resonant particle-field interaction, the last three are based on coherent processes. However, in the first case coherent processes sometimes result in both quantitative and qualitative effects. The purpose of the present work is to show an important role of coherent processes in nonlinear laser superhigh resolution spectroscopy.

3.1 Coherent Phenomena in Resonant Processes

A resonant response of probe wave absorption in the presence of other fields is of frequent interest in nonlinear laser high-resolution spectroscopy. Along with

[1]Here and below under Doppler free resonances we mean resonances with a homogeneous width whose frequency coincides with that of the corresponding transition of an immovable atom (or of an atom moving perpendicularly to an observer's direction).

coherent processes incoherent ones occur in the interaction of several waves. Among these, as a rule, are saturation effects that are due to the change in level population. It is well known that under the action of a strong field that can vary level population, a nonuniform velocity distribution of particles arises in a gas at each operating level. This phenomenon was noticed by BENNETT as early as 1962 [3.3] in analyzing the work of a gas laser. With inhomogeneous broadening the variation of velocity distribution of particles is of resonant nature. The population variation occurs in the particles whose velocities satisfy the conditions of resonant interaction (for details see [3.2]) $v = \omega - \omega_{21}/k$ (k is a wave number, v is the velocity projection onto the wave propagation direction, ω is the field frequency, ω_{21} is the transition frequency). If a probe wave with frequency ω travels in the same direction as a strong one with frequency ω', it will "feel" a decrease of the population difference at the interaction with those atoms that interact with the strong wave. This will take place when $\omega = \omega_{21} + kv$. So a resonance dip appears in absorption (gain) of a probe wave at the frequency $\omega = \omega'$. This scheme of resonant interaction of a probe wave with a medium corresponds to the conventional explanation of population effects.

The layout of radiative transitions may be represented as

$$\dot{n}_1 \xrightarrow{E_1} d_{12} \xrightarrow{E_1} \dot{n}_2 \xrightarrow{E_2} d_{12} \xrightarrow{E_2} \dot{n}_1 \quad . \tag{3.1}$$

The field E_1 produces polarization. The interaction of a polarized particle with the field of the same frequency results in field absorption and variation of level population. Then the process repeats itself but with a probe field E_2 and a population difference that arises under the action of field E_1. The similar phenomenon takes place, for instance, in a three-level system. The layout of radiative transitions in a three-level system may be represented as

$$\dot{n}_1 \xrightarrow{E_1} d_{01} \xrightarrow{E_1} \dot{n}_0 \xrightarrow{E_2} d_{02} \xrightarrow{E_2} \dot{n}_2 \quad . \tag{3.2}$$

Here the field E_1, when interacting with the $1 \rightarrow 0$ transition (Fig.3.1), produces a population of the level 0. The field E_2 that is resonant to the $0 \rightarrow 2$ transition interacts with this transition at a given population of the level 0.

The population approach to determination of the probe wave absorption results in that the solution of the problem on interaction of two waves with gas reduces to the solution of equations for diagonal elements of a density matrix. It is natural here to assume that a line shape of the probe wave absorption is not varied in the presence of fields, and changes in the probe wave absorption occur only due to population changes. The solution of the problem is therefore divided into two stages: firstly, the population difference in a strong field should be found, secondly, at a given population the response of a probe wave is found. A simplified system of equations has been referred to as rate equations that are frequently used

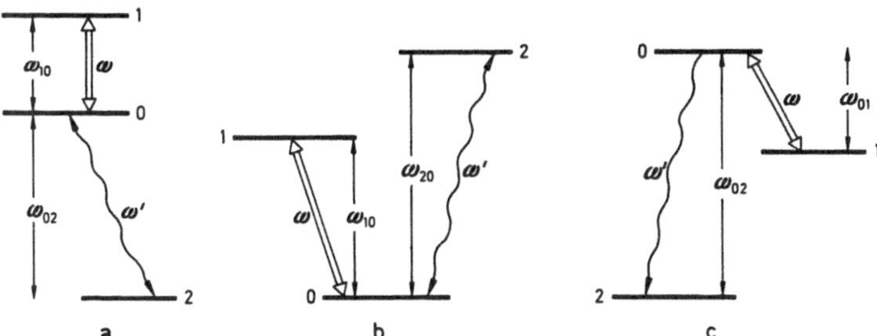

Fig.3.1. Energy level configurations for two coupled transitions: (a) cascade configuration; (b,c) bent configurations

to obtain some qualitative results. The use of these equations is reasoned in [3.4]. As the resulting effect of the probe wave absorption is determined by level population, the processes we have considered are irresponsive to the probability amplitude phases of atomic states that are in turn associated with the phases of fields. The schematic of the process (3.1) directly follows from strict equations for a density matrix with taking into account only a dipole moment arising under the action of a probe wave and of a strong field. At the nonlinear interaction of fields the dipole moment can arise not only under the action of a probe wave and with population of operating levels but also under the action of other fields. In this case the phase and frequency of the dipole moment will depend on field phases, frequencies, level structures and so on. In accordance with the equations for a density matrix, consideration of field coherence results in population modulation both in space and in time. Relevant spatial and time harmonics of population result in corresponding harmonics of the dipole moment of a particle. It is just the consideration of the spatial modulation of the medium population (e.g., in case of standing waves) or time modulation of the medium (unidirectional waves) that is frequently related to coherent effects. Depending on a given situation the phase of an induced dipole moment and its magnitude do not coincide with the phase and magnitude of the dipole moment arising in consideration of level population only. As a consequence, qualitative peculiarities arise in the line of probe signal absorption. Owing to an additional dipole moment that arises due to coherent effects, the process for two- and three-level systems, respectively, may be schematically represented as

$$\dot{n}_1 \xrightarrow{E_1} d_{12} \xrightarrow{E_2} \dot{n}_2 \xrightarrow{E_1} d_{12} \xrightarrow{E_2} \dot{n}_1 \quad . \tag{3.3}$$

$$\dot{n}_1 \xrightarrow{E_1} d_{01} \xrightarrow{E_2} d_{12} \xrightarrow{E_1} d_{02} \xrightarrow{E_2} \dot{n}_2 \quad . \tag{3.4}$$

In both cases the interaction of the dipole moment induced by one field with another field is considered. In the first case oscillations of absorption and level population take place due to the frequency difference. In the second case the dipole moment arises at the frequency of field E_2 without level population. From the given schemes one can readily extract coherent processes where the frequency of an induced dipole moment is a frequency combination, i.e., the dipole moment arises at the frequencies different from the field frequencies. An important circumstance resulting from coherent processes is the appearance of macroscopic polarization of a medium in averaging dipole moments of individual particles. Fulfilling the condition of synchronism, i.e., matching the phase velocities of propagation of polarization waves and of the field, may result in coherent radiation at new frequencies, e.g., harmonic generation, CARS, four-wave frequency conversion and others. The role of coherent processes in considering such nonsteady phenomena as a radiation echo would raise no doubts. Here the conservation of an induced dipole moment of individual particles in phasing them in time τ after the first pulse results in coherent radiation at the moments when some other fields do not act. It is sometimes difficult to determine the role of coherent phenomena in resonant processes due to the fact that the frequency of an additional dipole moment coincides with that of a probe field. It is therefore impossible to separate the radiation that is generated by induced polarization and the probe wave. Added to the probe wave it results in a change of the wave intensity which is generally interpreted as the change of a line shape of the probe wave absorption. In a given case, e.g., two-photon resonance, these changes in the absorption line are due to coherent phenomena only. The role of coherent phenomena is obvious in obtaining resonances by the method of separated optical fields. We are interested in the interaction with the field of the particles whose dipole moment arises in the interaction with the other fields in the other region of space. The appearance of resonances is due to only spatial transfer of the dipole moment of particles and to coherence conservation in particles.

3.2 Coherent Phenomena in Saturated Absorption Spectroscopy

3.2.1 Standing Wave

Of great importance in saturation spectroscopy is the formation of a LAMB dip [3.5] in an absorption line in a standing-wave field and a resonance of absorption of a weak oppositely traveling (probe) wave in the presence of a pumping wave [3.6]. From the point of view of population effects the physical picture of formation of a resonance dip is similar in both cases. In the line center both waves interact with the same atoms whose velocity projection onto the wave propagation axis is $v = 0$. Owing to saturation of the population difference in the line center $\Omega = \omega - \omega_{21} = 0$, a resonance dip arises in the absorption of an oppositely traveling wave. In weak fields the resonance dip shape found from the rate equations coincides with the solution derived from the equations for a density matrix.

Let us first consider the formation of a Lamp dip in an emission (absorption) line. It has been first observed in a He-Ne laser [3.7]. The Lamb dip in an absorption line in neon was independently observed in [3.8,9]. This extended the potentialities of the method and attracted much attention. The attainment of narrow resonance made it possible to study the influence of such phenomena as a second-order Doppler effect, a recoil effect, transit phenomena and so on [3.2].

The absorption of a standing wave is given by

$$\alpha(\omega) = \alpha_0(\omega)\left[1 + G/2\left(1 + \frac{\Gamma^2}{\Gamma^2 + \Omega^2}\right)\right], \quad G \ll 1 \quad . \tag{3.5}$$

$G = (d_{12}E/\hbar)^2 1/\gamma\Gamma$, d_{12} is the matrix element of the dipole moment, E is the traveling-wave field amplitude, $2/\gamma = 1/\gamma_1 + 1/\gamma_2$, γ_1 and γ_2 are the level relaxation rates, Γ is the transition halfwidth, $\alpha_0(\omega)$ is a nonsaturated absorption coefficient.

In consideration of the following terms of expansion in saturation parameter there appears some difference in the absorption line shape found from exact and rate equations. In the case of standing-wave field the solution of exact equations does not differ from the results obtained in a rate approximation. The quantiative contribution of coherent effects is insignificant even in very strong fields. With $G \gg 1$ and in the case where level relaxation constants are identical, the Lamb dip depth found from rate equations differs by about 10% from the corresponding value obtained from exact equations [3.10]. With different level relaxation constants the difference in results is less. This indicates a small contribution of coherent processes. It should be noted that in strong standing-wave fields the problem cannot be analytically solved. So the analysis of the influence of coherent phenomena presents difficulties. An approximate method of calculation of the absorption in standing-wave fields with large G providing that $\gamma/\Gamma G \ll 1$ was described in [3.11]. Reference [3.12] makes use of the smallness of $\gamma/\Gamma G$ to analyze the influence of the strong field on a nonlinear resonance shape in a three-level gas. Generally, in the optical band the level relaxation constants differ much. The above-mentioned approximation can therefore work at the saturation parameter $G > 1$. The physical sense of the parameter $\gamma/\Gamma G$ is as follows. It indicates the relation between a Rabi frequency and a homogeneous transition width. The condition $\gamma/\Gamma G \ll 1$ corresponds to $dE/\hbar\Gamma \ll 1$, i.e., an oscillation frequency of the probability amplitude is far less than a transition width. But in this case saturation effects may be significant. Note that an analytical solution of the problem is important in considering stable laser operation in a single-frequency regime.

In the line center the contribution of coherent effects is maximum. Providing $\gamma/\Gamma G \ll 1$ and $\Gamma_B \ll ku$ (Γ_B is a Bennett hole width, u is a thermal velocity, k is a

---

$\hbar = h/2\pi$ (normalized Planck's constant)

wave number), absorption in a standing-wave field for $\Omega = 0$ is given by

$$\alpha/\alpha_0 = (1 + 2G)^{-1/2} + (\gamma G/2\Gamma)(1 + 2G)^{-3/2} - (\gamma GA/2\Gamma)(1 + 2G)^{-2} \quad , \qquad (3.6)$$

where $A = 1 + 1/4P + 11/96P^2 + \ldots$, $P = 4G^2/(1 + 2G)^2$.

The first term describes the absorption conditioned by population effects. Two other terms may be related to coherence effects. These terms are of different signs. The last circumstance means that atoms with different velocities provide different contributions to absorption of oppositely traveling waves. For instance, for $\Omega = 0$ the atoms with velocity $v > \Gamma_B/k$ do not "feel" a spatial field amplitude. These atoms are responsive to an "average" field (they cross many nodes and loops of a standing wave). It is sufficient for these atoms to take into account the first spatial harmonic. For the atoms with a small velocity projection it is necessary to take into account all spatial harmonics of population and polarization.

Consideration of population effects corresponds to consideration of spatial medium inhomogeneity. It should be noted that in considering population effects and using rate equations, the spatial medium inhomogeneity in the strong field is not taken into account. To put it another way, neither the spatial structure of a standing-wave field is taken into consideration, nor consequently, the coherence between fields.

3.2.2 Probe Wave Resonances

Let us now consider probe wave resonances in the presence of a strong wave. For saturation spectroscopy the two following cases are important:

1) A probe and strong waves propagate in opposite directions.

2) A probe and strong waves propagate in the same direction. Firstly, we shall analyze the results that should be expected from the point of view of population saturation effects.

Oppositely Traveling Waves

In the line center both waves of the same frequency interact with the same atoms. Owing to the population difference saturation the absorption of a probe wave is decreased and tends to zero with an increase in the strong field intensity. With frequency detunings from the line center, the waves interact with different atoms. A probe wave does not "feel" the presence of a strong one, and the absorption of a probe wave is identical to the unsaturated absorption. Thus, a resonance dip arises in the line center. This case was first discussed in [3.6]. First experiments on observation of a resonance dip were reported in [3.13,14]. Important spectroscopic studies performed in Ne [3.15], PF_5 [3.16], and CH_4 [3.17] have shown the efficiency of this method. Together with the change of absorption a refractive index of the medium is resonantly changed [3.18] and one can observed rotation of a probe wave polarization plane [3.19]. The last methods permit recording saturation resonances at small absorptions.

It is not difficult to find the probe wave absorption without allowing for coherent effects [3.20]. The absorption of a probe wave of the same frequency is given by

$$\alpha(\omega) = \alpha_0[1 - (1 - 1/\sqrt{1 + G})\tilde{\Gamma}^2/(\Omega^2 + \tilde{\Gamma}^2)] \quad . \tag{3.7}$$

Here G is the saturation parameter for a strong wave, $\tilde{\Gamma} = (\Gamma/2)(1 + \sqrt{1 + G})$. The probe wave absorption in resonance is identical to the strong wave absorption. A strict treatment of the problem has been made in [3.20,21].

In weak fields ($G \ll 1$) the coherent effects introduce no changes in the line shape of probe wave absorption. The line shape of probe wave absorption is given in Fig.3.2. The absorption line shape that appears in considering the population effects only is denoted with dashed curves. Qualitative differences in the line shape appear in saturation $G \gg 1$. With allowing for coherent effects the probe wave absorption in the line center in a very strong field ($G \gg 1$) tends to a constant value dependent on level relaxation constants.

$$\alpha = \alpha_0(3/2)(\gamma/\Gamma)[1/(3 + \gamma/\Gamma)]\exp[-(1/3)(\gamma/\Gamma)(\tilde{\Gamma}/ku)^2] \quad , \tag{3.8}$$

where $2/\gamma = 1/\gamma_1 + 1/\gamma_2$.

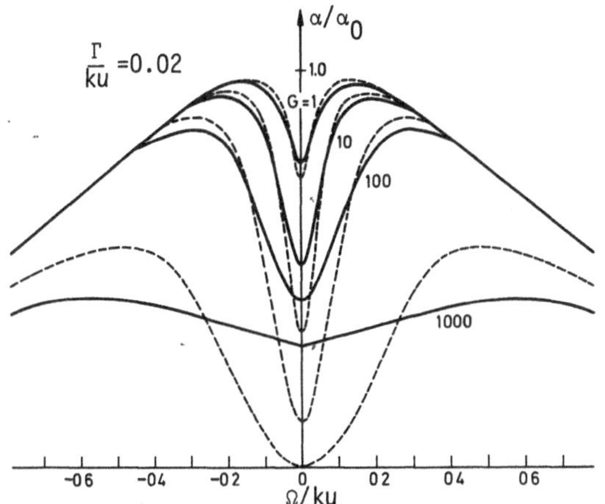

Fig.3.2. Shape of the absorption line of a weak probe wave in the presence of a strong counter-running wave at different degrees of saturation G. The calculation was carried out with the coherence effects (the solid curves, $\gamma = \Gamma$) and rate equations approximation (the dotted curves, $\gamma/\Gamma \rightarrow 0$)

The physical explanation of this effect turned out to be most readily related to a high-frequency Stark effect. The strong field affects not only the velocity distribution of particles but also the condition of their resonant interaction with a probe wave due to effective splitting of levels in a strong variable field (Townes-Outler effect). By an order of magnitude the splitting is equal to $\Delta \sim dE/\hbar$. The main contribution to the probe wave absorption is provided by the particles whose Doppler frequency shift cancels the magnitude of this splitting. So, for instance, for the detuning $\Omega = 0$ the resonance atoms are those whose velocities satisfy the condition $kv \sim dE/\hbar$.

The magnitude of absorption will be determined by the population difference for the particles with resonant velocities. The fraction of these particles will depend on the relation between a Bennett dip width (in units of frequency kv) and the magnitude of splitting Δ. The dip width is determined by the saturation parameter and with $G \gg 1$ equal to $\Gamma_B = \Gamma\sqrt{G}$, i.e., it is $\sqrt{\Gamma/\gamma}$ times as large as the magnitude of splitting. With equal level relaxation constants $\gamma_1 = \gamma_2$ the magnitude of splitting Δ is comparable with the dip width Γ_B. So the contribution of coherent effects is maximum. With $\gamma/\Gamma \ll 1$ the difference of populations of resonantly interacting particles tends to zero. Consequently, the contribution of coherent effects is scarcely noticeable. The oppositely traveling waves of the same frequency interact with the same atoms at $\Omega = 0$. The similar results may be obtained for oppositely traveling waves of different frequencies. In this case the resonances that are similar to those described arise at the frequency position symmetric relative to the line center.

Unidirectional Waves

In the interaction of unidirectional waves the coherent effects result in qualitative changes in the line shape of a probe signal even with low intensities of a pumping field. For arbitrary relaxation constants the line shape of probe wave absorption has been found in [3.22] and with low field intensities ($G \ll 1$) is given by

$$\alpha(\Delta)/\alpha_0 = \exp[-(\Omega/ku)^2]\Bigg\{1-(G/2)(2\Gamma)^2/[(2\Gamma)^2+\Delta^2]$$
$$\left[1+\left(\frac{\gamma_1}{\gamma_2^2+\Delta^2}+\frac{\gamma_2}{\gamma_2^2+\Delta^2}\right)\gamma/2\right.$$
$$\left.\left.+\gamma/4\Gamma\left(\frac{\Delta^2}{\gamma_1^2+\Delta^2}+\frac{\Delta^2}{\gamma_2^2+\Delta^2}\right)\right]\right\}\ , \qquad (3.9)$$

where Δ is the field frequency difference.

A resonant factor with a halfwidth 2Γ arises in consideration of population effects. It describes the appearance of a Bennett hole with a width 2Γ. In the line of a probe wave owing to a finite transition width 2Γ the dip width is doubled. The resonant factors with halfwidths γ_1 and γ_2 are due to coherent effects. In this case the coherent effects are directly associated with the coherent properties of fields. In the interaction of two unidirectional waves whose frequencies differ by a value Δ, owing to nonlinear properties of a medium, there appears polarization at frequencies ω, $\omega + \Delta$, and $\omega - \Delta$. An additional polarization appearing at the frequency $\omega - \Delta$ results in some peculiarities in the probe wave absorption. The additional medium polarization (in accordance with equations for a density matrix) is associated with the modulation of population of working levels. A relative modulation depth depends on the level relaxation constants and on the ratio of Δ, γ_1, and γ_2. With $\Delta \gg \gamma_1, \gamma_2$ the medium "has no time" to react to the changes in an instantaneous field amplitude.

In strong pumping fields the line shape of a probe signal undergoes considerable changes. With $dE/\hbar \gg \Gamma$ the probe wave absorption (with an accuracy up to $1/\sqrt{G}$) is given by

$$\frac{\alpha}{\alpha_0} = \begin{cases} 0 & \text{at} \quad |\Delta| < dE/\hbar \\ \dfrac{|\Delta|[\Delta^2 - (dE/\hbar)^2]^{\frac{1}{2}}}{\Delta^2 + \Gamma_B^2[1 - (\gamma/\Gamma)]} & \text{at} \quad |\Delta| \geqq dE/\hbar \quad . \end{cases} \tag{3.10}$$

In the region of frequency detunings $|\Delta| < dE/\hbar$ the probe wave absorption is zero. It may be interpreted as the absence of particles that resonantly interact with a probe wave. With $|\Delta| > dE/\hbar$ the probe wave absorption rapidly increases and becomes equal to an unsaturated one. The above peculiarities of the line shape of a probe wave in strong fields are due to a dynamic Stark effect on Doppler broadened transitions.

Figure 3.3 shows the line shape of probe signal absorption at various fields and relations between relaxation constants. As well as in the case of weak fields, of special importance for spectroscopy may be the case of strong fields.

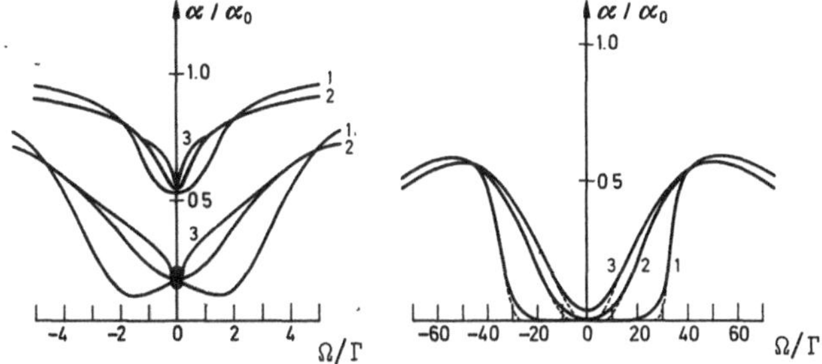

Fig.3.3. Shape of the absorption line of a probe wave in the presence of a strong parallel wave for the saturation parameters G = 1 and G = 10 (a) and G = 10^3 (b). Curves 1 correspond to $\gamma_1/\gamma_2 = 1$; 2 to $\gamma_1/\gamma_2 = 10$; 3 to $\gamma_1/\gamma_2 = 100$; $\gamma/ku = 10^{-2}$

High-Frequency Stark Effect on Doppler Broadened Transitions

Let the energies of an upper and lower levels be E_m and E_n, respectively, in the absence of a strong field. If an atom is affected by a strong monochromatic field detuned relative to the transition frequency ω_{mn} by a value $\delta = \Omega - kv$, the levels m and n are split into

$$E_m^{(1,2)} = E_m + \delta/2 \pm [(\delta/2)^2 + (dE/2\hbar)^2]^{\frac{1}{2}} \ ,$$

$$E_n^{(1,2)} = E_n - \delta/2 \pm [(\delta/2)^2 + (dE/2\hbar)^2]^{\frac{1}{2}} \ . \tag{3.11}$$

These splittings result in that instead of one resonance frequency ω_{mn} a moving atom in the strong field acquires three frequencies: $\omega_{1,2} = \omega_{mn} + \delta \pm [\delta^2 + (dE/\hbar)^2]^{\frac{1}{2}}$, $\omega_3 = \omega_{mn} + \delta$. If an atom has the velocity projection onto the wave propagation direction v, it will be affected by the field of frequency $\omega = \omega_{mn} + \Omega - kv$, and consequently, the atom turns out to be detuned with respect to the transition frequency by a value $\delta = \Omega - kv$. From the side of a weak oppositely traveling wave this atom will be affected by the field of frequency $\omega' = \omega_{mn} + \Omega + kv$. The condition of resonance $\omega' = \omega$ determines the velocities of the atoms at which an effective absorption of a weak oppositely traveling wave occurs. The conditions $\omega_{1,2} = \omega'$ give $2kv \pm \sqrt{(\Omega - kv)^2 + (dE/\hbar)^2} = 0$. Solving this equation we obtain

$$(kv)_{1,2} = -\Omega/3 \pm (2/3)\sqrt{\Omega^2 + (3/4)(dE/\hbar)^2} \ , \quad (kv)_3 = \Omega \ . \tag{3.12}$$

As has been shown in exact calculations, the regions of an order of Γ around the same velocities provided the main contribution into the probe wave absorption. This permits the conclusion that the atoms whose velocities satisfy the condition (3.12) are resonant for a weak oppositely traveling wave. For $\Omega = 0$ these velocities are

$$(kv)_{1,2} = \pm dE/\hbar\sqrt{3} \ , \quad (kv)_3 = 0 \ . \tag{3.13}$$

With no external field and in consideration of the Doppler shift the resonant frequency of these atoms for a weak field was

$$\omega = \omega_{mn} \pm (kv)_{1,2} \ . \tag{3.14}$$

Since these atoms are now in resonance, the field effect may be interpreted as cancellation of the Doppler shift with level splitting. Now it is not difficult to explain why with an increase of field the absorption coefficient α (3.8) tends to a constant value rather than to zero, as it follows from the consideration of population effects. With $\Omega = 0$ the contribution is provided only by the atom velocities which are equal to $dE/\hbar k\sqrt{3}$.

It is evident that the contribution of these atoms into absorption is proportional to the number of atoms. Then α/α_0 may be qualitatively estimated as the relation of atomic densities in the regions $\pm dE/\hbar\sqrt{3}$ in the presence of a strong field to those with $v = 0$ and no fields. For a very strong field the velocity distribution of atoms is

$$n(v) = \frac{(kv)^2}{(kv)^2 + \Gamma_B^2} \ , \quad \Gamma_B \ll ku \ . \tag{3.15}$$

Then for the absorption factor in a strong field we obtain

$$\alpha/\alpha_0 = 2\,\frac{\gamma}{\Gamma}\,\frac{1}{3 + (\gamma/\Gamma)} \quad . \tag{3.16}$$

Figure 3.4 illustrates the above and explains the sense of the result. The velocity of the atoms that resonantly interact with a weak wave is increased proportionally to the field. The dip width in velocities $2\sqrt{\Gamma/\gamma}dE/\hbar$ is also increased proportionally to the field so that the number of atoms is not changed. With equal relaxation constants of upper and lower levels $\gamma/\Gamma = 1$ the magnitude of splitting is of the same order as the dip width, and the splitting effects are more essential. With a decrease of γ/Γ (different level relaxation constants) the number of resonant atoms is decreased and with $\gamma/\Gamma \ll 1$ the contribution of splitting effects turns out to be small. Thus the parameter $\beta = \gamma/\Gamma$ acquires a quite definite meaning: it determines an additional contribution of coherent effects to absorption with allowing for the level splitting and the changes in the velocity distribution of atoms arising in a strong monochromatic field. When the splitting $dE/\hbar$ is far less than the dip width Γ_B, i.e., $\gamma/\Gamma \ll 1$, we draw an important conclusion: saturation effects in the interaction of two oppositely traveling waves may be described within the frames of population. In the other words, a gaseous medium with considerably different level relaxation constants is similar to that with a transverse Γ and longitudinal γ relaxation constant.

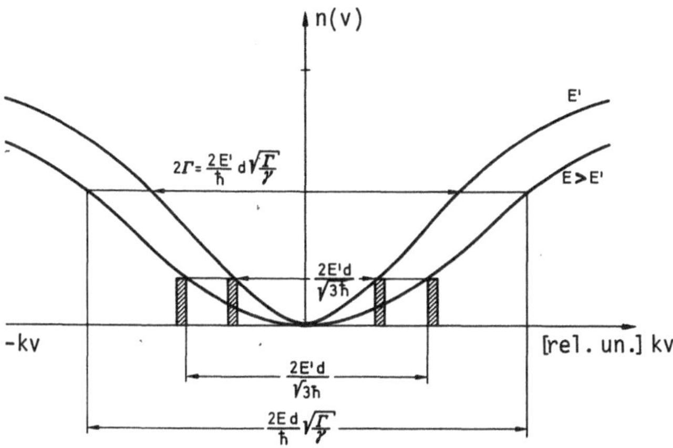

Fig.3.4. Velocity distribution of the population difference for two values of the strong field E and E' at $\Omega = 0$. The shaded regions correspond to the number of atoms resonantly interacting with the weak wave

The effects of splitting may be not taken into account in the approximation of a weak field $G \ll 1$. Here the splitting is always far less than the dip width Γ_B, in the first saturation order the effects do not manifest themselves. These effects become significant if $(\gamma/\Gamma)G \sim 1$.

For unidirectional waves the main contribution is provided by the atoms whose velocities satisfy the condition of resonance in splitting

$$(kv)_{1,2} = \pm \sqrt{\Delta^2 - (dE/\hbar)^2} \quad . \tag{3.17}$$

It is seen from here that there are no resonant atoms at $|\Delta| < dE/\hbar$. The probe wave absorption is zero. The resonant atoms appear at $|\Delta| \sim dE/\hbar$. Figure 3.5 shows the dependence of the resonance frequency of a probe signal on the atomic velocity. The dashed curves denote the dependence in the absence of the field. This dependence is described by a straight line corresponding to the Doppler shift $\Delta = \pm(kv)$. In the presence of a strong field considerable changes in this dependence are observed. For slow velocities that correspond to the condition of particle-field interaction the resonance frequency lies in the region $\Delta \sim \pm dE/\hbar$.

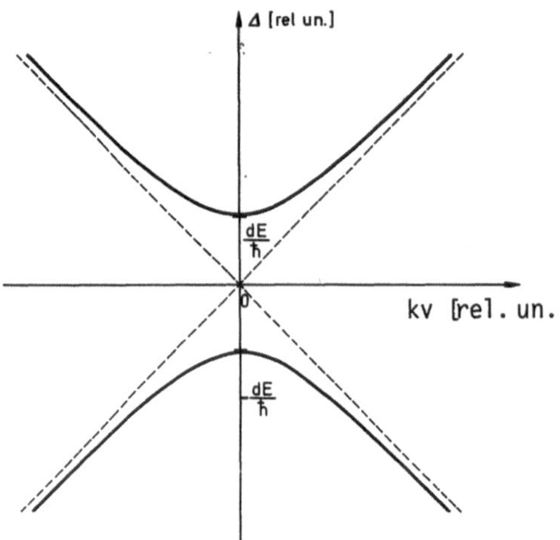

Fig.3.5. Dependence of the frequency of a probe signal on the velocity of resonantly interacting atoms at $\Omega = 0$. Dotted curves correspond to $E_1 = 0$

Spectroscopic Applications. Measurement of Relaxation Constants

The above discussed coherent phenomena open up qualitatively new possibilities in spectroscopy. In Lamb dip experiments much information is associated with measurements of a transition linewidth. Studies of an absorption line of a probe signal at the collinear wave propagation permit obtaining not only a linewidth but also relaxation constants of individual levels. This is very useful in studying vibrational-rotational molecular transitions. Figure 3.6 shows the dependence of probe wave absorption in SF_6 in different regions of detuning frequencies of a probe and pumping waves. In the frequency region of about 1 MHz a resonance width is due to rotational relaxation and collisional broadening. In the frequency region of about 1 kHz a resonance width is determined by the vibrational level relaxation. In the region of strong fields the absorption of a probe oppositely traveling wave is determined by a parameter γ/Γ that can depend on collisions in different ways. Studying the behavior of the parameter γ/Γ one can draw a conclusion about the

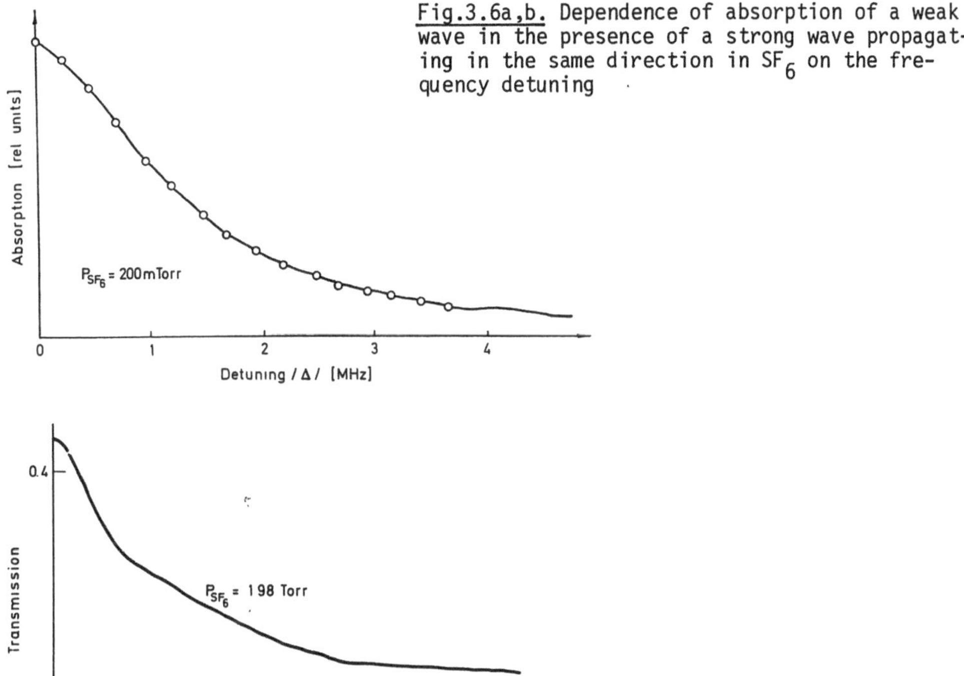

Fig.3.6a,b. Dependence of absorption of a weak wave in the presence of a strong wave propagating in the same direction in SF_6 on the frequency detuning

mechanism of collisions. It is this that very frequently presents a greater problem than a simple change of collisional line broadening.

Study of Level Structures and Separation of Weak Lines

Study into a line shape of a probe signal at the collinear wave propagation provides information on a fine line structure. With a constant field E the frequency region where absorption is zero depends on the dipole moment of a matrix element. If a line consists of several components with different dipole moments, the absorption line will comprise several regions where the absorption does not change. This may be seen better in recording a derivative of the probe wave absorption. Figure 3.7 shows the calculated dependence of the derivative of the probe wave absorption for the P(7) line of the ν_3 band in methane at λ = 3.39 μm. About 7 resonances are observed which correspond to the same number of transitions. The P(7) line is known to consist of three strong MHS components that correspond to the 5 → 6, 6 → 7, and 7 → 8 transitions with $\Delta F = -1$. The component spacing is about 10 kHz. Each component has 6-8 transitions between magnetic sublevels for a circularly polarized wave.

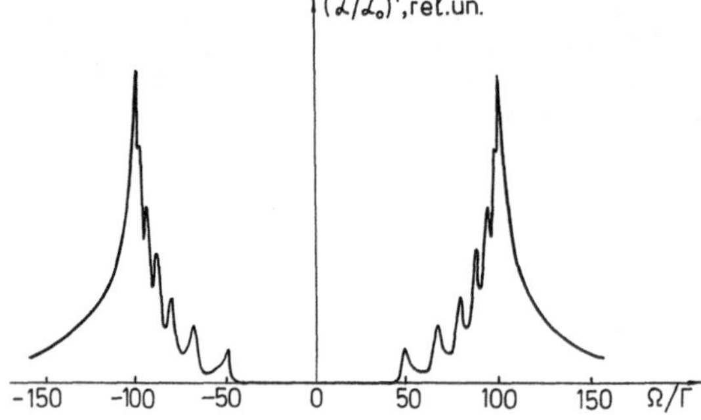

Fig.3.7. Frequency dependence of the derivative of an absorption coefficient of a weak wave in the presence of a strong one propagating in the same direction in CH_4. G = 10^4

The same effect may be used to separate weak lines from strong ones. In the strong field the line undergoes saturation while in a weak line the saturation is slight. This circumstance may be used to suppress absorption in strong lines and to separate weak lines. In scanning a frequency the modulation signal for a strong line is additionally suppressed due to line broadening.

Optical Instability. Generation Stability

Coherent phenomena in the interaction of several waves may be of significance in such important problems as instability, the condition for single-frequency operation in lasers with a nonlinear absorber. The physical nature of optical instability in a cavity with a nonlinear absorber is simple. The magnitude of saturated absorption depends on the field. At certain relations between the parameters of a cavity, absorption, and the magnitude of saturation this system with a positive feedback becomes unstable.

Optical instability and various hysteresis phenomena were first observed in a laser with nonlinear absorption [3.23]. If the rate of level decay is far less than the lifetime of photons in the laser cavity there appear stationary radiation pulsations [3.24].

Of great interest now are the works on bistability of interferometers filled with a nonlinear absorbent, see [3.25]. We shall restrict ourselves to analyzing the influence of coherent effects on stability of a laser with nonlinear absorption where the influence of coherent effects is determinant. The interaction of standing waves in the line center should be considered. With a given field in the cavity, saturated gain for unidirectional waves is always more than the gain of a weak signal due to coherent effects, see (3.9). So near frequency detunings comparable with level widths, the weak signal gain is less than losses in the cavity and generation turns out to be stable. Another situation is observed with nonlinear absorption. An effective gain of a weak signal at the field frequency is determined by the difference between coherent addition in absorption and gain, i.e.,

$$\alpha_1 \frac{G_1}{2(1 + G_1)^{3/2}} - \alpha_2 \frac{G_2}{2(1 + G_2)^{3/2}} \quad , \tag{3.18}$$

where α_1 is unsaturated gain, α_2 is unsaturated absorption, $G_{1,2}$ are the saturation parameters in an amplifying and absorbing media, respectively. With low saturation the condition of stability is

$$\alpha_1 G_1 > \alpha_2 G_2 \quad . \tag{3.19}$$

As should be expected, this condition coincides with that of the appearance of a hysteresis resonance [3.23].

The coherent phenomena are also important in analyzing single-frequency generation stability. Let us consider the possibility to achieve single-frequency generation in a laser with nonlinear absorption. The introduction of nonlinear absorption is known to result in mode selection and stable single-frequency generation [3.26]. Owing to a large saturation difference in absorption and gain, extra losses at the generation frequency caused by the introduction of absorption are insignificant, while at the other frequencies the absorption is identical to an unsaturated one and may be essential. The condition of generation is not fulfilled. This enabled us to select modes that by frequency are located at a distance much larger than a homogeneous linewidth (a detailed analysis is given in [3.27, 28]). In tuning the laser frequency there can appear a case where the cavity modes are symmetric with respect to the line center of absorption or gain. In this case the consideration of coherent effects is very important in solving the problems of stability.

Let the distance between the frequencies of the corresponding modes be much more than a homogeneous width. Then a standing-wave field can be readily represented as two oppositely traveling waves interacting with different atoms. The wave moving in the positive direction of the Z axis forms a dip in the vicinity of Ω/k, in the negative direction $-\Omega/k$. Let a weak wave field at an image frequency be also represented in the form of two traveling waves. We can notice that oppositely traveling waves with different frequencies interact with the same atoms. Thus a strong and weak wave are traveling oppositely which enables us to use the results obtained above. The strong wave gain and absorption in the line center are

$$\alpha_i^{(H)} = \alpha_i \left\{ \frac{1}{\sqrt{1+G_i}} + \frac{\gamma_i}{\Gamma_i} G_i \frac{a-1}{a} \right.$$
$$\left. \frac{(3a+1)(2a+1)}{(3a+1)(a+1)\left(2a+\frac{\gamma_{i1}}{\Gamma_i}\right)\left(2a+\frac{\gamma_{i2}}{\Gamma_i}\right)+(2a+1)^2 G_i \frac{\gamma_i}{\Gamma_i}} \right\} \tag{3.20}$$

where $i = 1,2$ corresponds to an amplifying and absorbing media, $a = \sqrt{1+G_i}$. The absorption of a weak oppositely traveling wave at the symmetric position of frequencies is given by

$$\tilde{\alpha} = \alpha_0 \left\{ \frac{\tilde{\alpha}^{(1)}}{\alpha_0} + 4\,\frac{(dE_1)^2}{\hbar^2\Gamma}\,(\Gamma_0 - \gamma) \right. \tag{3.21}$$

$$\left. \mathrm{Re}\left\{ - \frac{f(\Omega + \frac{\Delta}{2} + i\Gamma_0)}{2(\Omega + \frac{\Delta}{2}) + i(\Gamma_0 + \gamma)} \right\} \exp\left[-\left(\frac{\Omega}{kv_0}\right)^2\right] \right\} ,$$

where $\tilde{\alpha}^{(1)} = \alpha_0 \left\{ \exp\left[-\left(\frac{\Omega + \Delta}{kv_0}\right)^2\right] - \left(\frac{b\tilde{\Gamma}^2}{\left(\Omega + \frac{\Delta}{2}\right)^2 + \tilde{\Gamma}^2}\right) \exp\left[-\left(\frac{\Omega}{kv_0}\right)^2\right] \right\}$, $\Omega = \omega_1 - \omega_{21}$ is the frequency detuning in the strong field (E_1) with respect to the line center, $\Delta = \omega_2 - \omega_1$ is the frequency detuning of a weak wave with respect to a strong one, $b = \left(\frac{G}{1 + G + \sqrt{1 + G}}\right)$, $\tilde{\Gamma} = \Gamma\left(\frac{1 + \sqrt{1 + G}}{2}\right)$, $\Gamma_0 = \Gamma\sqrt{1 + G}$, $f(x)$ is determined by

$$f(x) = \frac{(3x - \Omega + i\Gamma)(x + i\Gamma)(2x + i\Gamma)}{(3x - \Omega + i\Gamma)(x + \Omega + i\Gamma)(2x + i\gamma_1)(2x + i\gamma_2) - (2x + i\Gamma)^2 4(dE_1/\hbar)^2} .$$

It can be readily noticed that an addition term in (3.21) that is due to the splitting effects is equal to the difference between the gain (absorption) of a weak signal and the saturated gain (absorption) of a strong one. So the relation between these addition terms in the weak signal gain and absorption mostly solves the problem on stability of modes of one type; if an additional contribution of the splitting effects to the weak signal absorption is more than that to the gain, the generation regime of the modes of one type is stable. Allowing for (3.20,21) one can write down this condition for symmetrically located modes

$$\left(\frac{\gamma}{\Gamma}\right)_1 \alpha_1 \varphi_1 < \left(\frac{\gamma}{\Gamma}\right)_2 \alpha_2 \varphi_2 , \tag{3.22}$$

where

$$\varphi_i = G_i\,\frac{a_i - 1}{a_i}\,\frac{(3a_i + 1)\left(2a_i + \frac{\gamma_1^{(i)} + \gamma_2^{(i)}}{\Gamma_i}\right)}{\left[(3a_i + 1)(a_i + 1)\left(2a_i + \frac{\gamma_1^{(i)}}{\Gamma_i}\right)\left(2a_i + \frac{\gamma_2^{(i)}}{\Gamma_i}\right) + \left(2a_i + \frac{\gamma_1^{(i)} + \gamma_2^{(i)}}{\Gamma_i}\right)(2a_i + 1)G_i\,\frac{\gamma_i}{\Gamma_i}\right]}$$

$$a_i = \sqrt{1 + G_i} .$$

With large saturation parameters in both media (3.22) takes a simple form

$$\left(\frac{\gamma}{\Gamma}\right)_1 \alpha_1 \frac{1}{\left[3 + \left(\frac{\gamma}{\Gamma}\right)_1\right]} < \left(\frac{\gamma}{\Gamma}\right)_2 \alpha_2 \frac{1}{\left[3 + \left(\frac{\gamma}{\Gamma}\right)_2\right]} . \tag{3.23}$$

Thus, with large saturation parameters the generation stability of the modes of one type is determined by the parameter $(\gamma/\Gamma)_i$ and by the magnitude of unsaturated absorption and gain. This condition also determines the stability of unidirectional generation at symmetrically located modes in lasers with ring cavities.

Recoil Effect

In absorption or emission of a photon there occur changes in the pulses of an atom with which the change of energy by a value $\Delta\varepsilon = \hbar\Delta$ is associated ($\Delta = \hbar k^2/2M$, k is the wave vector of a photon, M is an atomic mass). Owing to this effect (recoil effect) an absorption and emission line contour turn out to be shifted by 2Δ. This also results in splitting of a Lamb dip into two resonances spaced at 2Δ [3.29]. For the absorption factor of the standing-wave field we have

$$\frac{\alpha}{\alpha_0} = 1 - \frac{|V|^2}{\Gamma}\left[\frac{1}{\gamma_1}\left(1 + \frac{\Gamma^2}{(\Omega - \Delta)^2 + \Gamma^2}\right) + \frac{1}{\gamma_2}\left(1 + \frac{\Gamma^2}{(\Omega + \Delta)^2 + \Gamma^2}\right)\right] \quad , \quad V = \frac{dE}{\hbar} \tag{3.24}$$

where α_0 is an unsaturated absorption factor. The resonances described by this formula have a width 2Γ and are due to the saturation effects. With $\Omega = \Delta$ a resonance is due to saturation of level 1, with $\Omega = -\Delta$ due to saturation of level 2. The consideration of the saturation effects in higher orders of the parameter $|V|^2$ may be shown to introduce no qualitative changes in the results. As before we have two resonances at $\Omega = \pm\Delta$.

In the absorption factor of a standing wave the coherent effects arise in the terms proportional to $|V|^4$ [3.30]. Despite that their contribution to absorption amounts to about 10% of saturation effects, with allowing for the recoil effect the coherent effects result in qualitatively new peculiarities of a line shape: additional resonances at $\Omega = \pm 3\Delta$ with a width 2Γ. With the saturation of an order of unity the magnitude of these resonances amounts to several per cent. This follows from [3.30] where an addition to (3.24) proportional to $|V|^4$ has been found. The resonance at $\Omega = \pm 3\Delta$ is qualitatively due to multiphoton processes in a standing-wave field at which an atomic pulse is changed by a value larger than a photon pulse $\hbar k$. For example, the change of the pulse by $3\hbar k$ is due to photon absorption in the positive direction, emission of an oppositely moving photon, and absorption of a unidirectional photon.

When $\Gamma \gg \Delta$ the resonances are not resolved. However, the allowance for recoil results in such effects as the shift of the center of a Lamb dip from the magnitude of the field. According to [3.30] for the case $\gamma_1 \approx \gamma_2 \approx \Gamma$ the shift is

$$\Delta\Omega = \Omega_0\left(1 - \frac{8}{27}\frac{|V|^2}{\Gamma^2}\right) \quad , \tag{3.25}$$

where $\Omega_0 = [(\gamma_1 - \gamma_2)/(\gamma_1 + \gamma_2)]\Delta$. The saturation effects turned out to make no contribution to the Lamb dip shift. The shift is due to the coherent effects that are determined by the parameter $2\gamma_1\gamma_2/\Gamma(\gamma_1 + \gamma_2)$. In this problem it is necessary to take them into account. When the resonances are resolved $\Delta \ll \Gamma$, their shifts from field and pressure are determined by the influence of their "tails" upon each other.

3.2.3 Influence of Collisions on Coherent Processes

Study of Relaxation Processes

It is well known that collisions may disturb coherence between states. They can decrease particle lifetimes on levels. All the above processes result in broadening of both spectral line and resonances. Study of the influence of collisions on resonance broadening is a traditional trend in spectroscopy. Unfortunately, in many cases such studies do not permit one to say something definite about the mechanism of collisions that requires additional investigation.

Studies into the coherent processes enable one to elucidate their mechanism. For instance, as we have already seen, three resonances arise in the probe wave line shape: one with a homogeneous linewidth and two with level widths. The behavior of these resonances in the presence of collisions indicates the mechanism of relaxation processes of a two-level atom. In strong fields the influence of collisions is determined by the behavior of the parameter γ/Γ. If, for instance, collisions equally destroy upper and lower levels and do not randomize the phase of a dipole moment (this situation may take place for a line of vibrational-rotational molecular transitions), the parameter γ/Γ tends to 1 with an increase of collision frequency. Studying the behavior of the parameters γ and Γ one can detect the influence of collisions on various processes. With $\gamma/\Gamma \simeq 1$ the role of coherent effects becomes more essential. However, in a case of oppositely traveling waves strong fields are required for the coherent effects to be observed so that the condition $dE/\hbar \gg \Gamma$ can be realized.

Dipole Scattering

Of great importance in investigating the coherent processes is the study of the influence of collisions on an induced atomic dipole moment. Collisions can change the dipole velocity without phase randomization, change the dipole moment amplitude and phase. The behavior of narrow optical resonances in a gas largely depends on the dipole moment relaxation (changes in its velocity, phase and amplitude). This is unusual for optical spectroscopy, as it has been assumed that in the optical band along with the velocity change phase randomization and destruction of an induced dipole moment always take place in collisions. This is due to the fact that scattering amplitudes on upper and lower levels of optical transitions are different, so the coherence between states is always destroyed after collisions. Let us analyze the physical picture of destruction of coherence in elastic collisions from the point of view of scattering of a two-level system. Let scattering on an upper level occur at a potential U_2, on a lower level at a potential U_1 in the elastic collision of an atom with the immovable center. The difference between U_2 and U_1 makes it impossible to consider the motion of an atom as a whole in a mixed state, as it is not clear along what trajectory an atom determined by the potentials U_2 and U_1 will move. So the consideration of an induced dipole, i.e., an atom in a mixed state, requires the quantum treatment of its trajectory. In a general case a wave function may be written as

$$\Psi(\underline{r},t) = \sum_i \Psi_i(\underline{r},t)U_i(\xi)\exp\left(-\frac{iE_i t}{\hbar}\right) \quad , \tag{3.26}$$

where E_i is the ith level energy, $U_i(\xi)$ is the proper function, ξ is the totality of all internal atomic coordinates, $\underline{r}$ is the coordinate of the center of inertia, $\Psi_i(\underline{r},t)$ satisfies Schrödinger's equation with an atomic mass. An average value of the atomic dipole moment is found from the formula

$$\begin{aligned} \langle\underline{d}(t)\rangle &= \int d^3r \int d\xi\Psi^*(\underline{r},t)\underline{d}(t)\Psi(\underline{r},t) \\ &= \sum_{ik} \underline{d}_{ik} \exp(-i\omega_{ik}t) \int d^3r\Psi_i^*(\underline{r},t)\Psi_k(\underline{r},t) \quad , \end{aligned} \tag{3.27}$$

where $\underline{d}(t)$ is the atomic dipole moment operator, $d_{ik} = \int d\xi U_i^*(\xi)\underline{d}(\xi)U_k(\xi)$ is its matrix element, $\omega_{ik} = (E_i - E_k)/\hbar$.

For a two-level system

$$\langle\underline{d}(t)\rangle = \underline{d}_{12} \exp(-i\omega_{12}t) \int d^3r\Psi_1^*(\underline{r},t)\Psi_2(\underline{r},t) + \text{c.c.} \tag{3.28}$$

The obtained expression indicates that for the dipole moment to be different from zero it is necessary to have overlapping wave functions of upper and lower states. If an atomic dipole moment is induced at the $2 \rightarrow 1$ transition, we should consider this state as two overlapping wave packets on the upper and lower levels with the nonzero integral (3.28).

In a quasiclassical treatment of the scattering process the size of these packets a may be considered to be of an order of De Broglie's wave of an atom, i.e., $a = \lambda\!\!\!^{-} = \hbar/Mv$, where M is an atomic mass, v is its velocity. If U_1 and U_2 are different, according to Ehrenfest's theory the packets will move along the corresponding classical trajectories with probabilities $P_1 = \int \Psi_1^*(\underline{r},t)\Psi_2(\underline{r},t)d^3r, P_2 = \int\Psi_2^*(\underline{r},t)\Psi_1(\underline{r},t)d^3r$. In separation of the packets at the distance of an order of their size, i.e., $\lambda\!\!\!^{-}$, they may not be assumed to overlap each other, and according to (3.28) $\langle\underline{d}(t)\rangle = 0$, which means the destruction of a mixed state and, consequently, the destruction of a dipole. In the other words, deflection of an atom in the state 2 and 1, i.e., elastic scattering of an atom on any level, results in the loss of a dipole. The elastic scattering probability on the ith level is

$$W_i = nv\sigma_i \quad , \tag{3.29}$$

where n is the density of the dipole scattering centers, σ_i is the total elastic scattering cross section on the ith level. The dipole resolution probability is

$$\Gamma = \frac{W_1 + W_2}{2} = \frac{nv(\sigma_1 + \sigma_2)}{2} \quad . \tag{3.30}$$

In the sense Γ is a linewidth of impact broadening at collisions.

Let us find the time for which a dipole is destroyed after collision. If θ is the difference between the angles of the classical trajectories on the upper and lower levels, having traveled a distance $\ell = vt$ after scattering the packets are separated at about $\theta\ell$. When $\theta\ell = a$, the dipole is destroyed for the time

$$\tau \simeq \frac{a}{\theta v} \quad . \tag{3.31}$$

With the atomic velocity $v = 10^4$ cm/s, $M = 10^{-23}$ g we have $\lambdabar \sim 10^{-8}$ cm. If $a \simeq \lambdabar$

$$\tau = \frac{\lambdabar}{\theta v} \sim 10^{-12} \frac{1}{\theta} \quad . \tag{3.32}$$

This means that at $\theta = 1$ the dipole is destroyed immediately after collision. For the dipole to remain at scattering it is necessary that the trajectories coincide with each other to a high accuracy ($\theta \to 0$), i.e., the scattering potentials are almost identical.

Another situation is observed in considering the scattering of atoms at small angles both on an upper and a lower level, where a quasiclassical approximation is inapplicable. In this case an approximation of high energies which is also called diffraction scattering is generally used for scattering of atoms (see [3.31]). The region of impact parameters is of an order of Weiskopf's radius that determines the wave packet size.

The scattering angle $\theta \sim \lambdabar/k$, so

$$\tau = \frac{\lambdabar}{\theta^2 v} \sim 10^{-12} \frac{1}{\theta^2} \quad . \tag{3.33}$$

The lifetime is inverse to the square of difference of the trajectory angles θ along which the wave packets move, so may be considerably more than that in scattering at large angles. Even with considerably different scattering potentials (e.g., one of them is zero) and with $\theta \sim 10^{-2}$ we have $\tau \sim 10^{-8}$ s. If the potentials are close $U_2 - U_1 \sim 0.1U$, $\theta \sim 10^{-8}$ and, hence, $\tau \sim 10^{-6}$ s, i.e., the time is quite long. We should note that in the scattering at large angles with this difference in potentials we would have $\theta \sim 0.1$, i.e., $\tau \sim 10^{-11}$ s.

Influence of the Elastic Scattering Without Phase Randomization on Resonance Characteristics

The scattering at small angles without phase randomization results in qualitative peculiarities in the behavior of narrow resonances. In a low-pressure gas there has been discovered a nonlinear dependence of collisional broadening and shift of a resonance on the gas density [3.32]. Figure 3.8 shows the observed dependence of

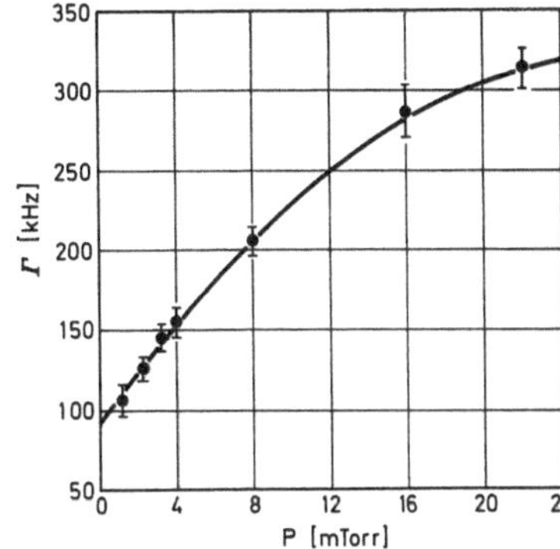

Fig.3.8. Dependence of the Lamb dip width in CH_4 at λ = 3.39 μm on pressure

a resonance width in CH_4 on the gas density. It should be particularly noted that this dependence has been observed in the conditions where the model of pair collisions is a priori true (gas pressure $\sim 10^{-3}$ Torr). Previous theories describing the influence of collisions on a resonance shape did not explain this dependence even qualitatively. The qualitative explanation of the nonlinear dependence of collisional broadening presented in [3.32] is as follows. If at scattering at an angle θ the typical Doppler shift equal to $ku\theta$ is more than a homogeneous linewidth Γ, then after collision a particle does not interact with the field. The interaction time is determined by the collision frequency. In this case it is not important whether the dipole moment phase is randomized or not. Under these conditions the resonance broadening is determined by the total scattering cross section.

When $ku\theta < \Gamma$ the dipole moment phase randomization is essential. Naturally, providing $ku\theta < \Gamma$ the collisions with velocity change and without dipole moment phase randomization result in no resonance broadening. Thus, here the resonance broadening will be determined by the same processes as the Doppler contour broadening. In this region the broadenings of resonance and Doppler contour coincide. It may be noted that the resonance broadening in the regions where $ku\theta \gg \Gamma$ and $ku\theta \ll \Gamma$ may largely differ. With the change of gas pressure a homogeneous halfwidth Γ is changed too, which results in the nonlinear run of the dependence of a resonance width in the presence of collisions with no phase randomization.

The theory of collisional resonance broadening based on kinetic equations for a density matrix in which arrival and departure terms are expressed through the scattering amplitudes of particles on the upper and lower levels is given in [3.33, 34]. The results are in qualitative agreement with experiment. The nonlinear dependences of broadening on gas density were also observed in NH_3 [3.35], CO_2 [3.36, 37], Xe [3.38]. The theoretical dependence calculated in [3.39] for CO_2 agrees satisfactorily with the experiment carried out in [3.37].

Providing $ku\theta \gg \Gamma$ a nonlinear resonance shape turns out to be complicated: it consists of an underlining of $ku\theta$ wide and a sharp part whose width is determined by the collision frequency, i.e., by the total elastic scattering cross section. Recently there has been obtained the direct experimental confirmation of this complicated shape of a resonance arising in particle scattering at small angles [3.40]. The experiments were performed with CH_4. Figure 3.9 shows the power resonance shape in a He-Ne laser with a methane absorption cell at $\lambda = 3.39$ μm with various helium pressures in the cell. With a helium pressure of more than 10^{-2} Torr an underlining is well pronounced, which is slowly broadened with an increase of pressure. Study of the underlining shape gives a qualitative concept of the characteristics of a differential scattering cross section.

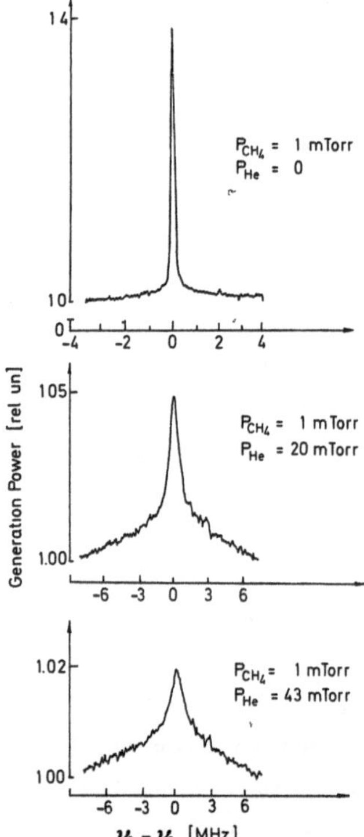

Fig.3.9. Record of the resonance shape in methane at various helium pressures: a - P_{CH4} = 1 mTorr, P_{He} = 0; b - P_{CH4} = 1 mTorr, P_{He} = 20 mTorr; c - P_{CH4} = 1 mTorr, P_{He} = 43 mTorr

3.3 Coherent Phenomena in Multilevel Systems

3.3.1 Resonant Processes in Three-Level Systems

A great many works are devoted to radiative transitions in three-level systems, studies of a line shape of radiation (emission) of fields that are resonant to adjacent transitions in atoms and molecules. The processes in three-level systems are closely related to such phenomena as resonances of absorption and gain, two-photon resonances, CARS, and four-wave processes. It is therefore impossible here to dwell even briefly on these various problems. We shall restrict ourselves to discussion of the role of coherent processes in multilevel systems. A general approach should be elaborated to explain the facts that would permit one to relate phenomena usually considered separately.

The layout of radiative transitions in three-level systems is given in Fig.3.1. We shall begin considering the processes in three-level systems at a case where both fields are in resonance with adjacent transitions. The radiative processes in this system were discussed in detail in [3.41,42]. So here we shall try to separate population and coherent processes. If a field E_1 at frequency ω is resonant to the $1 \rightarrow 0$ transition, and a field E_2 at frequency ω' is resonant to the $0 \rightarrow 2$ transition, and the frequency difference $\omega - \omega'$ is much more than a Doppler width, the interaction of the field E_1 with the $2 \rightarrow 0$ transition and of the field E_2 with the $1 \rightarrow 0$ transition may not be taken into account. In this case the probability of particle transition from the level 0 to the level 2 is given by the following expression (within the frames of second-order perturbation theory) [3.42]:

$$W_{1 \rightarrow 2} = \frac{2V_1^2V_2^2}{\left(\frac{\gamma_1 + \gamma_0}{2}\right)^2 + \Omega^2} \mathrm{Re}\left\{\frac{1}{\gamma_1} \frac{1}{\frac{\gamma_1 + \gamma_2}{2} + i(\Omega' - \Omega)} + \frac{1}{\gamma_0} \frac{1}{\frac{\gamma_0 + \gamma_2}{2} + i\Omega'} + \frac{1}{\left(\frac{\gamma_0 + \gamma_2}{2} + i\Omega'\right)\left[\frac{\gamma_1 + \gamma_2}{2} + i(\Omega' - \Omega)\right]}\right\}, \quad (3.34)$$

where $V_1 = d_{01}E_1/\hbar$, $V_2 = d_{02}E_2/\hbar$, $\Omega = \omega - \omega_{01}$, $\Omega' = \omega' - \omega_{02}$, γ_0, γ_1, γ_2 are the level relaxation constants. The expression (3.34) describes the behavior of the transition probability both in the conditions of resonance ($\Omega \sim \gamma_1 + \gamma_0$) and out of resonance.

The two processes determine the behavior of the probability of transition. The first one corresponds to a two-quantum process of simultaneous absorption of a quantum at the frequency ω and emission at the frequency ω'. The second process corresponds to a stepwise transition of an atom to the level 0 with consequent transitions to the level 2. This process is due to changes in population of an intermediate level. The first term in (3.34) corresponds to two-quantum processes. It has a width of the forbidden $1 \rightarrow 2$ transition, the line maximum is located at the frequency $\omega' = \omega_{12} + \omega$, i.e., when the conditions of two-photon resonance are fulfilled $\Omega' = \Omega$. The second term in (3.34) corresponds to the stepwise process associated with the population of the level 0. The linewidth that corresponds to this process is therefore determined by the width of the $0 \rightarrow 2$ transition, the line maximum is located at the frequency $\Omega' = 0$.

Far from resonance both lines and processes are readily separated. The line at the frequency $\omega' = \omega_{12} + \omega$ corresponds to a Raman scattering line. This line is related to coherent processes. That is to say, far from resonance separation of population and coherent processes is trivial: at $\omega' = \omega_{12} + \omega$ a particle goes to the level 2 due to a two-quantum process, at $\omega' = \omega_{02}$ of importance are the stepwise processes that are due to level population.

Another situation is observed in the conditions of resonance. With $\Omega = 0$, as can be readily seen from (3.34), the line of two-quantum process coincides with that of stepwise transition. It is therefore impossible in the conditions of resonance to unambiguously separate the processes. As these occur simultaneously, their interference takes place. The contribution of the interference is quantitatively described by the third term. There is a very important circumstance described in [3.43], that permits estimation of the contribution of coherent and stepwise population processes even in the conditions of resonance. Both in two- and in three-level systems the contribution of coherent processes in the conditions of resonance turns out to be largely dependent on the relation of level relaxation constants. As it is seen from (3.34), with $\gamma_0 \gg \gamma_1, \gamma_2$ the probability of transition $W_{1\rightarrow 2}$ is determined by two-quantum processes, with $\gamma_0 \sim \gamma_1$ one-quantum stepwise transitions play an essential role. With $\gamma_0 \ll \gamma_1$ the two processes are equivalent and a resonance shape is determined by both of them. Their interference is essential.

In strong fields new peculiarities appear in the probe wave absorption. These are essential at $k' < k$, where k' and k are the wave vectors of E_2 and E_1, respectively [3.44]. A wide region, where absorption is close to zero, appears in the probe wave absorption. Then the absorption sharply increases but has a resonance character. The resonances are spaced at $S = |V_1|\sqrt{k'(k-k')}/k^2$ between them the absorption tends to zero. The resonance shape $\Omega = 0$ is given in Fig.3.10. The behavior of the probe wave absorption in the frequency region where $|\Omega'| < S$ resembles that of absorption of a probe unidirectional wave for two-level atoms. As has been shown in [3.2], this behavior is due to a Stark effect as well as in the case of two-level atoms. Following the above described model one can show that the velocities of atoms that resonantly interact with a probe wave satisfy the relation at $\Omega = 0$

$$v_r = \frac{\Omega'(k' - k/2) \pm \sqrt{(k\Omega'/2)^2 + (k'^2 - k'k)V_1^2/4}}{k'^2 - k'k} \quad . \tag{3.35}$$

Now it can be readily seen that at $|\Omega'| < S$ there are no atoms that resonantly interact with a probe wave. So the probe wave absorption is zero. Figure 3.11 shows the dependence of probe wave frequency on the velocity of resonantly interacting atoms (for simplicity we assume that $\Omega = 0$). It is seen that with $|\Omega'| = S$ the number of atoms that can interact with the probe wave sharply increases, which results in the appearance of a resonance. To put it the other way round, one may talk

[2]In the layout of levels given in Fig.3.1a the lines under consideration correspond to unidirectional waves, for a cascade scheme to oppositely travelling ones.

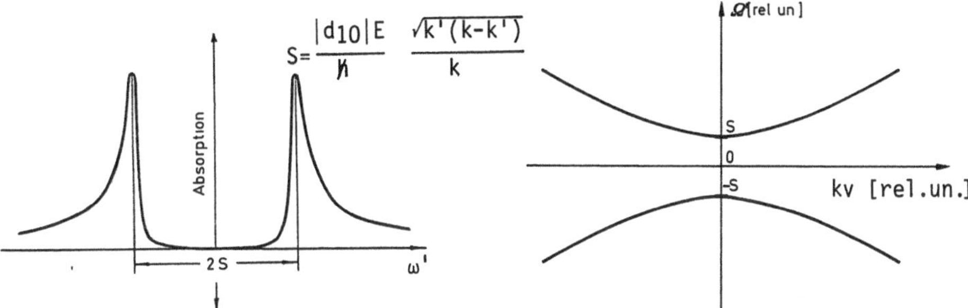

Fig.3.10. Absorption line shape on the $0 \rightarrow 2$ transition in the strong field on the $1 \rightarrow 0$ transition at $k' = k/2$

Fig.3.11. Dependence of the probe wave frequency on the velocity of resonantly interacting atoms at $\Omega = 0$. $k' = k/2$

about cancellation of the Doppler shift in most atoms. An observation of the Stark effect on Doppler broadened transitions has been discussed in detail in [3.45,46]. Under the action of several fields a qualitative analysis becomes more difficult. As has been shown in [3.47], the action of Stark splitting may be used in cancelling the Doppler shift and in obtaining a line of stimulated radiation without Doppler shift.

Let us dwell on another important effect of coherent phenomena in three-level systems. In stabilization of laser frequencies to a Lamp dip the influence of a closely located adjacent transition is appreciable. The presence of this transition results in the appearance of a so-called crossing resonance. It is located at the field frequency that is equal to the half-sum of frequencies of the $0 \rightarrow 1$ and $0 \rightarrow 2$ transitions. Its intensity is proportional to the square root of the probabilities of the $0 \rightarrow 2$ and $0 \rightarrow 1$ transitions [3.2]. Crossing resonances (we consider the case where the frequency difference $\omega_1 - \omega'$ is far less than a Doppler width) are of significance for a He-Ne laser at 3.39 μm stabilized to an absorption $F_2^{(2)}$ line in methane. As is known, the $F_2^{(2)}$ line in methane has a magnetic hyperfine structure that consists of three strong components ($\Delta F = -1$: $5 \rightarrow 6$, $6 \rightarrow 7$, and $7 \rightarrow 8$ transitions) and of two weak ones ($\Delta F = 0$: $6 \rightarrow 6$, $7 \rightarrow 7$ transitions). The probabilities of the $5 \rightarrow 6$, $6 \rightarrow 7$, and $7 \rightarrow 8$ transitions exceed those of the $6 \rightarrow 6$ and $7 \rightarrow 7$ transitions by a factor of 10^2. Since the $6 \rightarrow 7$, $7 \rightarrow 8$, and $6 \rightarrow 6$, $7 \rightarrow 7$ transitions are close to each other, we shall consider a simplified three-level scheme. In weak fields the Lamb dip intensity at the $1 \rightarrow 0$ and $0 \rightarrow 2$ transitions is extremely low as compared to the Lamb dip amplitude at the $0 \rightarrow 2$ transition. The resonance position at the $0 \rightarrow 2$ transition is largely affected by the processes in a three-level system. Two mechanisms affect the resonance position: the influence of the wing of a crossing resonance resulting in pulling the resonance maximum towards the center of transition. The magnitude of the pulling depends on a homogeneous width of the transition and increases with an increase of the width. Another process is due to multiquantum processes similar to those we have considered above. Owing to the closeness of the $0 \to 1$ transition, the field resonant to the $0 \to 2$ transition will interact with the $0 \to 1$ transition as well.

The resonance shifts in this system have been investigated in [3.48]. In the third order perturbation theory the shift is described by

$$\omega_r = \omega_{20} - \frac{F|\omega_{21}|}{2} \frac{d_{10}^2}{d_{20}^2} \quad , \tag{3.36}$$

where ω_r is the resonance position corresponding to the frequency ω_{20}:

$$F = \frac{1}{2(1+\delta^2)} \left[-1 + \frac{1}{1+\delta^2} - \frac{2(1-\delta^2)}{(1+\delta^2)^2} \right] + \frac{1}{(1+\delta^2/4)^2} \quad , \tag{3.37}$$

$$\delta = (\omega_{20} - \omega_{30})/\Gamma \quad ; \quad d_{10}^2 d_{20}^{-2} \ll 1 \quad .$$

The form of the function (3.37) is given in Fig.3.12.

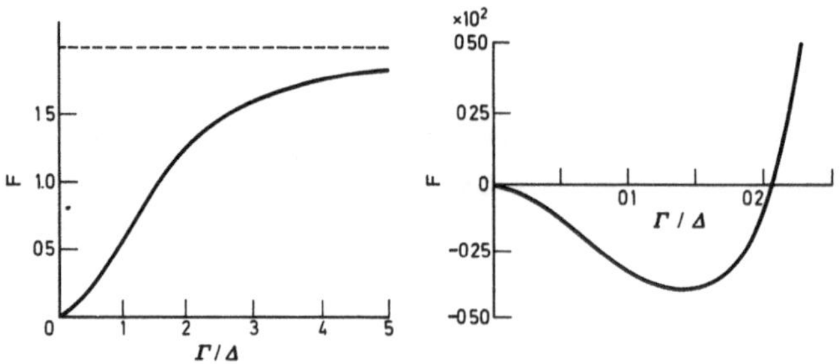

Fig.3.12. Shift of the maximum of the nonlinear power resonance formed by two components with a common level but largely different in intensity. 2Δ is the frequency distance between the components

In the region of small widths the coherent phenomena produce an effect that is in sign opposite to the influence of the crossing resonance, which results in a slight resulting shift. This effect proved to be significant in achieving a high frequency reproducibility of the He-Ne laser with a methane absorption cell [3.49]. The physical nature of the nonlinear resonance shift is similar to a well-known phenomena of an optical shift. The difference is as follows. A nonlinear resonance arises in the same order in the field as a shift. So it shows itself in the constant shift independent of field. For a field shift to be obtained in analogy with a usual optical shift it is necessary to consider higher orders of perturbation theory.

3.3.2 Two-Photon Resonances

We have been considering the situation where both fields are resonant to adjacent transitions. A new type of resonance phenomena arises if the sum or difference of frequencies of the applied fields is equal to the transition frequency. In this case some individual frequencies may be far from a resonance. Of great importance are so-called two-photon resonances in the standing-wave field. The resonance arises for oppositely travelling waves in the cascade layout of transitions when $\omega + \omega' = \omega_{21}$. It can be easily seen that in the standing-wave field the resonance takes place if $2\omega = \omega_{21}$. In the reference system of a moving atom it perceives one wave with the frequency $\omega + kv$, another wave with the frequency $\omega - kv$. In the absorption of two oppositely travelling photons of the same frequency the Doppler shift is completely eliminated. The resonance with a homogeneous width arises on the background of a Doppler underlining. The relation between the resonance amplitudes and the underlining is approximately identical with that between Doppler and homogeneous widths. This resonance has been predicted in [3.50] and first realized in [3.51,52]. The method proved to be very useful in spectroscopy. It was used in carrying out a great many experiments [3.53].

We shall not consider in detail the properties of the resonance, note only the similarity of the processes responsible for the formation of two-photon resonances and resonances on adjacent transitions. As before, the process may be schematically represented as

$$\dot{n}_1 \xrightarrow{E(\omega)} d_{01} \xrightarrow{E'(\omega)} d_{12} \xrightarrow{E(\omega)} d_{02} \xrightarrow{E'(\omega)} \dot{n}_2 \quad .$$

Here E and E' are two oppositely travelling waves. Since the field frequency is far from the resonance frequencies ω_{01}, ω_{20}, the stepwise processes may be completely ignored in considering the two-photon processes. Unlike the previously considered resonant cases, here both fields can interact with the $0 \to 1$ and $2 \to 0$ transitions. As a result, the probability of two-photon absorption increases by a factor of four as compared with the transition probability in the process (3.4). An additional increase by two times is due to coherence of the processes: the probability amplitude of detection of a particle in the level 2 increases twice, the probability by a factor of four. The complete elimination of the Doppler shift in the two-photon absorption permits experiments with immovable particles and observation of the phenomena that cannot be usually observed in a gas due to chaotic particle motion. Among these phenomena are self-emission, resonant transients, resonances in time and spatially separated fields. We shall consider some phenomena in detail.

3.3.3 Relation to Other Phenomena

We have considered the coherent processes in three-level systems responsible for the occurrence of resonances. At the action of two fields whose frequencies are close to those of adjacent transitions, we are interested in the dipole moment of a particle at the frequency of one field. The scheme of the process (3.4) is not exhaustive in our system as it does not take into account, for instance, the interaction of field E_1 with the $0 \to 2$ transition. With allowing for this interaction the following process may be represented

$$\dot{n}_1 \xrightarrow{E_1} d(\omega_1) \xrightarrow{E_2} d(\omega_1 - \omega_2) \xrightarrow{E_1} d(2\omega_1 - \omega_2) \quad . \tag{3.39}$$

Unlike the previously considered processes, the dipole moment of a particle arises at the frequency different from that of fields E_1 and E_2. The dipole moment of particles leads to the appearance of macroscopic polarization and coherent radiation that is known as CARS. It is seen that the dipole moment at the frequency $2\omega_1 - \omega_2$ is maximum if $\omega_1 - \omega_2 = \omega_{21}$. Following (3.39) one can consider the interaction with the third field at the frequency ω_3 instead of the interaction with the field E_1 at the third stage when there is a dipole moment at the frequency $\omega_1 \pm \omega_2$. Then the process can be readily represented as

$$\dot{n}_1 \xrightarrow{E_1} d_{10}(\omega_1) \xrightarrow{E_2} d_{12}(\omega_1 + \omega_2) \xrightarrow{E_3} d_{12}(\omega_1 + \omega_2 + \omega_3) \quad . \tag{3.40}$$

The process results in polarization and radiation at the resulting frequency. In the literature this process is known as a four-wave mixing in a gas. The efficiency of this process will sharply increase if the frequency of field E_3 is close to that of the $2 \to 3$ transition. If all fields are in resonance with transitions, the efficiency may be considerable even with weak fields. This permits obtaining an effective conversion of frequencies that lie in different ranges at continuous operation. The scheme of the first experiment on continuous frequency conversion in a gas is given in Fig.3.13 [3.54]. The experiment was performed in Ne on the cascade $3s_2 - 3p_4$, $3p_4 - 2s_2$, and $2s_2 - 2p_4$ transitions. Under certain conditions there occurs simultaneous cascade generation on the 3.39, 2.39, and 1.15 μm lines in a He-Ne mixture [3.55]. At this moment a continuous coherent radiation was observed at the frequency of the $3s_2 - 2p_4$ transition ($\lambda = 0.63$ μm). In this way the continuous mixing of frequencies in a gas was performed. The efficiency of this conversion is many orders as high as that in nonlinear optical crystals.

Let us come back once more to the scheme of (3.4). If the field E_1 has a spatial structure, the latter will take place in polarization at the frequency of the field E_2, i.e., we have wavefront reversal.

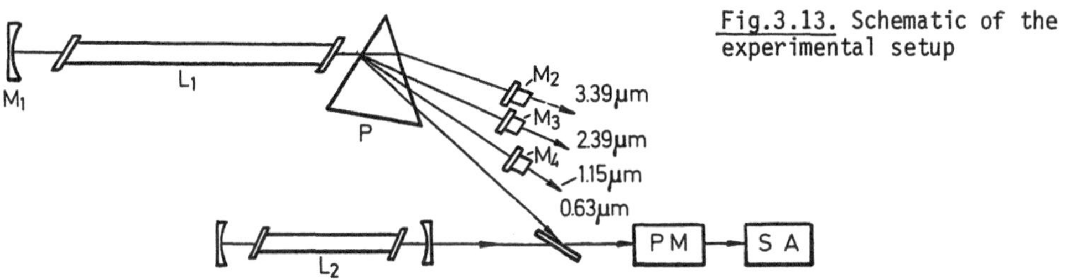

Fig.3.13. Schematic of the experimental setup

3.4 Method of Separated Optical Fields

Now let us go on to consider the method of separated optical fields that is new for the optical band. It is based on coherence transfer at large distances, which permits obtaining narrow optical resonances with a width that is inverse to the transit time of a particle between fields. Depending on the nature of particle-field interaction one can detect three ways to obtain resonances in separated fields:

1) three-beam system of separated fields (SF) to obtain resonances in two-level particles [3.56],
2) two-photon resonances in SF [3.57],
3) resonances in three-level particles [3.58].

In the course of investigations of the method of separated fields resonances of continuous coherent radiation in separated fields have been detected [3.59]. The peculiarities of this phenomenon are closely related to such well-known phenomena as photon echo and superradiance. A damping coherent radiation with the frequency located in the line center that arises in pulsed switching of a standing-wave field has been detected in studying transient resonant phenomena [3.60]. The frequency of the standing-wave field is detuned from a resonance. The totality of various resonant phenomena arising in spatially and time separated fields makes up the method of separated fields.

The method of separated fields is well known in a microwave range and has been developed by RAMSEY as early as the 1950s [3.61]. However, it proved to be impossible to use the method directly in an optical band. This is due to the influence of a Doppler effect. For better understanding let us consider the basic idea of this method in a SHF range. Figure 3.14 is a schematic of the method. A beam of particles interacts with two fields. After the resonant interaction with the first field a particle has the dipole moment at the transition frequency ω_{21} $d(t) = d_{21}V \exp(-i\omega_{21}t) + c.c.$, where d_{21} is the matrix element of the dipole moment, $V = id_{21}E/\hbar$, E is the field amplitude, τ is the interaction time. An energy absorbed in the second field periodically depends on the phase difference of the dipole moment and of the field.

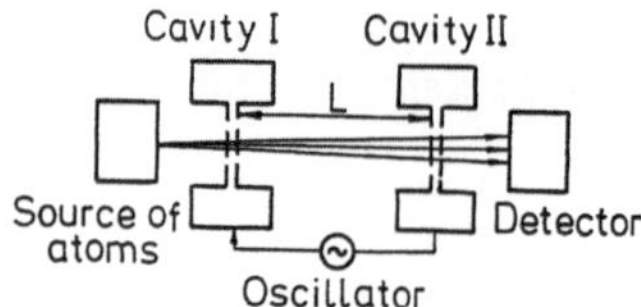

Fig.3.14. Schematic diagram of the method of separated fields in a microwave range

$$E = 2\hbar\omega_{21}\tau^2|V|^2 \cos(\Omega T - \varphi) \quad , \tag{3.41}$$

where T is the transit time of a particle, $\Omega = \omega - \omega_{21}$ is the field frequency detuning from the resonance frequency, φ is the phase difference between the fields. At the interaction of an ensemble of particles only the transit time of a particle between the fields is averaged. This results in that absorption oscillations begin to damp rapidly with an increase of the detuning frequency.

In the optical band we encounter a new important effect that does not permit a direct use of the method of separated fields (MSF) in optics: field dimensions are usually much larger than a wavelength, and the Doppler effect becomes essential. Coherent properties of an individual particle will show themselves if they are reflected in macroscopic polarization of a medium, i.e., of great importance in the optical band is the macroscopic polarization transfer. Let us consider two optical fields (Fig.3.15a,b) interacting with the beams of particles of finite divergence. The dipole moment of particles going out of the point z_1 is identical for all particles and determined by the field phase in the point z_1. However, if the distance between the fields is sufficiently large, the phase difference between the second field and the dipole moment will depend not only on the transit time of particles and the phase difference between fields but also on the velocity projection of particles. From Fig.3.15a one can see that the phase of the dipoles in the fields has a "Doppler" phase angle equal to kvT. Despite that some particles transfer coherence, the absorption of an ensemble of particles connected with the polarization transfer is zero. The same result is obtained in considering the dipole moment that is introduced by particles into the point z_2 of the second field after the interaction with the first field. The phases of the dipole moments depend on those of the fields in the points z_1, z_1', z_1'', and so on. If $\Delta z \gg \lambda$, an average dipole moment in the point z_2 will be zero. Thus, to observe the effects associated with coherence transfer requires elimination of the influence of a Doppler effect.

3.4.1 Two-photon Resonance in Separated Fields

As we have seen before, the complete compensation of the Doppler effect takes place in two-photon absorption of two oppositely traveling photons. The interaction with a beam also occurs as if the particle beam would have no divergence and fly perpendicularly to the direction of wave propagation. This means that the method of separated optical fields may be realized in the interaction with two fields for the case of two-photon absorption.

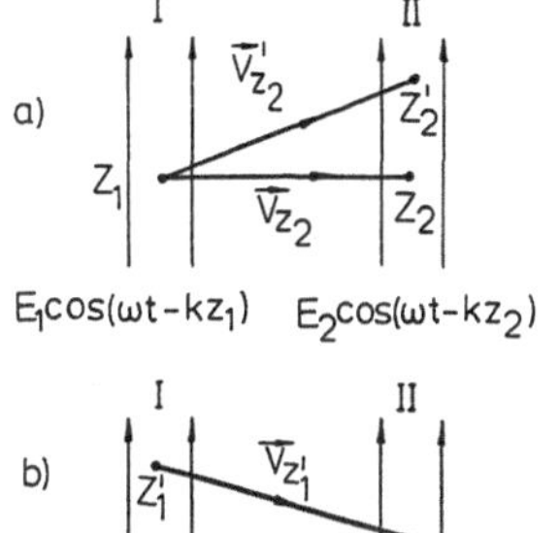

◄ Fig.3.15. Scheme of the interaction of particles with two separated optical fields

▼ Fig.3.16. Experimental arrangement for obtaining the two-photon resonance with spatially separated standing waves

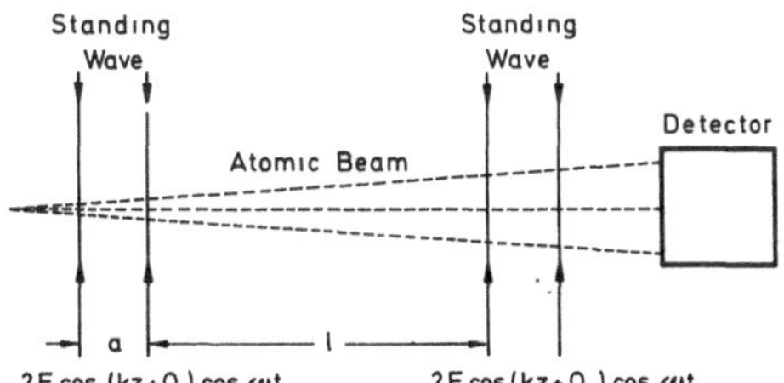

Figure 3.16 shows a simplified scheme for observing the resonance in separated fields. The particle flow passes through two standing-wave fields where the two-photon absorption occurs.

Let us consider an atom moving in the field of two standing waves. It is acted upon by the field

$$E(t) = 2E(t)\cos\omega t \quad ,$$

where

$$E(t) = E_1 g(t)\cos(kvt + Q_1) + E_2 g(t - T)\cos(kvt + Q_2)$$

$$g(x) = \begin{cases} 1 & \text{at } 0 < x < \tau \\ 0 & \text{at } x < 0,\quad x > \tau \end{cases} \quad .$$

Here $\tau = a/u$ is the time of flight of an atom with transverse velocity u through the field, $T = L/u$ is the time of flight between the light beams separated by L, and v is the velocity projection onto the light beam direction. The distance between beams is considered to be much larger than the beam dimension a, i.e., $L/a \ll 1$.

We assume that the doubled field frequency 2ω is close to the frequency of an atomic transition ω_{21}. Keeping only terms responsible for the two-photon resonance we find the equations of motion for the atom probability amplitudes

$$\hbar\dot{a}_n = id_{n2}E(t)\exp[i(\omega_{n2} + \omega)t]a_2 + id_{n1}E(t)\exp[i(\omega_{n1} - \omega)t]a_1 \quad , \tag{3.42}$$

$$\hbar a_1 = i\sum_n d_{1n}E(t)\exp[i(\omega_{1n} + \omega)t]a_n \quad , \tag{3.43}$$

$$\hbar a_2 = i\sum_n d_{2n}E(t)\exp[i(\omega_{2n} - \omega)t]a_n \quad , \tag{3.44}$$

where ω_{n2} and ω_{n1} are frequencies of the transitions $n \to 2$ and $n \to 1$, d_{n2} and d_{n1} are the corresponding matrix elements of a dipole moment. Since the detunings $\omega_{n2} + \omega$ and $\omega_{n1} + \omega$ are far from the resonance, one can derive the expression for a_n from (3.42). Inserting it into (3.43,44) we get a set of equations that determines a_1 and a_2 in a fashion similar to the Schrödinger equation for a two-level scheme

$$\dot{a}_1 = iD_{11}E^2(t)a_1 + iD_{12}E^2(t)e^{i\Omega t}a_2$$

$$\dot{a}_2 = iD_{22}E^2(t)a_2 + iD_{21}E^2(t)e^{-i\Omega t}a_1 \quad , \tag{3.45}$$

where $\Omega = 2\omega - \omega_{21}$, $D_{ik} = \sum_n d_{in}d_{nk}/\hbar^2(\omega_{n2} + \omega)$.

When solving (3.45) we consider the field to be weak. An initial condition at the moment of particle flight into the first beam ($t = 0$) is $a_1 = 1$, $a_2 = 0$. The probability amplitude at $t > T + 2\tau$ is

$$a_2 = a_2^{(1)} + a_2^{(2)} \tag{3.46}$$

where

$$a_2^{(1)} = \frac{i}{2} D_{21}E_1^2\tau \quad , \quad a_2^{(2)} = e^{-i\Omega T} \frac{i}{2} D_{21}E_2^2\tau \quad ,$$

$a_2^{(1)}$ and $a_2^{(2)}$ are the excitation amplitudes of particles in the level 2 in the first and second beams, respectively. In (3.46) we consider the region of detuning $\Omega \sim T^{-1} \ll \tau^{-1}$. The probability of detecting an atom in the level 2 after having passed through the two fields is equal to

$$n_2 = |a_2^{(1)}|^2 + |a_2^{(2)}|^2 + 2\mathrm{Re}\left\{a_2^{(1)}a_2^{(2)*}\right\}$$

$$= \frac{|D_{21}|^2\tau^2}{4}\left[E_1^4 + E_2^4 + 2E_1^2E_2^2 \cos(\Omega T)\right] \quad . \tag{3.47}$$

Besides the two-photon absorption resonances corresponding to the interaction of an atom with each of the fields, (3.47) has a term which is due to the interference of the amplitudes $a_2^{(1)}$ and $a_2^{(2)}$.

Now consider an atomic beam with the velocity distribution

$$f(u) = \frac{4}{\sqrt{\pi}} \frac{u^2}{v_0^3} \exp[-(u^2/v_0^2)] \quad , \tag{3.48}$$

where v_0 is the thermal velocity. We are interested in the third term in (3.47) responsible for the resonance. Averaging it with the distribution (3.48) we obtain the expression for the change in intensity of the excited particles

$$\Delta J_2 = J_1|D_{12}|^2\tau_0^2E_1^2E_2^2K(\Omega T_0) \quad ,$$

$$K(x) = \int_0^\infty ye^{-y^2} \cos\frac{x}{y}\,dy \quad , \tag{3.49}$$

where J_1 is the particle flow in the level 1, $T_0 = 1/v_0$, $\tau_0 = a/v_0$. The function $K(x)$ is given in Fig.3.17. It describes the resonance with a width which is reci-

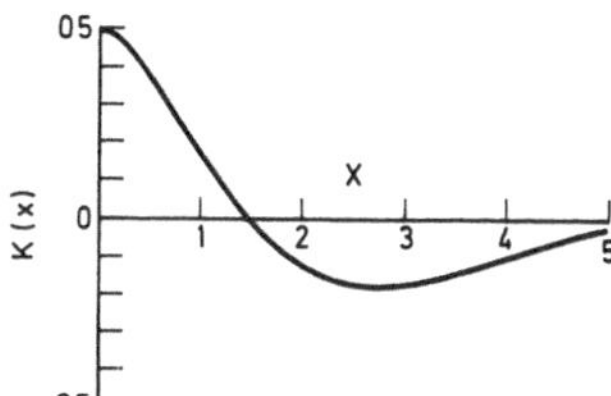

Fig.3.17. Line shape of the two-photon absorption resonance in spatially separated light beams

procal to the time of flight between fields for the number of particles excited in level 2.

Let us consider the interaction of an atomic beam with the field of two separated standing waves with a Gaussian profile [3.62]

$$g(x,y) = \exp\left(-\frac{y^2+x^2}{a^2}\right) + \exp\left(-\frac{y^2+(x-L)^2}{a^2}\right) . \tag{3.50}$$

L is assumed to be the distance between the beams which is much larger than the transverse beam dimension ($L \gg a$). Solving (3.42,43) and taking into account the level widths one can obtain the expression for the flow of particles excited in level 2 which consists of three terms

$$J_2 = J_2^{(1)} + J_2^{(2)} + J' .$$

$J_2^{(1)}$ and $J_2^{(2)}$ describe the resonance arising in the interaction of an atom with each of the beams with a width $\sim 1/\tau_0$. We are interested in the third term that can be written in the form

$$J'/J = Q2\pi|V|^2\tau_0^2 f(\Omega T_0) , \tag{3.51}$$

where

$$f(x) = 2\int_0^\infty d\xi\xi\exp[-(\Gamma T_0/\xi) - \xi^2]\cos(x/\xi) , \tag{3.52}$$

here ξ is an integration variable, $T_0 = L/v_0$, Γ is the homogeneous halfwidth of two-photon transitions. The function f(x) describes the two-photon absorption resonance in separated light fields. The resonance halfwidth is determined from the condition

$$\frac{f(\Gamma' T_0)}{f(0)} = \frac{1}{2} .$$

Figure 3.18 shows the dependence of the resonance halfwidth Γ' and its amplitude f(0) on ΓT_0. Figure 3.19 shows the function f(x) at different ΓT_0. When $\Gamma T_0 \ll 1$, $\Gamma' = 0.67/T_0$. When $\Gamma T_0 \gg 1$,

$$f(x) = 2(\pi/3)^{1/2}\beta e^{-3\beta^2} e^{-(1/12)(x/\beta^2)^2}\cos(x/\beta) ,$$

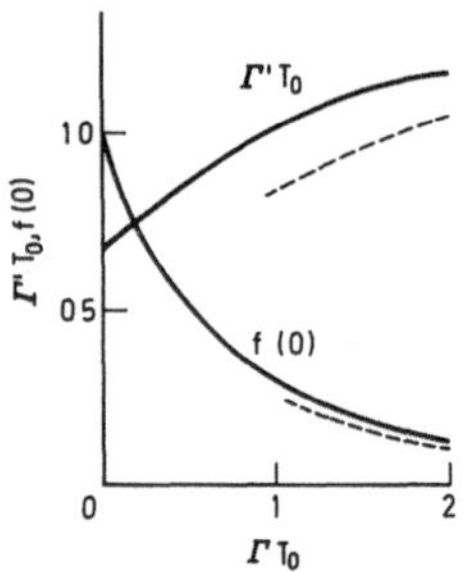

Fig.3.18. Dependence of the resonance width and amplitude in separated light beams on the parameter ΓT_0. Dotted curves are plotted according to (3.53,54)

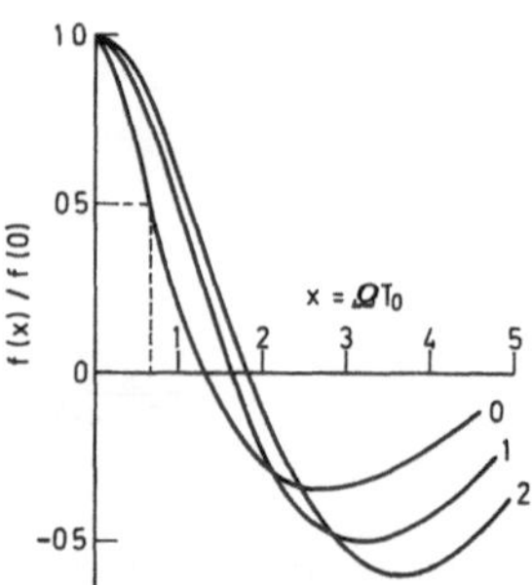

Fig.3.19. Line shape of the two-photon absorption resonance in spatially separated light beams at various values of the parameter ΓT_0. 0 - $\Gamma T_0 = 0$; 1 - $\Gamma T_0 = 1$; 2 - $\Gamma T_0 = 2$

where $\beta = (\Gamma T_0/2)^{1/3}$. The resonance halfwidth in this case is

$$\Gamma' = (\pi/3T_0)\beta^{1/3} \quad , \tag{3.53}$$

its amplitude is

$$f(0) = 2(\pi/3)^{1/2}\beta e^{-3\beta^2} \quad . \tag{3.54}$$

Note that the use of two-photon absorption in beams of such metals as Ca, Sr, Ba, Hg, and other permits one to obtain resonances in the optical region with widths of about 100 Hz. The distance between standing waves must be of an order of 1 m. In this case a limiting resonance width is determined by the second-order Doppler effect.

The resonances of two-photon absorption in separated fields were first observed in [3.63] on the 3s-4s transition of Na atoms. Since the 4s state was short lived, the distance between beams was about 100 μm. HALL [3.64] used transitions of Rb from the ground to the Rydberg state. The lifetime of highly excited states is quite long, which may be used to obtain narrow lines. Very promising in obtaining narrow lines is the use of transitions to metastable states in vapors of metals Ca, St and others.

In the system of coordinates of a moving atom the interaction with spatially separated fields is similar to the action of two pulsed fields with delay time T equal to the transit time of a particle between the fields. Some difference between the spatial and time pictures is observed at the interaction with an ensemble of particles. In spatially separated fields T depends on the particle velocity. In pulsed fields this time is identical for all atoms. So a typical interference pattern is well pronounced. Experiments on the interaction with two standing-wave fields were carried out in [3.63,66].

Narrow Two-Photon Absorption Resonances of the Sequence of Supershort Pulses in a Gas

Among the interesting applications of two-photon absorption in the standing-wave field [3.51] is the possibility to obtain narrow resonances on the 1s-2s transition of hydrogen [3.68]. This requires a source of continuous radiation at 2400 Å. In [3.67] we considered the case of two-photon absorption of the sequence of supershort pulses which is of interest for measuring the frequency of the 1s-2s transition of a hydrogen atom[3].

Let us show the role coherent effects play at the two-photon absorption of a periodic pulse sequence. After the interaction of an atom with one K pulse of the field

$$a_2^{(K)} = e^{-i\Omega KT} iD_{21}E_2^2\tau \quad . \tag{3.55}$$

After the interaction with pulses the probability of detecting a particle in level 2 is

$$|a_2|^2 = |\sum_{n=0}^{N} a_2^{(K)}|^2 = D_{21}^2 E^4 \tau^2 \frac{\sin^2 \frac{\Omega NT}{2}}{4 \sin^2 \frac{\Omega T}{2}} \quad . \tag{3.56}$$

Since the quantity NT can be equated with the total time of the coherent field-atom interaction, an effective resonance width is determined by the homogeneous transition width γ.

3.4.2 Resonance in Separated Fields for Two-Level Atoms

For two-level atoms, as we have seen earlier, the method of separated fields cannot be realized for two fields. The influence of the Doppler effect can be eliminated here in the interaction with three fields. After the nonlinear interaction with two fields the phase of the dipole moment of a particle undergoes a jump that compensates the Doppler phase angle. The interaction with three fields results in the elimination of the Doppler effect.

Let us consider a gas of atoms resonantly interacting with the field of three standing waves (Fig.3.20)

$$E(x, z, t) = 2E(x, z)\cos\omega t \quad , \tag{3.57}$$

$$E(x, z) = E_1 g(x)\cos(kz + \varphi_1) + E_2 g(x - L)\cos(kz + \varphi_2) + E_3 g(x - 2L)\cos(kz + \varphi_3) \quad ,$$

[3]The two-photon absorption resonance intensities for single-frequency and multimode runnings in the regime of supershort pulses are identical. However, the efficiency of frequency multiplication from the region of 4860 Å in the regime of supershort pulses is much higher.

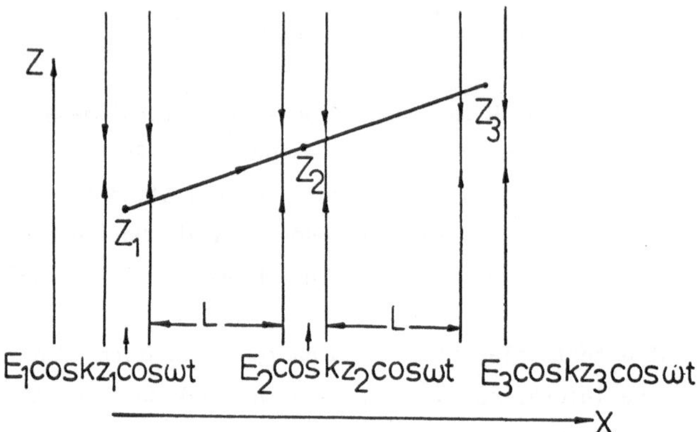

Fig.3.20. Scheme of the interaction of particles with three separated optical fields

where

$$g(x) = \begin{cases} 1 & \text{at} \quad 0 < x < a \\ 0 & \text{at} \quad x < 0 \ , \quad x > a \ . \end{cases}$$

The equations for the probability amplitudes of an atom are of the form

$$\dot{b} = \frac{i}{\hbar} d_{21} E(x,z) e^{-i\Omega t} a \ , \quad \dot{a} = \frac{i}{\hbar} d_{12} E(x,z) e^{-i\Omega t} b \ , \tag{3.58}$$

where a and b are the probability amplitudes for the lower 1 and upper 2 levels, respectively, d_{21} is the dipole moment of transition, $\Omega = \omega - \omega_{21}$ is the frequency detuning from the transition frequency ω_{21}.

We are interested in the energy absorbed by an atom after the interaction with three fields. In order to find the polarization we look for a dipole moment introduced by an atom into the point x, z and then average over velocities (averaging over velocities at the fixed z corresponds to that over the point z_1 of flight into the first beam). Following the conventional methods we shall find the probability amplitude after having passed through three beams.

For the sake of simplicity we shall consider the beams to be narrow, i.e., $kv\tau \ll 1$, $\Omega\tau \ll 1$, where $\tau = a/u$ is the time of flight through a beam, u is the atomic velocity along the x axis. The interaction with the first beam is considered in the first order of perturbation theory, with the second in the third order. The initial condition at the moment of atomic entrance into the first beam (t = 0) is b = 0, a = 1. The probability amplitudes after the atomic flight through the first beam are of the form

$$b_1 = V_1 \tau \cos(kz_1 + \varphi_1) e^{-i\Omega t_1} \ , \quad a_1 = 1 \ . \tag{3.59}$$

After the interaction with the second field the probability amplitude is increased by a value

$$b_2 = V_2\tau\cos(kz_2 + \varphi_2)e^{-i\Omega T} - \frac{|V_2|^2\tau^2}{2}\cos^2(kz_2 + \varphi_2)b_1 \quad , \tag{3.60}$$

in the third field the probability amplitude is increased by a value

$$b_3 = V_3\tau\cos(kz_3 + \varphi_3)\exp(-i\Omega L/u)$$

$$[1 - V_2^*\tau\cos(kz_2 + \varphi_2)\exp(i\Omega T)b_1 - |V_2|^2(\tau^2/2)\cos^2(kz_2 + \varphi_2)] \quad . \tag{3.61}$$

After the interaction with three fields the probability amplitude is $b = b_1 + b_2 + b_3$. The probability of detecting a particle in level 2 after the interaction with three fields is

$$|b|^2 = |b_1|^2 + |b_2|^2 + |b_3|^2 + 2\mathrm{Re}\{b_1b_2^*\} + 2\mathrm{Re}\{b_1b_3^*\} + 2\mathrm{Re}\{b_2b_3^*\} \quad . \tag{3.62}$$

The squares of the modules b_1, b_2, and b_3 respond to an independent absorption in each beam. Three last terms respond to the interference of wave functions after the interaction with individual fields. The fourth term corresponds to the interaction of an atom with two fields. As has been already shown, in averaging over velocities v the interference effect disappears. We shall be therefore interested only in the two last terms that meet the condition of interaction of a particle with three fields.

$$\begin{aligned} &2\mathrm{Re}\{(b_1 + b_2)b_3^*\} \\ &= 2\mathrm{Re}\Big\{b_1\tau\cos(kz_1 + \varphi_1) + V_2\tau\cos(kz_2 + \varphi_2)\exp(-i\Omega T) \\ &\quad -|V_2|^2(\tau^2/2)\cos^2(kz_2 + \varphi_2)V_1\tau\cos(kz_1 + \varphi_1)\Big\} \\ &\quad \times V_3^*\tau\cos(kz_3 + \varphi_3)\exp(i\Omega L/u)\Big[1 - V_2\cos(kz_2 + \varphi_2) \\ &\quad \times \exp(-i\Omega T)V_1\cos(kz_1 + \varphi_1) - |V_2|^2(\tau^2/2)\cos^2(kz_2 + \varphi_2)\Big] \quad . \end{aligned} \tag{3.63}$$

At the fixed coordinate z_1 the coordinates z_2 and z_3 depend on the velocity v in the following way

$$z_2 = z_1 + vT \quad ,$$

$$z_3 = z_1 + vL/u \quad ,$$

where L is the distance between the first and second fields. Thus, the oscillating terms in (3.63) contain the velocity v. These terms can yield zero average values in averaging over v. In (3.63) we shall be interested in the terms that permit elimination of the dependence on v. Keeping only these terms we obtain

$$2\mathrm{Re}\{(b_1 + b_2)b_3^*\} = \mathrm{Re}V_1V_3^*|V_2|^2\tau^4 \cos(kz_1 + \varphi_1)$$

$$\times \cos[2(kz_2 + \varphi_2)]\cos(kz_3 + \varphi_3)\exp(i\Omega L/u)$$

$$\times [1 + \exp(-2i\Omega T)] \ . \qquad (3.64)$$

Analyzing (3.64) one can notice that the Doppler phase angle that is stored during the drift between the first and third beams is cancelled in the third field at $z_3 - z_1 = 2(z_2 - z_1)$ due to the phase jump $\Delta\varphi = \pm 2kz_2$ at the nonlinear interaction in the second field.

The energy absorbed by a particle $E = \hbar\omega|b|^2$ after averaging (3.64) over z_1 and v is

$$E = \hbar\omega \frac{|V_2|^2\tau^4}{2} |V_2||V_1 \cos^2\Omega T\cos(\varphi_1 + \varphi_3 - 2\varphi_2) \ . \qquad (3.65)$$

Averaging over velocities gives the resonance that is in its shape close to the Ramsey resonance.

We shall give the results of the calculation made in [3.69] for three standing waves of arbitrary intensity with a finite beam width and allowing for decay of atomic levels in a gas with Maxwell distribution of velocities. It follows from the calculation that the resonance has the largest relative amplitude of about $(\Phi/L)\times 10^{-2}$ at the field $V_{opt}\tau_0 \sim 1$ ($\tau_0 = \Phi/v_0$, v_0 is the thermal velocity, $V_1 = V_2 = V_3 = V_{opt}$). The factor Φ/L is connected with a relatively small number of atoms flying simultaneously through three light beams. The resonance shape is given in Fig.3.21. The characteristic resonance width is $1/T_0$ ($T_0 = L/v_0$).

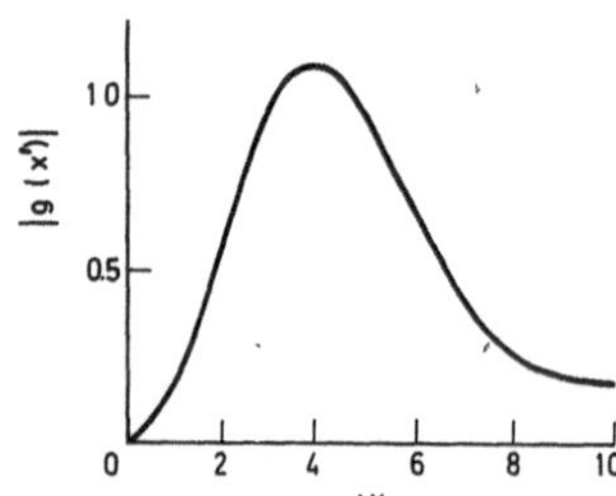

Fig.3.21. Resonance shape in a gas with three separated optical fields. $x' = 2V\tau_0$

The first experimental observation of the resonance in three separated fields was carried out by BERGQUIST et al.[3.70], who used a monokinetic beam of metastable neon atoms interacting with spatially separated light beams of a single-mode frequency-stabilized dye laser. The experiment is schematically represented in Fig. 3.22. Several μA ions of $^{20}Ne^+$ from a hot cathode that served as a source of ions were accelerated up to 5-50 keV and focused in a Na oven. An effective charge exchange occurred through the reaction $Ne^+ + Na \rightarrow Ne$ with an energy deficiency of

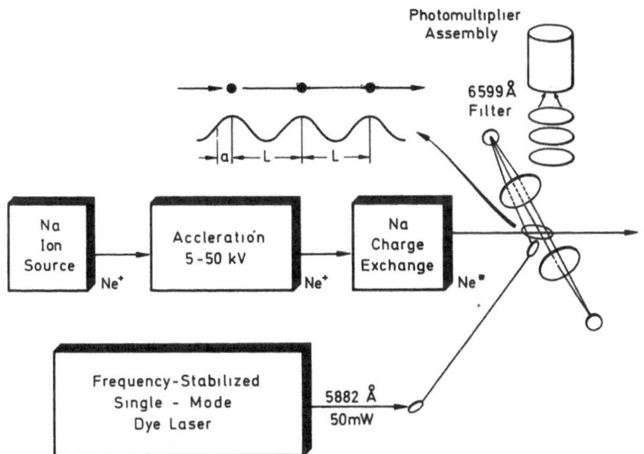

Fig.3.22. Schematic of the experiment. Three standing-wave interaction regions are formed by two well-corrected cat's-eye retroreflectors. Typical values are the following: $i(Ne^+) = 3\ \mu A$, $V = 20$ kV, and the laser power is 50 mW. Fluorescence signals $\sim 10^8$ photons/s reach the multiplier through the f/2 collection optics and filter [3.70]

about 0.2 eV. The beam of metastable atoms had a diameter of about 2 mm, divergence of 2 mrad, and velocity spread $\Delta v/v \lesssim 10^{-4}$. The laser power of about 50 mW sufficed in saturating the $1s_2$-$2p_2$ transition even for fast beams ($v/c \sim 10^{-3}$).

The interaction was recorded with a high sensitivity through strong fluorescence on the $2p_2$-$1s_2$ transition (6599 Å) by an appropriately filtered photomultiplier. The atomic system ^{20}Ne free of hyperfine structure allowed a clear interpretation of the results. The distance between neighboring light beams was 0.5 cm. At an average beam energy ($\sim$20 keV) the lifetime of an upper state of about 18.7 ± 0.3 ns gives the population decay length of about 0.8 cm and the phase coherence length, or the dipole decay length, of about 1.6 cm. A symmetric structure in absorption is obtained provided that spatially identical phases in three beams lie in the same plane. This condition permits the production of the system of two opposite "cat eyes" [3.70].

Recently, BARGER et al. [3.71] carried out investigations of resonances on the 6573 Å line in Ca. In Ca there have been recorded recoil effects and obtained the line with a relative width of about 10^{-11}.

3.4.3 Coherent Radiation and Macroscopic Polarization Transfer in Separated Fields

The resonances in three beams can be obtained in another way. Since we are interested in coherent effects, let us see what happens to the dipole moment of a particle in the interaction with optical fields. This approach permits an easy explanation of both absorption resonances in three fields and coherent radiation from the region spaced from the second field at the distance equal to that between fields. After the interaction with the first field an atom has the dipole moment

$d(t) = d_0 \cos(\omega t + kz_1)$. The field in the point z_3 is $E_3 \cos(\omega t + kz_3)$. The phase difference between the dipole moment of a particle d and the field E_3 determines the energy absorbed by a particle in the field E_3. This difference depends on the particle velocity v and amounts to $\varphi = kv2T = kv2L/U$, where 2T is the transit time of a particle between fields E_1 and E_3, 2L is the distance between fields. Now let us place the standing-wave field $E_2 \cos\omega t \cos kz$ between the fields E_1 and E_3 at the distance L.

As has been shown in Sect.3.4.2, in the interaction with the standing-wave field the dipole moment of a particle undergoes a phase jump equal to the doubled field phase in the point z_2. Thus, having passed through the second field the dipole moment of a particle is $d \sim \cos(\omega t + kz_1 + 2kz_2)$. It is seen that the phase difference between the dipole moment and the field in the point z_3 is independent of the point position as the phase jump of the dipole moment $\Delta\varphi = 2kz_2 = 2kvT$ is exactly equal to the field phase delay $kz_3 = 2kvT$.

As a result, the absorbed energy is determined by the field frequency detuning from the resonance $\varphi = 2\Omega T$ and independent of the velocity v. All the particles arriving at the point z_3 have a spatial phase equal to kz_3.

Now let us consider this process in more detail. At first we shall consider the case of traveling waves. After the interaction with the second field the dipole moment of a particle is

$$\begin{aligned} d_2(t,|V_2|^2) = -\Bigg\{ & d_1 \frac{|V_2|^2}{4} \tau^2 \left[2 + \exp(2ikz_2 + 2i\varphi_2) + \exp(-2ikz_2 - 2i\varphi_2)\right] \\ & + d_{12} b_1^* \frac{|V_2|^2}{4} \tau^2 \left[2 + \exp(2ikz_2 + 2i\varphi_2) \right. \\ & \left. + \exp(-2ikz_2 - 2i\varphi_2)\right] \times \exp(2i\Omega t_2)\Bigg\} \exp(-i\omega_{21} t) + \mathrm{c.c.} \quad , \end{aligned} \tag{3.66}$$

where $d_1(t) = d_1 \exp(-i\omega_{21}t) + \mathrm{c.c.} = d_{12} b_1 \exp(-i\omega_{21}t) + \mathrm{c.c.}$ is the dipole moment after the interaction with the first field, $t_2 = t_1 + T$. The dipole moment nonlinear in V_2 contains two terms: the first one is connected with one-quantum processes, the second with two-quantum processes. A very important fact is that two-quantum processes do not lead to changes in the dipole moment phase and change only its influence. The one-quantum processes are responsible for the phase jump of the dipole moment that is equal to

$$\Delta\varphi = 2k(z_2 - z_1) + 2\Omega T \quad . \tag{3.67}$$

Figure 3.23 depicts the diagram of the dependence of the phase difference between the dipole moment of a particle induced by the first field and the field in the point where the particle is located at a given moment. The phase difference contains a time part connected with the frequency detuning and a spatial part (Doppler effect). They are shown separately in Fig.3.23. After the interaction the phase

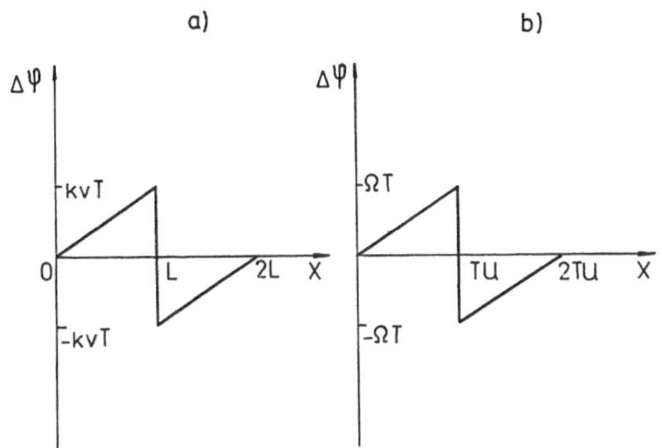

Fig.3.23. Diagram of the dependence of the phase difference between the dipole moment of a particle and the field in a given point: (a) spatial part; (b) time part

difference is increased. This is due to the transverse particle velocity u. After the phase jump at $x = L$ the phase difference is zero at $x = 2L$. The time part also grows linearly. The slope of the straight line depends on the frequency detuning from the line center $\Delta\varphi = \Omega \cdot x/u$. It can be readily seen that at the distance L from the second beam the phase of the dipole moment of a particle will completely coincide with the field phase. The phase jump contains the spatial and time parts. At the distance L from the second beam the phase of the dipole moment of a particle is

$$\varphi = kz_1 + 2k(z_2 - z_1) + 2\Omega T + 2\omega_{21}T \quad . \tag{3.68}$$

It is seen that the phase of the dipole moment of the particle arriving in the point z_3 will coincide with the field phase in this point, i.e., the particle will be in spatial and time synchronism with the wave. In the same way we may consider the other points z_3', z_3'', and so on located at the distance L away from the second field. In these points the phase of the dipole moment will, respectively, kz_3', kz_3'', and so on. Thus, we have a polarization wave and, accordingly, coherent radiation in the same direction. Note that the time phase jump $2\Omega T$ exactly cancels the phase difference between the dipole and the field at the frequency detuning Ω. The interference effects and frequency dependence of absorption or coherent radiation are not therefore observed. The mechanism of elimination of the Doppler effect in separated traveling waves is close to a photon echo in its nature. The cancellation of the time phase explains why in numerous photon echo experiments resonance effects have not been observed.

Now we consider the case where the second field is a standing wave. Two mechanisms of elimination of the Doppler effect work here. The first one corresponds to the interaction with traveling waves. (We should recall that a standing wave may be represented as the sum of two oppositely traveling waves). This mechanism has been described above. Qualitatively new peculiarities arise in the interaction of standing waves. The phenomena that occur here are typical of a gas only and due to the two-photon process of absorption and emission from oppositely traveling waves. It is evident that these processes may occur only in the line center where particles may interact simultaneously with the two waves. In the interaction with oppositely traveling waves the phase jump of the dipole moment is

$$\Delta\varphi = \pm 2kz_2 \quad .$$

Since the dipole moment phase of a particle before the interaction was kz_1, after the interaction it will be $-2kz_2 + kz_1$. At the distance $x = 2L$ from the first beam $2z_2 - z_1 = z_3$ and the dipole moment phase is $-kz_3$. When considering the other points in the same way one can notice that a polarization wave will propagate in the direction opposite to that of the first traveling wave.

The difference from the case of traveling waves is that the phase jump $\Delta\varphi = -2kz_2$ that is necessary for polarization transfer arises at the expense of the two-photon process of absorption and emission of oppositely traveling waves. At the distance 2L there occurs the shift of the field phase and of the particle dipole moment by a value $2\Omega T$. This is the circumstance that is responsible for the occurrence of resonance phenomena.

3.4.4 Properties of Coherent Radiation in Separated Fields

Macroscopic polarization transferred at a distance may be recorded over absorption of a probe wave or over coherent radiation in separated fields (CRSF). We shall consider the latter in more detail.

The principal properties of CRSF are as follows.

1) Coherent radiation is of resonance character, the manifestation of which depends on the conditions of observation.
2) Coherent radiation shows the properties of continuous superradiance. Its intensity is proportional to the square of particle density. When using an inverted beam one can observe other phenomena inherent of superradiance.
3) In the system there occurs wavefront reversal of a traveling wave.
4) Coherent radiation is sensitive to changes in the dipole moment phase during drift.
5) Coherent radiation that is due to stimulated processes takes place at the field frequency rather than the transition frequency.
6) Since coherent radiation arises with delay with respect to an exciting field, the system operates as a coherent delay line or a system with record and restoration of an optical image.
7) The radiation line is very sensitive to deflection of particles caused by the action of various physical factors (gravitational, electric and magnetic fields, collisions, recoil effects and others).
8) Under the action of fields with different frequencies the polarization arises at combination frequencies.

9) In strong standing-wave fields the polarization is transferred at the distances multiple to L.

The resonance phenomena that are due to the interference ones can take place only in the interaction of standing waves. At the two-photon process of absorption and emission from oppositely traveling waves there occurs cancellation of the Doppler shift in a standing wave. The polarization phase for the time of drift of a particle T will therefore differ from the field phase at the distance 2L by a value $2\Omega T$. The occurrence of interference phenomena largely depends on the properties of beams, field configurations and other experimental conditions. This is the circumstance that gives rise to so-called Ramsey fringes in observing the probe wave absorption. A somewhat different situation is observed in the CRSF. As an example we consider the resonance phenomena when the first field is a traveling wave, the second one is a standing wave (Fig.3.24). Let the beam of particles that transfers the polarization be monokinetic. It is a good analog of the interaction of the system with two pulses. As it is seen, in this case the coherent radiation arises in two directions. In the same direction the radiation phase coincides with the field phase, in the opposite direction differs by a value $2\Omega T$. However, in the direct observation of the radiation intensity versus frequency this circumstance is not appreciable. In the first case the radiation is observed within the limits of a Doppler width, in the second within the limits of $1/\tau$. If the first field is a standing wave, the CRSF will also occur in both directions. In each direction there will be two waves caused by the interaction of traveling and standing waves. In this case there arises the interference between the CRSF components associated with traveling and standing waves, and in each direction there are observed intensity beats. These are better pronounced for a monokinetic beam, as in this case all particles provide a contribution to the radiation with identical phases. In pulsed fields all particles interact with the field in the same time interval and have the same phases. Figure 3.25 shows the dependence of the CRSF intensity in the standing-wave field in SF_6. The interference pattern is clear.

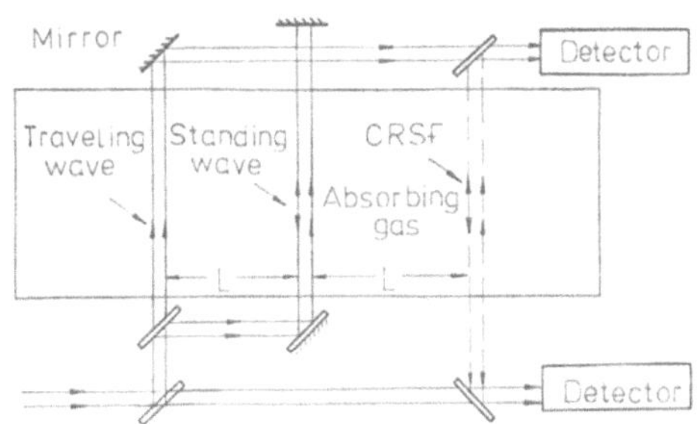

Fig.3.24. Coherent radiation in separated optical fields

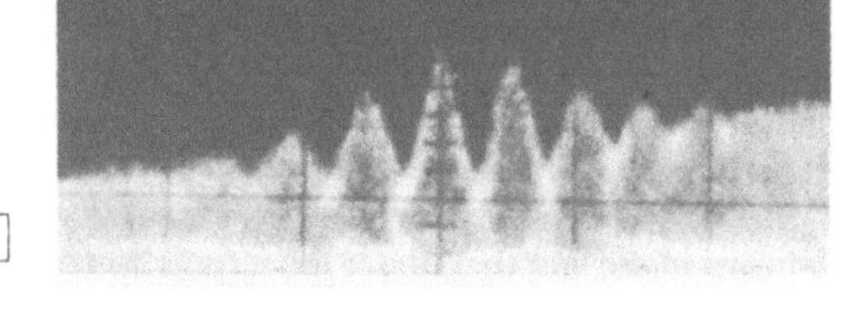

Fig.3.25. Intensity change of the coherent radiation with the laser frequency tuning. The time between pulses is $T = 0.4\ \mu s$

Destruction of an Interference Structure and Attainment of Resonances with a Radiative Width

An interference pattern makes it difficult to gain spectroscopic information associated, for example, with the presence of a hyperfine line structure. There is a simple method to reduce and possibly to exclude its influence that is based on variation of the time delay between fields. In this case oscillating terms with the detuning Ω will periodically depend on T. The simplest way is to vary T in time separated fields. Averaging the signal over the delay time depending on frequency we obtain an ordinary resonance curve with a width equal to a line width. As we have already seen, the signal intensity in time T is described by

$$I = I_0(1 + \cos 2\Omega T)e^{-\gamma T} \quad . \tag{3.69}$$

Averaging over T gives

$$\langle I\rangle = I_0\left(1 + \frac{\gamma^2}{\gamma^2 + \Omega^2}\right) \quad . \tag{3.70}$$

In spatially separated fields T may be changed by varying the distance between beams and the beam velocity. For this purpose it is convenient to use ion beams whose velocity is readily changed. With atomic and molecular beams the case is more complicated. Here a particle velocity can be varied by using effects of light pressure. Choosing temperatures of various beams one can reduce oscillations and destroy Ramsey fringes. For example, the second oscillation amplitude was reduced by a factor of 10 as compared with the use of one beam. Figure 3.26 shows the resonance shape for beams with two different velocity distributions of particles.

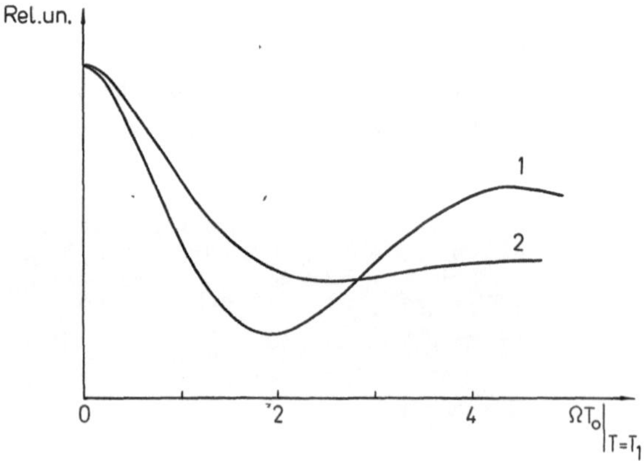

Fig.3.26. Resonance in the number of excited particles arising in the interaction with separated fields of a particle beam with temperature T_1 (curve 1) and of a two-temperature beam with temperatures T_1 and T_2 (curve 2). $T_2/T_1 = 5.2$

Particle Scattering

In the interaction of particles with the standing-wave field there takes place scattering of particles due to a recoil effect. This may result in variation of the resonance properties in separated fields. Here an effect of resonance splitting appears. However, its mechanism differs from that of saturation resonance splitting. The influence of the recoil effect on the CRSF properties has been studied in [3.72]. Since the recoil should be taken into account when Δ is comparable with the width of these resonances $\gamma \sim T^{-1}$ that is far less than a transit linewidth $1/\tau$, $\Delta \ll 1/\tau$. This means that splitting of line contours of emission and absorption may be neglected. However, the particle velocity is changed due to the recoil effect (Fig.3.27). A very important fact for us is that in the interaction of an atom with the first wave the recoil effect may be neglected. Actually, for the CRSF to appear it is necessary that after the interaction with the first wave an ensemble of polarized atoms with random velocity spread should arise in a medium. The presence of recoil does not change this assumption. The variation of the phase for the time of interaction with the wave τ may be neglected, as it is $\Delta\tau \ll 1$.

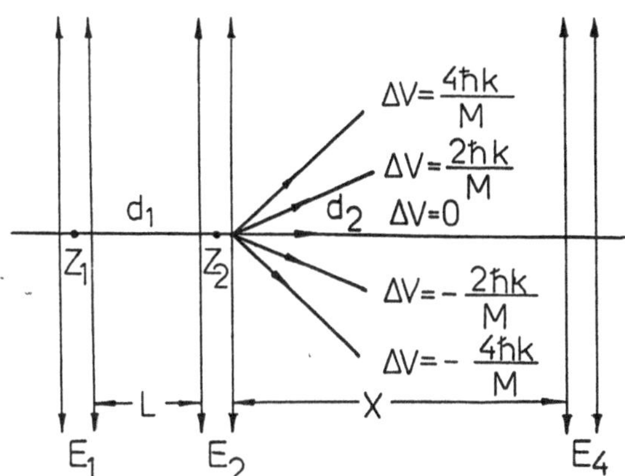

Fig.3.27. Scheme of the interaction of particles with three separated fields in consideration of a recoil effect

In the interaction with the second field particles acquire an additional velocity change due to the recoil effect. During the drift between the second and third fields particles undergo an extra Doppler shift equal to $2\Delta vkT$, where Δv is the particle velocity variation in absorption (emission) of a photon. In order to cancel this shift it is necessary to vary the field frequency by a value $2\Omega T = 2\Delta vkT$. Hence $\Omega = \pm\Delta vk$, i.e., the resonance is split. An interesting situation arises at the distances $x_n = (2n - 1)$. Here the coherent radiation arises in even orders of the second wave field. This means that an even number of photons participates in the interaction, and hence, there are processes where an atomic velocity is not changed at all ($\Delta v = 0$). In these processes the allowance for recoil does not change the conditions of phase cancellation. These processes are responsible for the resonance components in the line center. The total theory of the CRSF effect that simultaneously takes into account the influence of recoil and strong field has been elaborated in [3.72]. We should point out that the component in the line center arises not only in the CRSF line but also in the absorption resonances in separated fields [3.73]. If in the region of drift a particle is affected by some forces, particle deflection produces an extra phase shift, which may be compensated by frequency detuning. With the drift region length of about 50 cm, the described system turns out to be very sensitive. For example, the gravitational field deflects particle trajectories, which results in the resonance shift by a value of about 1 kHz (for particles with M = 10, T = 300 K the axis of fields is vertical). The systems with ion beams may turn out to be very sensitive to electric and magnetic fields.

In the electric field of about 10^{-8} V/cm the frequency shift for an ion with M = 10 will be about 1 kHz with the distance between fields of 50 cm. The sensitivity to the magnetic field is also very high.

Since the considered phenomena are associated with the action of time-limited fields, here the phenomena may occur which are in their physical sense similar to the action of a $\pi/2$ pulse. These phenomena are more or less clear. We shall consider qualitatively new phenomena that appear in standing-wave fields and are due to polarization transfer. In considering the effects in the higher orders perturbation theory, the polarization transfer appears at the distances 3L, 4L only in standing-wave fields. The polarization transfer at the other distances may be also explained by using the model of phase jumps. The dipole moment that arises in a particle after the interaction with the field contains the signal "n" corresponding to the number of an order of perturbation theory $d \sim |V_1|^n \cos^n kz_1$. The interaction with the second field should be considered in the even orders of perturbation theory. The dipole moment of interest is $\sim \cos^n kz_1 \cos^m kz_3$. The perturbation theory orders should be combined so that the dipole moment at the distance kL away from the first beam would have the first spatial harmonic. After the interaction with two fields the dipole moment will contain the term

$$k(nz_1 - mz_2) \quad .$$

Since the coordinate z at the distance x = kL away from the first beam is

$$z_3 = z_1 + k(z_2 - z_1)$$

it is evident that the perturbation theory orders, i.e., the numbers m, n, and ℓ, should be combined so that the following equality would take place, for example,

$$nz_1 - mz_2 = z_1 + \ell(z_2 - z_1) \quad .$$

In order to obtain the polarization at the distance 4L away from the first beam, n should be equal to 5, m = 4. This case is of interest for obtaining recoil free resonances.

3.4.5 Coherent Raman Scattering in Separated Fields

Here we consider a new phenomenon, coherent Raman scattering (CRS) in separated fields [3.58]. It arises in the linear approximation of an incident field. Its physical nature largely differs from the stimulated Raman scattering (SRS) process. The CRS may be observed in an absorbing medium and, hence, not connected with the process of amplification. Under the action of two fields resonant to the $1 \rightarrow 2$ transition there takes place the polarization at the distance $x = 2L$ from the first beam. If in this region there is a field at the frequency ω_3 and combination transitions are resolved, it is natural to expect the coherent Raman scattering at the frequencies $\omega_3 \pm \omega_{21}$.

If a one-photon transition between levels 1 and 2 is forbidden, the coherence between these levels may be produced by acting upon the system with two fields with frequencies ω_1 and ω_2 whose difference is equal to the frequency $\omega_{21} = \omega_1 - \omega_2$ (Fig.3.28).

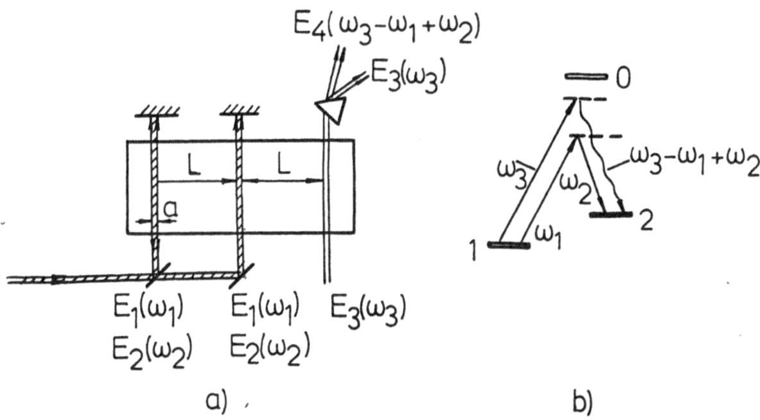

Fig.3.28. Scheme for observing the coherent Raman scattering of three-level atoms (a). Transition layout in a three-level scheme (b)

We have considered spatially separated fields. The similar phenomena may be observed in pulsed fields. In this case in time 2T after the first pulse a signal arises at combination frequencies, which may be interpreted as an effect of the CARS echo.

3.4.6 Transient Resonant Coherent Effects

The method of separated optical fields is based on the changes in the phase of the dipole moment of a particle in the nonlinear interaction of other fields, which permits elimination of the influence of a Doppler effect. Here the decay of an induced dipole moment is neglected. Let us see what will happen to the induced dipole moment in the presence of other fields. In other words, unlike the previous studies of free induction decay [3.74], we are interested in free induction decay in the presence of other fields.

As has been shown in [3.60], the free induction decay produced by a traveling wave pulse in the presence of a standing-wave field has new interesting properties: coherent damping radiation arises in the direction opposite to that of pulse propagation. It has the transition frequency. The last circumstance permits the use of free induction decay not only in studies of relaxation processes but also in superhigh resolution spectroscopy. At the time $t_1 = 0$ a gas cell is affected by a traveling wave pulse of τ duration that polarizes the medium (Fig.3.29). Then the medium is acted upon by the standing-wave field. After the pulse action the damping radiation arises in the opposite direction (this may be interpreted as wavefront reversal of induction). The radiation frequency is located in the line center.

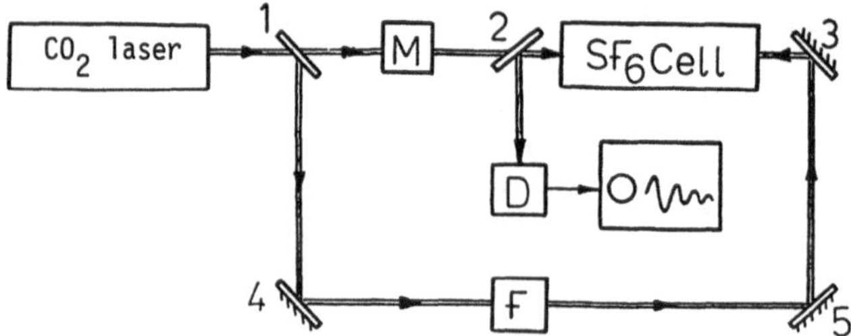

Fig.3.29. Schematic of the experiment for observing nonstationary resonance effects. 1-5 - mirrors, M - modulator, F - filter, D - detector

The interference of this radiation with the traveling component of the standing wave results in radiation pulsation. Figure 3.29 shows the scheme for observing this phenomenon. The radiation of a CO_2 laser passes through a SF_6 cell and through a modulator M. After the action of the pulse there is a standing-wave field in the cell which is formed by mirrors 2 and 3 and the modulator M. The radiation is explored in the direction opposite to that of radiation pulse propagation. Figure 3.30 depicts the beat signal at the detuning frequency.

In short the physical picture of the radiation occurrence may be represented as follows. After the interaction with the traveling wave, we have the distribution of damping dipoles over velocities described by

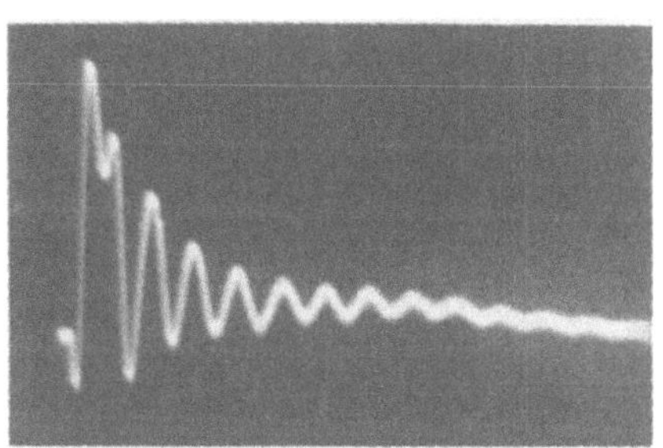

Fig.3.30. Oscillogram of the coherent transient in SF_6

$$d^{\ell}(v,t) = \frac{i|d_{21}|^2}{\hbar} E\tau \exp[-(\Gamma + i\omega_{21} + ikv)t]\exp(ikz) \quad .$$

The wave traveling in the same direction produces population pulsations with the frequency (Ω - kv). The second wave interacting with the particles whose population is modulated produces the polarization equal to

$$P^{n\ell} = \frac{i|d_{21}|^4}{\hbar^3} E_2 E_1^* E\tau e^{ikz} \int dvF(v)$$
$$\frac{\exp[-(\Gamma + i\omega_{21} + ikv)t] - \exp[-(\Gamma + i\omega_{21} - ikv)t]}{2ikv(\Gamma - \gamma - i\Omega + ikv)} \quad .$$

Averaging over velocities gives the damping radiation at the frequencies ω_{21} and ω that decay with constants Γ and γ, respectively.

This method may be useful in studying collisional shifts and structures.

References

3.1 H. Walther (ed.): *Laser Spectroscopy*, Topics in Applied Physics, Vol.2 (Springer, Berlin, Heidelberg, New York 1976)
3.2 V.S. Letokhov, V.P. Chebotayev: *Nonlinear Laser Spectroscopy*, Springer Series in Optical Sciences, Vol.4 (Springer, Berlin, Heidelberg, New York 1977)
3.3 W.R. Bennett, Jr.: Phys. Rev. *126*, 580 (1962)
3.4 S. Stenholm: Appl. Phys. *16*, 159 (1978)
3.5 W.E. Lamb, Jr.: Phys. Rev. *134*A, 1429 (1964)
3.6 V.S. Letokhov, V.P. Chebotayev: Zh. Eksp. Teor. Fiz. Pisma *9*, 364 (1969)
3.7 R.A. McFarlane, W.R. Bennett, Jr., W.E. Lamb, Jr.: Appl. Phys. Lett. *2*, 189 (1963)
A. Szöke, A. Javan: Phys. Rev. Lett. *10*, 521 (1963)
3.8 P.H. Lee, M.L. Skolnick: Appl. Phys. Lett. *10*, 303 (1967)
3.9 V.N. Lisitsyn, V.P. Chebotayev: Zh. Eksp. Teor. Fiz. *54*, 419 (1968)
3.10 B.J. Feldman, M.S. Feld: Phys. Rev. *A1*, 1375 (1970)
3.11 E.V. Baklanov, V.P. Chebotayev: Zh. Eksp. Teor. Fiz. *62*, 541 (1972)
3.12 B.Y. Dubetsky: Ph.D. Thesis, Novosibirsk (1979)
3.13 N.G. Basov, I.N. Kompanets, O.N. Kompanets, V.S. Letokhov, V.V. Nikitin: Zh. Eksp. Teor. Fiz. Pisma *9*, 568 (1969)
3.14 Yu.A. Matyugin, B.I. Troshin, V.P. Chebotayev: Abstracts of the All-Union Symposium on Gas Laser Physics, Novosibirsk, USSR (June 1969) p.56
3.15 P.W. Smith, T. Hänsch: Phys. Lett. *26*, 740 (1971)
3.16 C. Borde: C.R. Acad. Sci. Paris *271*, 371 (1970)

3.17 E.E. Uzgiris, J.L. Hall, R.L. Barger: Phys. Rev. Lett. *26*, 289 (1971)
3.18 S.N. Bagayev, L.S. Vasilenko, A.K. Dmitriyev, V.G. Goldort, M.N. Skvortsov, V.P. Chebotayev: Appl. Phys. *10*, 231 (1976)
3.19 C. Wieman, T.W. Hänsch: Phys. Rev. Lett. *36*, 1170 (1976)
3.20 E.V. Baklanov, V.P. Chebotayev: Zh. Eksp. Teor. Fiz. *60*, 551 (1971)
3.21 S. Haroche, F. Hartmann: Phys. Rev. *6*A, 1280 (1972)
3.22 E.V. Baklanov, V.P. Chebotayev: Zh. Eksp. Teor. Fiz. *61*, 922 (1971)
3.23 V.N. Lisitsyn, V.P. Chebotayev: Zh. Eksp. Teor. Fiz. Pisma *7*, 3 (1968)
3.24 Yu.V. Brzhazovsky, L.S. Vasilenko, V.P. Chebotayev: IEEE J. *QE-4*, 23 (1968); *QE-5*, 146 (1969)
3.25 H.M. Gibbs, S.L. McCall, T.N.C. Venkatesam: In *Coherence in Spectroscopy and Modern Physics*, ed. by F.T. Arecchi, R. Bonifacio, M.O. Scully (Plenum Press, New York 1978)
3.26 V.P. Chebotayev, I.M. Beterov, V.N. Lisitsyn: IEEE J. *QE-4*, 788 (1968)
3.27 I.M. Beterov, V.N. Lisitsyn, V.P. Chebotayev: Opt. Spektrosk. *30*, 932, 1108 (1971)
3.28 W.R. Bennett, Jr.: Comments Atomic Molec. Phys. *2*, 10 (1970)
3.29 A.P. Kolchenko, S.G. Rautian, R.I. Sokolovsky: Zh. Eksp. Teor. Fiz. *55*, 1864 (1968)
3.30 E.V. Baklanov: Opt. Spektrosk. *38*, 622 (1975)
3.31 L.D. Landau, E.M. Lifshits: *Quantum Mechanics* (Izd. Nauka, Moscow 1974)
3.32 S.N. Bagayev, E.V. Baklanov, V.P. Chebotayev: Zh. Eksp. Teor. Fiz. Pisma *16*, 15 (1972)
3.33 V.A. Alexeyev, T.L. Andreyeva, I.I. Sobelman: Zh. Eksp. Teor. Fiz. *64*, 813 (1973)
3.34 E.V. Baklanov: Opt. Spektrosk. *38*, 24 (1975)
3.35 A.T. Mattick, N.A. Kurnit, A. Javan: Chem. Phys. Lett. *38*, 176 (1976)
3.36 T.W. Meyer, C.K. Rhodes, H.A. Haus: Phys. Rev. A*12*, 1993 (1975)
3.37 L.S. Vasilenko, V.P. Kochanov, V.P. Chebotayev: Opt. Commun. *20*, 409 (1977)
3.38 Ph. Cahuzac, E. Marie, O. Robaux, R. Vetter, P.R. Berman: J. Phys. B*11*, 645 (1978)
3.39 V.P. Kochanov, S.G. Rautian, A.G. Shalagin: Zh. Eksp. Teor. Fiz. *72*, 1358 (1977)
3.40 S.N. Bagayev, A.S. Dychkov, V.P. Chebotayev: Zh. Eksp. Teor. Fiz. Pisma *29*, 570 (1979)
3.41 I.M. Beterov, V.P. Chebotayev: In *Progress in Quantum Electronics*, Vol.3, ed. by J.H. Sanders, S. Stenholm (Pergamon Press, Oxford 1974)
3.42 V.P. Chebotayev: In *High-Resolution Laser Spectroscopy*, Topics in Applied Physics, Vol.13, ed. by K. Shimoda (Springer, Berlin, Heidelberg, New York 1976)
3.43 I.M. Beterov, Yu.A. Matyugin, V.P. Chebotayev: Zh. Eksp. Teor. Fiz. *64*, 1495 (1973)
3.44 N. Skribanowitz, M.J. Kelly, M.S. Feld: Phys. Rev A*6*, 2302 (1972)
3.45 B.J. Feldman, M.S. Feld: Phys. Rev. A*5*, 899 (1972)
3.46 N. Skribanowitz, M.J. Kelly, M.S. Feld: Phys. Rev. A*6*, 2302 (1972)
3.47 C. Cohen-Tannoudji, F. Hoffbeck, S. Reynand: Opt. Commun. *27*, 71 (1978)
3.48 E.V. Baklanov, E.A. Titov: Kvantovaya Elektron *2*, 1893 (1975)
3.49 V.P. Chebotayev: Report at the XIX General Assembly of URSI (Helsinki, Finland 1978); Preprint No. 42-79 (Institute of Thermophysics, Sib. Branch of USSR Acad. of Sci., Novosibirsk 1979)
3.50 L.S. Vasilenko, V.P. Chebotayev, A.V. Shishayev: Zh. Eksp. Teor. Fiz. Pisma *12*, 161 (1970)
3.51 B. Cagnac, G. Grynberg, F. Biraben: Phys. Rev. Lett. *32*, 643 (1974)
3.52 M.D. Levenson, N. Bloembergen: Phys. Rev. Lett. *32*, 645 (1974)
3.53 N. Bloembergen, M.D. Levenson: In *High-Resolution Laser Spectroscopy*, Topics in Applied Physics, Vol.13, ed. by K. Shimoda (Springer, Berlin, Heidelberg, New York 1976) p.315
3.54 V.M. Klementyev, Yu.A. Matyugin, V.P. Chebotayev: Zh. Eksp. Teor. Fiz. Pisma *24*, 8 (1976)
3.55 W.R. Bennett, Jr.: *Some Aspects of the Physics of Gas Lasers*, 2nd ed. (Gordon and Breach Science Publ. Inc., New York 1973)
3.56 E.V. Baklanov, B.Y. Dubetsky, V.P. Chebotayev: Appl. Phys. *9*, 171 (1976)

3.57 E.V. Baklanov, V.P. Chebotayev, B.Y. Dubetsky: Appl. Phys. *11*, 201 (1976)
3.58 V.P. Chebotayev: Vestnik Moscow State University *19*, 159 (1978)
V.P. Chebotayev, B.Y. Dubetsky: Appl. Phys. *18*, 217 (1979)
3.59 V.P. Chebotayev: Appl. Phys. *15*, 219 (1978)
3.60 L.S. Vasilenko, M.N. Skvortsov, V.P. Chebotayev: Zh. Tekhn. Fiz. Pisma *4*, 1120 (1978)
3.61 N.F. Ramsey: *Molecular Beams* (Oxford University Press, New York 1956)
3.62 E.V. Baklanov, B.Y. Dubetsky: Kvantovaya Elektron. *5*, 99 (1978)
3.63 V.P. Chebotayev, A.V. Shishayev, B.Ya. Yurshin, L.S. Vasilenko: Appl. Phys. *15*, 43 (1978)
3.64 J.L. Hall, S.A. Lee, J. Helmcke: In *Laser Spectroscopy IV*, ed. by H. Walther, K.W. Rothe (Springer, Berlin, Heidelberg, New York 1979) p.130
3.65 M.M. Salour, C. Cohen-Tannoudji: Phys. Rev. Lett. *38*, 757 (1977)
3.66 R. Teets, J. Eckstein, T.W. Hänsch: Phys. Rev. Lett. *38*, 760 (1977)
3.67 E.V. Baklanov, V.P. Chebotayev: Appl. Phys. *12*, 97 (1977)
3.68 S.N. Bagayev, V.P. Chebotayev: Nonlinear Processes in Optics, Proc. 3rd Vavilov Conference (Novosibirsk, USSR 1973) p.107
E.V. Baklanov: Nonlinear Processes in Optics, Proc. 3rd Vavilov Conference (Novosibirsk, USSR 1973) p.114
E.V. Baklanov, V.P. Chebotayev: Opt. Commun. *12*, 312 (1974)
3.69 B.Y. Dubetsky: Kvantovaya Elektron. *6*, 682 (1976)
3.70 J.C. Bergquist, S.A. Lee, J.L. Hall: Phys. Rev. Lett. *38*, 159 (1977)
3.71 R.L. Barger, J.C. Bergquist, T.C. English, D.J. Glaze: Appl. Phys. Lett. *34*, 850 (1979)
3.72 E.V. Baklanov, B.Y. Dubetsky, V.M. Semibalamut: Zh. Eksp. Teor. Fiz. *76*, 482 (1979)
3.73 B.Y. Dubetsky: Kvantovaya Elektron. *3*, 1258 (1976)
B.Y. Dubetsky, V.M. Semibalamut: Kvantovaya Elektron. *5*, 176 (1976)
3.74 R.G. Brewer, R.L. Shoemaker: Phys. Rev. A*6*, 2001 (1972)

3.57 E.V. Baklanov, V.P. Chebotayev, B.Ya. Dubetsky: [illegible] 312 (1976)
3.58 V.S. Letokhov: [illegible] (1977)
V.P. Chebotayev, [illegible]: Appl. Phys. [illegible] (1979)
3.59 V.P. Chebotayev: Appl. Phys. 15, 219 (1978)
3.60 L.S. Vasilenko, N.N. Rubtsova, V.P. Chebotayev: Zh. Eksp. Teor. Fiz. [illegible] (1978)
[illegible]
[illegible]
[illegible] (1978)
[illegible]
[illegible]
[illegible]
[illegible]
[illegible]
3.65 [illegible], V.P. Chebotayev: Nonlinear Processes in Optics, Proc. 3rd Novosibirsk Conference [illegible]
[illegible] Conference [illegible]
[illegible] p. 174
[illegible]
[illegible]

4. Multiphoton Resonant Processes in Atoms

G. Grynberg, B. Cagnac, and F. Biraben

With 23 Figures

In this chapter, we are dealing with the processes, where transitions between discrete levels of atoms are produced by the simultaneous absorption of n photons. They are resonant in the sense that the sum of the energies of the n photons is equal to the energy difference between the two concerned levels. But we exclude the case where an intermediate level would be resonantly excited during the process (its energy difference from the ground state being equal to the sum of the energies of n-1 photons, or n-2 photons, ...). Many experiments have been done with two such successive resonant processes: these involve, for example, the well-known stepwise excitation or the three-level resonance technique [4.1]. Nevertheless, the detailed study of these two successive processes raises quite different problems; and we restrict our topic to resonant n-photon transitions without a resonant intermediate step.

The principle and the theory of resonant two-photon transition has been known since the pioneering work of GÖPPERT-MAYER [4.2]. If we consider two levels, ground level g and excited level e, of the same parity (see Fig.4.1), it is possible to induce a direct transition from g to e, the atom absorbing two photons. In such a process, the absorption of one photon is nonresonant, the important point is that at the end of the process, the energy is conserved which means that the energy difference between the levels g and e: E_{eg} is equal to $2\hbar\omega$.

The probability of two-photon absorption at low intensity is proportional to the square of the intensity of the light field I^2, the probability of absorption of each photon being proportional to I. In the case of a n-photon transition, the probability of excitation is proportional to I^n.

Even if it is not necessary to have a real intermediate state exactly between g and e, the transition probability is very sensitive to the detuning from the single-photon transitions. For instance, if we consider a three-level atom g, j, e, (Fig.4.1) j being of a parity opposite to g and e, we can qualitatively understand why the transition probability depends on the detuning from the single-photon transition $\hbar\Delta\omega_j = E_{jg} - \hbar\omega$ where E_{jg} is the energy difference between the levels g and j. If we consider the first photon absorbed, the atom can stay in the j level

$\hbar = h/2\pi$ (normalized Planck's constant)

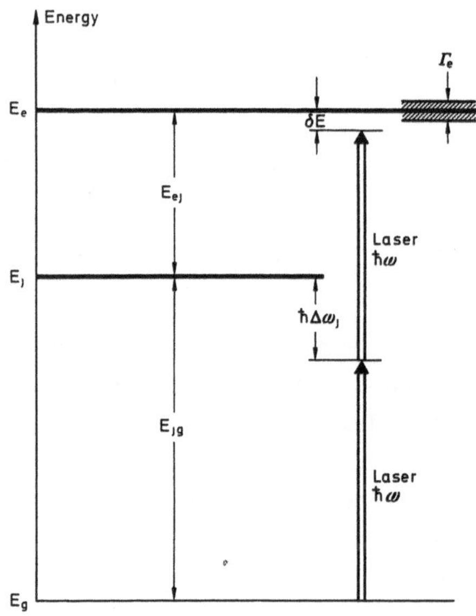

Fig.4.1. Schematic energy diagram for two-photon transition

a time of the order of $(\Delta\omega_j)^{-1}$ because of the uncertainty principle. The two-photon transition can only take place if a second photon is absorbed during the time spent by the atom in the j state. If $\Delta\omega_j$ increases, this time becomes shorter and the transition probability is reduced.

We will first present in the two following sections experimental aspects of these multiphoton transitions, with special interest in Doppler-free experiments. Afterwards we will develop separately the theoretical problems involved in two-photon transitions and in many-photon transitions.

4.1 Various Experimental Aspects of Resonant Multiphoton Transitions in Atoms

Because we, generally, need a large radiation intensity to induce the multiphotonic absorption, these transitions were first observed in the radiofrequency range [4.3]. We do not intend to discuss these experiments in the present paper, we only want to present more recent experiments which take place in the optical range. For a two-photon absorption between states g and e (Fig.4.1), the use of a tunable laser is essential because $(E_e - E_g)/2\hbar$ does not generally correspond to an atomic laser frequency. The first evidence of two-photon absorption in the optical range between discrete levels of atoms was performed by ABELLA [4.4]. The transition between the $6S_{1/2}$ ground state and the $9D_{3/2}$ excited state of caesium was induced by absorption of two photons at 6535.5 Å produced by a thermally tuned ruby laser. The detection of a two-photon resonance is obvious. The photons spontaneously emitted from the excited state e have wavelengths different from the exciting wavelength of the laser

(Fig.4.1). By detecting those emitted photons, the observation of the resonance is sensitive and unambiguous.

Resonant multiphoton processes in atoms have been studied by many groups of physicists, but in most cases their ultimate goal was not the study of the process of multiphoton transition itself. We describe in this section some situations where resonant multiphoton excitation plays an important role. For instance, it has been known for a long time that it is an important step in problems such as multiphoton inonization [4.5,6] and nonlinear generation of vacuum ultraviolet light [4.7]. Multiphoton excitation can also be used to excite a precise level (for instance a metastable state which cannot be attained by a single-photon transition). In that case, multiphoton excitation is much more precise than other techniques such as discharges.

The results obtained in recent years have amply demonstrated that multiphoton transition is a very powerful tool in spectroscopy. The power of the method has increased by many orders of magnitude when it becomes clear that it is possible to obtain Doppler-free spectra with multiphoton excitation [4.8,9]. Owing to the considerable number of experiments performed in the past five years which use this Doppler-free technique, Sect.4.2 will be devoted to them. In this section we discuss the applications in experiments where elimination of the Doppler broadening is not necessary.

4.1.1 Selective Pumping of an Excited Level with Multiphoton Transition

An important aspect of multiphoton transition which has not been greatly developed up to now is the utilization of these transitions to populate excited states which are difficult to populate with a sufficiently high selectivity by the classical methods. For instance, multiphoton excitation can be used in order to create the atom in a particular excited state and to study the collision processes and (or) the chemical reactions in this state. The three-photon excitation of X_e in the s^3P_1 resonant state [4.10] and the three-photon excitation of H_g in np^3P_1, np^1P_1 and nf^1F_3 excited levels [4.11] are good examples of the selectivity of the multiphoton excitation in a range of wavelength where the most common way to excite these levels is to use a discharge.

In the case of two-photon excitation, BIRABEN et al. [4.12] have shown what information on the interatomic potentials can be obtained by measuring the collisional transfer of population (and other multipoles) from the selected excited levels to other excited levels. Another possibility with two-photon excitation is to create the atoms in a metastable state by a clean method and to study the relaxation of this metastable atom. Good candidates for such experiments are the metastable $^1D_2$ states of Ca, Sr ... .

Another subject for which the pumping with two-photon transition may be interesting is the verification of the Bell inequality and the possible existence of hidden variables [4.13]. In the corresponding experiments, the important quantity is the correlation function between two successive spontaneous photons emitted in a cascade. It is important in these experiments to be extremely careful with the type of excitation of the atom in order to avoid parasitic signals. It seems that two-photon excitation is clearly a very good method to perform the excitation: the intermediate states (j in Fig.4.1) are not populated by the excitation, the quantum numbers of the excited state can be accurately determined from the knowledge of the polarization of the exciting light and of the quantum numbers of the ground state.

4.1.2 Intermediate Step in Other Processes

Resonant multiphoton excitation often occurs as an intermediate step in other processes. In the case of multiphoton ionization the number of ions increases by a huge factor when the wavelength of the exciting laser is adjusted to obtain a resonant multiphoton transition with an intermediate level (see Fig.4.2). This effect has been the subject of many experiments [4.6] because some features of the multiphoton ionization probability when an intermediate state in the ionization process is resonantly excited were unclear: the experimental variation of the ions—yield versus the laser intensity I did not appear to follow the theoretical law, when the laser wavelength is close to the value corresponding to the intermediate multiphoton resonance. This effect seems now to be understood; the explanation, which relies on the light shift of the levels, is given in Sect.4.4.

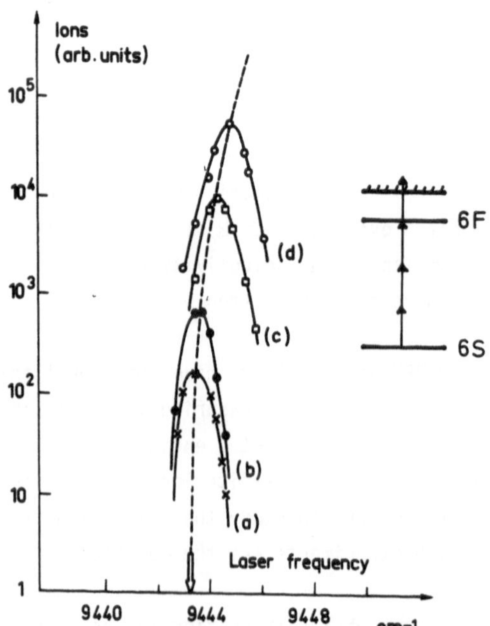

Fig.4.2. Four-photon ionization of caesium with an intermediate resonant level: Variation of the number of atomic caesium ions as a function of the laser frequency in the neighborhood of the three-photon transition 6S→6F, whose exact position is indicated by the arrow. Each curve is obtained for a fixed laser intensity (units 10^8 W/cm^2): (a) 1.8; (b) 5.2; (c) 11; (d) 15,7. [4.6]

An intermediate resonant multiphoton transition is also important in order to magnify the nonlinear generation of light in an atomic vapor. Let us, for instance, discuss the generation of vacuum ultraviolet light by frequency mixing in Sr vapor [4.7]. The principle of the experiment is to mix in Sr vapor two photons at frequency ν_1 and one photon at frequency ν_2 in order to obtain a third harmonic generation at $2\nu_1 + \nu_2$. A resonant enhancement of this generation occurs when $2h\nu_1$ corresponds to a resonant two-photon transition in Sr. In that case an enhancement of uv light by four orders of magnitude is obtained. By scanning ν_2, ν_1 being fixed, it is thus possible to obtain a tunable coherent radiation in the uv range. More precisely in the case of Sr, HODGSON et al. [4.7] were able to continuously tune their source from 1778 to 1817 Å and from 1833 to 1957 Å. Let us also notice that in the course of tuning ν_2, a further enhancement of the uv light has been observed when $h(2\nu_1 + \nu_2)$ corresponds to the position of an autoionizing state of Sr.

4.1.3 Spectroscopy Using Broadband Lasers

We describe the experiments which do not intend to obtain the ultimate linewidth (e.g., the natural width) which can be obtained in multiphoton transition. For all these experiments, the linewidth is larger than the Doppler width or equal to it. The first experiment of this type was done by ABELLA [4.4] and was presented in the introduction of Sect.4.1. Many experiments have been done since this time on various elements K [4.14], Ca [4.15], Sr [4.16], Ba [4.17] ... . The aim of most experiments was to determine the positions of the excited levels. As an example, we reproduced in Fig.4.3 an experimental recording in the case of Sr.

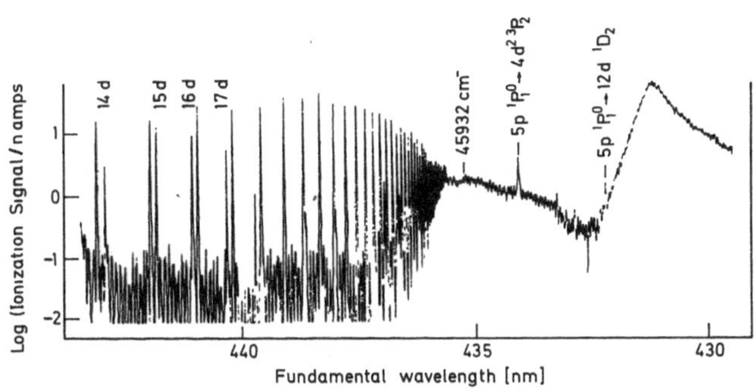

Fig.4.3. Two-photon spectroscopy of strontium. The highly excited levels are detected using a thermoionic detector. Only J = 2 levels are excited from the $^1S_0$ ground level when the light is circularly polarized (the experiment is performed with a heated pipe at 680 C. The Sr presure is 0.2 Torr and the buffer gas is Ne). [4.16]

In the case of two-photon excitation, the levels which can be reached are of the same parity as the ground state and cannot therefore be observed by single-photon excitation. If the ground level is a $^1S_0$ level, it is possible to show (see Sect. 4.3.6) that only levels of angular momenta $J = 0$ and $J = 2$ can be excited in a two-photon excitation (with photons of the same frequency). To distinguish the spectrum corresponding to the $J = 0$ levels from the one corresponding to $J = 2$, which is important because of the large number of resonances observed (see Fig.4.3) one can record two spectra: one with linearly polarized light, the second with circularly polarized light. In the case of linearly polarized light all the levels ($J = 0$ and $J = 2$) are excited while with a circularly polarized light, only the $J = 2$ levels are excited (because the selection rule $\Delta m = 2$ cannot be respected in a $J = 0 \rightarrow J = 0$ transition).

As can be seen on the list of the elements reported above, many of them correspond to atoms with two valence electrons. In several atoms, many levels with two excited valence electrons are localized below the ionization limit. When the energy of such a level is close to the energy of a level of a Rydberg series with only one electron excited, there may be a perturbation on the Rydberg series. The study of these perturbations is of particular interest to analyze the interaction between the different configurations.

The interest of multiphoton spectroscopy is notably increased by the Doppler-free technique, which will be described in the following section.

4.2 Doppler-Free Two-Photon Experiments

We present in this section the principle of Doppler-free multiphoton transition and some experiments which demonstrate the powerfulness of this method. Even though the method is extremely recent, the first experimental demonstrations [4.18,19] being performed in 1974, it is not possible to describe here the numerous experiments which have been done with this method during the five last years. We only want to point out here some of the striking applications of the method. An exhaustive list of the applications of the method can be found in a recent review paper [4.20].

4.2.1 Principle of Doppler-Free Multiphoton Transitions

The Doppler broadening is due to the thermal velocities of the atoms in the vapor. More precisely, if $\underline{V}$ is the velocity of the atom and $\underline{k}$ is the wave vector of the light beam, the first-order Doppler shift is $\underline{k} \cdot \underline{V}$. If one reverses the sense of the propagation of light ($\underline{k} \rightarrow -\underline{k}$), the first-order Doppler shift is reversed in sign.

We suppose that a two-photon transition can occur between the levels E_g and E_e of an atom in a standing electromagnetic wave of angular frequency ω. (The standing wave is produced, for instance, by reflecting a laser beam onto itself using a mirror). In its rest frame, the atom interacts with two oppositely travelling waves of angular frequencies $\omega + \underline{k} \cdot \underline{V}$ and $\omega - \underline{k} \cdot \underline{V}$. If the atom absorbs one photon from each travelling wave, the conservation of energy implies

$$E_e - E_g = E_{eg} = \hbar(\omega + \underline{k} \cdot \underline{V}) + \hbar(\omega - \underline{k} \cdot \underline{V}) = 2\hbar\omega \quad . \tag{4.1}$$

The term depending on the velocity of the atom disappear, indicating that, at resonance, all the atoms, irrespective of their velocities, can absorb two photons [4.8,9].

Theoretically the Doppler-free two-photon absorption resonance must have a Lorentzian shape, the width of the resonance being the natural one. But the wings of the resonance generally differ from the Lorentzian curve for the following reason: if the frequency ω of the laser does not fulfill the resonant condition (4.1) but is still close to it, the atoms cannot absorb two photons propagating in opposite directions, although some atoms of definite velocity can absorb two photons propagating in the same direction, provided that the energy defect $(E_{eg} - 2\hbar\omega)$ is equal to the Doppler shift $\pm 2\underline{k} \cdot \underline{V}$. For each value of ω, there is only one group of velocities which contribute to this signal, whereas at resonance (due to the absorption of photons propagating in opposite directions) all the atoms contribute. It follows that the intensity of the wings is much smaller than the intensity at resonance. More precisely, the two-photon line shape appears as the superposition of two curves (Fig.4.4):

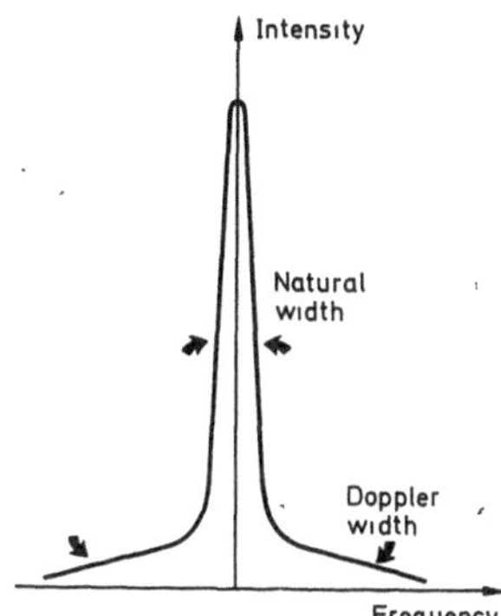

Fig.4.4. Theoretical two-photon line shape in a standing wave

A Lorentzian curve of large intensity and narrow width (natural width) corresponding to the absorption of photons from the oppositely travelling waves.

A Gaussian curve of small intensity and broad width (Doppler width) corresponding to the absorption of photons from the same travelling wave.

If the two travelling waves have the same polarization, a simple discussion shows that the area of the narrow curve is twice the area of the broad curve. Currently the Doppler width of the Gaussian curve is 100 or 1000 times larger than the natural width of the Lorentzian curve; and the Gaussian curve will appear as a very small background. In some cases, the choice of different polarizations permits one to suppress completely the Doppler background, using the different selection rules corresponding to different polarization [4.18].

As was noticed in [4.9], it is also possible to observe the cancellation of the Doppler broadening in transitions which involve more than two photons. We consider an atom of velocity V interacting with several plane waves, each of these waves being characterized by its wave vector $\underline{k}_i$. The first-order Doppler shift for each wave is $\underline{k}_i \cdot \underline{V}$. If the atom absorbs n photons $k_1, k_2, \ldots, k_n$, the n-photon transition is shifted by a quantity equal to $\sum_{i=1}^{n} \underline{k}_i \cdot \underline{V}$. It appears that if

$$\sum_{i=1}^{n} \underline{k}_i = 0 \tag{4.2}$$

the n-photon absorption is not Doppler shifted. In the particular case of n = 2, the two wave vectors must be equal in magnitude and opposite in sign. This is the case of the standing wave which was first investigated. If there are more than two photons, it is possible for the same amount of energy brought to the atomic system to choose the exciting frequencies within wide limits. For instance, for n = 3, the only limitations to the frequencies are the triangle relations.

Equation (4.2) can be interpreted as a *momentum conservation equation*: The momentum of each absorbed photon is $\hbar \underline{k}_i$. The total momentum of all the photons absorbed by the atom being equal to 0, its velocity does not change and there is no modification of its kinetic energy. *All of the photon energy is then transferred into internal energy*:

$$E_{eg} = \sum_{i=1}^{n} \hbar c k_i \quad . \tag{4.3}$$

If (4.2) is fulfilled, the n-photon transition is free of Doppler broadening. Furthermore, this transition is free of the recoil effect because the momentum transferred from the photons to the atom is equal to 0. It appears that the only "kinematical" limitation of the method is the second-order Doppler shift, whose magnitude is $E_{eg}V^2/2c^2$. In most cases, this is too small to be observed.

It is also noticed that the principle of Doppler-free multiphoton transition is not limited to a multiphoton absorption. We can, for instance, imagine a process where two photons, $\underline{k}_1$ and $\underline{k}_2$, are absorbed and another, $\underline{k}_3$, is emitted. In that case, to eliminate the Doppler broadening in the three-photon process, we have to change the sign of k_3 in (4.2,3). The first experiment of Doppler-free three-photon spectroscopy [4.21] was realized with that geometry. This experiment will be discussed in more detail in Sect.4.4.

Nevertheless, it is clear that up to now most of the experiments have been performed using the technique of two-photon absorption. That is why we shall present the two-photon case in more detail.

4.2.2 Experimental Observation of Doppler-Free Two-Photon Transitions

Two-photon experiments using the Doppler-free technique raise two contradictory requirements, as they need simultaneously very high light power and extreme spectral purity. The first experimental demonstrations of Doppler-free two-photon transitions in Paris [4.18] and in Harvard [4.19] were performed with pulsed dye lasers on the 3S - 5S transition in sodium (see Fig.4.6). But the precision of the measurements was increased by the use of CW dye lasers in single mode operation. The light power available in this case is rather small (currently 100 mW) and permits one to work only in favorable cases; but the number of these favorable cases increases with the improvement of the dye lasers (a CW power of the order of 1 W will be available now with the new ring laser [4.22]).

Experimental setups using pulsed dye lasers are described elsewhere [4.23]; and we describe here a typical setup using a CW laser (Fig.4.5).

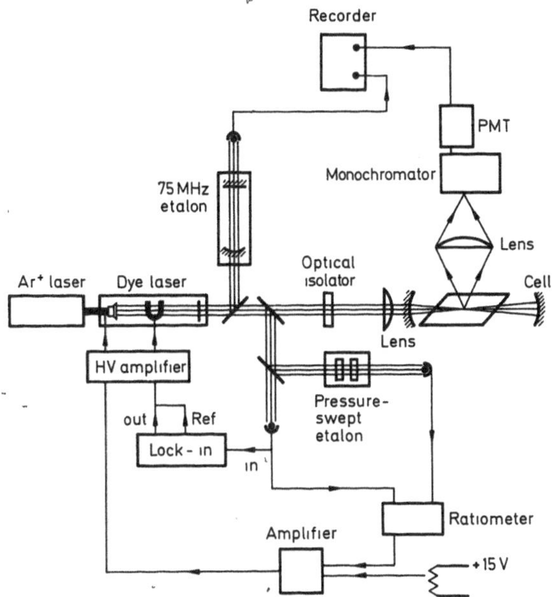

Fig.4.5. Experimental setup for two-photon spectroscopy with a CW dye laser. The bottom of the figure shows the two servo loops used for stabilization and control of the laser frequency

The CW laser is pumped by an argon ion laser. In order to obtain good control of the laser frequency, one uses two servo loops:

I) the purpose of the first one is to maintain the single-frequency oscillation of the dye laser. The Fabry-Pérot etalon inside the laser cavity selects one particular longitudinal mode of the cavity. One has to avoid possible mode hopping, and to attain this goal, the length of the Fabry-Pérot etalon is modulated piezoelectrically, which produces a small modulation of the laser intensity. This intensity modulation is detected with a lock-in amplifier and is used to control piezoelectrically the length of the internal Fabry-Pérot etalon. With the appropriate phase of the lock-in, one maintains the internal etalon centered on the same mode of the laser cavity.

II) the second servo loop is used to control the frequency of the laser cavity and does not include any modulation. Some of the laser light passes through an external Fabry-Pérot etalon. The transmitted signal is compared with the laser intensity; then the ratio or the difference of the two signals is used as an error signal and amplified, and applied to the piezoceramic carrying one of the mirrors of the laser cavity. The working point is at the mid-height of the transmission peak of this external etalon. The spacers of the etalon are made in Cervit and the etalon is enclosed in a vacuum box. One thus obtains very good thermal and pressure stability: the drift in frequency is less then 10 MHz over a few hours. This servo loop also reduces the jitter of the laser frequency due to acoustical vibrations of the laser cavity. It is easy to reduce the frequency jitter to about 1 MHz and it is possible to reduce it still more [4.24]. In most experiments one has to scan the laser frequency. One obtains an almost linear frequency scan (plotted against time) by pressure sweeping the external etalon. A small part of the laser light passes through a second external etalon which is about 1 m long; The transmitted signal gives very closely spaced transmission peaks which are used to scale the frequency axis.

After the problems involved in the light sources we describe now the experiment itself. The light coming from the laser is focused into the experimental cell (Fig.4.5) in order to increase the energy density. The transmitted light is refocused from the other side into the cell using a concave mirror whose center coincides with the focus of the lens. To obtain a precise coincidence of the narrow beam waists, which are often less than 50 μm, we need a good mechanical mounting.

In some experiments the energy density has been increased by placing the experimental cell in a spherical concentric Fabry-Pérot cavity [4.25]. The lens is chosen to match the radius of curvature of the wavefront to the radius of the first mirror. The windows of the experimental cell are tilted to the Brewster angle in order to reduce the losses in the cavity. The length of this cavity is locked to the laser frequency in order to transmit the maximum signal. Another solution would be to place the cell inside the laser cavity, but this technique requires a larger laser cavity and very low losses due to the cell.

When the return beam coincides exactly with the incident one, that produces interference effects which lead to mode hopping. Mode hopping is a good indicator of exact alignment of the return beam with the incident beam, but it presents, of course, many problems for the experiment itself. We must place an optical isolator between the laser and the experimental region in order to prevent these instabilities. The optical isolator can use either the birefringence of a quarter-wave plate or the Faraday rotation in a flint glass in a longitudinal magnetic field [4.26]. In both cases the return beam comes back into the laser with a polarization at right angles to that of the incident beam and does not interfere with it. These two techniques are not equivalent if the two-photon excitation is sensitive to the polarization of light; in the first case, the atoms are excited with circularly polarized light and in the second case with linearly polarized light.

The two-photon resonance is detected by collecting photons emitted from the excited state at a wavelength λ_{ej} (see Fig.4.1) different from the exciting wavelength λ (it is sometimes more convenient to detect the resonance on another wavelength λ_{ab} emitted by the atom in a cascade). The characteristic wavelength λ_{ej} is selected with an interference filter or a monochromator. The difference between λ_{ej} and λ allows the complete elimination of the stray light of the laser, despite its high intensity, and the observation of very small signals on a black background.

Typical Experiment in Sodium

As an example, we present the results obtained in the case of the 3S-4D two-photon transition in sodium [4.27,28]. Figure 4.6 shows this transition on the energy diagram of sodium. The laser wavelength λ is equal to 5787.3 Å. The two-photon transition probability is high because the oscillator strengths f_{3S-3P} and f_{3P-4D} are strong and because the energy detuning $\hbar\Delta\omega_j$ from the single photon transition to 3P level is small ($\Delta\omega_j/\omega \sim 0.02$). In fact, the 3S-4D line is split into four components because of the fine structure in the excited 4D level and of the hyperfine interaction in the 3S level (see Fig.4.6). Figure 4.7 shows a typical recording obtained when the photomultiplier current is plotted versus the laser frequency. The four components correspond to the transitions: (a) $F = 2 \rightarrow J' = 5/2$; (b) $F = 2 \rightarrow J' = 3/2$; (c) $F = 1 \rightarrow J' = 5/2$; (d) $F = 1 \rightarrow J' = 3/2$. It is possible to observe in this recording the small Doppler background due to the absorption of two photons propagating in the same direction. The width of the narrow lines is of the order of 10 MHz and represents principally the uncertainty of the light frequency due to residual jitter of the laser. This width can be reduced now to less than 5 MHz. Anyway it is much smaller than the Doppler width which is about 2000 MHz on the laser frequency scale (all the values must be doubled on the atomic energy scale).

As can be seen in Fig.4.7, the hyperfine interaction in the excited state is not optically resolved. However, by measuring very carefully the positions of the four different components, BURGHARDT et al. [4.29] noticed that their spacings were different from those expected from the known values of the hyperfine structure in the ground state and of the fine structure in the excited state. They deduced from their measurements that the shifts of the spacings can be interpreted if one takes into account the hyperfine interaction in the excited state. They found for the hyperfine A factor: $A(4^2D_{3/2}) = 215 \pm 15$ kHz and $A(4^2D_{5/2}) = 29 \pm 6$ kHz. (Both

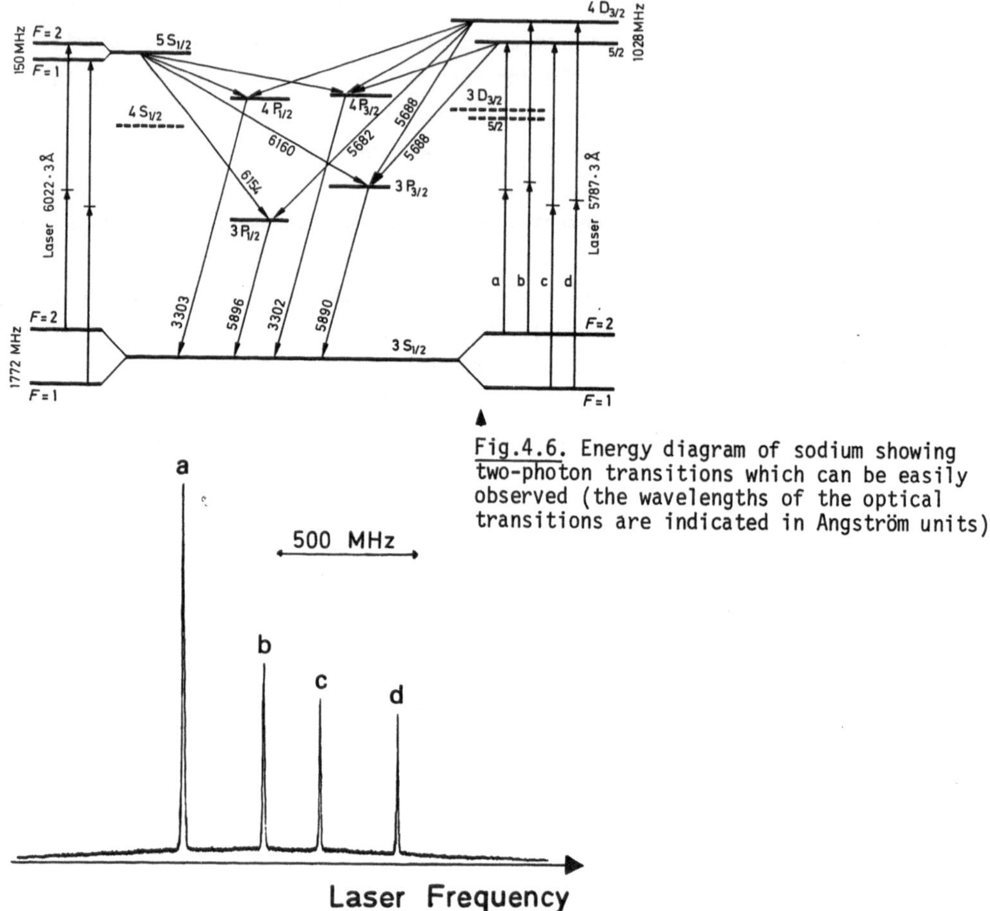

Fig.4.6. Energy diagram of sodium showing two-photon transitions which can be easily observed (the wavelengths of the optical transitions are indicated in Angström units)

Fig.4.7. Experimental recording of the Doppler-free two-photon transition 3S → 4D in sodium (fluorescence intensity vs the laser frequency). It shows the four components predicted in Fig.4.6. The small Doppler background produced by absorption of two photons of the same travelling wave can also be observed [4.28]

values agree with an earlier measurement [4.30] which was done using the polarization rate of the fluorescence after a Doppler-free two-photon excitation). In the same paper [4.29], the 4D fine structure was found to be: 1027.989 ± 0.030 MHz. These values are not important in themselves, nevertheless they do show the very high accuracy which can be attained now with Doppler-free two-photon spectroscopy.

Thermoionic Detection

In the preceding experiment the detection of the resonance was done by observing the fluorescent light emitted by the atoms from the excited state. Nevertheless, this method is not efficient for the study of highly excited levels near the ionization limit called Rydberg states, because the fluorescence is weak and distributed on many various wavelengths from infrared to ultraviolet. It has thus been proposed in [4.31] to use a modified type of the thermoionic detector already used in other experiments [4.32]. The experimental setup with the usual CW stabilized dye laser is shown in Fig.4.8. The thermoionic detector may be thought of as a simple diode working in the space-charge regime: it consists of a cylindrical anode surrounding a wire cathode. The highly excited atoms produced by the laser excitation are ionized by thermal collisions, and their presence in the space charge affects the flow current. The trapping of the ions greatly enhances the sensibility of the device; almost 100% efficiency is achieved, with gains of the order of 10^5. Since the polarizabilities of the states under investigation are large, an electrostatically shielded volume is made inside the detector: the anode is separated into two compartments by a nickel mesh, one of them containing the cathode tungsten wire, the second one being free of electric field. The laser is focused into the second compartment, and the excited atoms diffuse into the region containing the cathode.

It was possible with this detector to observe the Doppler-free two-photon transitions 5S → nD in rubidium with $25 < n < 85$ [4.31]. This large number of levels which can be excited is another indication of the possibilities of Doppler-free two-photon excitation.

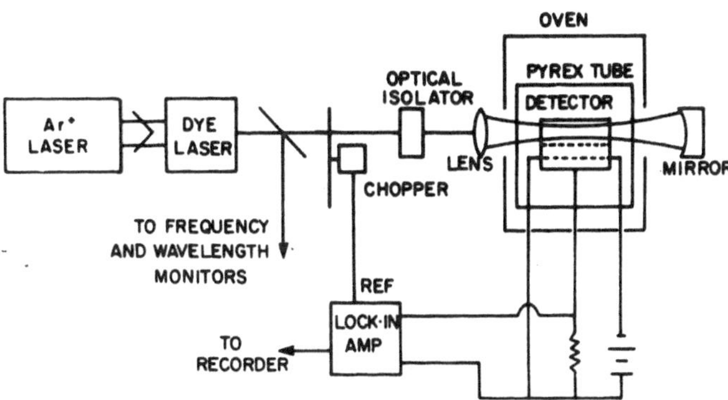

Fig.4.8. Experimental setup for Doppler-free two-photon spectroscopy of highly excited Rydberg states, using the thermoionic detection [4.31]

4.2.3 Doppler-Free Two-Photon Transitions in Hydrogen

The interest of the 1S-2S Doppler-free two-photon transition in hydrogen has been pointed out by various authors [4.9,27,33]. Due to the very long lifetime of the 2S state, the width of the two-photon line can be extremely narrow. Furthermore the measurement of the 1S-2S frequency permits the determination of the Lamb shift of the ground state 1S of hydrogen. We now describe the experiment performed in Stanford [4.34,35].

The wavelength used to realize the two-photon excitation is 2430 Å (two times the wavelength of the Lyman α line). To obtain it, the frequency of a nitrogen pumped dye laser operating near 4860 Å was doubled with a crystal of lithium formate monohydrate. The pressure tuned dye laser system operates at 17 pps and provides 10 ns long pulses of 15 kW peak power and 120 MHz linewidth at 4860 Å. The ultraviolet pulses have a power of about 300 W. Part of the 4860 Å output is sent through a plane Fabry-Pérot interferometer to provide frequency calibration markers (see Fig.4.9). Another part of the 4860 Å output allows the simultaneous observation of the Balmer β line by saturated absorption spectroscopy and permits an absolute calibration.

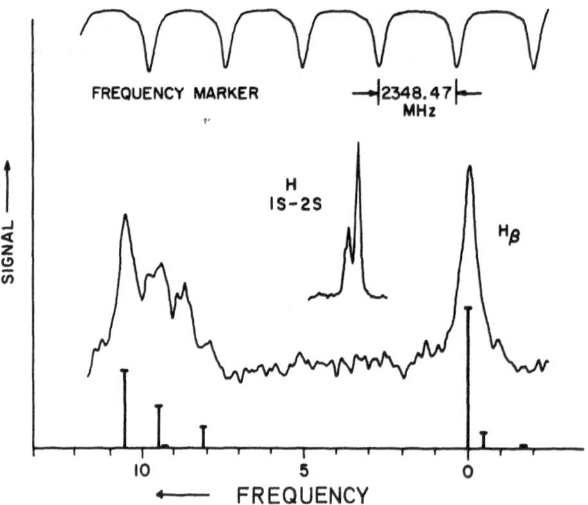

Fig.4.9. Doppler-free two-photon transition 1S → 2S in a hydrogen atom (middle curve) compared to the saturation spectrum of the hydrogen Balmer line H_β (bottom curve), which is simultaneously recorded. The vertical straight lines in the bottom indicate the theoretical position of the fine structure components of the H_β spectrum. The top curve gives the resonances of the calibration interferometer. [4.35]

The atoms are generated in a gas discharge in a mixture of H_2 and D_2 at 0.2 Torr. The two-photon transition is monitored by observing the 2P-1S fluorescence at 1215 Å with a solar blind photomultiplier through a MgF_2 window (the transitions from 2S to 2P are induced by collisions). Figure 4.9 shows the Doppler-free two-photon spectrum (middle curve) compared to saturated absorption of the Balmer line (bottom curve) and to frequency markers (upper curve). The two-photon spectrum of H consists of two peaks corresponding to the F = 0 → F' = 0 and F = 1 → F' = 1 hyperfine components of the two-photon transition. Only transitions between states of the same F value are allowed in that case. The reason is explained in Sect.4.3.6. It can be understood from the fact that the electric dipole operator only acts on the orbital part of the wave function, and it can modify the electronic and nuclear spin of the atom only by coupling with orbital momentum. Since both the initial and final states are S levels (L = 0), we have no coupling with zero orbital momentum, and thus conservation of F and m_F.

The simultaneous recording of the two-photon transition and of the saturation spectrum of the Balmer line allows one to measure the 1S Lamb shift by avoiding the technical difficulties of vacuum ultraviolet spectroscopy. If Bohr's formula were correct, the n = 1 - 2 interval would be exactly four times larger than the n = 2 - 4 interval, and the two observed transitions would occur at exactly the same dye laser frequency. The actual displacement is due to relativistic and quantum electrodynamic corrections plus small nuclear-structure effects. It is thus possible to deduce the 1S Lamb shift, the other corrections being evaluated theoretically. The experimental values of the 1S Lamb shift are 8.20 ± 0.10 GHz for H and 8.25 ± 0.11 GHz for D. Both values agree within their error limits with the theoretical values.

In the same experiment, the 1S-2S isotope shift has been measured, its value is 670.933 ± 0.056 GHz, in fair agreement with the theoretical predictions. Several improvements can still be performed on this experiment. The first limitation is the linewidth of the laser, the linewidth of the nitrogen pumped dye laser presently used here is about one hundred times larger than the one of a commercial CW dye laser. Even if the use of a CW dye laser is not considered for the present, the linewidth can be improved by using a pulsed dye laser synchronized on a CW dye laser [4.36]. Another possible improvement would be to compare the 1S-2S and the 2S-4S two-photon transitions. In that case the linewidth obtained on the Balmer β line will not be limited by a short living P state as is the case with saturated absorption spectroscopy.

4.2.4 Other Possibilities of Doppler-Free Two-Photon Transitions

A lot of experiments have been done following the scheme of Sect.4.2.2, in order to measure fine or hyperfine splittings. But many other applications of the Doppler-free technique have been performed.

The application of *small external fields* permits one to observe easily the separation of electric or magnetic components of an optical line. Measurements have been done for Zeeman splittings [4.37] and Stark splittings [4.38]. The experiments give not only the splitting between the components, as radio frequency or level-crossing experiments, but also the absolute shift of the components: they permit for example the measurement of diamagnetic shifts [4.39]; in the case of the Stark effect, they permit to deduce both the scalar and the tensorial polarizabilities of the atomic states [4.38].

Special attention must be paid to *isotopic shift* measurements, as they cannot be obtained with the older Doppler-free method such as radio frequency or level crossings. The preceding section has shown an example of measurement of isotopic shift in hydrogen. Figure 4.10 shows another recording of isotopic shift in the case of neon [4.40]. The better precision due to the use of a CW dye laser is clearly evident; small frequency differences of the order of 10 MHz have been measured and theoretically interpreted [4.41]. The separation between the two-photon lines of the two even isotopes of neon is smaller than the Doppler width and would not be visible without the Doppler-free technique. Another interesting feature of this recording is that the lower level of the two-photon transition is not the ground state of the atom, but the metastable level 3s (J = 2) populated in a weak discharge. (The discharge is chopped; and the two-photon transition is detected in the afterglow, in order to avoid the parasitic light of the discharge).

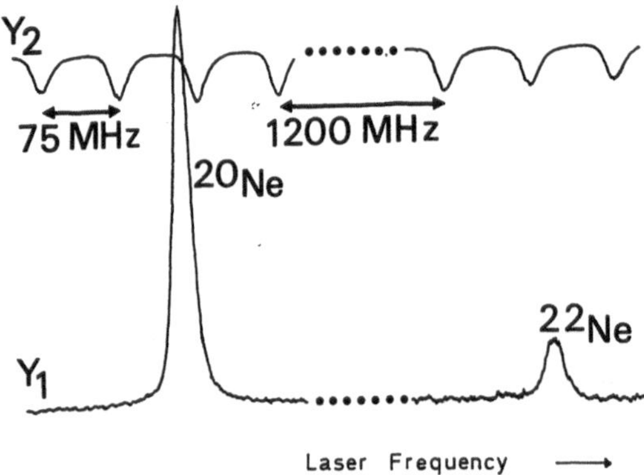

Fig.4.10. Recording of the Doppler-free two-photon transition 3s (J = 2) → 4d' [5/2] J = 3 in natural neon (bottom curve Y_1) showing the isotopic shift between the two even isotopes ^{20}Ne and ^{22}Ne. The upper curve Y_2 shows the frequency markers from a stable interferometer. The distance between two peaks is 75 MHz, which corresponds to 150 MHz for the atomic frequency [4.40]

Other isotopic shift measurements using the Doppler-free two-photon method have been performed in rubidium [4.42,43], in thallium [4.44], and in helium [4.79].

Using Doppler-free two-photon excitation, it is also possible to obtain much information on *collision processes*. In particular, it must be noticed that *all* the atoms interact with both waves; this implies that, in conditions which are usually satisfied, velocity changing collisions do not modify the line shape, which usually remains a Lorentzian curve (this not being true with saturated absorption). The influence of collisions on the line shape can thus be presented only in terms of shift and broadening of the line. The broadening of the line corresponds to the relaxation of the optical multipole coherence between the ground and excited levels. One of the clear advantages of Doppler-free two-photon excitation is that the shift and the broadening can be observed at very low pressure of gas. Figure 4.11 displays a good example [4.45,46]; it shows, on an expanded

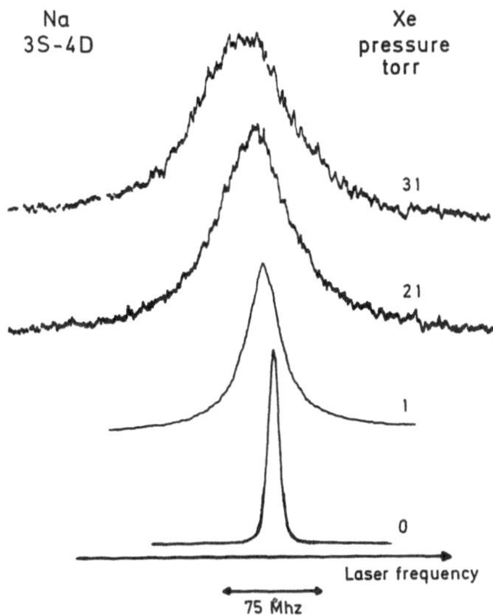

Fig.4.11. Collisional broadening and shift of a Doppler-free two-photon transition: the same component (a) of the 3S → 4D two-photon line, shown in Fig.4.7, is recorded with various pressures of Xenon buffer gas from 0 to 3.1 Torr. (in abcissa laser frequency) [4.45]

scale, various recordings of the first component of the two-photon line in sodium presented in Fig.4.7. The bottom curve is recorded in a cell containing only sodium; the upper curves are obtained when the experimental cell is filled with various pressures of Xenon, from 1 to 3 Torr, showing the broadening and the shift. Numerous experiments have been performed using this scheme [4.47-49].

Another application of Doppler-free two-photon transition in collision experiments is to excite only one sublevel inside the Doppler width and to detect the transfer from this sublevel to other levels [4.12].

Owing to the very small linewidth obtained in a Doppler-free two-photon transition, one may consider the possibilities of this method for *metrological purposes*. It has been recently shown that Doppler-free two-photon transition to Rydberg levels can be used as a very interesting calibration spectrum [4.24]. In theory these lines can be extremely narrow because the lifetimes of the Rydberg levels vary as n^{*3} versus the effective quantum number n^*. Furthermore, these transitions permit one to cover a relatively large spectral range.

We hope that these few examples have convinced the reader of the interest of the Doppler-free two-photon excitation method.

4.2.5 Experiments with Two Different Light Sources

If one uses two different light sources of frequencies ω_1 and ω_2 instead of one, it is possible, keeping $\hbar\omega_1 + \hbar\omega_2$ constant and equal to the energy difference $E_e - E_g$, to increase by a large factor the transition probability by reducing the energy defect $\hbar\Delta\omega_j$ (see Fig.4.1). The gain in transition probability is compensated for by the presence of a residual Doppler broadening which is proportional to $\omega_2 - \omega_1$. Such an experiment has been performed on the 3S-4D two-photon transition of sodium [4.50]. In this case, it has been possible to increase the transition probability by about seven orders of magnitude while the residual Doppler broadening attains a value of the order of 60 MHz.

The two-photon spectrum is recorded by fixing the wavelength λ_1 of the first absorbed photon and scanning the wavelength λ_2 of the second. The intensity of the two-photon line depends on the wavelength λ_1. The variation of the two-photon transition rate plotted against λ_1 is represented in Fig.4.12. When λ_1 is close to one of the resonance lines of sodium, there is of course a sharp increase in the transition probability. The variation of the transition rates with λ_1 are different in the cases of the excited states $4D_{3/2}$ and $4D_{5/2}$. For the transition to $4D_{5/2}$, only the $3P_{3/2}$ state can act as in intermediate state since the $3P_{1/2} \to 4D_{5/2}$ single-photon transition is forbidden. It follows that the corresponding transition rate presents only one maximum. On the other hand, both $3P_{1/2}$ and $3P_{3/2}$ are possible intermediate states in the case of the two-photon transition to $4D_{3/2}$ and the corresponding transition rate presents two maxima. Moreover, there is a destructive interference between the two possible paths in that second case for a value of λ_1 lying between 5890 and 5896 Å (see Fig.4.12).

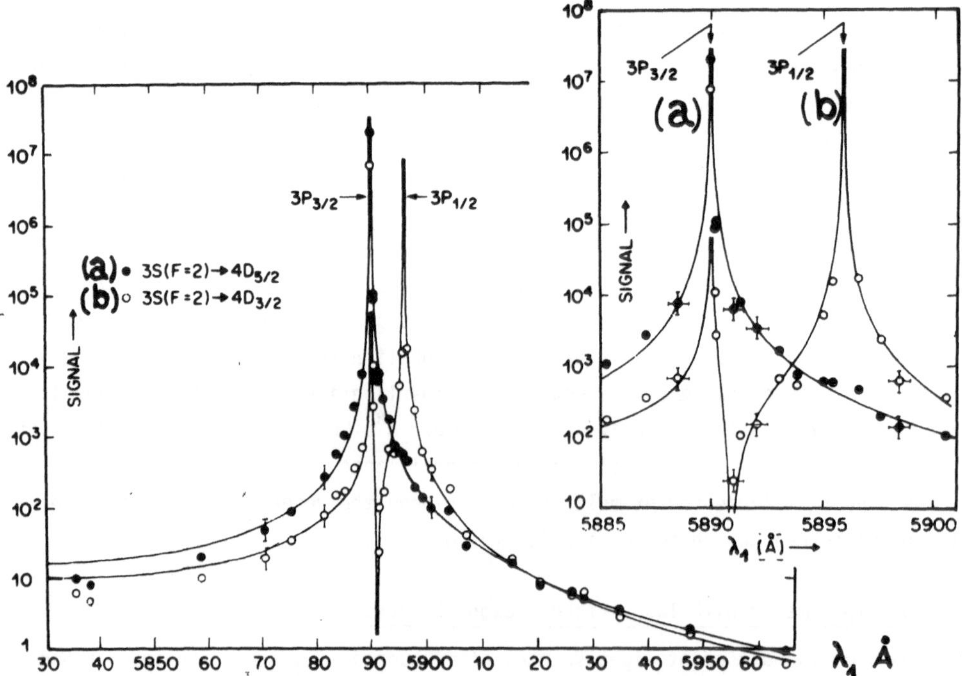

Fig.4.12. Influence of the energy defect $\hbar\Delta\omega_r$ of the intermediate relay level on the two-photon transitions produced by two lasers of different wavelengths: Normalized transition rates as a function of the wavelength λ_1 of the first absorbed photon for the following two-photon lines in sodium (see Fig.4.6): (a) 3S (F=2) $\to 4D_{5/2}$ black points; (b) 3S (F=2) $\to 4D_{3/2}$ open circles. The points are experimental and the curves are theoretical. The inset shows the behavior in the region from 5885 to 5900 Å with an expanded horizontal scale [4.50]

The limiting situation of such experiments corresponds to the case where the energy defect $\hbar\Delta\omega_j$ is of the same order as the Doppler width of the single-photon transition g-j. The two-photon transition probability is then different from one velocity group to another. If $\Delta\omega_j$ is small enough, atoms of a particular velocity group are excited in the intermediate state, and the two-photon excitation appears for these atoms as a real two-step excitation. In that case, the two-photon absorption line generally appears as the sum of two curves: a Doppler-broadened curve whose width is proportional to $(\omega_2 - \omega_1)$ and a narrow peak corresponding to the real two-step excitation concerning a particular velocity group [4.51]. Figure 4.13 shows an example of such a recording. The ratio of the intensities of the two curves depends on the energy detuning, $\hbar\Delta\omega_j$ from the intermediate state. When $\Delta\omega_j$ is large compared to the Doppler width of the g-j transition, the intensity of the Doppler broadened curve is predominant. When $\Delta\omega_j$ becomes close to zero, the opposite situation prevails. Let us finally notice that if $\Delta\omega_j$ is small, the situation is closely related to the one encountered in laser-induced line-narrowing techniques in three-level resonances [4.1].

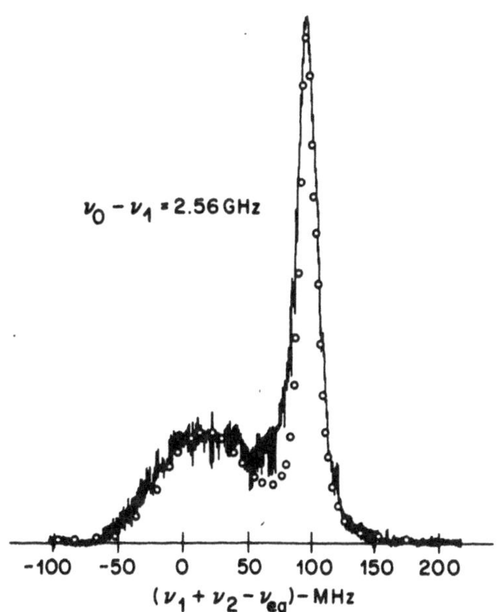

Fig.4.13. Two-photon transition produced by two counter propagating waves of different frequencies with a nearly resonant intermediate level [transition (b) 3S(F=2)→$4D_{3/2}$ in sodium]. The frequency ν_1 of the first absorbed photon is fixed at $\nu_0 - \nu_1 = 2.56$ GHz, where ν_0 is the frequency of the sodium D_1 line (relay level $3P_{1/2}$ F=1); and the frequency of the second photon ν_2 is scanned through the two-photon resonance at $\nu_{eg} = E_{eg}/h$. The open circles are theoretical. The high and narrow peak correspond to a stepwise excitation (two successive single-photon transitions) of a particular velocity group and is Doppler shifted. The smaller and broader peak corresponds to the "true" two-photon transition produced by all the atoms; the width is partly due to the residual Doppler broadening proportional to $\nu_2 - \nu_1$ [4.51]

4.3 Theory of Two-Photon Transitions in Atoms

In Sect.4.2 we have presented several experiments about two-photon excitation. In order to interpret the experimental observations, we present in this section a theory of two-photon excitation, which will be generalized to n-photon excitation in Sect.4.4. We are mainly interested in the following aspects: the frequency displacement of the transition due to light shifts and the selection rules. The light shift [4.52] is certainly one of the mot important problems connected with multi-photon excitation. It is the first evident difficulty encountered in the application of Doppler-free two-photon excitation for metrological purposes [4.9]. It leads to a strange dependence of the order of nonlinearity in the case of resonant multi-photon ionization [4.6]. It is one of the basic limitations in attempts to generalized the Doppler-free experiments to n-photon transitions (for $n > 2$) [4.9,53]. It is thus very important to have a good understanding of the importance of the light shift.

We first introduce an effective Hamiltonian (Sect.4.3.1) and solve the density matrix equation (Sect.4.3.2). Then we consider the case of two different waves (Sect.4.3.3) thus allowing the calculation of the two-photon line shape in a standing wave (Sect.4.3.4). Finally we discuss the importance of the light shifts (Sect.4.3.5) and the selection rules (Sect.4.3.6).

4.3.1 The Effective Hamiltonian

We consider an atom, whose energy levels g, e, j_1, j_2, ... are charaterized by the energies E_g, E_e, E_{j_1}, E_{j_2}, ... In practice we will use the energy differences from the ground state $E_{jg} = E_j - E_g$.

The atom interacts with an electric field, which is written *in the rest frame of the atom* in the general form

$$\underline{E}(t) = \mathrm{Re}\left\{\underline{\varepsilon} \cdot \mathcal{E} \cdot e^{-i\omega t}\right\} = \frac{1}{2}\left(\underline{\varepsilon}\mathcal{E} \cdot e^{-i\omega t} + \underline{\varepsilon}^*\mathcal{E}^* \cdot e^{+i\omega t}\right) \tag{4.4}$$

where the unit vector $\underline{\varepsilon}$ represents the polarization of the electromagnetic wave. Complex vectors $\underline{\varepsilon}$ permit representation of circular polarizations; *but in this section we suppose a linear polarization and $\underline{\varepsilon}$ is real.*

We assume that the level e can be excited from level g by two-photon absorption, and that $2\hbar\omega$ is close to E_{eg}. On the other hand, we make the following assumption: all the energy defects $\hbar\Delta\omega_j = |\hbar\omega - E_{jg}|$ are very large compared with

- the natural width of the g-j and j-e transitions,
- the saturation width of the g-j and j-e transitions,
- the two-photon detuning $|2\hbar\omega - E_{eg}|$. (4.5)

The interaction Hamiltonian is $-\underline{D} \cdot \underline{E}$, where $\underline{D}$ is the electric dipole moment of the atom for the choice between the interaction Hamiltonian $-\underline{D} \cdot \underline{E}$ and $-\underline{p} \cdot \underline{A}$, see for example [4.54,55] and references therein). We call D_{gj} the matrix element of the component $\underline{D} \cdot \underline{\varepsilon}$ between the levels g and j, and so on ... . The levels e and g have the same parity and the matrix elements D_{eg} is zero; on the other hand, we will neglect the coupling between levels j and j' different from g and e.

The Schrödinger equation is applied to the wave function $|\psi(t)\rangle$

$$|\psi(t)\rangle = b_g(t)|g\rangle + b_e(t)\left[\exp\left(-\,i\,\frac{E_{eg}t}{\hbar}\right)\right]|e\rangle + \sum_j b_j(t)\left[\exp\left(-\,i\,\frac{E_{jg}t}{\hbar}\right)\right]|j\rangle \qquad (4.6)$$

(the energy E_g of level g is taken equal to 0). It gives

$$i\hbar\dot{b}_g = -\sum_j b_j\left[\exp\left(-\,i\,\frac{E_{jg}t}{\hbar}\right)\right]D_{gj}\ \mathrm{Re}\{\mathcal{E}\ e^{-i\omega t}\} \qquad (4.7)$$

$$i\hbar\dot{b}_e = -\sum_j b_j\left[\exp\left(i\,\frac{E_{ej}t}{\hbar}\right)\right]D_{ej}\ \mathrm{Re}\{\mathcal{E}\ e^{-i\omega t}\} \qquad (4.8)$$

$$i\hbar\dot{b}_j = -\,b_g\left[\exp\left(i\,\frac{E_{jg}t}{\hbar}\right)\right]D_{jg}\ \mathrm{Re}\{\mathcal{E}\ e^{-i\omega t}\} - b_e\left[\exp\left(-\,i\,\frac{E_{ej}t}{\hbar}\right)\right]D_{je}\ \mathrm{Re}\{\mathcal{E}\ e^{-i\omega t}\}\ . \qquad (4.9)$$

If the time evolution of b_g and b_e is slow compared with $1/\Delta\omega_j$ we can easily integrate (4.9) and we obtain

$$b_j = \frac{D_{jg}}{2}\,b_g\left\{\frac{\mathcal{E}^*\exp[i(\omega + E_{jg}/\hbar)t]}{\hbar\omega + E_{jg}} - \frac{\mathcal{E}\exp[i(-\,\omega + E_{jg}/\hbar)t]}{\hbar\omega - E_{jg}}\right\}$$
$$+\,\frac{D_{je}}{2}\,b_e\left\{\frac{\mathcal{E}^*\exp[i(\omega - E_{ej}/\hbar)t]}{\hbar\omega - E_{ej}} - \frac{\mathcal{E}\exp[-i(\omega + E_{ej}/\hbar)t]}{\hbar\omega + E_{ej}}\right\}\ . \qquad (4.10)$$

If we substitute the value of b_j in (4.7,8), we see that the interaction between levels g and e can be represented by an effective Hamiltonian. We neglect the antiresonant terms, that is to say, we keep in the final expressions giving b_e and b_g only the terms whose modulation frequency is $(2\omega - E_{eg}/\hbar)$. Under that condition the matrix V which represents the effective coupling between g and e is

$$V = \begin{pmatrix} \frac{1}{4}\langle e|Q'|e\rangle|\mathcal{E}|^2 & \frac{1}{4}\langle e|Q|g\rangle\mathcal{E}^2\ e^{-2i\omega t} \\ \\ \frac{1}{4}\langle g|Q|e\rangle\mathcal{E}^{*2}\ e^{2i\omega t} & \frac{1}{4}\langle g|Q|g\rangle|\mathcal{E}|^2 \end{pmatrix}, \qquad (4.11)$$

$\langle e|Q|g\rangle$ is the matrix element of the two-photon operator Q between g and e [4.9,56]

$$Q = D\cdot\underline{\varepsilon}\,\frac{1}{\hbar\omega - H_0}\,D\cdot\underline{\varepsilon} \qquad (4.12)$$

(H_0 is the Hamiltonian of the free atom).

The diagonal matrix elements of V correspond to the light shift of the levels (dynamic Stark effect) [4.52]. In the light shift of the ground level, we have retained only the terms containing $1/\hbar\Delta\omega_j$ and neglected those with $1/(\hbar\omega + E_{jg})$ because E_{jg} is always positive. It is not so simple for the light shift of the excited level, because E_{ej} can be negative and we must use a more complicated operator Q'

$$Q' = \underline{D}\cdot\varepsilon \frac{1}{E_e - \hbar\omega - H_0} \underline{D}\cdot\underline{\varepsilon} + D\cdot\underline{\varepsilon} \frac{1}{E_e + \hbar\omega - H_0} \underline{D}\cdot\underline{\varepsilon} \quad . \tag{4.13}$$

4.3.2 Solution of the Density Matrix Equation

The density matrix equation is written

$$\frac{d}{dt}\rho = \frac{1}{i\hbar}[(H_0 + V), \rho] - \left(\frac{d}{dt}\rho\right)_{rel.} \quad . \tag{4.14}$$

The last term represents the relaxation, and its matrix elements are

$$\left.\begin{aligned} \left(\frac{d}{dt}\rho_{ee}\right)_{rel} &= -\Gamma_e\rho_{ee} \\ \left(\frac{d}{dt}\rho_{gg}\right)_{rel} &= \Gamma_e\rho_{ee} \\ \left(\frac{d}{dt}\rho_{eg}\right)_{rel} &= -\Gamma_{eg}\rho_{eg} \end{aligned}\right\} \tag{4.15}$$

We suppose that all the atoms leaving the excited level return to the level g. In the absence of collisions Γ_e is the inverse of the radiative lifetime of e level, and $\Gamma_{eg} = \Gamma_e/2$.

All experiments described in Sect.4.2 are performed in the steady state regime; and the detection techniques give a signal proportional to the population of the excited level. This is the reason why we now calculate the steady state solution of (4.14). Using the effective Hamiltonian (4.11) we find the fraction of the atoms in the e level

$$\frac{N_e}{N_0} = \rho_{ee}(\omega) = \frac{\hbar^2\Gamma_{eg}^2 a^2/2}{(E_{eg} + s - 2\hbar\omega)^2 + \hbar^2\Gamma_{eg}^2(1 + a^2)} \tag{4.16}$$

$$\text{with } a^2 = |\langle e|Q|g\rangle\ ^2|^2/4\hbar^2\Gamma_e\Gamma_{eg} \tag{4.17}$$

$$s = \frac{1}{4}\left(\langle e|Q'|e\rangle - \langle g|Q|g\rangle\right) |\mathcal{E}|^2 \quad . \tag{4.18}$$

In these equations a represents the saturation parameter (dimensionless) of the two-photon transition. The width of the two-photon absorption line shape is equal to $\Gamma_{eg}\sqrt{1 + a^2}$. At high intensity, it is proportional to the saturation parameter a, i.e., to the intensity of the field $|\mathcal{E}|^2$. The shift of the resonance is s, which is proportional also to the intensity of the field $|\mathcal{E}|^2$ [4.52].

The fact that both the saturation parameter a and the light shift s are proportional to the intensity is important for the discussion on the light shift which will be given in Sect.4.3.5.

We must not forget the validity condition of the effective Hamiltonian (4.11). We have assumed in Sect.4.3.1 that the time evolution of b_g and b_e have to be slow compared with the various $1/\Delta\omega_j$. It is clear, from the matrix elements of V, that this implies the following conditions:

1) concerning the term which expresses the transition rate

$$|\langle e|Q|g\rangle\mathcal{E}^2| \ll \hbar\Delta\omega_j \quad , \quad \text{i.e.,} \quad a \ll \Delta\omega_j/\sqrt{\Gamma_e\Gamma_{eg}} \quad . \tag{4.19}$$

2) concerning the effective detuning of the two-photon transition

$$\delta E = E_{eg} + s - 2\hbar\omega \ll \hbar\Delta\omega_j \quad . \tag{4.20}$$

These two conditions are contained in the general assumption (4.5).

4.3.3 Case of Two Waves with Complex Polarizations

We have seen in Sect.4.2 why it is interesting to produce two-photon transitions with two electromagnetic waves differing either with respect to direction or to frequency. We suppose now that the atom, in its rest frame, is submitted to two electric fields analogous to (4.4) but with distinct complex polarizations $\underline{\varepsilon}_1$ and $\underline{\varepsilon}_2$, and also distinct frequencies ω_1 and ω_2. The frequency difference $|\omega_1 - \omega_2|$ is assumed to be small compared with the $\Delta\omega_j$, in such a way that we have now three small two-photon detunings:

$$|2\hbar\omega_1 - E_{eg}|, \quad |2\hbar\omega_2 - E_{eg}| \quad \text{and} \quad |\hbar(\omega_1 + \omega_2) - E_{eg}| \ll \hbar\Delta\omega \quad . \tag{4.21}$$

The generalization of the calculation of Sect.4.3.1 permits easy attainment of the matrix elements of the effective Hamiltonian V

$$\langle e|V|e\rangle = \frac{1}{4}\sum_p\sum_q \langle e|Q'_{\varepsilon_p\varepsilon_q^*}|e\rangle\mathcal{E}_p\mathcal{E}_q^* \exp[i(\omega_q - \omega_p)t]$$

$$\langle e|V|g\rangle = \frac{1}{4}\sum_p\sum_q \langle e|Q_{\varepsilon_p\varepsilon_q}|g\rangle\mathcal{E}_p\mathcal{E}_q \exp[-i(\omega_p + \omega_q)t]$$

$$\langle g|V|g\rangle = \frac{1}{4}\sum_p\sum_q \langle g|Q_{\varepsilon_p^*\varepsilon_q}|g\rangle\mathcal{E}_p^*\mathcal{E}_q \exp[i(\omega_p - \omega_q)t] \quad . \tag{4.22}$$

In these expressions the indices p and q take the values 1 and 2; and each matrix element contains four terms, which will be discussed below. The two-photon operators Q are simply straightforward generalizations of (4.12,13), for instance,

$$Q_{\varepsilon_p^*\varepsilon_q} = \underline{D}\cdot\underline{\varepsilon}_p^* \frac{1}{\hbar\omega - H_0} \underline{D}\cdot\underline{\varepsilon}_q \tag{4.23}$$

$$Q'_{\varepsilon_p\varepsilon_q^*} = \underline{D}\cdot\underline{\varepsilon}_p \frac{1}{E_e - \hbar\omega - H_0} \underline{D}\cdot\underline{\varepsilon}_q^* + \underline{D}\cdot\underline{\varepsilon}_q^* \frac{1}{E_e + \hbar\omega - H_0} \underline{D}\cdot\underline{\varepsilon}_p \quad . \tag{4.24}$$

Using (4.21), ω can represent either ω_1 or ω_2. In the diagonal matrix elements, corresponding to the light shifts, we find two terms constant (for p = q) and two imaginary conjugate terms depending on the time (for $p \neq q$). This variation with time can be well understood: the light shift is produced by the total electric field $E_1 + E_2$, which is the sum of two waves with different frequencies; and its instantaneous value varies with the beat frequency of the two waves. Nevertheless in most cases, the amplitude of the light shift is small and its time variation is very fast

$$\left|\frac{\langle e|V/\hbar|e\rangle}{\omega_1 - \omega_2}\right| \quad \text{and} \quad \left|\frac{\langle g|V/\hbar|g\rangle}{\omega_1 - \omega_2}\right| \ll 1 \quad . \tag{4.25}$$

The condition of motional narrowing is thus satisfied [4.57,53] and we can keep only the two constant terms. The light shift is given by the sum of the light shifts of the two waves

$$s = \left(\langle e|Q'_{\varepsilon_1\varepsilon_1}|e\rangle - \langle g|Q_{\varepsilon_1\varepsilon_1}|g\rangle\right)\frac{|\mathcal{E}_1|^2}{4} + \left(\langle e|Q'_{\varepsilon_2\varepsilon_2}|e\rangle - \langle g|Q_{\varepsilon_2\varepsilon_2}|g\rangle\right)\frac{|\mathcal{E}_2|^2}{4} \quad . \tag{4.26}$$

In the off-diagonal matrix element $\langle e|V|g\rangle$ we find:

1) Two terms corresponding to $p = q$, which are resonant when $2\hbar\omega_1 = E_{eg}$ and $2\hbar\omega_2 = E_{eg}$, respectively. They correspond to the absorption of two photons of the same wave, either wave 1, or wave 2.

2) Two symmetrical terms for $p \neq q$, which are resonant under the same condition $\hbar(\omega_1 + \omega_2) = E_{eg}$, and which correspond to the absorption of one photon of each wave. These two terms are explained by the fact that this absorption can be produced by two distinct ways: the atom absorbs first a photon 1 and then a photon 2, or vice versa.

These two symmetrical terms are responsible for the narrow line observed in Doppler-free experiments; and it is convenient to consider them together by introducing the symmetrical two-photon operator

$$Q^S_{\varepsilon_1\varepsilon_2} = \frac{1}{2}\left(\underline{D}\cdot\underline{\varepsilon}_1 \frac{1}{\hbar\omega - H_0} \underline{D}\cdot\underline{\varepsilon}_2 + \underline{D}\cdot\underline{\varepsilon}_2 \frac{1}{\hbar\omega - H_0} \underline{D}\cdot\underline{\varepsilon}_1\right) \quad . \tag{4.27}$$

This new operator will be particularly useful in Sect.4.3.6 in order to obtain the selection rules.

If we suppose that the frequency difference $|\omega_1 - \omega_2|$ is greater than the linewidth

$$|\omega_2 - \omega_1| \gg \Gamma_{eg}\sqrt{1 + a^2} \tag{4.28}$$

then we obtain a steady-state solution of the density matrix containing three independent terms analogous to (4.16) corresponding to the three resonance frequencies. These three terms differ only in the saturation parameters, which are calculated from different operators. The terms resonant at $2\omega_1$, $2\omega_2$ and $\omega_1 + \omega_2$ are represented by the saturation parameters a_{11}, a_{22} and a_{12}, respectively, such that

$$a_{11} = \left| \langle e|Q_{\varepsilon_1\varepsilon_1}|g\rangle \mathcal{E}_1^2 \right| / 2\hbar\sqrt{\Gamma_e\Gamma_{eg}}$$

$$a_{22} = \left| \langle e|Q_{\varepsilon_2\varepsilon_2}|g\rangle \mathcal{E}_2^2 \right| / 2\hbar\sqrt{\Gamma_e\Gamma_{eg}}$$

$$a_{12} = \left| \langle e|Q^S_{\varepsilon_1\varepsilon_2}|g\rangle \mathcal{E}_1\mathcal{E}_2 \right| / \hbar\sqrt{\Gamma_e\Gamma_{eg}} \quad . \tag{4.29}$$

4.3.4 Two-Photon Line Shape in Vapors

Most experiments are performed in vapors, and we must take into account the thermal motion of the atoms. We assume that the motion of each atom is described by a straight path $z = z_0 + v_z t$.

We first study the case of atoms *in a travelling wave*, of wave vector $\underline{k}$ parallel to the z axis, whose electric field can be written in the laboratory frame

$$E(z,t) = \underline{\varepsilon}\mathrm{Re}\left\{\mathcal{E}\cdot e^{i(kz-\omega t)}\right\} \quad . \tag{4.30}$$

Substituting $z_0 + v_z t$ for z in this expression, we obtain the electric field in the atomic frame and we see that its frequency is now $\omega - kv_z$. The density matrix calculated for each atom is exactly the same as (4.16) except for the value of ω which becomes $\omega - kv_z$. The two-photon absorption line shape $A(\omega)$ is obtained by averaging over the velocity distribution $f(v_z)$.

$$A(\omega) = \int_{-\infty}^{+\infty} dv_z f(v_z)\rho_{ee}(\omega - kv_z) \tag{4.31}$$

We assume in the following that $f(v_z)$ is the classical Maxwell-Boltzmann distribution

$$f(v_z) = \frac{1}{u\sqrt{\pi}} \exp\left(-\frac{v_z^2}{u^2}\right) \quad \text{with} \quad u^2 = \frac{2kT}{M} \quad , \tag{4.32}$$

(k is the Boltzmann constant, M the mass of the atoms and T the absolute temperature of the vapor). If we suppose that the natural width and the saturation width are much smaller than the Doppler width, we can integrate (4.31) by replacing $\rho_{ee}(\omega - kv_z)$ by a Dirac δ function

$$A(\omega) = \frac{\sqrt{\pi}}{4} \frac{a^2}{\sqrt{1 + a^2}} \exp\left[-\left(\frac{E_{eg} + s - 2\hbar\omega}{2\hbar ku}\right)^2\right] \quad , \tag{4.33}$$

where the saturation parameter a is given by (4.17) and the light shift s by (4.18). We obtain, as expected for a travelling wave, a Gaussian curve whose width is the Doppler width ($\sim ku$) and whose center is shifted by an amount equal to the light shift of the transition. We remark that at low intensity ($a \ll 1$), the number of atoms excited is proportional to the square of intensity $|\mathcal{E}|^4$, while at higher intensities ($a \ll 1$) the number of atoms increases in proportion to the light intensity $|\mathcal{E}|^2$.

Let us now consider the case of atoms *in a standing wave*, obtained by reflecting a laser beam back on itself, as is commonly done in Doppler-free experiments (see Sect.4.2). In fact it is possible to have different polarizations for the two oppositely travelling waves; and the electric field can be written in the laboratory frame

$$E(z,t) = \mathrm{Re}\left\{\underline{\varepsilon}_1 \cdot \mathcal{E}_1\, e^{i(kz-\omega t)} + \underline{\varepsilon}_2 \cdot \mathcal{E}_2\, e^{-i(kz+\omega t)}\right\} \quad . \tag{4.34}$$

Taking into account the motion of the atom, we obtain in its rest frame two waves of different frequencies $\omega_1 = \omega - kv_z$ and $\omega_2 = \omega + kv_z$. Using the results of Sect.4.3.3, we find the steady-state density matrix of this particular atom

$$\rho_{ee}(\omega,v_z) = \frac{\Gamma_{eg}^2}{4}\left[\frac{a_{12}^2}{\left(\frac{\delta E}{\hbar}\right)^2 + \Gamma_{eg}^2(1 + a_{12}^2)} + \frac{a_{11}^2}{\left[\left(\frac{\delta E}{\hbar}\right) + kv_z\right]^2 + \Gamma_{eg}^2(1 + a_{11}^2)} + \frac{a_{22}^2}{\left[\left(\frac{\delta E}{\hbar}\right) - kv_z\right]^2 + \Gamma_{eg}^2(1 + a_{22}^2)}\right] \quad . \tag{4.35}$$

The right-hand side of this formula depends on the light frequency ω through the effective energy detuning δE (taking into account the light shift), which is given by (4.20,26). The saturation parameters a_{12}, a_{11} and a_{22} are given by (4.29). We do not forget the validity conditions (4.25,28) of the calculations leading to this formula. They are verified as long as the light intensity is not too high, in such a way that the saturation width and the light shifts are much smaller than the Doppler width ku; under that condition we can exclude the few slow atoms, whose individual Doppler shifts kv_z are less than the saturation or natural widths. It is important to notice that the first term of (4.35) does not depend on the velocity. It is precisely this term which corresponds to the absorption of two photons of opposite directions, leading to the cancellation of the Doppler broadening.

We must now average (4.35) over the velocity distribution (4.32). Using a formula similar to (4.31) we deduce the line shape

$$A(\delta E) = \frac{\Gamma_{eg}^2 a_{12}^2}{\left(\frac{\delta E}{\hbar}\right)^2 + \Gamma_{eg}^2(1 + a_{12}^2)} + \frac{\sqrt{\pi}\Gamma_{eg}}{4ku}\left(\frac{a_{11}^2}{\sqrt{1 + a_{11}^2}} + \frac{a_{22}^2}{\sqrt{1 + a_{22}^2}}\right)\exp\left[-\left(\frac{\delta E}{2\hbar ku}\right)^2\right] \quad . \tag{4.36}$$

The line shape appears thus as a superposition of a narrow Lorentzian peak and a broad Gaussian Doppler line. The narrow peak comes from the absorption of two photons propagating in opposite directions while the broad Gaussian curve corresponds to the absorption of two photons propagating in the same direction.

In the low intensity limit (parameters $a \ll 1$) we find the classical formulas of [4.8,9]. In order to compare the two curves it is convenient to deduce from (4.36) the ratio of their areas (see Fig.4.4): the ratio of the area of the Lorentzian curve to that of the Gaussian is equal to $a_{12}^2/(a_{11}^2 + a_{22}^2)$. If the two waves have the same polarization and the same intensity, ($a = \frac{1}{2} a_{12} = a_{11} = a_{22}$), the area of the Lorentzian peak is two times larger than the area of the Gaussian curve. That can be understood easily [4.9]: as we have noted in the preceding section, the absorption of two oppositely propagating photons (corresponding to the narrow Lorentzian peak) may occur in two ways: the atom absorbs first one photon 1 and then a photon 2; or in the reverse order. These two processes are not independent: we must first add their equal amplitudes a, and then square to obtain the probability $4a^2$. On the contrary, the two processes corresponding to the Gaussian curve are independent and we add their probabilities a^2.

When the light intensity is above the saturation limit ($a \gg 1$), (with the same polarization for the two waves) the two areas become equal. But the two curves do not evolve in an identical way: the intensity of the Lorentzian peak does not vary while its width increases proportionally to the intensity. On the other hand, the width of the Doppler curve does not change while its intensity increases proportionally to the intensity.

If the two waves have different polarizations, it is possible to find particular situations where $a_{11} = a_{22} = 0$ while a_{12} is not zero; this leads to the elimination of the Doppler background [4.9]. An example of this effect has been demonstrated with two light beams of opposite circular polarizations σ^+ and σ^-, producing a two-photon transition between two S levels of zero orbital angular momentum ($L = 0$): it is not possible to absorb two photons of the same circular polarization because such a process corresponds to a selection rule $\Delta m_L = \pm 2$ which cannot be satisfied (see remark at the end of Sect.4.3.6). It implies that the only process which is allowed is the absorption of two photons σ^+ and σ^- coming from the two counter-propagating waves because it permits the conservation of angular momentum. This situation permits one to observe only the narrow Lorentzian peak [4.18,19].

We have assumed in this section, that the atoms interact all the time with electromagnetic waves. But this is not true when the light beam is focused in order to increase the energy density (see Sect.4.2.2). In that case, the transit time of the atom through the light beam can be less than the radiative lifetime $1/\Gamma_e$; a more precise treatment is necessary, taking into account the spatial variation of the Gaussian light beam: the exact line shape obtained for the Doppler-free narrow peak is then the convolution of the Lorentzian curve in (4.36) with a double exponential curve [4.58,59] in good agreement with experiment [4.59].

4.3.5 Light Shifts

In this section we discuss in more detail the problem due to light shifts, using the formulae obtained in Sects.4.3.2 and 3. The fact that both the saturation parameter a and the shift s are proportional to the light intensity has the following important consequence: it is generally possible to find experimental conditions far below the saturation limit, where the light shift is much smaller than the natural width of the transition [4.9]. This result is of course fundamental because it implies that for most spectroscopic applications the method can be used without troubles.

Far below saturation ($a \ll 1$) the relative population of the excited level N_e/N_0 (number of atoms in the e level over the total number of atoms) at resonance is obtained, using (4.16,17)

$$\frac{N_e}{N_0} = \frac{a^2}{2} = \frac{|\langle e|Q|g\rangle \mathcal{E}^2|^2}{8\hbar^2 \Gamma_e \Gamma_{eg}} \quad . \tag{4.37}$$

Combining this expression and (4.18) giving the shift (for the sake of simplicity, we assume that the relaxation is only due to spontaneous emission so that $\Gamma_{eg} = \Gamma_e/2$) we find

$$\left(\frac{s}{\hbar\Gamma_{eg}}\right)^2 = \left|\frac{\langle e|Q'|e\rangle - \langle g|Q|g\rangle}{\langle e|Q|g\rangle}\right|^2 \frac{N_e}{N_0} = R \cdot \frac{N_e}{N_0} \quad . \tag{4.38}$$

The dimensionless ratio R depends on the wave functions of the atom. We can estimate its order of magnitude in the particular case where the two-photon transition probability can be calculated *using the one-intermediate-state-approximation*, i.e., where one particular relay level r (between g and e) gives the essential contribution to the transition probability. If we assume that this level gives also the main contribution to the light shifts of level e ang g, using (4.12,13) we obtain

$$\langle e|Q'|e\rangle - \langle g|Q|g\rangle \simeq \frac{|\langle e|\underline{D}\cdot\underline{\varepsilon}|r\rangle|^2}{E_e - E_r - \hbar\omega} - \frac{|\langle r|\underline{D}\cdot\underline{\varepsilon}|g\rangle|^2}{\hbar\omega - E_{rg}} \simeq \frac{|\langle e|\underline{D}\cdot\underline{\varepsilon}|r\rangle|^2}{\hbar\omega - E_{rg}} - \frac{|\langle r|\underline{D}\cdot\underline{\varepsilon}|g\rangle|^2}{\hbar\omega - E_{rg}}$$

$$\langle e|Q|g\rangle = \frac{\langle e|\underline{D}\cdot\underline{\varepsilon}|r\rangle\langle r|\underline{D}\cdot\underline{\varepsilon}|g\rangle}{\hbar\omega - E_{rg}} \quad .$$

We can express the matrix elements of the electric dipole moment in terms of the oscillator strengths f_{rg} and f_{er} of the one-photon transitions $g \to r$ and $r \to e$. The formula (4.38) becomes

$$\left(\frac{s}{\hbar\Gamma_{eg}}\right)^2 = R \cdot \frac{N_e}{N_0} = \frac{(E_{rg}f_{er} - E_{er}f_{rg})^2}{E_{er}E_{rg}f_{er}f_{rg}} \cdot \frac{N_e}{N_0} \quad . \tag{4.39}$$

If the two oscillator strengths f_{er} and f_{rg} of the one-photon transitions are of the same order of magnitude and close to each other, the ratio R can be very close to zero: the light shift of the excited level compensates that of the ground level; and the resulting shift for the two-photon transition is very small. Such cases can be especially useful in metrological applications.

If the two oscillator strengths f_{er} and f_{rg} are quite different, it is easy to show that the ratio R is almost equal to the ratio of the greatest f value over the smallest one. On the other hand, the experiments show that two-photon transitions are easily detected, with a good signal-to-noise ratio, in conditions where the relative population of the excited level N_e/N_0 is as small as 10^{-6}, or even 10^{-8}. In that condition we see that the light shift will be extremely small (compared to the linewidth) even in experiments where the ratio R of the oscillator strengths is of the order of 10^3.

The formulae used in this discussion are valid if the light wave is strictly monochromatic, that is to say if the spectral width of the laser $\Delta\omega_L$ is much smaller than the natural linewidth ($\Delta\omega_L \ll \Gamma_e$). In the contrary case ($\Delta\omega_L \gg \Gamma_e$) the above conclusions do not hold. It can be shown that, for the same light intensity, (that is to say the same light shift s), the transition rate N_e/N_0 is divided by the ratio $\Delta\omega_L/\Gamma_e$ of the laser spectral linewidth to the natural linewidth (case of incoherent light); we deduce that the ratio R of (4.39) must be then multiplied by $\Delta\omega_L/\Gamma_e$.

Comparison with Experiments

Up to now, there have been no systematic measurements of the light shift for Doppler-free two-photon transitions in a standing wave. Even in the more precise experiments performed recently [4.24,29], the problem of the dynamic Stark effect was just mentioned and it seems that its value was below the limit of resolution of the apparatus. The experiments confirm the preceding conclusions.

To observe important light shifts in a two-photon transition it is necessary to make the system asymmetrical. For instance, if the two counterpropagating waves have very different intensities ($\mathcal{E}_2 \gg \mathcal{E}_1$), we obtain in the effective Hamiltonian (4.22) very large diagonal terms $\langle e|Q'_{\varepsilon_2\varepsilon_2}|e\rangle|\mathcal{E}_2|^2$ and $\langle g|Q_{\varepsilon_2\varepsilon_2}|g\rangle|\mathcal{E}_2|^2$, leading to an important shift (4.26), while the nondiagonal term which leads to the Doppler free line line $\langle e|Q^S_{\varepsilon_1\varepsilon_2}|g\rangle\mathcal{E}_1\mathcal{E}_2$ remains rather small. This implies that the ratio R in (4.38) is multiplied by $|\mathcal{E}_1|^2/|\mathcal{E}_2|^2$; and for the same value of N_e/N_0 the light shift is $|\mathcal{E}_1|/|\mathcal{E}_2|$ times larger.

A similar device has been used by LIAO and BJORKHOLM to observe the light shift [4.60]. In order to increase the transition probability, the frequencies of the beams 1 and 2 were slightly different to obtain the energy $\hbar\omega_2$ very close to the energy level of the intermediate state of the two-photon transition. The experimental setup is very similar to the one described in Sect.4.2.5; the beams coming from two lasers have frequencies ω_1 and ω_2 so that $\hbar\omega_1 + \hbar\omega_2$ coincide with the energy of the $3S_{1/2}(F = 2) - 4D_{5/2}$ transition in sodium. The energy defect $\hbar\Delta\nu$ between $\hbar\omega_2$ and $(E_{3P_{3/2}} - E_{3S_{1/2}})$ was kept very small (5 GHz$<|\Delta\nu|<$30 GHz).

The power of laser 1 is kept constant and has a very low value, while the power of laser 2 varies in the range ($2\cdot10^2 - 2\cdot10^3$ W/cm^2). The shift of the two-photon transition is clearly observed up to a maximum value of the order of 1 GHz. For a constant energy defect $\hbar\Delta\nu$, the shift is proportional to the intensity of laser 2; with constant intensities of the two lasers, it is proportional to the inverse of the energy defect $\hbar\Delta\nu$.

4.3.6 Selection Rules for Two-Photon Transitions

In this section, we present the selection rules for two-photon excitation. The knowledge of the selection rules is important in order to know which levels can be populated by multiphoton excitation. We obtain these selection rules by using the tensorial formalism technique. This tensorial formalism is also of great interest for the calculation of the line intensities and of the two-photon pumping matrix. These aspects will not be presented here; they can be found in the original papers [4.44,56] and in the review papers [4.20,61]. The probability of two-photon excitation depends on the value of the matrix element of the symmetrical operator $Q^S_{\varepsilon 1 \varepsilon 2}$ (4.27) between the levels e ang g. If the value of the matrix element is 0, it is clear that there is no transition. We analyze the tensorial properties of the $Q^S_{\varepsilon 1 \varepsilon 2}$ operator and we deduce which transitions are forbidden.

As we know, the operator $Q^S_{\varepsilon 1 \varepsilon 2}$ is symmetrical with respect to the exchange of the two polarizations $\underline{\varepsilon}_1$ and $\underline{\varepsilon}_2$. This property is associated with the choice of the same frequency for the two oppositely travelling waves. If one uses waves of different frequencies (Sect.4.2.5), it is necessary to define a different operator which is not always symmetrical. The rank of the operator $Q^S_{\varepsilon 1 \varepsilon 2}$ is less than or equal to 2 (one obtains $Q^S_{\varepsilon 1 \varepsilon 2}$ by coupling the vectorial operator $\underline{D}$ twice and the scalar operator $[\hbar\omega - \mathcal{H}_0]^{-1}$ once). But there is no tensor of rank 1 in the decomposition of $Q^S_{\varepsilon 1 \varepsilon 2}$ because it is symmetrical: $Q^S_{\varepsilon 1 \varepsilon 2}$ is the sum of a scalar operator and of a quadrupolar operator.

The result on the rank of $Q^S_{\varepsilon 1 \varepsilon 2}$ demonstrated here is theoretically only true if the angular momentum under consideration is the total angular momentum F = I + J (where I and J are the nuclear and electronic angular momenta): $[\hbar\omega - \mathcal{H}_0]^{-1}$ is invariant in a rotation of both nuclear and electronic variables. Nevertheless, in most usual cases, these conclusions are equally true if $Q^S_{\varepsilon 1 \varepsilon 2}$ is based on J. This is true if the hyperfine structure of the intermediate state r is much smaller than the energy defect $\hbar\Delta\omega_r$ of the single-quantum transition. That result can be understood as follows: owing to the uncertainty principle, the atom stays in the state r a time of the order of $1/\Delta\omega_r$; if this quantity is small enough, the nuclear and electronic momenta have not enough time to be coupled. In the following, this assumption will always be considered as being true. In a similar way, for an atom in LS coupling, if the fine structure of the state is small compared to $\hbar\Delta\omega_r$, the preceding results are equally valid, based on L.

In order to find the selection rule, we apply the Wigner-Eckart theorem to the $\langle e\ J_e\ m_e | Q^S_{\varepsilon 1 \varepsilon 2} | g\ J_g\ m_g \rangle$ matrix element. Because of the decomposition of $Q^S_{\varepsilon 1 \varepsilon 2}$ in terms of tensors of rank 0 and 2, we obtain [4.9]

$$|J_e - J_g| \leq 2 \qquad (4.40)$$

$$J_\alpha = 0 \leftrightarrow J_\beta = 1 \quad \text{forbidden} \quad (\alpha,\beta = g,\ e \text{ or } e,g) \quad .$$

These two last selection rules may look surprising because it is well known that, for instance, the 7^3S_1 level of mercury can be excited from the ground state 6^1S_0 by a two-step excitation (see Fig.4.14). There is no two-photon transition between these two states because there is always a destructive interference between the various possible paths. For instance, if one wave is polarized σ^+ and the other one σ^-, there are two possible paths (Fig.4.14b). If all the photons have the same frequency, it is easy to show that the probability amplitude of path $(\sigma^-\sigma^+)$ is just opposite to the one of path $(\sigma^+\sigma^-)$ [4.56].

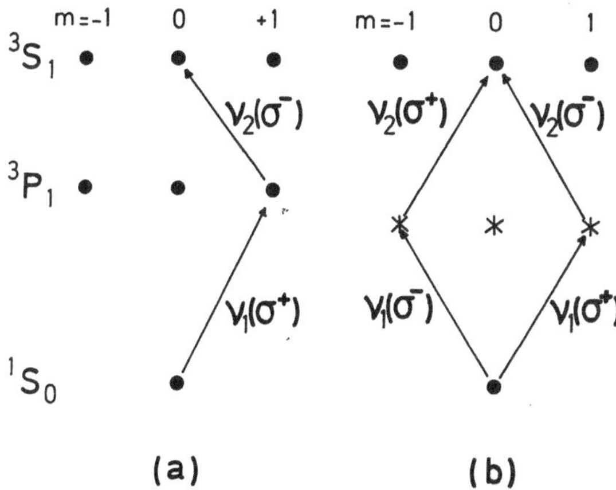

Fig.4.14. Schematic energy diagram showing the comparison between the theoretical calculations of stepwise excitation, in a transition from J = 0 to J = 1: (a) in the case of stepwise excitation, there is a resonant path with a photon σ^+ first absorbed and then a photon σ^-; (b) in the case of two-photon excitation, there are two paths of equal amplitude, but opposite in sign

That property is a consequence of the fact that the photons have the same frequency. If the photons σ^+ and σ^- have different frequencies ω_+ and ω_-, the energy denominators in the transition probability formula will be different ($\omega_+ - \omega_{rg}$ and $\omega_- - \omega_{rg}$); one of the paths is then favored. Furthermore, this example shows that if the two opposite waves do not have the same frequency, a tensorial component of rank 1 may appear in the decomposition of the two-photon operator.

Remark: The selection rules demonstrated for J are, of course, equally true for F. Nevertheless it is sometimes possible to find more precise selection rules. For instance, if the two levels have the same J value, and if $J_g = J_e = 0$ or $\frac{1}{2}$, then only a scalar operator can couple the ground state and the excited state. This implies the following selection rules for F [4.9]

$$\Delta F = 0 \quad (F_e = F_g)$$
$$\Delta m_F = 0 \quad (m_{Fe} = m_{Fg}) \quad . \tag{4.41}$$

These selection rules have been verified for a large number of S-S two-photon transitions in sodium [4.18,19,26,62] and in hydrogen [4.34,35] (see also Fig.4.9). If the levels have different J values, the selection rules are those of an electric quadrupolar transition [4.40].

4.4 Multiphoton Transitions

In Sect.4.3 we have restricted ourselves to the theory of two-photon transitions, presenting a treatment in some detail. We now generalize the ideas introduced previously to the case of n-photon transitions with $n \geq 3$. The complexity of the problems encountered does not allow us to give here any detailed calculation. However, some general ideas will be enough to make comparison and to evaluate orders of magnitude.

4.4.1 Generalization of the Effective Hamiltonian

We generalize here the calculations of Sect.4.3.1, using the same notation. For the sake of simplicity, most of the formulae will be developed for a three-photon transition, however, the generalization to n photons is obvious. We assume as before that the levels g and e are nondegenerate. The atom interacts with three waves, which are written in the rest frame of the atom

$$\underline{E}_p(t) = \mathrm{Re}\left\{\underline{\varepsilon}_p \mathcal{E}_p \exp(-i\omega_p t)\right\} \quad p = 1,2,3 \tag{4.42}$$

and we assume that neither one photon nor two photons correspond to a resonant transition for the atom. If this assumption is true, we can derive an effective Hamiltonian V coupling the levels g and e, similar to (4.11,22), whose matrix elements are

$$\langle e|V|e\rangle = \frac{1}{4}\sum_p\sum_q \langle e|Q'_{\varepsilon_p\varepsilon_q^*}|e\rangle \mathcal{E}_p\mathcal{E}_q^* \exp[i(\omega_q - \omega_p)t]$$

$$\langle e|V|g\rangle = \frac{1}{2^3}\sum_p\sum_q\sum_r \langle e|Q^{(3)}_{\varepsilon_p\varepsilon_q\varepsilon_r}|g\rangle \mathcal{E}_p\mathcal{E}_q\mathcal{E}_r \exp[-i(\omega_p + \omega_q + \omega_r)t]$$

$$\langle g|V|g\rangle = \frac{1}{4}\sum_p\sum_q \langle g|Q_{\varepsilon_p^*\varepsilon_q}|g\rangle \mathcal{E}_p^*\mathcal{E}_q \exp[i(\omega_p - \omega_q)t] \quad . \tag{4.43}$$

The diagonal matrix elements are very similar to those of (4.22) and they contain the same operators $Q_{\varepsilon_p^*\varepsilon_q}$ and $Q'_{\varepsilon_p\varepsilon_q^*}$ defined by (4.23,24).

The off-diagonal matrix elements involve a new operator

$$Q^{(3)}_{\varepsilon_p\varepsilon_q\varepsilon_r} = \underline{D}\cdot\underline{\varepsilon}_p \frac{1}{\hbar(\omega_q + \omega_r) - H_0} \underline{D}\cdot\underline{\varepsilon}_q \frac{1}{\hbar\omega_r - H_0} \underline{D}\cdot\underline{\varepsilon}_r \quad . \tag{4.44}$$

This off-diagonal term comprises a summation over various paths. For the sake of simplicity we suppose that only the path 1-2-3 has a siginificant contribution since it is resonant ($E_{eg}/\hbar \simeq \omega_1 + \omega_2 + \omega_3$) and, also the intermediate relay levels favor the order 3-2-1. In this case we can calculate the line shape as done in Sect.4.3.2. We obtain for the fraction N_e/N_0 of the excited atoms a formula similar to (4.16) except that the energy detuning in the denominator is now $[E_{eg} + s - \hbar(\omega_1 + \omega_2 + \omega_3)]$, and the saturation parameter a^2 becomes

$$a^2_{123} = 4\left|\langle e\, Q^{(3)}_{\varepsilon_1\varepsilon_2\varepsilon_3}\, g\rangle \frac{\mathcal{E}_1\mathcal{E}_2\mathcal{E}_3}{2^3}\right|^2 \Big/ \hbar^2\Gamma_e\Gamma_{eg} \quad . \tag{4.45}$$

(In case of comparable contributions from several paths, we must add the various amplitudes before squaring). The generalization of (4.44,45) to the case of n photons is obvious.

The light shift s, which occurs in the energy detuning, is given by a straightforward generalization of (4.27). It must not be forgotten that this simple expression (4.27) is valid only if the condition of motional narrowing (4.25) is satisfied; if this condition is not satisfied, we must include the time-dependent terms in the diagonal matrix elements (4.43).

4.4.2 Discussion of the Light Shifts

It may be surprising to find us again discussing light shifts, since the light shift formula presented above is not different from that found in the case of two-photon transitions. The important change, however, lies in the off-diagonal terms of (4.43), and consequently in the light intensity required to produce observable transitions.

In the case of an n quantum transition, the diagonal matrix elements of the effective Hamiltonian V (light shift) vary as the intensity $\mathcal{E}^2$, while the off-diagonal elements (saturation) are proportional to $\mathcal{E}^n$. It follows that, contrary to the case of two-photon absorption, where the light shift and the broadening are of the same order of magnitude, the light shifts, in the n-photon case, can be very important even if this transition is far below the saturation [4.9]. In practice, it is often necessary to tolerate a sizeable light shift in order to obtain a nonnegligible transition rate N_e/N_0. It is not possilbe in the general n-photons case to give a precise discussion of these effects as was done in Sect. 4.3.5 for two photons; but we can still present some general ideas.

At resonance and in the case of nonsaturation the fraction of atoms in the excited state produced by an n-photon transition can be deduced from (4.16,45)

$$\frac{N_e}{N_0}(n) = \frac{a^2}{2} = 2\left|\langle e|Q^{(n)}_{\varepsilon_1\ldots\varepsilon_n}|g\rangle\left(\frac{\mathcal{E}}{2}\right)^n\right|^2 \Big/ \hbar^2\Gamma_e\Gamma_{eg} \quad . \tag{4.46}$$

Let $\hbar\mathcal{H}$ denote a typical matrix element of the coupling Hamiltonian ($-\underline{E}\cdot\underline{D}$) between the atom and the electromagnetic field, and $\hbar\Delta\omega$ a typical energy defect for the various intermediate levels E_j which are involved in one, two, ... photon transitions. Then an order of magnitude calculation of $N_e/N_0(n)$ gives

$$\frac{N_e}{N_0}(n) \sim \frac{\mathcal{H}^{2n}}{\Delta\omega^{2n-2}\Gamma_e^2} \quad . \tag{4.47}$$

Similarly, we obtain a rough expression for the light shift

$$\frac{s}{\hbar} \sim \frac{\mathcal{H}^2}{\Delta\omega} \tag{4.48}$$

and we deduce

$$\left(\frac{s}{\hbar\Gamma_e}\right)^n \sim \left(\frac{\Delta\omega}{\Gamma_e}\right)^{n-2} \frac{N_e}{N_0}(n) \quad . \tag{4.49}$$

In the case of two photons, the ratio $s/\hbar\Gamma_e$ of the light shift to the natural width is of the order of the transition rate N_e/N_0 (with this rough approximation, the R factor of (4.38) is equal to unity). In an n-photon transition, for the same transition rate, the light shift is multiplied by a huge factor, $(\Delta\omega/\Gamma_e)^{(1-2/n)}$ which increases with the number n of photons up to $\Delta\omega/\Gamma_e \sim 10^{+7}$ or 10^{+8} (in the sort currently investigated).

Assuming that the relative population of the excited state has the same small value as in Sect.4.3.5, i.e., $N_e/N_0 \sim 10^{-6}$ to 10^{-8}, we find by calculation for 3 photons a shift of the order of the natural width, and for 4 photons a shift one hundred times larger, i.e., of the order of the Doppler width.

Case of a Standing Wave

The comparison of the shift with the Doppler width is of particular interest when the n-photon transitions are produced by a standing light wave; this is the case if we try to observe Doppler-free n-photon transitions by using the same laser to produce two or more photons among the n photons involved in the transition. As explained in Sect.4.2.1, we must use several light beams having different directions of propagation, chosen in such a manner that $\sum\underline{k}_i = 0$ (4.2); two beams with different wave vectors $\underline{k}_1$ and $\underline{k}_2$, but the same frequency $\omega = \hbar|\underline{k}_1| = \hbar|\underline{k}_2|$ create a standing wave.

In that case, an atom of velocity $\underline{V}$ is submitted in its rest frame to two waves whose frequencies are slightly different

$$\omega_2 - \omega_1 = (\omega - \underline{k}_2\cdot\underline{V}) - (\omega - k_1\underline{V}) = (\underline{k}_1 - \underline{k}_2)\cdot\underline{V} \quad .$$

This frequency difference is smaller than or of the same order of magnitude as the Doppler width. If the light shift is of the order of or larger than the Doppler width, the condition of motional narrowing (4.25) is no longer satisfied; and we must take into account the time-dependent terms in the diagonal matrix elements of the effective Hamiltonian (4.43). These terms produced then, not only a shift but also a broadening and an asymmetrical shape of the spectral line.

The problem may also be discussed in the laboratory frame: the intensity of the standing wave appears to be variable in space with a spatial periodicity of the order of the wavelength $\lambda = 2\pi/k$. The moving atom sees a light shift variable in time with a temporal period of the order of $T \sim \lambda/V \sim 1/kV$; and the light shift is modulated with the same period. If this modulation of the atomic energy has a small amplitude, the moving atom is sensitive only to the mean value (motional narrowing). This is no longer true, however, if the modulation amplitude is larger than the modulation frequency kV. In this case the slow atoms will contribute more to the line shape than the fast atoms, thus resulting in a broadening and an asymmetry of the line shape [4.53].

4.4.3 Application to Multiphoton Ionization

The importance of light shifts in n-photon transitions is well illustrated by the multiphoton ionization experiments *using an intermediate resonant step*, which have been mentioned in Sect.4.1.2. Figure 4.2 shows, as an example, the number of caesium ions produced by the absorption of four photons produced by a Nd glass laser ($\lambda = 1,06\mu$) as a function of the laser frequency. (The experiment is performed with a travelling wave). It shows a big enhancement in the neighborhood of the three-photon transition 6S → 6F. A comparison of the various curves corresponding to different intensities shows that the center of the resonance is shifted by an amount which is proportional to the intensity. For an intensity I = 1 GW/cm^2, the wavelength of excitation for the 6S - 6F three-photon transition in caesium is reduced by an amount larger than 1 Å.

This important light shift explains the strange behavior of the order of nonlinearity of a multiphoton ionization process near an n-photon resonance [4.63]. The order of nonlinearity K is defined by

$$K = \frac{\partial \log Ni}{\partial \log I} \, , \tag{4.50}$$

where Ni is the number of ions obtained in the multiphoton ionization and I is the light intensity. Far away from any intermediate resonance, K is equal to the number of photons Ko which is needed for photoionization (do not confuse Ko for ionization with n for the intermediate step). But close to a resonance, this is not true. Because of the light shift, the effective detuning from the resonance, analogous to (4.20), increases or decreases, depending on the sign of the laser detuning ($E_{eg} - n\hbar\omega$). This explains why, on one side of the resonance ($E_{eg} = n\hbar\omega$), K is much larger than Ko while, on the other side, K is smaller than Ko. Figure 4.15 shows the variation of K as a function of the laser detuning $E_{eg} - n\hbar\omega$ under the same

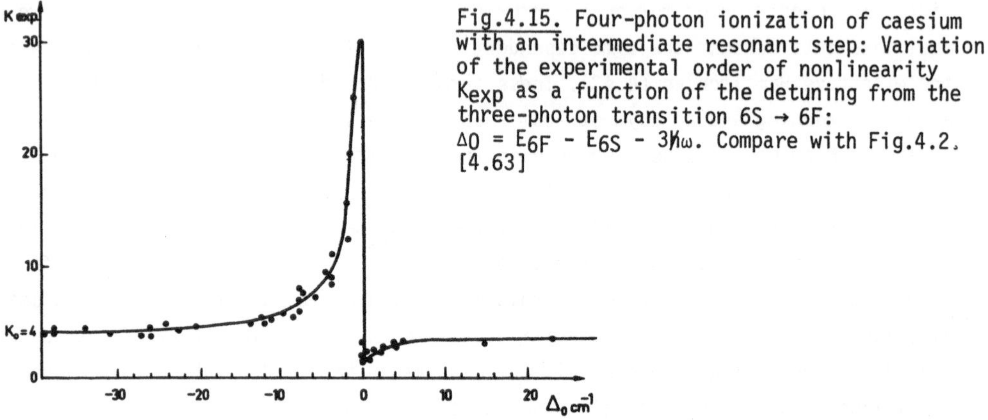

Fig.4.15. Four-photon ionization of caesium with an intermediate resonant step: Variation of the experimental order of nonlinearity K_{exp} as a function of the detuning from the three-photon transition 6S → 6F: $\Delta_0 = E_{6F} - E_{6S} - 3\hbar\omega$. Compare with Fig.4.2. [4.63]

experimental conditions as for Fig.4.2. Figure 4.15 is quite well explained by Fig.4.2.

Remark: An account of the exact shape of ionization curves exemplified in Fig.4.2 is beyond the scope of this chapter. But, the asymmetry shown in the figure can be well interpreted [4.64] as a consequence of the interference between two possible paths for photoionization: for instance, in the case of the four-photon ionization of C_S the first path takes as virtual intermediate states all the levels of the atom except the quasi-resonant, 6F level, and we call this the nonresonant path. The second path, which we call the resonant path, takes as the third intermediate level, the 6F level. If the final state of the ionized atom is the same, the two paths interfere and since the sign of the resonant path is different on either side of the resonance, we expect to find interference which is constructive on one side and destructive on the other. Such asymmetries are observed on resonant curves in many nonlinear processes (see, e.g., Fig.4.12).

4.4.4 Doppler-free Three-Photon Transition

As we have shown in Sect.4.2.1 the cancellation of Doppler broadening is not limited to two-photon transitions and the method can, in theory, be extended to any n-photon transition ($n > 2$). The interest of such an extension is obvious: using two-photon absorption one can investigate states of the same parity as the ground state, whereas in an n-photon absorption (with n odd) it is possible to study states of the opposite parity. Nevertheless, a three-photon experiment is generally more difficult to realize than a two-photon experiment due to a smaller transition probability. To avoid the use of high-power lasers we have performed an experiment with small values of the energy defects $\Delta\omega_j$. That is why we have performed a three-photon excitation of the $3P_{1/2}$ resonance level of sodium with two photons absorbed and one photon emitted [4.21]. The principle of the experiment is shown in Fig.4.16. The three

Fig.4.16 ▼ Fig.4.17 ►

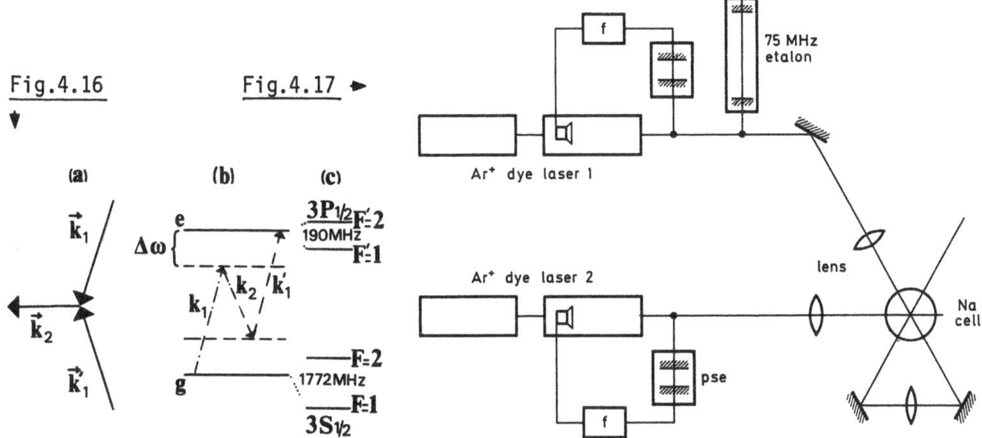

Fig.4.16. Principle of the Doppler-free three-photon experiment in sodium. (a) Spatial orientation of the wave vectors in order to eliminate the Doppler broadening; (b) schematic diagram of energy (two photons absorbed $\hbar k_1$ and $\hbar k_1'$; one photon emitted $\hbar k_2$; (c) real energy levels for the sodium experiment

Fig.4.17. Experimental setup for Doppler-fre three-photon transition following the mechanism of Fig.4.16

wave vectors $\underline{k}_1$, $\underline{k}_1'$ are disposed in such a way that the total momentum exchanged between the atom and photons is 0

$$\hbar\underline{k}_1 + \hbar\underline{k}_1' - \hbar\underline{k}_2 = 0 \qquad \text{(see Fig.4.16a)} \quad . \tag{4.51}$$

The conservation of energy requires

$$E_e - E_g = \hbar ck_1 + \hbar ck_1' - \hbar ck_2 \qquad \text{(see Fig.4.16b)} \quad . \tag{4.52}$$

In this case, the intermediate states involved in the three-photon process are e and g. (Due to the smallness of $\Delta\omega$, we do not need to take into account other levels in the calculation).

In the experiment, the light beams come from two different lasers (see Fig. 4.17), so that $\hbar ck_1 = \hbar ck_1' \neq \hbar ck_2$. By scanning the frequency of one laser, the frequency of the second being fixed, we observe a Doppler-free three-photon spectrum. The resonance is detected by collecting photons spontaneously emitted from the $3P_{1/2}$ level.

Figure 4.17 shows the experimental setup. The two tunable light sources are CW dye lasers similar to those described in Sect.4.2.2. Each laser is locked onto the side of the transmission peak of a pressure-swept Fabry-Pérot etalon. The laser powers are about 60 mW (laser 1) and 30 mW (laser 2) and the linewidth is the same for both lasers (about 7 MHz). The light coming from laser 1 is focused into the sodium cell; then, using a set of mirrors and a lens, the transmitted light is refocused at the same point of the sodium cell, the angle between the two beams being equal to 120°. The beam coming from laser 2 is parallel to the bisectrix of the preceding beams and is focused at the same point. To select the fluorescent light at the atomic frequency in the presence of Rayleigh scattering at the laser frequency, a high resolution monochromator is used.

Figure 4.18 shows the signal recorded when the frequency of laser 1 is scanned, while the frequency of the laser 2 is fixed at 30 GHz (or 1 cm^{-1}) below the sodium resonance frequency ($3S_{1/2}$ - $3P_{1/2}$). The four peaks correspond to the transitions from the two hyperfine sublevels (F = 1 and F = 2) of the ground state to the two hyperfine sublevels (F' = 1 and F' = 2) of the excited state. When one of the three beams is interrupted, the resonances disappear. Even a slight modification of the relative position of the focus leads to an abrupt decrease of the signal. The width of each resonance is about 60 MHz. This is much less than the Doppler width (2000 MHz) but it is more than one expects from the linewidth of the lasers and from the width of the excited level. The two main sources of broadening seem to be:

I) A residual Doppler effect due to the fact that the angles between the laser beams may be slightly different from 60° (the residual Doppler effect due to the focusing is much smaller, of the order of 2 MHz).

II) The light shifts: in Sect.4.4.2 we have shown that the light shifts, which are generally small in the case of two-photon transitions, may be large in the case of three photons, and the light shifts contribute to the width as we will see below.

These light shifts have been studied in more detail [4.45] by choosing an energy defect three times smaller, $\Delta\nu = \Delta\omega/2\pi = 5$ GHz (i.e., the laser 2 frequency is fixed 10 GHz below the sodium resonance frequency). The power of laser 2 is kept constant while the Doppler-free three-photon spectrum is recorded for different values of the power of laser 1. For instance, we show in Fig.4.19 the two low-frequency components of the three-photon transition recorded for $I_1 \sim 250$ Watt/cm^2 (upper curve) and for $I_1 \sim 450$ Watt/cm^2 (lower curve). There is an evident shift of the resonance. The value of the shift is of the order of 0.1 GHz; this is still small compared to the Doppler width ($\sim$2 GHz), and is why a narrow line is observed.

Apart from the shift there is also an obvious broadening of the line. There are several reasons for this broadening: 1) the intensity is not uniform within the region of observation because of the Gaussian spatial character of the exciting beams; 2) the light shift changes from one velocity group to another because the Doppler broadening is not very small compared to the detuning $\Delta\nu$; 3) the various Zeeman sublevels of the excited state can experience different light shifts, because the hyperfine splitting of the ground level (1.77 GHz) is not very small compared to $\Delta\nu$ (when $\Delta\nu$ is large compared to 1.77 GHz we can neglect the hyperfine splitting and the tensor which represents the light shift has a spherical symmetry [4.65]);4) the light shift is perhaps not small enough to neglect the influence of the atoms of very small velocity for which there is no motional narrowing (see Sect.4.4.2).

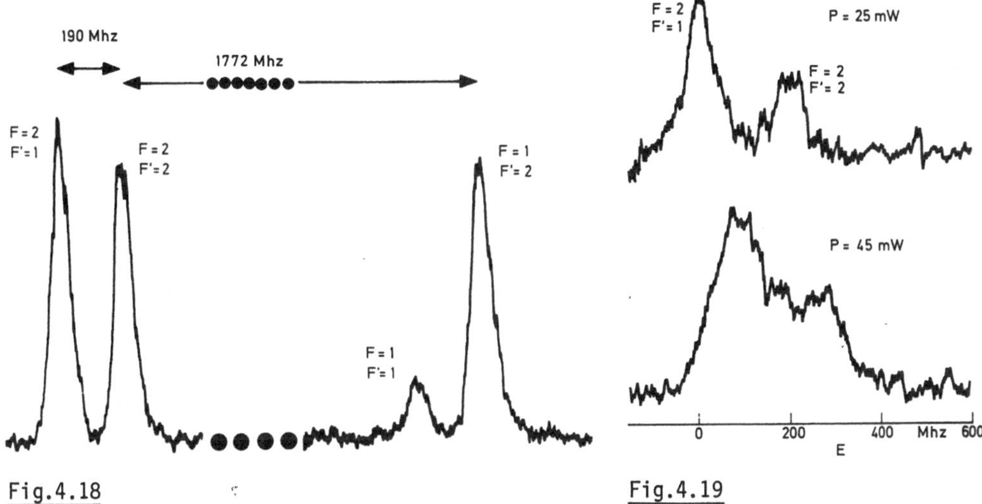

Fig.4.18 Fig.4.19

Fig.4.18. Experimental recording of the Doppler-free three-photon spectrum, following the mechanism of Fig.4.16. The frequency ν_2 of the emitted photon is fixed at $\nu_0 - \nu_2 = 2(\Delta\omega/2\pi) = 30$ GHz where ν_0 is the frequency of the sodium D_1 line. The resonances are observed by scanning the common frequency ν_1 of the two absorbed photons. The separation of the lines are indicated for the atomic frequency. [4.21]

Fig.4.19. Recording of two components of the three-photon transition $3S \rightarrow 3P_{1/2}$ in sodium ($F = 2 \rightarrow F' = 1$ and $F = 2 \rightarrow F' = 2$) for a smaller detuning of laser 2: $\nu_0 - \nu_2 = 2(\Delta\omega/2\pi) = 10$ GHz. The two curves are obtained for the same power of laser 2 but different powers of laser 1; at higher power the resonances are shifted to higher frequencies (in abcissa: the atomic frequency $2\nu_1$). [4.45]

This experiment demonstrates that Doppler-free multiphoton spectroscopy is not restricted to two-photon transition. Three-photon spectroscopy is complementary to two-photon spectroscopy because it permits the study of states of opposite parity. But one must be cautious with the light shifts.

4.4.5 Three-Photon Selection Rules

The probability of three-photon excitation depends on the matrix element of the operator $Q^{(3)}_{\varepsilon_p\varepsilon_q\varepsilon_r}$ (4.44) between the levels e and g. More precisely, *if the three photons have the same frequency*, the three-photon operator is completely symmetric, it is invariant by any permutation of the polarizations $\underline{\varepsilon}_1$, $\underline{\varepsilon}_2$, and $\underline{\varepsilon}_3$. This property is important in finding the tensorial properties of the operator.

We want to find the standard components of the totally symmetric operator $Q^S_{\varepsilon_1\varepsilon_2\varepsilon_3}$. The rank of this operator is less than or equal to 3 [one obtains $Q^S_{\varepsilon_1\varepsilon_2\varepsilon_3}$ by coupling the vectorial operator $\underline{D}$ three times and the scalar operator $(\hbar\omega - H_0)^{-1}$ twice]. In order to make the rank precise, we remark that the present problem is similar to making explicit the symmetric tensorial product of three spins 1 as a function of the eigenstates of the total angular momentum J. This latter problem is easy to solve: only those states corresponding to the values 3 and 1 of J are completely symmetric. It follows that the totally symmetric operator $Q^S_{\varepsilon_1\varepsilon_2\varepsilon_3}$ is the sum of two operators: one of rank 3 and one of rank 1. This implies the following selection rules [4.78]:

$$|J_e - J_g| \leq 3$$

$$\left.\begin{array}{l} J_\alpha = 0 \leftrightarrow J_\beta = 2 \\ J_\alpha = 0 \leftrightarrow J_\beta = 0 \end{array}\right\} \text{forbidden } (\alpha,\beta = g,e \text{ or } e,g) \quad . \tag{4.53}$$

These selection rules show that it is not possible to populate the metastable levels $^3P_0$ and $^3P_2$ of the noble gas atoms or of mercury by three-photon absorption (three photons of the same frequency). This result is not obvious because, as in the case of the $6^1S_0 - 7^3S_1$ two-photon transition in Hg, it is possible to find several paths which couple for instance the levels 6^1S_0 and 6^3P_2: however, all the paths interfere destructively as long as the photons have the same frequency.

4.5 Dispersion near a Two-Photon Resonance

In this section the problem of dispersion near a resonant multiphoton transition is discussed. Despite the small number of experiments which have investigated this phenomenon [4.66], the subject presents a number of interesting features and, moreover, is intrinsically related to the phenomenon of resonance absorption. We consider the calculation of the refractive index near a two-photon absorption resonance for a travelling wave (Sect.4.5.1); for two oppositely travelling waves of different frequency (Sect.4.5.2); and for a standing wave (Sect.4.5.3). We give in the following section the experimental results which are at present available and some remarks about the possible developments of experiments.

4.5.1 Refractive Index for a Travelling Wave

The first step in finding an expression for the refractive index is to calculate the mean value of the electric dipole moment of an atom, using the density matrix ρ

$$\langle \underline{D} \cdot \underline{\varepsilon} \rangle = \mathrm{tr}\{\rho \underline{D} \cdot \underline{\varepsilon}\} = 2\mathrm{Re}\left\{\sum_j (\rho_{je} D_{ej} + \rho_{jg} D_{gj})\right\} \quad . \tag{4.54}$$

We use again the calculations of Sect.4.3.1. The formula (4.10) giving the amplitude b_j of the intermediate states $|j\rangle$ allows the coherences ρ_{je} and ρ_{jg} to be easily expressed in terms of the matrix elements ρ_{gg}, ρ_{ee}, ρ_{eg}. Substituting these expressions in (4.54) we obtain

$$\langle \underline{D} \cdot \underline{\varepsilon} \rangle = \mathrm{Re}\left\{(-\rho_{gg} \langle g|Q|g\rangle - \rho_{ee} \langle e|Q'|e\rangle - 2\rho_{ge} \langle e|Q|g\rangle)\mathcal{E}\, e^{-i\omega t}\right\} \tag{4.55}$$

(the terms giving a modulation at a frequency 3ω have been neglected). We can now use the steady state solution of the reduced density matrix equation (4.14) in

Sect.4.3.2; using the relation $\rho_{gg} = 1 - \rho_{ee}$, (4.16) for ρ_{ee}, and a similar formula for ρ_{ge}, we obtain a final expression for $\langle \underline{D} \cdot \underline{\varepsilon} \rangle$. The complex susceptibility $\chi' + i\chi''$ is defined in terms of the electric dipole moment per unit volume, which is obtained by multiplying $\langle \underline{D} \cdot \underline{\varepsilon} \rangle$ by the atomic density N_0

$$P = N_0 \langle \underline{D} \cdot \underline{\varepsilon} \rangle = \mathrm{Re}[\{\chi' + i\chi''\}\mathcal{E}\, e^{-i\omega t}] \tag{4.56}$$

and we deduce

$$\chi' + i\chi'' = -\langle g|Q|g\rangle N_0 + \left(\delta E - \frac{\Gamma_{eg}}{\Gamma_e} s + i\hbar\Gamma_{eg}\right) \frac{|\langle g|Q|e\rangle|^2 \mathcal{E}\mathcal{E}^*}{\delta E^2 + \hbar^2\Gamma_{eg}^2(1 + a^2)} \cdot \frac{N_0}{2} \ . \tag{4.57}$$

The notation used in this formula can be found in Sect.4.3.2: the saturation parameter a is given by (4.17); the light shift s by (4.18); and the effective detuning δE from the two-photon transition (taking into account the light shift) is given by (4.20). The present derivation is valid only for small detunings δE, because if the detuning becomes larger, the effects of saturation on the single-photon transitions, which are not taken into account in the present derivation, will be of similar or greater importance than the terms which have been retained.

The first term of (4.57), $\langle g|Q|g\rangle N_0$, is independent of the field intensity; it is the first order susceptibility, which corresponds to the ordinary refractive index when only single quantum transitions from g to the level j are taken into account. As is well known, this ordinary refractive index is strictly proportional to the light shift of the ground state [4.67]. It will be neglected in the following.

The second term of (4.57) comes from the two quantum absorption processes from level g to level e and is proportional, in the low intensity regime, to the field intensity $|\mathcal{E}|^2$. This term is what is referred to as the third order susceptibility. As usual, the imaginary part χ'' of the susceptibility is proportional to the relaxation rate Γ_{eg}, while the real part χ' is proportional to the detuning from the resonance, in the first approximation where the shift s is small. Additionally, however, we notice that there is a correction terms, $(\Gamma_{eg}/\Gamma_e)s$, which arises from the variation of the populations ρ_{ee} and ρ_{gg} under the action of two-photon transitions. It follows that at high intensities the real susceptibility χ' is zero, for a value of the light frequency which is shifted from the resonance at E_{eg} by an amount equal to $s\ (1 - \Gamma_{eg}/\Gamma_e)$ (i.e., currently s/2), contrary to the absorption line whose center is shifted by the amount s.

The imaginary part χ'' does not give us any additional information, as it is directly related to the population ρ_{ee} of the excited state which has been calculated in Sect.4.3.2 and is given by (4.16). It is easy to verify the relation between ρ_{ee} and χ'', being simply deduced from the energy balance between the power absorbed from the exciting beam and the power reemitted in the spontaneous decay from the excited state

$$\chi''|\mathcal{E}|^2 = 4N_0\hbar\Gamma_e\rho_{ee} \quad . \tag{4.58}$$

We consider in the following the low-intensity regime which is defined by the condition $s \ll \Gamma_e$. It can be noticed that this condition is not always fulfilled in experimental situations [4.80]. Quite often, the observation of dispersion needs an important intensity and we have to take into account the exact expression (4.57). From the real part χ', we deduce the refractive index $n(\omega)$ of the atomic vapor, which is given *in the low-intensity regime* by the expression

$$n(\omega) - 1 \simeq \frac{\chi'}{2\varepsilon_0} \simeq \frac{N_0}{4\varepsilon_0} \cdot \frac{|\langle g|Q|e\rangle|^2|\mathcal{E}|^2\delta E}{\delta E^2 + \Gamma_{eg}^2} \tag{4.59}$$

(the variation of n with ω is contained in the detuning $\delta E \simeq E_{eg} - 2\hbar\omega$). In the case where the detuning is much larger than the natural width ($\delta E \gg \hbar\Gamma_{eg}$) this formula can be simplified

$$n(\omega) - 1 \simeq \frac{N_0}{4\varepsilon_0} \cdot \frac{|\langle g|Q|e\rangle|^2|\mathcal{E}|^2}{\delta E} \quad . \tag{4.60}$$

The refractive index n depends on the value of the intensity $|\mathcal{E}|^2$. This effect, which is intrinsic to a two-quantum transition (without saturation), may be used to observe self-focusing, self-defocusing and also bistable behavior when the atoms are inside a Fabry-Pérot cavity [4.68]. A calculation of the order of magnitude of the intensity-dependent part of n shows that these effects can be observed with moderate laser powers. In vapor cell experiments, in which we must consider the velocity distribution $f(v_z)$ of the atoms, the refractive index is obtained by averaging the index for each velocity group by a formula similar to (4.31)

$$\bar{n}(\omega) = \int_{-\infty}^{+\infty} dv_z \; f(v_z) n(\omega - kv_z) \quad . \tag{4.61}$$

4.5.2 Refractive Indices for Two Waves of Different Frequencies

We consider two waves of frequencies ω_1 and ω_2 such that $\hbar(\omega_1 + \omega_2)$ is close to a two-photon resonance. For the sake of simplicity, we suppose that the detuning from the resonance at $\hbar(\omega_1 + \omega_2)$ is much smaller than the detunings from the resonances at $2\hbar\omega_1$ or $2\hbar\omega_2$

$$|\delta E_{12}| = |E_{eg} - \hbar(\omega_1 + \omega_2)| \ll |E_{eg} - 2\hbar\omega_1| \quad \text{and} \quad |E_{eg} - 2\hbar\omega_2| \quad . \tag{4.62}$$

In this case we can neglect the terms resonant at $2\omega_1$ and $2\omega_2$ and take into account only those terms resonant at $(\omega_1 + \omega_2)$: the refractive index n_1 of wave 1, due to the proximity of the two-photon resonance, depends on the intensity $|\mathcal{E}_2|^2$ of the wave 2. A similar behavior occurs for the index n_2 of wave 2 which is dependent on the intensity $|\mathcal{E}_1|^2$. The indices n_1 and n_2 are given by formulae derived from (4.59) with two modifications:

I) $|\mathcal{E}|^2$ is replaced by $|\mathcal{E}_2|^2$ in the formula for n_1 and by $|\mathcal{E}_1|^2$ in the formula for n_2; II) each formula is two times (4.59) since the two-photon process is now possible by two different ways (first $\hbar\omega_1$ and then $\hbar\omega_2$ or vice versa).

Let us now assume that *the two waves propagate in opposite directions*, in order to reduce the Doppler broadening to an amount proportional to $(\omega_2 - \omega_1)$ [4.50].

In the averaging formula (4.61) kv_z is replaced by $(k_1 - k_2)v_z = (\omega_1 - \omega_2)v_z/c$. The frequency interval over which we must average $n(\omega)$ becomes very small; and we can usually replace the velocity distribution $f(v_z)$ by a Dirac δ function. This is the case if the energy detuning from the two-photon resonance is larger than the residual Doppler width (and also larger than the natural width)

$$|\delta E_{12}| = |E_{eg} - \hbar(\omega_1 + \omega_2)| \gg \hbar|\omega_1 - \omega_2|\frac{v_z}{c} \gg \hbar\Gamma_{eg} \quad . \tag{4.63}$$

The indices n_1 and n_2 are then given by formulae similar to (4.60), but which now take into account the velocity distribution

$$\begin{cases} n_1 - 1 = \dfrac{N_0}{2\varepsilon_0} \dfrac{|\langle e|Q^S_{\varepsilon_1\varepsilon_2}|g\rangle|^2}{E_{eg} - \hbar(\omega_1 + \omega_2)} |\mathcal{E}_2|^2 \\[2ex] n_2 - 1 = \dfrac{N_0}{2\varepsilon_0} \dfrac{|\langle e|Q^S_{\varepsilon_1\varepsilon_2}|g\rangle|^2}{E_{eg} - \hbar(\omega_1 + \omega_2)} |\mathcal{E}_1|^2 \end{cases} \tag{4.64}$$

where the operator $Q^S_{\varepsilon_1\varepsilon_2}$ defined by (4.27) depends on the polarizations $\underline{\varepsilon}_1$ and $\underline{\varepsilon}_2$ of the two light beams.

The expressions (4.64) of the refractive index have been tested experimentally [4.66]. The experiment was performed on the 3S - 5S two-photon transition in sodium. The beam 2 was circularly polarized σ^+. The selection rule $\Delta m = 0$, previously demonstrated for a S-S transition (see Sect.4.3.6), implies that the refractive index is not the same for the left-hand as for the right-hand circularly polarized wave 1. More precisely, the refractive index deduced from (4.64) is different from unity for a beam 1 polarized σ^-, while for a beam 1 polarized σ^+ it is equal to unity. If one uses a linearly polarized light beam 1, its component σ^+ and σ^- are transmitted with different indices; it follows that the linear polarization of beam 1 rotates under the action of beam 2.

We show, in Fig.4.20, the rotation angle as a function of the energy detuning from the two-photon resonance. Near the resonance, the shape of the curve corresponds to a dispersion curve, but there is an asymmetry between the short and long wavelength sides of the line. This asymmetry is a consequence of an interference between the rotation induced by the two-photon transition and the rotation induced by first order processes. These first order processes, which give rise to an ordinary refractive index proportional to the light shift and to saturation terms of the single quantum processes (which were neglected in the calculations of

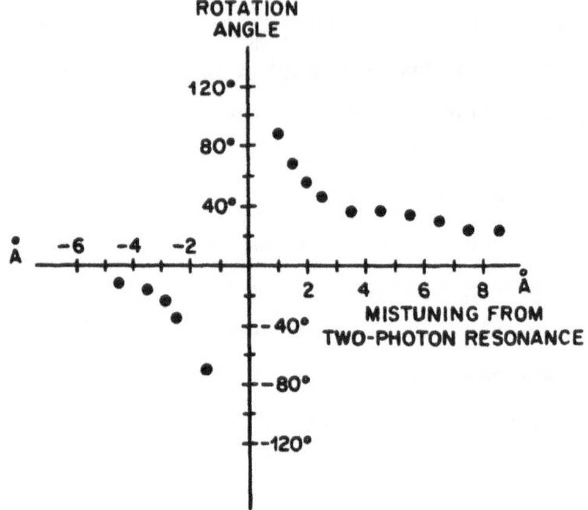

Fig.4.20. Experimental measurement of two-photon dispersion, in the neighborhood of a two-photon resonance (sodium 3S→5S). The light beam 2 is circularly polarized and its wavelength is fixed at 1 Å from the D_2 sodium line (pulsed power of the order of 1 MW/cm^2). The linear polarization of beam 1 rotates when it passes through the sodium vapor (density$\simeq 3\cdot 10^{14}$ cm^{-3}): the rotation angle is measured versus the laser 1 wavelength. [4.66]

Sect.4.5.1), are particularly important in the present case and can no longer be neglected because the energy detuning from the single-photon is of the same order of magnitude as the energy detuning from the two-photon transition.

4.5.3 Refractive Index for a Standing Wave

The problem is more complicated when the two oppositely travelling waves have the same frequency ω, so forming a standing wave.

The atom of velocity v_z is then submitted in its rest frame to two waves of frequencies $\omega_1 = \omega - kv_z$ and $\omega_2 = \omega + kv_z$; and the problem is to know whether or not these two frequencies ω_1 and ω_2 satisfy (4.62) for the three resonances at $(\omega_1 + \omega_2)$, $2\omega_1$ or $2\omega_2$ to be resolved.

If the detuning of the laser is small compared to the Doppler width

$$|E_{eg} - 2\hbar\omega| \ll \hbar ku \tag{4.65}$$

[u characterizes the velocity distribution, see (4.32)].

The condition (4.62) is satisfied for most of the atoms since they have large enough velocities, and we can exclude the small number of slow atoms for which (4.62) is not true. In this case the terms resonant at $2\omega_1 = 2\omega - 2kv_z$ or $2\omega_2 = 2\omega + 2kv_z$ (corresponding to the absorption of two photons of the same travelling wave) are much smaller than the term resonant at $\omega_1 + \omega_2 = 2\omega$ (corresponding to the absorption of two counterpropagating photons). It follows that we can generalize the results of the preceding section—with this simplification that the remaining terms do not depend at all on the atomic velocities—and the result can be given *without any approximation concerning the velocity averaging*

$$n_1 - 1 = \frac{N_0}{2\varepsilon_0}\,|\langle e|Q^S_{\varepsilon_1\varepsilon_2}|g\rangle|^2|\mathcal{E}_2|^2\,\frac{E_{eg} - 2\hbar\omega}{(E_{eg} - 2\hbar\omega)^2 + \hbar^2\Gamma_{eg}^2}$$

$$n_2 - 1 = \frac{N_0}{2\varepsilon_0} |\langle e|Q^S_{\varepsilon_1\varepsilon_2}|g\rangle|^2 |\mathcal{E}_1|^2 \frac{E_{eg} - 2\hbar\omega}{(E_{eg} - 2\hbar\omega)^2 + \hbar^2\Gamma^2_{eg}} \quad . \tag{4.66}$$

The fact that the discussion is related to the absorption of two photons from waves travelling in opposite directions implies that the width of the dispersion line is the natural width. This suggests the possible use of Doppler-free two-photon dispersion for spectroscopy application. Nevertheless, the detection of dispersion is generally less convenient than the observation of the fluorescent light in a two-photon experiment. Another possible application of (4.66) is to use the two-photon dispersion to observe a bistable behavior if the experimental cell is placed inside a cavity. If $E_{eg} - 2\hbar\omega \sim \hbar\Gamma_{eg}$, we can obtain a very considerable effect on the dispersion because all the atoms (or almost all) have the same contribution to the dispersion. Such an experiment has been recently realized in rubidium vapor [4.80].

We now consider larger detunings, such that

$$|E_{eg} - 2\hbar\omega| \gtrsim ku \quad . \tag{4.67}$$

The condition (4.62) is no longer satisfied because the energy detunings for the resonance at 2ω and for the resonances at $2(\omega \pm kv_z)$ are of the same order of magnitude, and the three resonances cannot be resolved (all the atoms are slow in the sense of the preceding discussion). The calculation becomes then much more complicated because new terms appear from interference between the three resonances. In the expression of the electric dipole moment of the vapor, $N_0\langle \underline{D} \cdot \underline{\varepsilon}\rangle$, we obtain new terms whose spatial variation is different from that of the electric field; and strictly speaking, it is no longer possible to define rigorously a refractive index.

Nevertheless, in the case where the detuning is much larger than the Doppler width

$$|E_{eg} - 2\hbar\omega| >> ku \tag{4.68}$$

it is possible to avoid the velocity averaging and to define approximately a refractive index which is obtained by summing the quantities n-1 determined for the travelling wave (4.59) and for the oppositely travelling waves (4.66,68).

In other words, we can discuss the behavior of the atom in the laboratory frame (see Sect.4.4.2): the atom moving in the standing wave is submitted to an electric field whose intensity is modulated with a temporal period $T \simeq \lambda/v_z \simeq 1/kv_z \simeq 1/ku$ and this period must be compared with the evolution time of the atom $(E_{eg}/\hbar - 2\omega)^{-1}$. In the case of (4.65) the period T is much shorter than the evolution time and the atom cannot follow the change of the light intensity; it sees only its mean value. On the contrary, for (4.68), the atom has the time to follow the field intensity for each value of z; it is then possible to calculate the value of the dipole moment for each value of z. In many cases, it is possible to attain a mean effect on the light beam by retaining the spatial component of the dipole which has the same spatial variation as the electric field.

4.6 Transient Processes Involving Doppler-Free Two-Photon Excitation

Two-photon excitation is also a very powerful tool to investigate transient effects. More precisely, Doppler-free two-photon excitation allows the excitation of all the atoms exactly on resonance, or off resonance with the same energy detuning, whatever their velocities may be. It follows that the Doppler dephasing which exists with single-photon excitation and which results in destructive interference between different velocity classes does not occur with Doppler-free two-photon excitation. As we shall see below this has a lot of important consequences for future experiments. This chapter is divided in two parts, the first one concerning free induction experiments and their use in relaxation-time studies, the second one dealing with transients in the driven regime and their dependence on the energy detuning of the two-photon resonance.

4.6.1 Free Induction Transients

The observation of two-photon free induction decay has been reported [4.69-71]. We discuss here in more details the experiment of LIAO et al. [4.71] who observed the evolution of a two-photon coherence by a delayed pulsed technique. Two identical counterpropagating simultaneous laser pulses were used to produce a two-photon coherent state. The decay of this coherent state is monitored by a delayed probe pulse. In the same way as for the exciting pulses, the phase-matching condition is such that the probe pulse produces a backward emission whose amplitude is proportional to the coherence. As long as the atoms remain coherently excited the probe will continue to generate a signal; measurements of the decrease of the backward-wave intensity with increasing probe delay give the lifetime T_2 of the optical coherence. The interesting point is that observation of this phenomenon *does not require narrow linewidth pulses* as the exciting (and probing) frequency does not need to be measured: I) the finite bandwidth of the pump pulse only leads to a very slight Doppler dephasing. This dephasing occurs because the two-photon transition will not be excited by frequency components at exactly $\omega = E_{eg}/2\hbar$ but at a combination of the frequencies $(E_{eg}/2\hbar) + \delta$ and $(E_{eg}/2\hbar) - \delta$. Because the frequencies are unequal, a residual Doppler width of $2\delta \cdot \bar{v}/c$ is obtained (where $\bar{v}$ is the mean quadratic velocity). For a laser linewidth of $\Delta\nu_L/c = 0.1\ \mathrm{cm}^{-1}$ this additional linewidth is only of a few kHz; II) the finite bandwidth of the probe gives a finite bandwidth to the generated signal but does not affect its intensity. Nevertheless, the bandwidth of the laser should be kept reasonably narrow since the backward-wave signal can only be generated along the laser's coherence length $c/\Delta\nu_L$.

The experiment has been performed on sodium and the pressure-broadening coefficients for the 3S-4D two-photon transition for collision with neon, argon and helium have been determined. The sodium atoms are pumped with the 5 ns, 15 kW pulses from a nitrogen-laser-pumped dye laser; the bandwidth is approximately 4 GHz. The output of the laser is divided into two beams having orthogonal polarizations. One of them is reflected back on itself to pump the two-photon transition, the second one is used to probe the two-photon coherence. As the backward reemitted pulse has the same polarization as the probe pulse, it can easily be isolated from the pump pulse. Moreover the probe pulse is not necessarily colinear with the pump pulse, and the signal can also be isolated spatially. The sodium vapor pressure is maintained at approximately 0.3 Torr. Figure 4.21 shows the dependence of the signal intensity on the probe delay when 0.8 Torr of neon buffer gas is added. The pressure-broadening coefficients, measured here at 400°C, agree with those of [4.46], which have been measured directly from the broadening of the Doppler-free two-photon line.

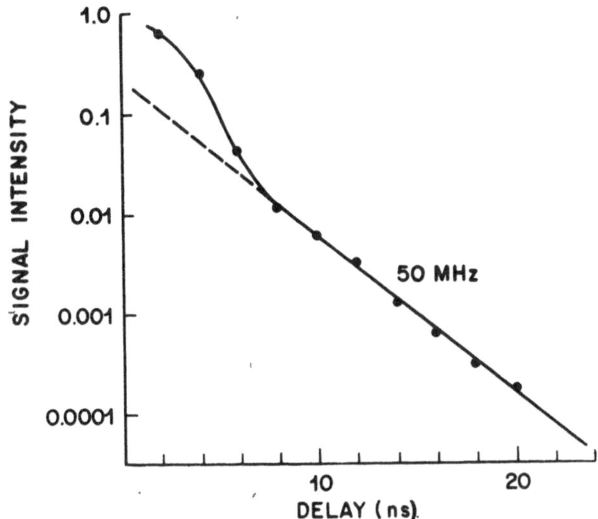

Fig.4.21. Optical two-photon free-induction decay in sodium. The two-photon coherence (3S→4D) is excited by two counterpropagating laser pulses, and is monitored with a delayed probe pulse which produces a backward emission: Intensity of the backward emission vs the delay of the probe pulse. (With 0.8 Torr of neon buffer gas). (The initial part of the curve corresponds to a situation where the probe pulse overlaps the pump pulse. [4.71]

The experiments of [4.69,70] are quite similar in their principle, except for the following features: I) the excitation is done by two counterpropagating laser beams at slightly different frequencies ω_1 and ω_2, which gives a small residual Doppler broadening, but increases the transition probability and allows easier saturation of two-photon transition; II) the Stark switching technique is used to produce the transient process; the sudden application of an electric field shifts the studied energy level by the Stark effect and brings the two-photon transition widely off resonance. Then the two-photon optical coherence begins a free precession with the shifted atomic frequency $E_{eg}/\hbar$, whereas the laser frequencies remain constant; a beat signal can be observed between the shifted atomic frequency $E_{eg}/\hbar$ and "the two-photon frequency" $\omega_1 + \omega_2$. So in these experiments the laser beams successively play two different roles: first they resonantly excite the optical coherence; second, after the Stark pulse, they probe the evolution of the optical coherence; the beam at frequency ω_1 generates the backward emission at the complementary frequency $(E_{eg}/\hbar) - \omega_1$.

4.6.2 Transients in the Driven Regime

The important point is here that all the atoms can be excited with the same energy detuning. It is thus possible to study very precisely the off-resonant transient effects, particularly for weak detunings from resonance. A quite simple experiment has been performed by BASSINI et al. [4.72] on the 3S-4D two-photon transition of sodium. They observed the transient fluorescence signal when the light from a CW dye laser is suddenly switched off or on, for various fixed detunings of the laser frequency.

The theory of these two-photon transients is *identical to the theory of the one-photon transients* (when the Doppler effect is ignored) provided that the energy defects $\hbar\Delta\omega_j$ with the intermediate levels are large enough [see (4.5)]. The laser frequency detuning $\hbar\delta\omega_L$ must be replaced by the effective detuning of the two-photon transition $\delta E = E_{eg} + s - 2\hbar\omega$ [see (4.20)] and the electric dipole interaction, replaced by the effective interaction (4.11) or (4.22) between the e and g levels introduced in Sect.4.3.1.

This differs from the case of one-photon transient experiments in that it is possible to obtain a well-defined energy detuning δE with this Doppler-free technique. The experiment was performed with a CW dye laser focused into the sodium cell and reflected back on itself to provide Doppler-free excitation. An acousto-optic modulator is used to deflect the light beam; the rise time and the cut-off time of the system are smaller than the lifetime (50 ns) of the 4D state. The repetition rate of the light pulses is of the order of 10^6 s^{-1}. The number of fluorescence photons with CW excitation is small (10^3 s^{-1} to 10^6 s^{-1} at resonance), and the probability of observing more than one photon during the observation time following each pulse (150 ns) is negligible. A time to amplitude converter measures the arrival time of each photon after the leading edge or the trailing edge of the pulse. A multichannel analyzer gives the plot of the number of photons versus the arrival time (see Fig.4.21 and 22).

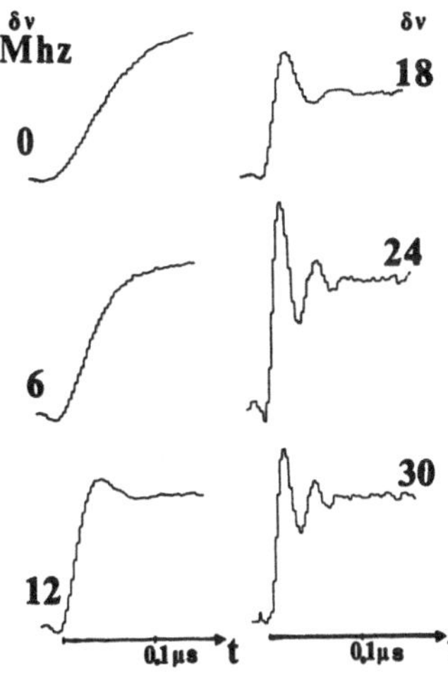

Fig.4.22. Optical two-photon transients observed by sudden irradiation, for various values of the two-photon energy detuning $\delta E = h\delta\nu$ (in sodium 3S→4D): number of spontaneously reemitted photons recorded as a function of the time after the sudden switching on of the laser light. [4.72]

I) The sudden irradiation of atoms causes a ringing corresponding to optical nutation. Very similar effects have been studied for a long time in the radio frequency domain [4.73]. It is well known from elementary quantum mechanical calculations that in such cases it is possible to observe oscillations at the Rabi frequency. In our case, the general solution of the density matrix equation (4.14) depending on the time, leads to the same result, and we obtain the Rabi frequency

$$\Omega = \sqrt{(W_{eg}/\hbar)^2 + (\delta E/\hbar)^2}$$

where

$$W_{eg} = 2\langle e|V|g\rangle e^{+2i\omega t} = \langle e|Q^S_{\varepsilon_1\varepsilon_2}|g\rangle \mathcal{E}_1\mathcal{E}_2 \quad . \tag{4.69}$$

The two-photon operator $Q^S_{\varepsilon_1\varepsilon_2}$ is defined by (4.27). We have retained in the effective Hamiltonian (4.22) only the Doppler-free terms (see Sect.4.3.4). The oscillations at this Rabi frequency are observed in the transient regime, and their amplitude decreases with a time constant equal to $T_2 = 1/\Gamma_{eg}$, the relaxation time of the optical coherences. They can be observed only when the rise time of the pulse is short compared to the period of the Rabi oscillation. Otherwise, the amplitude of the oscillation is reduced compared to the theoretical prediction given by a square shaped pulse.

This can be easily understood if one considers the Fourier transform of the pulse: when off-resonance electromagnetic radiation is turned on, the atomic resonance frequency $E_{eg}/\hbar$ is contained in the excitation Fourier spectrum if the edge of the pulse is steep enough compared to $\hbar/\delta E$. Under this condition beats will be observed between the resonance excitation at the atomic frequency $E_{eg}/\hbar$ and the Raman excitation at 2ω, twice the laser frequency. If the rise time of the pulse is not short enough, the component at the atomic frequency $E_{eg}/\hbar$ is less intense and can even disappear causing a decrease in the observed interference between resonant and Raman reemission.

In the experiment of BASSINI et al. [4.72] the frequency of the Rabi nutation is $\Omega \simeq \delta E/\hbar$, because the two-photon excitation is very far from saturation, and thus $W_{eg}/\hbar \ll \Gamma_{eg}$ while $\delta E/\hbar$ is of the order of Γ_{eg}. These damped Rabi oscillations can be seen in Fig.4.22, which shows various transients (fluorescence versus the time) obtained for different values of the energy detuning $\delta\nu = \delta E/\hbar$ expressed in MHz. The largest detuning of 30 MHz, is almost equal to ten times the linewidth $\Gamma_{eg}/2\pi$. (The steps in the experimental curves correspond to the 3.3 ns duration of each channel of the analyzer).

If the light intensity is higher and attains either a saturation width or a light shift larger than the natural width Γ_e of the transition (but still smaller than the Doppler width) the damped Rabi oscillations can be observed also at exact resonance $\delta E = 0$. The frequency of the oscillations is proportional to the light intensity and the damping is observed with a time constant T_2. A similar behavior has been observed by LOY [4.74]; but his results are rather complicated because the experiment was performed with a smooth-shaped laser pulse and the shift as well as the saturation could vary with time.

In the case where the light intensity is still higher and attains either saturation or a light shift larger than the Doppler width (4.69) is no longer valid, the Rabi oscillation must completely disappear. The origin of the cancellation of the oscillation is that the Rabi frequency depends on the position of the atom and the velocities of the atoms are not large enough to average the values of the Rabi frequency (see Sect.4.4.2).

II) If the light pulse is cut off instantaneously, the emitted light simply decays following an exponential function with a time constant equal to $T_1 = 1/\Gamma_e$ which is the relaxation time of the level population. But here the imperfection of the shutter considerably influences the shape of the signal. There has been a large theoretical interest in these effects [4.75], after the work of WILLIAMS et al. [4.76] who observed the different time-dependent decays between pure resonance fluorescence and pure Raman scattering after an exciting pulse, in a one-photon transition.

Suppose that the light is not cut off instantaneously but with a time constant θ shorter than T_1. We are interested in the behavior of the atoms during this time θ. Then the decay of the scattered light depends on the energy detuning $\delta E/\hbar$:

- if $\delta E/\hbar \ll 1/\theta$, there is only one exponential curve with time constant T_1,
- if $\delta E/\hbar \gg 1/\theta$, the scattered light decreases in the same way as the light pulse,

- in the intermediate cases ($\delta E \cdot \theta/\hbar \sim 1$) the decay is described by two successive curves; the first one follows the exciting pulse, the second one is an exponential curve with time constant T_1. The relative amplitude of the two successive curves depends on the product $(\delta E/\hbar) \cdot \theta$.

The experiment by WILLIAMS et al. was performed on a one-photon transition in iodine and the phenomena were blurred because of average over the Doppler shifts of different velocity classes. In the two-photon experiment by BASSINI et al. the energy detuning $\delta E/\hbar$ is well defined and can easily be compared with θ. Figure 4.23 shows the experimental curves representing the decay of the fluorescence light at the trailing edge of the pulse (decay time of the modulator θ = 22 ns). One clearly sees the progressive appearance of the sharp curve, which follows the light pulse, and the decrease of the amplitude of the slow T_1 exponential curve when the energy detuning $\delta E = \hbar\delta\nu$ is increased. The amplitude of the two successive curves is almost equal for the case (c) corresponding to the product $(\delta E/\hbar)\theta = 2\pi\ \delta\nu \cdot \theta = 1.5$.

Another very important effect which involves transient behavior is the Ramsey technique which has been recently extended to include Doppler-free two-photon excitation [4.77]. The various aspects of this problem are related to coherent phenomena in separated optical fields, which are discussed by Chebotayev in Chap.3.

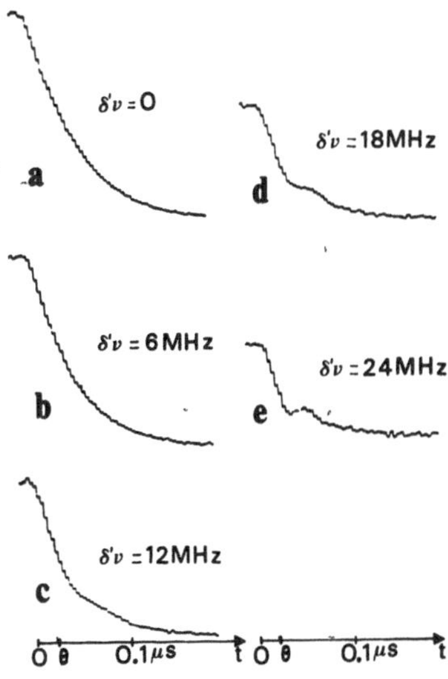

Fig.4.23. Optical two-photon transients observed by the cut-off of the light, for various values of the two-photon energy detunign $\delta E = \hbar\delta\nu$. (In sodium $3S \rightarrow 4D$): number of spontaneously reemitted photons recorded as a function of the time: 0 start of the cut-off; θ = 20 ns end of the cut-off time. (The small "ringing" observed on figures d and e is a parasitic effect). [4.72]

References

4.1 M.S. Feld, A. Javan: Phys. Rev. *117*, 540 (1969)
T.W. Hänsch, P. Toschek: Z. Phys. *236*, 213 (1970)
4.2 M. Göppert Mayer: Ann. Phys. *9*, 273 (1931)
4.3 V. Hughes, L. Grabner: Phys. Rev. *79*, 314 (1950)
V. Hughes, L. Grabner: Phys. Rev. *79*, 819 (1950)
J. Brossel, B. Cagnac, A. Kastler: C.R. Acad. Sc. Paris *237*, 984 (1953)
J. Brossel, B. Cagnac, A. Kastler: J. de Phys. *15*, 6 (1954)
P. Kusch: Phys. Rev. *93*, 1022 (1954)
P. Kusch: Phys. Rev. *101*, 627 (1956)
J.M. Winter: Ann. de Phys. *4*, 745 (1959)
4.4 I.D. Abella: Phys. Rev. Lett. *9*, 453 (1962)
4.5 B. Held, G. Mainfray, C. Manus, J. Morellec, F. Sanchez: Phys. Rev. Lett. *30*, 423 (1973)
G. Mainfray, C. Manus: J. Physique *39*, C1.1 (1978)
P. Lambropoulos: In *Advances in Atomic and Molecular Physics*, Vol.12, ed. by D.R. Bates (Academic Press, New York 1976) pp.87-164 and references therein
4.6 L.A. Lompre, G. Mainfray, C. Manus, J. Thebault: J. de Phys. *39*, 610 (1978)
4.7 R.T. Hodgson, P.P. Sorokin, J.J. Wynne: Phys. Rev. Lett. *32*, 343 (1974)
4.8 L.S. Vasilenko, V.P. Chebotayev, A.V. Shishaev: JETP Lett. *12*, 161 (1970)
4.9 B. Cagnac, G. Grynberg, F. Biraben: J. Phys. *34*, 845 (1973)
4.10 S.V. Filseth, R. Wallenstein, H. Zacharias: Opt. Commun. *23*, 231 (1977)
4.11 C. Tai, F.W. Dalby: Can. J. Phys. *55*, 434 (1977)
4.12 F. Biraben, B. Cagnac, G. Grynberg: C.R. Acad. Sc. Paris *208*B, 235 (1975)
F. Biraben, K. Beroff, E. Giacobino, G. Grynberg: J. de Phys. *39*, L108 (1978)
F. Biraben, K. Beroff, G. Grynberg, E. Giacobino: J. de Phys. *40*, 519 (1979)
G. Grynberg, E. Giacobino, F. Biraben, K. Beroff: J. de Phys. *40*, 533 (1979)
4.13 C.A. Kocher, E.D. Commins: Phys. Rev. Lett. *18*, 575 (1967)
J.F. Clauser: Phys. Rev. Lett. *36*, 1223 (1976)
E.S. Fry, R.C. Thompson: Phys. Rev. Lett. *37*, 465 (1976)
4.14 P. Agostini, P. Bensoussan, J.C. Boulassier: Opt. Commun. *5*, 293 (1972)
P. Bensoussan: Phys. Rev. *11*A, 1787 (1975)
C.D. Harper, S.E. Wheatley, M.D. Levenson: J. Opt. Soc. Am. *67*, 579 (1977)
4.15 J.A. Armstrong, P. Esherick, J.J. Wynne: Phys. Rev. A*15*, 180 (1977)
P. Esheric, J.A. Armstrong, R.W. Dreyfus, J.J. Wynne: Phys. Rev. Lett. *36*, 1296 (1976)
4.16 P. Esheric: Phys. Rev. *15*A, 1920 (1977)
4.17 P. Camus, C. Morillon: J. Phys. B*10*, L133 (1977)
M. Aymar, P. Camus, M. Dieulin, C. Morillon: Phys. Rev. A*18*, 2173 (1978)
J.J. Wynne, J.P. Hermann: Opt. Lett. *4*, 106 (1979)
4.18 F. Biraben, B. Cagnac, G. Grynberg: Phys. Rev. Lett. *32*, 643 (1974)
4.19 M.D. Levenson, N. Bloembergen: Phys. Rev. Lett. *32*, 645 (1974)
4.20 E. Giacobino, B. Cagnac: In *Progress in Optics*, Vol.17, ed. by E. Wolf (North-Holland, Amsterdam 1979)
4.21 G. Grynberg, F. Biraben, M. Bassini, B. Cagnac: Phys. Rev. Lett. *37*, 283 (1976)
4.22 H.W. Schröder, L. Stein, D. Fröhlich, B. Fugger, H. Welling: Appl. Phys. *14*, 377 (1977)
4.23 N. Bloembergen, M.D. Levenson: In *High-Resolution Laser Spectroscopy*, Topics in Applied Physics, Vol.13, ed. by K. Shimoda (Springer, Berlin, Heidelberg, New York 1976)
4.24 S.A. Lee, H. Helmcke, J.L. Hall, B.P. Stoicheff: Opt. Lett. *3*, 141 (1978)
4.25 E. Giacobino, F. Biraben, G. Grynberg, B. Cagnac: J. Physique *38*, 623 (1977)
4.26 F. Biraben, B. Cagnac, G. Grynberg: Phys. Lett. *49*A, 71 (1974)
4.27 T.W. Hansch, K. Harvey, G. Meisel, A.L. Schalow: Opt. Commun. *11*, 50 (1974)
4.28 F. Biraben, B. Cagnac, G. Grynberg: Phys. Lett. *48*A, 469 (1974)
4.29 B. Burghardt, M. Dubke, W. Jitschin, G. Meisel: Phys. Lett. *69*A, 93 (1978)
4.30 F. Biraben, K. Beroff: Phys. Lett. *65*A, 209 (1978)
4.31 K.C. Harvey, B.P. Stoicheff: Phys. Rev. Lett. *38*, 537 (1977)
4.32 S.M. Curry, C.B. Collins, M.Y. Mirza, D. Popescu, I. Popescu: Opt. Commun. *16*, 251 (1976)

4.33 E.V. Baklanov, V.P. Chebotayev: Opt. Commun. *12*, 312 (1974)
4.34 T.W. Hänsch, S.A. Lee, R. Wallenstein, C. Wieman: Phys. Rev. Lett. *34*, 307 (1975)
4.35 S.A. Lee, R. Wallenstein, T.W. Hänsch: Phys. Rev. Lett. *35*, 1262 (1975)
4.36 S. Blit, U. Ganiel, D. Treves: Appl. Phys. *12*, 69 (1977)
F. Trehin, G. Grynberg, B. Cagnac: Rev. Phys. Appl. *13*, 307 (1978)
4.37 C.D. Harper, M.D. Levenson: Opt. Commun. *20*, 107 (1977)
4.38 C.K. Harvey, R.T. Hawkins, G. Meisel, A.L. Schawlow: Phys. Rev. Lett. *34*, 1073 (1975)
4.39 F. Biraben, B. Cagnac, G. Grynberg: C.R. Acad. Sc. Paris *279*B, 51 (1974)
4.40 F. Biraben, E. Giacobino, G. Grynberg: Phys. Rev. A*12*, 2444 (1975)
4.41 F. Biraben, G. Grynberg, E. Giacobino, J. Bauche: Phys. Lett. *56*A, 441 (1976)
E. Giacobino, F. Biraben, E. de Clercq, K. Wohrer-Béroff, C. Grynberg: J. de Phys. *40*, 1139 (1979)
4.42 Y. Kato, B.P. Stoicheff: J. Opt. Soc. Am. *66*, 490 (1976)
4.43 D.E. Roberts, E.N. Forston: Opt. Commun. *14*, 332 (1975)
4.44 A. Flusberg, T. Mossberg, S.R. Hartmann: Phys. Rev. A*14*, 2146 (1976)
4.45 F. Biraben: Thesis, Paris (1977) unpublished
4.46 F. Biraben, B. Cagnac, G. Grynberg: J. de Phys. *36*, L41 (1975)
F. Biraben, B. Cagnac, E. Giacobino, G. Grynberg: J. de Phys. B*10*, 2369 (1977)
4.47 M.M. Salour: Phys. Rev. A*17*, 614 (1978)
4.48 J.P. Woerdman: Opt. Commun. *28*, 69 (1979)
4.49 K.H. Weber, K. Nemax: Opt. Commun. *31*, 52 (1979)
4.50 J.E. Bjorkholm, P.F. Liao: Phys. Rev. Lett. *33*, 128 (1974)
4.51 J.E. Bjorkholm, P.F. Liao: Phys. Rev. A*14*, 751 (1976)
4.52 C. Cohen-Tannoudji: Ann. de Phys. *7*, 423 (1962)
4.53 G. Grynberg: J. de Phys. *40*, 657 (1979)
4.54 F. Bassani, J.J. Forney, A. Quattropani: Phys. Rev. Lett. *39*, 1070 (1977)
4.55 G. Grynberg, E. Giacobino: J. Phys. B*12*, L93 (1979)
4.56 G. Grynberg: Thesis, Paris (1976) CNRS - AO 12497
G. Grynberg, F. Biraben, E. Giacobino, B. Cagnac: Opt. Commun. *18*, 374 (1976)
G. Grynberg, F. Biraben, E. Giacobino, B. Cagnac: J. de Phys. *38*, 629 (1977)
4.57 P.W. Anderson, P.R. Weiss: Rev. Mod. Phys. *25*, 269 (1953)
A. Abragam: *The Principles of Nuclear Magnetism* (The Clarendon Press, Oxford 1961)
C.P. Slichter: *Principles of Magnetic Resonance*, 2nd ed., Springer Series in Solid-State Sciences, Vol.1 (Springer, Berlin, Heidelberg, New York 1980)
J.M. Winter: J. Quantum Optics and Electronics. Les Houches 1964 (Gordon and Breach, New York, London, Paris)
4.58 C. Borde: C.R. Acad. Sc. Paris *282*B, 241 (1976)
4.59 F. Biraben, M. Bassini, B. Cagnac: J. de Phys. (1979) to be published
4.60 P.F. Liao, J.E. Bjorkholm: Phys. Rev. Lett. *34* , (1) 1540 (1975)
4.61 G. Grynberg, B. Cagnac: Rept. Progr. Phys. *40*, 791 (1977)
4.62 N. Bloembergen, M.D. Levenson, M.M. Salour: Phys. Rev. Lett. *32*, 867 (1974)
M.D. Levenson, M.M. Salour: Phys. Lett. *48*A, 331 (1974)
4.63 J. Morellec, D. Normand, G. Petite: Phys. Rev. A*14*, 300 (1976)
4.64 S. Feneuille, L. Armstrong, Jr.: J. de Phys. *36*, L235 (1975)
4.65 C. Cohen-Tannoudji, J. Dupont-Roc: Phys. Rev. A*5*, 968 (1972)
W. Happer: Rev. Mod. Phys. *44*, 169 (1972)
4.66 F. Liao, G.C. Bjorklund: Phys. Rev. A*15*, 2009 (1977)
4.67 C. Cohen-Tannoudji, F. Laloë: J. de Phys. *28*, 505, 722 (1967)
4.68 C. Grynberg, M. Devaud, C. Flytzanis, B. Cagnac: J. de Phys. (to be published)
4.69 M.M.T. Loy: Phys. Rev. Lett. *39*, 187 (1977)
4.70 P.F. Liao, J.E. Bjorkholm, J.P. Gordon: Phys. Rev. Lett. *39*, 15 (1977)
4.71 P.F. Liao, N.P. Economou, R.R. Freeman: Phys. Rev. Lett. *39*, 1473 (1977)
4.72 M. Bassini, F. Biraben, B. Cagnac, G. Grynberg: Opt. Commun. *21*, 263 (1977)
B. Cagnac, M. Bassini, F. Biraben, G. Grynberg: In *Laser Spectroscopy III*, ed. by J.L. Hall, J.L. Carlsten, Springer Series in Optical Sciences, Vol.7 (Springer, Berlin, Heidelberg, New York 1977) p.258
4.73 H.C. Torrey: Phys. Rev. *76*, 1059 (1949)
4.74 M.M.T. Loy: Phys. Rev. Lett. *36*, 1454 (1976)

4.75 J.M. Friedman, R.M. Hochstrasser: Chem. Phys. *6*, 155 (1974)
S. Mukamel, J. Jortner: J. Chem. Phys. *62*, 3609 (1975)
J.O. Berg, C.A. Langhoff, G.W. Robinson: Chem. Phys. Lett. *29*, 305 (1974)
R.C. Hilborn: Chem. Phys. Lett. *32*, 76 (1975)
A. Szöke, E. Courtens: Phys. Rev. Lett. *34*, 1053 (1975)
H. H. Metiu, J. Ross, A. Nitzan: J. Chem. Phys. *63*, 1289 (1975)
4.76 P.F. Williams, D.L. Rousseau, S.H. Dworetsky: Phys. Rev. Lett. *32*, 196 (1974)
4.77 M.M. Salour, C. Cohen-Tannoudji: Phys. Rev. Lett. *38*, 757 (1977)
4.78 G. Grynberg: J. de Phys. *40*, 965 (1979)
4.79 E. Giacobino, E. de Clercq, F. Biraben, G. Grynberg, B. Cagnac: In *Laser Spectroscopy IV*, ed. by H. Walther, K.W. Rothe, Springer Series in Optical Sciences, Vol.21 (Springer, Berlin, Heidelberg, New York 1979) p.626
4.80 E. Giacobino, M. Devaud, F. Biraben, G. Grynberg: To be published

5. Coherent Excitation of Multilevel Systems by Laser Light*

C. D. Cantrell,** V. S. Letokhov, and A. A. Makarov

With 25 Figures

The excitation and dissociation of polyatomic molecules have attracted much attention since the demonstration of isotopic selectivity in the multiple-photon dissociation of BCl_3 [5.1] and SF_6 [5.2], which has been shown to occur in the absence of collisions [5.3]. Whether the infrared-laser-induced dissociation of SF_6 and other polyatomic molecules must be described as a coherent process obeying the Schrödinger equation, or as a unimolecular chemical reaction obeying usual chemical kinetic equations, is currently the subject of much discussion in the scientific community. Certain effects such as collisions, which may be important in practical applications of laser isotope separation, may tend to reduce the importance of coherent effects. However, regardless of practical considerations, there remains the fundamental question of the process of excitation of polyatomic molecules at low laser intensities or fluences, under collisionless conditions. It is clear from experimental high-resolution spectra (for example, the Doppler-limited $3\nu_3$ spectrum of SF_6 [5.4]) that the vibration-rotation states of polyatomic molecules at excitations well below the dissociation threshold are discrete and free from perceptible lifetime broadening. One would therefore expect the dynamics of laser excitation of these discrete states to possess distinctively quantum-mechanical features, such as multiphoton resonances. In this chapter we review the theoretical tools and concepts needed for a first-principles quantum-mechanical study of the coherent laser excitation of polyatomic molecules under collision-free conditions, and illustrate these methods with selected numerical results.

5.1 Multilevel Molecular Systems

In an isolated polyatomic molecule the vibration-rotation states are always degenerate (by at least a factor of $2J + 1$). Different degenerate energy levels may be split from one another by a variety of interactions, and radiative transitions

---

*The Los Alamos portion of this research was supported by the United States Department of Energy.
**Former address: Theoretical Division, Los Alamos Scientific Laboratory, University of California, Los Alamos, NM 87545.

of different strengths may occur to several other levels. It is logical to begin a systematic study of the dynamics of excitation of such a complex structure by studying a simple ladder of nondegenerate states, which are supposed to represent some of the low-lying states of a polyatomic molecule (Fig.5.1a). Two distinctive features of the coherent excitation of such a system are: 1) Multiphoton resonances [5.5,6], in which the population cycles coherently between two states separated by an integer multiple of the photon energy. 2) the anharmonicity barrier [5.7,8], which owes its existence to the decrease of spacing of successive vibrational levels due to anharmonicity in a real molecule. At a fixed laser frequency, little excitation occurs beyond an energy level where the frequency of transitions up the ladder differs from the laser frequency by much more than the multilevel Rabi frequency. At the laser frequencies corresponding to multiphoton resonances, and at fixed laser intensity, the cycling of population between upper and lower states occurs with a longer and longer period as the level of excitation is increased, so that at some point the multiphoton Rabi period greatly exceeds the laser pulse length, making the degree of excitation negligible. How the anharmonicity barrier is overcome at low laser intensities is one of the central theoretical problems of the multiphoton excitation of polyatomic molecules.

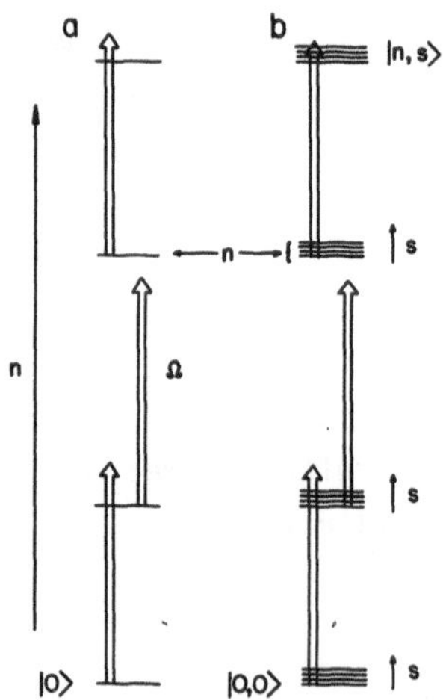

Fig.5.1a,b. Schematic level diagrams for (a) nondegenerate systems; (b) multiplet systems

Following our discussion of the simplest multilevel systems, we turn to a model of laser excitation which includes the so-called quasicontinuum [5.9], where the molecular energy levels are assumed to be so closely spaced that resonant excitation can occur over a range of laser frequencies. In this chapter we model the quasicontinuum as a set of equally spaced nondegenerate states [5.8]; this "quasicontinuum" is coupled to a lower system consisting of a nonequidistant ladder of (nondegenerate) states, of one of the type already studied. Our reasons for representing the quasicontinuum in this way, and not as a continuum, are the following: First, although the density of all states of a polyatomic molecule is high at high vibrational excitation, the density of states which possess substantial dipole strength at laser frequencies near the frequency of a fundamental ($v = 0 \rightarrow v = 1$) vibration-rotation band is much smaller. (The theoretical problem of determining the transition-dipole distribution in highly excited states of real polyatomic molecules is not yet fully solved). Second, there is no reason to believe that there are any physical intramolecular processes which lead to a true continuum of energy levels at excitation energies below the dissociation threshold.

Since we regard the quasicontinuum as consisting of a large number of closely spaced discrete states, the process of multiple-photon excitation is the process of coherent excitation of such a manifold of states. Since a laser beam of finite intensity or pulse duration will pump a coherent superpostion of these states, we study analytically the similarities and differences between coherent excitation from a lower nondegenerate level in this model, and excitation of a true continuum. We find that only in some cases is the Fermi golden rule, which applies to the true continuum, valid for a manifold of discrete states.

After studying some of the fundamental aspects of coherent excitation of nondegenerate systems, we turn to more realistic models of the energy levels and transition moments of polyatomic molecules, including the effects of vibrational degeneracy [5.10,11] and molecular rotation [5.12-15]. We review the theoretical foundation of vibrational and vibration-rotation bases, energy levels and transition moments in spherical-top molecules before discussing some of the qualitative features of multiphoton excitation using numerical techniques. We choose spherical-top molecules because many of the molecules of interest in multiple-infrared photon excitation are spherical tops, and because the high symmetry of spherical tops allows one to make significant analytical and computational simplifications without sacrificing complexity of vibrational structure. Certain qualitative features found in our analytical and numerical calculations for spherical-top molecules should be true in general for polyatomic molecules, and not only for the particular models we have considered. These generally important features are: 1) Vibrational anharmonic splitting, which occurs in the excited vibrational states of degenerate modes of vibration, can, if large, partially or completely compensate the anharmonic shift of energy levels [5.10,11]. In less symmetric molecules, the role of anharmonic splitting may be played by the splitting of combination vibrational states involving at least one infrared-active vibrational mode. In any case, regardless of the magnitude of the anharmonic splitting, the regions of laser frequency where N-photon resonances occur (for a specific N) are determined by the anharmonicity properties of the laser-pumped vibrational mode. 2) Molecular rotation leads to the presence of many near-resonant (but not exactly resonant) intermediate states in multiphoton excitation. These near-resonant intermediate states have a significant effect on the dependence upon laser intensity of the probability of excitation [5.14].

3) For each N, the presence of a dense set of N-photon resonances due to rotational structure leads to a nearly continuous excitation spectrum, with broad "bands" of 2-photon, 3-photon, etc., resonances centered about the purely vibrational multiphoton frequencies [5.13-15]. Thus, even the presence of many rotational states and splittings does not destroy the multiphoton resonances, which are among the most characteristic features of coherent excitation of a set of discrete levels.

5.1.1 The Schrödinger Equation for Multilevel Systems in the Rotating-Wave Approximation

The rotating-wave approximation (RWA) is well known for two-level systems [5.16]. the first application having been in the field of nuclear magnetic resonance [5.17]. This approximation considerably simplifies the solution of the Schrödinger equation by removing all exponential terms with rapidly varying phase from the equations for the probability amplitudes. The criteria for using the RWA in the two-level case are expressed by the following conditions:

$$|\Omega - \omega_{ji}| \ll \Omega \quad , \tag{5.1.1}$$ [1]

$$(\mathcal{E}/h)\ |\langle i|\hat{d}|j\rangle| \ll \Omega \quad . \tag{5.1.2}$$

Here $\mathcal{E}$ is the field amplitude, Ω is the field frequency, ω_{ji} is the transition frequency between the upper state $|j\rangle$ and lower state $|i\rangle$, and $\hat{d}$ denotes the dipole moment operator. Condition (5.1.1) means that the field frequency Ω is relatively close to the transition frequency ω_{ji}. As a result, the terms in these equations may be subdivided into resonant and nonresonant ones. Condition (5.1.2) means that the field is relatively weak, so that one may neglect the nonresonant terms. It should be noted that (5.1.2) is valid in the optical region up to very strong laser intensities which are near the maximum attainable with present experimental techniques.

Let us formulate the RWA for multilevel quantum systems, beginning with the simplest case (see Fig.5.1a) involving only one allowed quasi-resonant transition at each successive step. We call such system "nondegenerate" ones, and as will be seen below, the RWA Schrödinger equation allows rather complete and clear analysis of this case. In addition, many conclusions for nondegenerate systems remain true for more complicated systems (see Fig.5.1b) in which a few transitions are allowed at each step. We shall call the latter "multiplet systems".

Let the spectrum of a quantum systems with Hamiltonian $\hat{H}_0$ consist of discrete nondegenerate states $|n\rangle$, the energies $\hbar\varepsilon_n$ of which are read from the ground state $|0\rangle$ (see Fig.5.1a), and let a field $\mathcal{E}(t)\cos\Omega t$ act on the system. (Of course, in general the field amplitude $\mathcal{E}$ is a slowly varying function of time.) It is convenient

[1]Due to the numerous equations in this chapter they are numbered by section.

to try to find a solution of the Schrödinger equation

$$i\hbar \frac{\partial \Psi}{\partial t} = \hat{H}_0 \Psi - (\hat{d}\mathcal{E}\cos\Omega t)\Psi \tag{5.1.3}$$

in the form of an expansion using the stationary eigenfunctions of the Hamiltonian $\hat{H}_0$

$$\Psi = \sum_n a_n(t)\psi_n \exp(-in\Omega t) \quad . \tag{5.1.4}$$

In this expansion the probability amplitudes a_n in the interaction picture differ from those in the Schrödinger or Heisenberg pictures by phase factors. A more complete discussion is given in Sect.5.3. Since $\hat{H}_0\psi_n = \hbar\varepsilon_n\psi_n$, one arrives at the following set of equations for a_n by putting (5.1.4) into (5.1.3) and using the orthogonality property of stationary wave functions:

$$\frac{da_n}{dt} + i(\varepsilon_n - n\Omega)a_n = \frac{i\mathcal{E}}{2\hbar}\sum_\ell \langle n|\hat{d}|\ell\rangle a_\ell \{\exp[i(n - \ell + 1)\Omega t] + \exp[i(n - \ell - 1)\Omega t]\} \quad . \tag{5.1.5}$$

The conditions similar to (5.1.1,2) which make it possible to simplify (5.1.5) are the following:

$$|\varepsilon_n/n - \Omega| \ll \Omega, \quad \kappa = (\mathcal{E}/\hbar\Omega)\,|\langle n|\hat{d}|\ell\rangle| \ll 1 \quad . \tag{5.1.6}$$

These conditions permit us to identify in (5.1.5) the principal resonant terms (with zero exponential factor) which connect adjacent levels with $\ell = n \pm 1$. Keeping only these principal terms we arrive at the RWA equations for the probability amplitudes

$$\frac{da_n}{dt} + i(\varepsilon_n - n\Omega)a_n = (i\mathcal{E}/2\hbar)(\langle n|\hat{d}|n - 1\rangle a_{n-1} + \langle n|\hat{d}|n + 1\rangle a_{n+1}) \quad . \tag{5.1.7}$$

The physical meaning of the RWA is clear since every exponential term with rapidly changing phase gives a correction to the solution which is smaller than the slowly varying terms by the parameter κ. This fully substantiates the RWA for two-level systems (see, e.g. [5.18]). But, in the general case of a multilevel system, any of the equations (5.1.5) may involve a number of exponential terms, so that a correct substantiation of the equivalence of (5.1.5,7) requires additional cumbersome estimates. (Such estimates can be obtained using the methods given, for example, in [5.19]).

Of course, the situation is simpler for calculations of the effect of quasi-resonant radiation on molecular vibrations since molecular vibrational transition frequencies are close to the harmonic ones to a great extent, and since the tran-

---

$\hbar = h/2\pi$ (normalized Planck's constant)

sition dipole moments with $|\ell - n| > 1$ are small in comparison with those for $\ell - n = \pm 1$ and quickly decrease with increasing $|\ell - n|$. This circumstance certainly makes the RWA correct without any additional requirements, except (5.1.6).

Nondegenerate systems are considered below in Sects.5.2, 5.3.1, 3 and 4, and in part of Sect.5.3.2. Examples of more complicated multiplet (degenerate) systems are considered in part of Sect.5.3.2, and also in Sects.5.3.5 and 5.4. For multiplet systems (see Fig.5.1b) the generalization of the RWA offers no difficulty. Denoting the states of a given multiplet by the index s (for a fixed n) and repeating exactly the procedure given above for passing from the Schrödinger equation to equations for the probability amplitudes, we arrive at the following equations which are more general than (5.1.7):

$$\frac{da_{ns}}{dt} + i(\varepsilon_{ns} - n\Omega)a_{ns} = \frac{i\mathcal{E}}{2\hbar}\sum_q [\langle ns|\hat{d}|n-1,q\rangle\, a_{n-1,q} + \langle ns|\hat{d}|n+1,q\rangle a_{n+1,q}] \quad . \tag{5.1.8}$$

The terms in (5.1.8) connect in pairs all components of adjacent multiplets. The criterion for validity of these equations is expressed by the natural conditions

$$|\varepsilon_{ns}/n - \Omega| \ll \Omega,\ (\mathcal{E}/\hbar\Omega)|\langle ns|\hat{d}|\ell q\rangle| \ll 1 \quad . \tag{5.1.9}$$

Equations (5.1.7,8) may be investigated using a highly developed method based on the so-called "quasi-energy" or "dressed-states" approach. We shall consider this approach in Sect.5.1.2, where we shall cite all formulas and equations in a form appropriate for multiplet systems. The transformation of the latter results to the simpler case of nondegenerate systems should be obvious.

5.1.2 "Quasi-Energy" or "Dressed-States" Approach for Multilevel Systems

The "quasi-energy" or "dressed-states" approach may be formulated for any quantum system, the total Hamiltonian of which is a periodic function of time [5.20]. We shall consider this approach only for the important case of equations describing the dynamics of a multilevel system in the RWA.

Constant Optical Electric Field

When the field amplitude $\mathcal{E}$ in (5.1.8) is constant, the linearly independent solutions of (5.1.8) may be written in the form

$$a_{ns} = p_{ns}\exp(-i\lambda t) \quad , \tag{5.1.10}$$

where λ is any eigenvalue of the following system of time-independent linear equations:

$$(\varepsilon_{ns} - n\Omega - \lambda)p_{ns} = (\mathcal{E}/2\hbar)\Big[\sum_q \langle ns|\hat{d}|n-1,q\rangle p_{n-1,q}$$

$$+ \sum_q \langle ns|\hat{d}|n+1,q\rangle p_{n+1,q}\Big] \quad . \tag{5.1.11}$$

Because of the strict conservation law

$$\frac{d}{dt} \sum_{n,s} a^*_{ns}\, a_{ns} = 0 \quad , \tag{5.1.12}$$

all eigenvalues must be real. Note that (5.1.12) can be obtained directly from (5.1.8) by use of the Hermiticity of the dipole operator $\hat{d}$

$$\langle ns|\hat{d}|\ell q\rangle = \langle \ell q|\hat{d}|ns\rangle^* \quad . \tag{5.1.13}$$

The eigenvalues λ are the so-called "quasi-energies". The coefficients $\|p_{ns}\|$ corresponding to a given eigenvalue λ we shall call the "dressed-state eigenvector." This set of coefficients determines the structure of the dressed state $\Psi(\lambda)$

$$\Psi(\lambda) = \exp(-i\lambda t) \sum_{n,s} p_{ns}(\lambda)\psi_{ns} \exp(-in\Omega t) \quad . \tag{5.1.14}$$

A more detailed discussion of time dependence and excitation probabilities in terms of dressed states is given in Sect.5.3.2.

Any two eigenvectors which correspond to different quasi-energies λ_1 and λ_2 possess a very important orthogonality property, which can be obtained if one takes the scalar product of $p^*_{ns}(\lambda_2)$ with the left-hand sides of (5.1.11) and its complex conjugate. Using (5.1.13) we arrive at

$$(\lambda_1 - \lambda_2) \sum_{n,s} p^*_{ns}(\lambda_1)p_{ns}(\lambda_2) = 0 \quad , \tag{5.1.15}$$

which is equivalent to the orthogonality relation

$$\sum_{n,s} p^*_{ns}(\lambda_1)p_{ns}(\lambda_2) = 0 \quad \text{if} \quad \lambda_1 \neq \lambda_2 \quad . \tag{5.1.16}$$

We next consider the normalization of the eigenvectors. As in time-independent quantum mechanics there is a difference between the discrete quasi-energy spectrum and the continuous spectrum. For a discrete spectrum the eigenvectors can be normalized to unity, that is

$$\sum_{n,s} p^*_{ns}(\lambda)p_{ns}(\lambda) = 1 \quad , \tag{5.1.17}$$

but for a continuous spectrum they should be normalized to a δ function

$$\sum_{n,s} p^*_{ns}(\lambda)p_{ns}(\lambda') = \delta(\lambda - \lambda') \quad . \tag{5.1.18}$$

For either (5.1.17) or (5.1.18), the solution of (5.1.8) for given initial conditions can be obtained automatically. For example, let the quasi-energy spectrum be continuous, and let the system be in the stationary state $|n_0,s_0\rangle$ at time $t = 0$, when the laser field is switched on. Then the solution of (5.1.8) is

$$a_{ns}(t) = \int u(\lambda)p_{ns}(\lambda)\exp(-i\lambda t)d\lambda \quad , \tag{5.1.19}$$

where the weight function $u(\lambda)$ must obey

$$\int u(\lambda')p_{ns}(\lambda')d\lambda' = \delta_{n,n_0}\delta_{s,s_0} \quad . \tag{5.1.20}$$

Multiplying each equation by $p^*_{ns}(\lambda)$, summing on n, and using the normalization relation (5.1.18) we obtain the following relation for the weight function:

$$u(\lambda) = p^*_{n_0,s_0}(\lambda) \quad . \tag{5.1.21}$$

The same relation is valid for a discrete spectrum.

For any system with a finite number of levels the quasi-energy spectrum is always discrete, because the finite set of equations (5.1.11) gives a finite number of eigenvalues. For multilevel systems with an infinite number of levels the quasi-energy spectrum may be either discrete or continuous. It is also possible for the quasi-energy spectrum to contain both a discrete and a continuous part. In fact, excitation into the continuous spectrum means the escape of the system to an unbound state, just as in time-independent quantum mechanics. For multilevel systems, escape to an unbound state should be understood in such a way that the probability amplitude of any fixed level decays to zero as t becomes infinite. This is clear, for example, from (5.1.19). Of course, as can be seen from (5.1.12), the sum of populations $a^*_{ns}a_{ns}$ remains equal to unity. Practically, escape to an unbound state is equivalent to the possibility of unlimited excitation of the system.

Adiabatic Switching on of the Field

In addition to the properties of (5.1.8) in the case of a constant field amplitude as considered above, it is also worthwhile to consider the case when the field is switched on very slowly (adiabatically). Here too the quasi-energy approach appears to be very useful.

Before the laser pulse is switched on, let the system be in the stationary state $|n_0,s_0\rangle$, which we assume is not connected with other states through single-photon or multiphoton resonant transitions. At first the quasi-energy is equal to

$$(\varepsilon_{n_0,s_0} - n_0\Omega) \quad .$$

During the increase of the field amplitude from zero to the maximum value this quasi-energy changes continuously. If the increase of $\mathcal{E}$ occurs sufficiently slowly and if the level does not cross others, then at every instant of time the system is in that dressed state which corresponds to the current value

λ_{n_0,s_0} .

If the pulse is also switched off adiabatically, then the system will return to the initial state. If the pulse is switched off instantaneously, then the system will be in a superposition of the states $|ns\rangle$ for zero field, as determined by the resolution of $|n_0s_0\rangle$ into zero-field eigenstates just prior to switching off the field.

It is difficult to formulate a general criterion for adiabaticity. In most situations the following requirement is sufficient:

$$\tau_p^{-1} \ll |\Delta\lambda| , \tag{5.1.22}$$

where τ_p is the pulse duration, and $|\Delta\lambda|$ is the distance to the next quasi-energy level. A more exact criterion can be written in the following form:

$$\frac{d\mathcal{E}}{dt} \ll \mathcal{E}|\Delta\lambda| \quad . \tag{5.1.23}$$

Of course, the criterion for adiabaticity is violated either when a crossing of quasi-energy levels occurs or when the quasi-energy level $|n_0s_0\rangle$ enters the continuous spectrum.

5.2 Interaction of Equidistant Nondegenerate Multilevel Systems with a Quasi-Resonant Field

In this section we shall consider the excitation of multilevel systems with equidistant, nondegenerate levels (Fig.5.1a). In the notation of the previous section, we shall consider the excitation of a system as described by (5.1.7), using the energy spectrum

$$\varepsilon_n = n\omega (n \geq 0 \quad \text{and} \quad \omega = \text{const.}) \quad . \tag{5.2.1}$$

Since in the examples under consideration only one transition takes place at each successive step, one can for simplicity regard the transition dipole moments as positive. In what follows we shall use the following notation for the matrix elements of the dipole operator $\hat{d}\mathcal{E}/2\hbar$:

$$(\mathcal{E}/2\hbar)\langle n - 1|\hat{d}|n\rangle = \gamma_n = (\mathcal{E}/2\hbar)\langle n|\hat{d}|n - 1\rangle \quad . \tag{5.2.2}$$

The parameters γ_n correspond to the well-known Rabi frequency [5.21] which in the case of an exactly resonant two-level system gives the oscillation frequency of the probability amplitudes.[2]

---

[2]Sometimes the doubled value $2\gamma_n$ is called the Rabi frequency; in that case it is equal to the oscillation frequency of the probabilities for an exactly resonant system.

The systems which we shall consider differ either in the dependence of the dipole moments on the transition number or in the number of levels (finite or infinite). An additional parameter is the detuning of the field frequency Ω from exact resonance. In comparing the analytical solutions for specific systems, the parameters of which are given in Table 5.1, we shall investigate the following questions:

Table 5.1. Particular cases of equidistant multilevel systems, for the dynamics of which analytical solutions have been obtained

N	Number of levels	$\Omega - \omega$	γ_n	Sect.	Ref.
1	Infinite	0	$\gamma n^{1/2}$	5.2.1	[5.8,24-26,29]
2	Infinite	0	γ = constant	5.2.1	[5.8,26,28,29]
3	Infinite	0	$\gamma(n+1)^{-1/2}$	5.2.1	[5.28]
4	N	0	γ = constant	5.2.1	[5.26,27,29]
5	Infinite	0	γ = constant for $n \geqq 2$ and $\beta = \gamma_1/\gamma > \sqrt{2}$	5.2.2	[5.28]
6	Infinite	0	γ = constant for $n \geqq 2$ and $\beta = \gamma_1/\gamma \leqq \sqrt{2}$	5.2.2	[5.28]
7	Infinite	$\neq 0$	$\gamma n^{1/2}$	5.2.3	[5.24,25]

1) The sensitivity of the dynamics of excitation to the dependence of γ_n on n.
2) The specific character of excitation for systems with a finite number of levels.
3) The specific character of excitation for systems with a jump in the dependence of γ_n on n.
4) The role of detuning from exact resonance.

Also in Table 5.1 references are given to papers in which either the analytical solution was originally obtained or a method which is close to that of the present article was formulated.

The quasi-energy spectra and the analytical forms of the dressed-state eigenvectors for several examples are compiled in Table 5.2. The systematic method for obtaining the solution of (5.1.11) in a specific case is the reduction of these equations to the recurrence relations for a known family of orthogonal polynomials. Since the literature on orthogonal polynomials is both rich and accessible (see, e.g., [5.22,23]), we shall not give the details of most of the calculations. In those cases when in our opinion some comment is needed, a corresponding note is given in Table 5.2. For the sake of uniformity we give an unnormalized form of the eigenvectors such that $p_0 = 1$.

Table 5.2. Summary of analytical solutions for the dynamics of equidistant multilevel systems

N[a]	Quasi-energy spectrum	Unnormalized eigenvectors[b]	Remarks
1	$-\infty < \lambda < +\infty$	$(-1)^n(2^n n!)^{-1/2} H_n(\lambda/\gamma\sqrt{2})$	H are Hermite polynomials
2	$-1 \leq \lambda/2\gamma \leq 1$	$(-1)^n U_n(\lambda/2\gamma)$	U_n are Chebyshev polynomials of the second kind
3[c]	$\lambda_{(\pm m)} = \pm\gamma m^{-1/2} (m = 1,2,\dots)$	$(\mp 1)^n[(n+1)m^n/n!]^{1/2} C_{m-1}(n;m)$	C_{m-1} are Charlier polynomials
4[c]	$\lambda_m = -2\gamma\cos[m\pi/(N+1)]$ $m = (1,2,\dots N)$	$\sin[(n+1)m\pi/(N+1]/\sin[m\pi/(N+1)]$	
5[c]	$-1 \leq \lambda/2\gamma \leq 1;\ \lambda_{(\pm)} = \pm\gamma\beta^2(\beta^2-1)^{-1/2}$	$(-1)^n\beta^{-1}[2(\beta^2-1)T_n(\lambda/2\gamma) + (2-\beta^2)U_n(\lambda/2\gamma)]$ for $n \geq 1$	T_n and U_n are, respectively, Chebyshev polynomials of the first and second kinds
6	$-1 \leq \lambda/2\gamma \leq 1$		
7	$\lambda_m = m(\omega-\Omega) + \gamma^2/(\Omega-\omega)$ $(m = 0,1,\dots)$	$\left[\frac{\Omega-\omega}{\gamma}\right]^n (n!)^{1/2} L_n^{m-n}\left[\left(\frac{\gamma}{\Omega-\omega}\right)\right]^2$	L_n^{m-n} are Laguerre polynomials

[a]The numbers correspond to those in Table 5.1
[b]The unnormalized eigenvectors are chosen so that $p_0 = 1$
[c]For details see the text

5.2.1 Analytical Solutions for an Exactly Resonant Field

The equations of motion for a multilevel system with equidistant levels and an exactly resonant field have the following form:

$$\frac{da_0}{dt} = i\gamma_1 a_1$$

$$\frac{da_n}{dt} = i(\gamma_n a_{n-1} + \gamma_{n+1}a_{n+1})(n \geq 1) \quad . \tag{5.2.3}$$

If a system consists of a finite number of levels N, one should set $a_N \equiv 0$ (or $\gamma_N = 0$) in (5.2.3).

For simplicity we shall assume below that the field is switched on at time $t = 0$ and that the field amplitude remains constant thereafter. However, the assumption of resonance at every transition makes our solutions valid for an arbitrary (slow) time dependence of $\mathcal{E}$. It follows from (5.2.3) (after dividing by $\mathcal{E}$ and defining a new variable $du = \mathcal{E}dt$) that the solution for a step pulse may be modified to include any time dependence $\mathcal{E}(t)$ by the following substitution:

$$\mathcal{E}t \to \int_{-\infty}^{t} \mathcal{E}(\tau)d\tau \quad . \tag{5.2.4}$$

Below we shall be interested in the solution subjected to initial conditions for which the system is in the ground state at $t = 0$, that is

$$a_n(0) = \delta_{n0} \quad . \tag{5.2.5}$$

The first system under consideration is the harmonic oscillator, for which the Rabi frequency γ_n increases proportionally to $n^{1/2}$.

Harmonic Oscillator ($\gamma_n = \gamma n^{1/2}$)

The quasi-energy eigenvalues and eigenvectors are given for this case in the first line of Table 5.2. The continuity of the quasi-energy spectrum is in agreement with the well-known fact that the amplitude of the vibrations of a harmonic oscillator under the influence of a resonant external force is, in principle, unlimited.

To obtain the solution which satisfies the initial conditions (5.2.5) one need not follow precisely the procedure of Sect.5.1.2. From the known orthogonality relations of the Hermite polynomials the solution may be written immediately as

$$a_n(t) = (\pi 2^n n!)^{-1/2} \int_{-\infty}^{+\infty} H_n(z)\exp(-z^2 + i\sqrt{2}\gamma tz)dz \quad . \tag{5.2.6}$$

Performing the integral, we obtain the following expression for the probability amplitudes:

$$a_n(t) = (n!)^{-1/2}(i\gamma t)^n \exp(-\gamma^2 t^2/2) \quad . \tag{5.2.7}$$

Hence we have a Poisson distribution for the populations

$$a_n^* a_n = (n!)^{-1}(\gamma t)^{2n} \exp(-\gamma^2 t^2) \quad . \tag{5.2.8}$$

Since the expressions for the mean value and the variance of the Poisson distribution are well known, we give only the results

$$\bar{n} = \sum_{n=1}^{\infty} n a_n^* a_n = \gamma^2 t^2 \quad , \tag{5.2.9}$$

$$D = (\bar{n^2} - \bar{n}^2)^{1/2} = \left(\sum_{n=1}^{\infty} n^2 a_n^* a_n - \bar{n}^2\right)^{1/2} = \gamma t \quad . \tag{5.2.10}$$

Thus the mean energy of a harmonic oscillator excited by a resonant field increases in proportion to $\&^2 t^2$. For high excitation ($\gamma t \gg 1$) the population distribution has a form close to a Gaussian distribution and is concentrated near the mean value ($D \ll \bar{n}$). Hence it is clear that the function

$$S_k(t) = \sum_{n=k}^{\infty} a_n^* a_n = 1 - \sum_{n=0}^{k-1} a_n^* a_n \quad , \tag{5.2.11}$$

which characterizes the passage of the system through a fixed (k - th) level, changes from nearly zero to nearly unity within a relatively short interval of time. Such a dependence is usually classified as a "threshold-like" dependence on time. For the harmonic oscillator a typical dependence of $S_k(t)$ on time is given in Fig.5.2. The abrupt form of the curve justifies the use of the terminology "threshold-like".

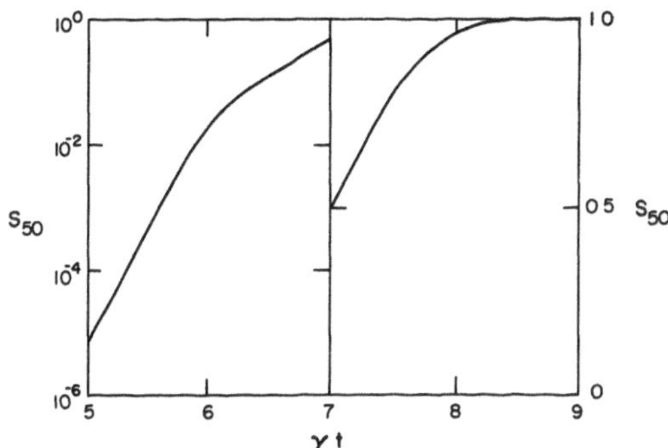

Fig.5.2. Time dependence of the total population for all levels with $k \geqq 50$ for the harmonic oscillator

We turn now to a multilevel system with equal dipole moments. It is interesting that the excitation dynamics for this system have the same main features as for the harmonic oscillator, although there are significant differences of detail.

Infinite System with Equal Dipole Moments ($\gamma_n = \gamma =$ constant)

For this case the quasi-energy spectrum is continuous, like the previous one (see the second line of Table 5.2), but is confined to a finite segment of the real line.

The solution which satisfies the initial conditions (5.2.5) may be obtained by using the orthogonality relation for Chebyshev polynomials of the second kind. The solution is given by the following integral, which can be reduced to ordinary Bessel functions:

$$\begin{aligned} a_n(t) &= \frac{2}{\pi}\int_{-1}^{1} (1 - z^2)^{\frac{1}{2}} U_n(z)\exp(2i\gamma tz)dz \\ &= i^n \frac{n+1}{\gamma t} J_{n+1}(2\gamma t) = i^n[J_n(2\gamma t) + J_{n+2}(2\gamma t)] \quad . \end{aligned} \tag{5.2.12}$$

When $t \to \infty$ the probability amplitudes $a_n(t)$ decay to zero; that is, the system escapes to infinity (see Sect.5.1.2).

Let us study the properties of the population distribution $a_n^* a_n$. From the well-known asymptotic expansion of the Bessel functions [5.30] one can conclude that at each instant of time for $\gamma t \gg 1$ the maximum of the distribution corresponds to levels with numbers n_{max} occurring within the interval

$$2\gamma t - (2\gamma t)^{1/3} < n_{max} < 2\gamma t + (2\gamma t)^{1/3} \quad . \tag{5.2.13}$$

Outside of this interval the populations $a_n^* a_n$ fall off exponentially in the region $n > 2\gamma t$ and oscillate in the region $n < 2\gamma t$, with the mean value decreasing smoothly toward smaller values of n according to the law

$$a_n^* a_n \sim (n + 1)^2/\pi\gamma^2 t^2[4\gamma^2 t^2 - (n + 1)^2]^{\frac{1}{2}} \quad . \tag{5.2.14}$$

It is clear from what has been said that the passage of the system through a given high-lying level has a clear "threshold-like" character (in the sense defined above). For $k \gg 1$ the time for attainment of the kth level is $t_k \sim k/2\gamma$.

For a harmonic oscillator the passage of the system through a given level can be characterized by the quantity S_k (5.2.11). One can obtain a rough expression for $S_k(t)$ above threshold (that is, when $k + 1 < 2\gamma t$) by substituting (5.2.14) for levels below the kth one into (5.2.11) and replacing the summation by an integration. We find that for arbitrary k the quantity S_k is the following universal function of the variable

$$\eta = 2\gamma t/(k + 1):$$

$$S_k(t) \sim g(\eta) = 1 - \frac{2}{\pi}\left[\arcsin\frac{1}{\eta} - \frac{1}{\eta^2}(\eta^2 - 1)^{\frac{1}{2}}\right] \quad (\eta > 1) \quad . \tag{5.2.15}$$

The function $g(\eta)$ increases rapidly, reaching a value $\cong 0.94$ when γt is twice threshold.

More accurate estimates are needed near the threshold of excitation of the kth level. The required results can be obtained from the following exact equation which may be derived directly from (5.2.3,12)

$$\frac{dS_k}{dt} = i\gamma(a_k^* a_{k-1} - a_k a_{k-1}^*) = [2k(k + 1)/\gamma t^2]J_k(2\gamma t)J_{k+1}(2\gamma t) \quad . \tag{5.2.16}$$

The typical dependence of $S_k(t)$ on time for a system with equal dipole moments is given in Fig.5.3. From that curve one can see that, above threshold, relatively rapid oscillations are superimposed on the gradual growth of the function $S_k(t)$, but the average behavior of $S_k(t)$ is well described by (5.2.15). Comparing the curves $S_k(t)$ in Fig.5.2 and 3 one notes that both display a "threshold-like" behavior in the above-mentioned sense.

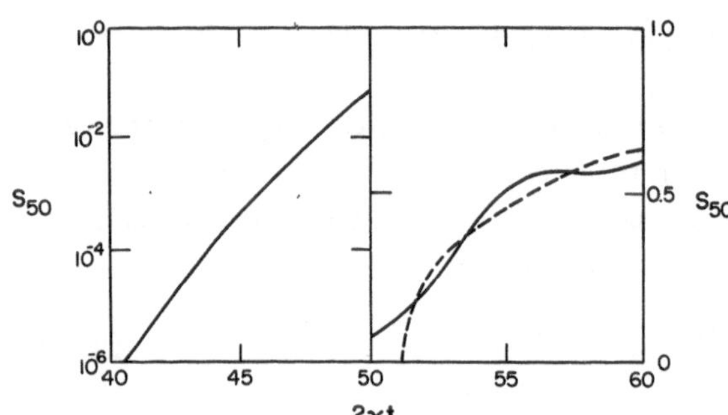

Fig.5.3. Time dependence of the total population for levels with $k \geq 50$ for a system with equal dipole moments

However, passage through a given level proceeds much more slowly for a system with equal dipole moments than for a harmonic oscillator. This fact can be illustrated by calculations of the mean value and the variance of the population distribution. The equation for the mean value follows directly from (5.2.3)

$$\frac{d^2\bar{n}}{dt^2} = 2\gamma^2 a_0^* a_0 = (2\gamma/t)J_1^2(2\gamma t) \ , \ \bar{n}(0) = 0 \ , \ \frac{d\bar{n}}{dt}(0) = 0 \ . \tag{5.2.17}$$

After integrating we obtain the following exact expression for $\bar{n}$:

$$\bar{n} = \frac{16}{3}\gamma^2 t^2\left[J_0^2(2\gamma t) + J_1^2(2\gamma t)\right] - \frac{8}{3}\gamma t\, J_0(2\gamma t)J_1(2\gamma t)$$
$$+ J_0^2(2\gamma t) + \frac{1}{3}J_1^2(2\gamma t) - 1 \ . \tag{5.2.18}$$

The asymptotic expansion of the Bessel functions for $\gamma t \gg 1$ reduces this expression to the form

$$\bar{n} \sim (16/3\pi)\gamma t \ , \tag{5.2.19}$$

that is, $\bar{n} \sim 0.85\, n_{max}$.

A formula for the standard deviation may be easily obtained by direct summation. After some simple transformations we find the following expression

$$D = (3\gamma^2 t^2 - 2\bar{n} - \bar{n}^2)^{\frac{1}{2}} \ , \tag{5.2.20}$$

which for $\gamma t \gg 1$ reduces to

$$D \sim (3 - 256/9\pi^2)^{\frac{1}{2}}\gamma t \ . \tag{5.2.21}$$

From this we conclude that the standard deviation characterizing the halfwidth of the distribution is approximately $D \sim 0.2\,\bar{n}$. Thus, for $\gamma t \gg 1$ the relative width of the distribution is constant, in contrast to the harmonic oscillator where it decreases with time [see (5.2.9,10)]. This difference is reflected in the curves for $S_k(t)$ in Figs.5.2,3.

Despite some quantitative differences between the dynamics of excitation of a system with equal dipole moments and a harmonic oscillator, the main qualitative feature (the possibility of unlimited excitation) is common to both cases. We turn now to an example for which γ_n decreases with n, for which this feature no longer exists.

A System with Decreasing Dipole Moments $[\gamma_n = \gamma(n+1)^{-\frac{1}{2}}]$

For this case (5.1.11) can be reduced with the help of a simple transformation to the recurrence relations for Laguerre polynomials of the following special form

$$L_n^{x-n-1}(x) \ , \ x = (\gamma/\lambda)^2 \tag{5.2.22}$$

(see the third line of Table 5.2). Note that the variable x is part of the upper index of the polynomials (5.2.22).

When $n \to \infty$ the polynomials (5.2.22) tend to zero only if x is equal to a positive integer, and diverge otherwise. Therefore, the spectrum of eigenvalues λ is discrete and the components of the eigenvectors are a special case of Charlier polynomials (see Table 5.2). Normalization factors can be obtained from the known orthogonality property of these polynomials, which may be expressed in the following form [5.23]

$$\sum_{n=0}^{\infty} \frac{n+1}{n!} m^n [C_{m-1}(n;m)]^2 = 2e^m m^{1-m} m! \quad . \tag{5.2.23}$$

From here, and taking (5.1.20) into account, we find after a simple transformation that the solution of the primary set of equations (5.2.3) satisfying the initial conditions (5.2.5) is the following

$$a_n(t) = i^n \left(\frac{n+1}{n!}\right)^{\frac{1}{2}} \sum_{m=1}^{\infty} \frac{m^{m-1+n/2}}{m!} e^{-m} C_{m-1}(n;m) \cos\left(\frac{\gamma t}{m^{\frac{1}{2}}} - \frac{n\pi}{2}\right) \quad . \tag{5.2.24}$$

Since the quasi-energy spectrum of the system under consideration is discrete, the probability amplitudes do not decay to zero with time, and the system does not escape to infinity. Thus, to some extent, the reduction of the dipole moment is analogous to an indefinitely increasing potential in the time-independent Schrödinger equation, which also results in a discrete energy spectrum.

By averaging the distribution over an infinite time interval one can estimate how effectively the high-lying levels may be populated. Such averaging gives the following expression

$$\begin{aligned} V_n &= \lim_{T\to\infty} \frac{1}{T} \int_0^T a_n^*(t) a_n(t) dt \\ &= \frac{1}{2} \frac{n+1}{n!} \sum_{m=1}^{\infty} \frac{m^{2m-2+n}}{(m!)^2} e^{-2m} [C_{m-1}(n;m)]^2 \quad . \end{aligned} \tag{5.2.25}$$

The mean populations V_n are partially tabulated in Table 5.3. Hence one can see that, at the lowest transition, an inversion exists on average. For higher transitions, the populations gradually decrease. This decrease, however, is slow and the probability that the system will be found at very high levels is appreciable.

A characteristic feature of functions which, like (5.2.24), consist of the sum of a large number of periodic terms was commented upon in [5.27], where the dynamics of an N-level system were investigated. The authors of this paper have appropriately named this feature a "flickering" of populations. In a condition of "flickering" populations the values $a_n^* a_n$ mainly fluctuate near the means V_n, but random spikes occur from time to time. Of course, if the system is initially in the ground state, the attainment of a condition of "flickering" populations requires a definite time interval, which is longer the higher the level considered (at least, $t > n/\gamma$).

A condition of "flickering" populations is also typical in the excitation of systems consisting of a large (but finite) number of levels. We shall consider such a case below.

Table 5.3. Distribution of average populations V_n for resonant excitation of an equidistant system with Rabi frequency decreasing as $\gamma_n = \gamma(n+1)^{-1/2}$

n	V_n	n	V_n	n_1	n_2	$\sum_{n=n_1}^{n_2} V_n$
0	0.08	3	0.07	6	10	0.12
1	0.15	4	0.05	11	20	0.10
2	0.10	5	0.04	21	∞	0.29

N-level System with Equal Dipole Moments ($\gamma_n = \gamma$ = constant)

As for an infinite system with equal dipole moments, the eigenvectors for an N-level system can be expressed in terms of Chebyshev polynomials of the second kind. The eigenvalues are determined by

$$a_N = 0 \quad \text{or} \quad U_N(\lambda/2\gamma) = 0 \quad , \tag{5.2.26}$$

which gives the results shown in the fourth line of Table 5.2. We omit the simple trigonometric calculations required for normalization of the eigenvectors. The solution satisfying the initial conditions (5.2.5) has the following form

$$a_n(t) = \frac{2}{N+1}\sum_{m=1}^{N} \sin[m\pi/(N+1)]\sin[(n+1)m\pi/(N+1)]$$
$$\exp\{2i\gamma t\cos[m\pi/(N+1)]\} \quad . \tag{5.2.27}$$

As in the previous example, it is of interest to average the population distribution over an infinite time interval. We obtain the following result for the mean populations V_n

$$V_0 = V_{N-1} = 3[2(N+1)]^{-1} \; , \quad V_n = (N+1)^{-1} \quad \text{for} \quad n \neq 0 \; , \quad N-1 \quad . \tag{5.2.28}$$

Thus the mean populations are equal for all levels, except the ground and the highest levels.

The averaging which gives (5.2.28) corresponds to the steady-state regime of the "flickering" populations. We now estimate the time required to attain this steady state, using the following Bessel-function representation of (5.2.27)

$$a_n(t) = i^n \left\{\sum_{r=0}^{\infty} \left[J_{2r(N+1)+n}(2\gamma t) + J_{2r(N+1)+n+2}(2\gamma t)\right]\right.$$
$$\left. - \sum_{r=1}^{\infty} \left[J_{2r(N+1)-n-2}(2\gamma t) + J_{2r(N+1)-n}(2\gamma t)\right]\right\} \quad . \tag{5.2.29}$$

The validity of (5.2.29) may be easily examined by direct substitution into (5.2.3). The representation (5.2.29) has a clear physical interpretation. The term with r = 0 in the first sum corresponds to the solution (5.2.12) for an infinite system and plays the main role until the distribution reaches the upper levels ($n \sim N-1$). The term with $r = 1$ in the second sum describes "reflection" of the distribution from the upper boundary ($n = N-1$); the term with $r = 1$ in the first sum describes the same from the lower boundary ($n = 0$), etc. In general, the first sum describes the upwards flow of the population distribution and the second sum describes the downwards flow.

It is convenient to study the transition to equilibrium of the total populations in the lower levels ($0 \leq n \leq [N/2]-1$) and in the upper levels ($[N/2] \leq n \leq N-1$) using the function $S_{[N/2]}(t)$ [see (5.2.11)]. To obtain a rough estimate of the time dependence of $S_{[N/2]}(t)$ it is sufficient to use (5.2.15) for every term in (5.2.29). [Of course, by limiting ourselves to this expression, we neglect the interference of different terms in the series (5.2.29)]. In this way the following estimate can be obtained

$$S_{[N/2]}(t) \sim \sum_{r=0}^{\infty} g\left(n_r^{(+)}\right) - \sum_{r=1}^{\infty} g\left(n_r^{(-)}\right) \quad . \tag{5.2.30}$$

Here the universal function g is determined by (5.2.15), and the variables $n_r^{(\pm)}$ have the following form

$$n_r^{(+)} = 4\gamma t/N(4r+1) \quad , \quad n_r^{(-)} = 4\gamma t/N(4r-1) \tag{5.2.31}$$

where we have assumed $N \gg 1$.

The results of a numerical calculation of the function $S_{[N/2]}(t)$ using (5.2.30) are given in Fig.5.4. From this calculation one can see that the typical time t_{eq} which is required to attain equilibrium is approximately one order of magnitude longer than the typical time $t_N \sim N/2\gamma$ which is required for distribution to reach the upper levels.

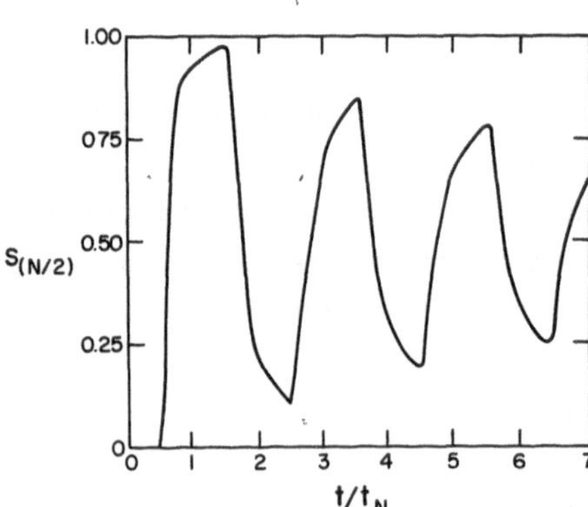

Fig.5.4. Time dependence of the total population of the upper levels for an N-level system with equal dipole moments

Having considered the example of an N-level system we do not investigate resonant dynamics further. We turn now to the question which is the subject of the next section. At first sight this question may seem to be irrelevant, since one is inclined to believe that resonance on successive transitions always results in the most effective possible excitation, and that the time for excitation of a given level may be estimated as the sum of the time intervals $\Delta t_n \sim 1/\gamma_n$ required to take each lower step. However, we shall show that such estimates may not always be correct.

5.2.2 Does Resonance Always Result in Effective Excitation?

In Sect.5.2.1 we have considered the dynamics of resonant excitation of an infinite system with equal dipole moments for successive transitions. In this section we shall show that a considerable change in the dipole moment of a single transition results in a considerable change in the dynamics of the populations. We limit ourselves to a situation in which the lowest transition $0 \to 1$ is singled out, that is $\gamma_1 \neq \gamma_n = \gamma (n \geq 2)$.

Thus, we shall obtain the solution of (5.2.3) with $\gamma_n = \gamma = \text{const.}$ for $n \geq 2$ and $\gamma_1 \neq \gamma$. Letting $\gamma_1/\gamma = \beta$ we arrive at the following form for (5.1.11) which determine the quasi-energy spectrum and the structure of the dressed-state eigenvectors

$$\begin{aligned} -\lambda p_0 &= \beta\gamma p_1 \\ -\lambda p_1 &= \gamma\beta p_0 + \gamma p_2 \\ -\lambda p_n &= \gamma(p_{n-1} + p_{n+1})(n \geq 2) \quad . \end{aligned} \tag{5.2.32}$$

In order to find the solution of (5.2.32) it is convenient to consider a truncated set from which the first two equations are excluded. Such a truncated set obviously has two linearly independent solutions, and its general solution can be represented as a superposition of Chebyshev polynomials of the first and second kind

$$p_n = C_1 T_n(z) + C_2 U_n(z) \ , \ n \geq 1 \ , \ z = -\lambda/2\gamma \quad . \tag{5.2.33}$$

Using the explicit form of the Chebyshev polynomials for $n = 1,2$ we find the constants C_1 and C_2 from the first two equations of (5.2.32) (see the fifth and sixth lines of Table 5.2).

Just as for a system with equal dipole moments, the eigenvalue spectrum contains a continuous region, the segment $-1 \leq z \leq 1$. Let us obtain the normalization of the eigenvectors belonging to the continuous spectrum by writing (5.1.18) in the following form

$$\begin{aligned} 1 + \sum_{n=1}^{\infty} [2\beta^{-1}(\beta^2 - 1)T_n(z) + \beta^{-1}(2 - \beta^2)U_n(z)][2\beta^{-1}(\beta^2 - 1)T_n(z') \\ + \beta^{-1}(2 - \beta^2)U_n(z')] = A(z)\delta(z - z') \quad . \end{aligned} \tag{5.2.34}$$

Multiplying both sides of (5.2.34) by $(1 - z'^2)^{\frac{1}{2}}$, integrating by z' in the segment $-1 \leq z' \leq 1$, and noting that only the integral with $T_2(z')$ gives a nonzero result we arrive at the following expression

$$A(z) = (\pi/2)\beta^{-2}[\beta^4 + 4(1 - \beta^2)z^2](1 - z^2)^{-\frac{1}{2}} \quad . \tag{5.2.35}$$

Besides the continuous spectrum, for certain values of β there are also discrete eigenvalues λ which can easily be determined from the following requirement

$$\lim_{n\to\infty} p_n(\lambda) = 0 \quad . \tag{5.2.36}$$

Using the explicit form of the Chebyshev polynomials, that is, $T_n(\cos\theta) = \cos\theta n$, $U_n(\cos\theta) = \sin\theta(n + 1)/\sin\theta$, and regarding $\theta = x + iy$ as a complex variable we find that (5.2.36) reduces to the following equations

$$[\beta^2 - 1 - \exp(2|y|)] \cos x = 0$$

$$[\beta^2 - 1 + \exp(2|y|)] \sin x = 0 \quad . \tag{5.2.37}$$

Equations (5.2.37) have simultaneous solutions if and only if the condition

$$\beta > \sqrt{2} \tag{5.2.38}$$

is valid. In the plane of the complex variable $\lambda = -2\gamma \cos(x + iy)$ these solutions reduce to the two eigenvalues which are given in the fifth line of Table 5.2. The corresponding normalized eigenvectors are

$$p_0^{(\pm)} = [(\beta^2 - 2)/2(\beta^2 - 1)]^{\frac{1}{2}} \quad , \tag{5.2.39}$$

$$p_n^{(\pm)} = (\mp 1)^n \beta[(\beta^2 - 2)/2(\beta^2 - 1)^{n+1}]^{\frac{1}{2}} \text{ for } \quad n \geq 1 \quad .$$

Thus, if the dipole moment of the lowest transition exceeds that of the others, in correspondence with (5.2.38), then a peculiarity appears in the dynamics of excitation. The solution $a_n(t)$ contains a decaying term $a_n^{(c)}(t)$ which comes from the continuous spectrum, and an oscillating term $a_n^{(d)}(t)$ which comes from the discrete spectrum. The ratio of the quantities

$$W_c = \sum_{n=0}^{\infty} |a_n^{(c)}|^2 \quad \text{and} \quad W_d = \sum_{n=0}^{\infty} |a_n^{(d)}|^2 \tag{5.2.40}$$

is the key to estimating the efficiency of excitation of the system. Since $W_c + W_d = 1$, this ratio can be obtained from either part (discrete or continuous) of the solution.

The complete solution for the case under consideration is described by formulas which follow from the expressions for the eigenvectors [see Table 5.2 and (5.2.39)] together with (5.1.21,34,35)

$$a_n(t) = a_n^{(c)}(t) + a_n^{(d)}(t) \tag{5.2.41}$$

$$a_0^{(c)}(t) = \frac{2\beta^2}{\pi} \int_{-1}^{1} \frac{(1 - z^2)^{\frac{1}{2}} \exp(2i\gamma tz)}{\beta^4 + 4(1 - \beta^2)z^2} \, dz \tag{5.2.42}$$

$$a_n^{(c)}(t) = \frac{2\beta}{\pi}\int_{-1}^{1}\frac{(1-z^2)^{\frac{1}{2}}[2(\beta^2-1)T_n(z)+(2-\beta^2)U_n(z)]}{\beta^4+4(1-\beta^2)z^2}\exp(2i\gamma tz)dz \quad (5.2.42)$$

$$a_0^{(d)}(t) = \frac{\beta^2-2}{\beta^2-1}\cos\left[\gamma\frac{\beta^2}{(\beta^2-1)^{\frac{1}{2}}}t\right] \quad (5.2.43)$$

$$a_n^{(d)}(t) = i^n\frac{\beta(\beta^2-2)}{(\beta^2-1)^{n/2+1}}\cos\left[\gamma\frac{\beta^2}{(\beta^2-1)^{\frac{1}{2}}}t-\frac{n\pi}{2}\right] \quad \text{for} \quad n \geqq 1 \ .$$

Expressions for the probabilities W_c and W_d can be obtained, for example, from the discrete part (5.2.43)

$$W_d = (\beta^2-2)/(\beta^2-1) \ , \ W_c = 1 - W_d = (\beta^2-1)^{-1} \ . \quad (5.2.44)$$

It follows that for $\beta \gg 1$, that is, when the dipole moment of the lowest transition is considerably greater than that of the subsequent ones, the probability W_c of escape of the system to an unbound state is considerably less than unity.

Thus a relatively strong transition which is inserted into a sequence of weaker transitions may be a barrier for further excitation. The physical reason for this effect may be clarified by conceptually dividing the problem into two stages, using the ideas of perturbation theory. First, we take into account only the strong resonant interaction between the field and the two levels coupled by the strong transition. As is well known, this interaction creates two dressed states, the quasi-energies of which differ from the zero-field energy eigenvalues by $\pm\gamma_1$. When we turn to the next approximation to consider the weaker transitions, we see that it is possible that the transition 1 → 2 has been taken out of resonance (see Fig.5.5). The induced detuning may be compensated only by changing the transition frequency or the laser frequency. This effect is also of importance for more complicated multilevel systems (see Sect.5.3.3).

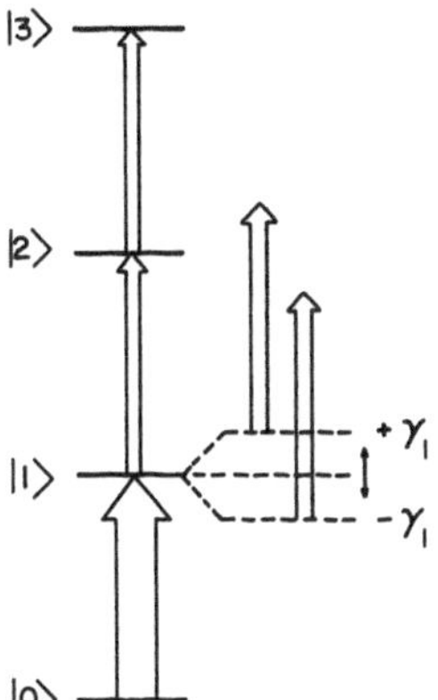

Fig.5.5. Change in the frequency of the transition 1 → 2 because of the effect of the field on the relatively strong transition 0 → 1

A relatively weak transition which is inserted into a sequence of stronger transitions need not result in so radical a change of the dynamics of excitation. When the condition

$$\beta \leq \sqrt{2} \tag{5.2.45}$$

which is the opposite of (5.2.38), is met, there is only a continuous quasi-energy spectrum. Here a specific effect arises, however, which results in an increase of the time required for passage through the weak transition.

Under the condition (5.2.45) the solution is given by the expression

$$a_n(t) = a_n^{(c)}(t) \tag{5.2.46}$$

where the functions $a_n^{(c)}(t)$ are defined by (5.2.42). For $\beta \sim 1$ the solution obviously will not differ essentially from that for a system with equal dipole moments (see Sect.5.2.1). For $\beta \ll 1$ the excitation rate of the system as a whole is determined by the rate of decay of the ground-state population. The neighborhood of the point $z = 0$ plays a dominating role in the integral (5.2.42) for $a_0(t)$, so that we may retain only the oscillating term in the numerator of the integrand, and then extend the integration to the entire real axis. As a result we obtain

$$a_0(t) \cong \exp(-\beta^2\gamma t) = \exp\left(-\frac{\gamma_1^2}{\gamma}t\right) \quad . \tag{5.2.47}$$

Thus, for $\beta \ll 1$ the rate of decay of the ground-state probability amplitude is considerably less than γ_1. According to (5.2.47) the probability amplitude decays exponentially, without oscillation, as is typical for a two-level system having an upper-level width exceeding the Rabi frequency γ_1. In the case under consideration the role of the excited-level width is played by the parameter γ which characterizes the decay due to transitions to higher levels.

We have shown that the resonant excitation of equidistant multilevel systems may be hindered by a sudden jump in the transition dipole moment. Another hindrance to the efficient excitation of equidistant multilevel systems is, of course, detuning of the laser frequency from the transition frequencies, which we consider in the next section.

5.2.3 Nonexact Resonance and Its Compensation by Power Broadening

To give a more complete picture of the coherent interaction between a monochromatic field and a multilevel quantum system with equidistant energy levels, let us consider a case in which the field frequency Ω is detuned from the transition frequency ω. It is obvious that for low field intensities such a detuning results in a considerable suppression of excitation. The question then arises, how strong must the field be in order to achieve substantial excitation of a detuned transition?

The answer is well known for a two-level system. Detuning in the two-level case may be compensated if the condition

$$2\gamma_{ij} = (\mathcal{E}/\hbar)|\langle i|\hat{d}|j\rangle| \gtrsim |\Omega - \omega_{ji}| \tag{5.2.48}$$

is met, where γ_{ij} is the Rabi frequency. Under the condition (5.2.48), the dynamics of excitation cannot be described within the framework of perturbation theory, but must be calculated by the same methods used for exact resonance. The upper-level occupation probability has the following form for the initial conditions $a_i = 1$ and $a_j = 0$:

$$a_j^* a_j = \frac{4\gamma_{ij}^2}{(\Omega - \omega_{ji})^2 + 4\gamma_{ij}^2} \sin^2\left\{\frac{1}{2}\left[(\Omega - \omega_{ji})^2 + 4\gamma_{ij}^2\right]^{\frac{1}{2}} t\right. \quad . \tag{5.2.49}$$

Of course, the criterion (5.2.48) follows directly from (5.2.49).

Naturally, the question arises how to generalize this criterion for equidistant multilevel systems. An estimate for the harmonic oscillator is the most easily obtained. Using the quasi-energy eigenvalues and the expressions for the dressed-state eigenvectors in the seventh line of Table 5.2, one may investigate two different limits of the time dependence of the laser pulse.

The solution in the case of instantaneous switching on of a laser pulse of constant amplitude can be obtained by the general method which has been used above for resonant excitation. However, in contrast to the case of resonance for all successive transitions, this solution may no longer be extended to an arbitrary $\mathcal{E}(t)$ by the simple substitution (5.2.4). As follows from Sect.5.1.2, a simple procedure is available when the field-pulse switching on is adiabatically slow. In this case the structure of the solution satisfying the initial conditions (5.2.5) is the same as that of a fixed dressed state which approaches the ground state for $\mathcal{E} \to 0$.

Step-Function Laser Pulse (Harmonic Oscillator)

The solution under the initial conditions (5.2.5) is the following superposition of dressed states

$$a_n(t) = \left(\frac{\Omega - \omega}{\gamma}\right)^n (n!)^{\frac{1}{2}} \exp\left[-\, i\,\frac{\gamma^2}{\Omega - \omega}\, t - \left(\frac{\gamma}{\Omega - \omega}\right)^2\right] \quad . \tag{5.2.50}$$

$$\sum_{m=0}^{\infty} \frac{1}{m!}\left[\left(\frac{\gamma}{\Omega - \omega}\right)^2 e^{i(\Omega-\omega)t}\right]^m L_n^{m-n}\left[\left(\frac{\gamma}{\Omega - \omega}\right)^2\right] \quad .$$

After resummation this formula reduces to

$$a_n(t) = (n!)^{-\frac{1}{2}}\left[\frac{\gamma}{\Omega - \omega}\left(e^{i(\Omega-\omega)t} - 1\right)\right]^n \exp\left[-\, i\,\frac{\gamma^2}{\Omega - \omega}\, t \right.$$
$$\left. + \left(\frac{\gamma^2}{\Omega - \omega}\right)^2\left(e^{i(\Omega-\omega)t} - 1\right)\right] \quad , \tag{5.2.51}$$

which transforms into (5.2.7) when $\Omega = \omega$. Thus we have a Poisson distribution for the populations

$$a_n^* a_n = (n!)^{-1}[\bar{n}(t)]^n \exp[-\bar{n}(t)] \quad , \tag{5.2.52}$$

where the mean number of photons absorbed is

$$\bar{n}(t) = 2[\gamma/(\Omega - \omega)]^2[1 - \cos(\Omega - \omega)t] \quad . \tag{5.2.53}$$

In contrast to the exactly resonant case, the mean value $\bar{n}$ does not increase without limit as $t \to \infty$, but oscillates between zero and the maximum value

$$n_{max} = 4[\gamma/(\Omega - \omega)]^2 \quad . \tag{5.2.54}$$

It is a well-known fact [5.31] that the time dependence of the mean energy is in exact correspondence with the time dependence of the energy of a classical harmonic oscillator.

Thus, effective excitation of the nth level of a harmonic oscillator under a step-function pulse may be accomplished when the relation

$$|\gamma/(\Omega - \omega)| \gtrsim n^{\frac{1}{2}}/2 \tag{5.2.55}$$

between the Rabi frequency γ and the detuning from resonance Ω is obeyed.

Adiabatically Switched-On Pulse (Harmonic Oscillator)

In this case an oscillator which was initially in the ground state is in the dressed state with m = 0 at any instant of time. The population distribution is also Poisson

$$a_n^* a_n = (n!)^{-1}[\gamma/(\Omega - \omega)]^{2n} \exp\left[-\left(\frac{\gamma}{\Omega - \omega}\right)^2\right] \quad . \tag{5.2.56}$$

From this it is easy to show that the maximum energy is four times smaller than for a step-function pulse of the same field amplitude. This property is also true of the classical solution.

General Estimates for Maximum Detuning

Let us define Δ_n as the absolute value of the maximum detuning under which the excitation of the nth level is still appreciable. For a harmonic oscillator there is a simple relation between Δ_n and the time t_n required for resonant excitation of the nth level (see Sect.5.2.1)

$$\Delta_n t_n \cong 2 \quad . \tag{5.2.57}$$

For an equidistant multilevel system with equal dipole moments the following estimate is applicable

$$\Delta_n \cong 4\gamma/n \quad . \tag{5.2.58}$$

We shall not discuss this estimate in detail, but will justify it by giving the solution for a system which is infinite in both directions, that is $-\infty < n < +\infty$ ($\gamma_n = \gamma$ = constant). In this case the probability amplitudes have the simple form

$$a_n(t) = i^n \exp\left[\frac{i}{2} n(\Omega - \omega)t\right] J_n\left[\frac{4\gamma}{\Omega - \omega} \sin\left(\frac{\Omega - \omega}{2}\right)t\right] \tag{5.2.59}$$

which may be easily verified by direct substitution into the equation of motion. The estimate (5.2.58) follows from (5.2.59).

A more rigid restriction of the allowed detuning for equidistant multilevel systems than (5.2.48), which was derived for a two-level system, results from an increase of detuning from resonance for subsequent multiphoton transitions from the ground level. If the system is not equidistant, and its levels are not more than a fixed value Δ away from the positions of multiphoton resonances (see Fig. 5.6), then a lower limit for the field amplitude is again given by (5.2.48). An analytical solution for a simple example has been given in [5.32]. The lower limit for the field amplitude is reduced further if one or more intermediate levels are nearly in multiphoton resonance. The dynamics of multiphoton transitions is considered at the beginning of Sect.5.3, which deals with the excitation of nonequidistant multilevel systems.

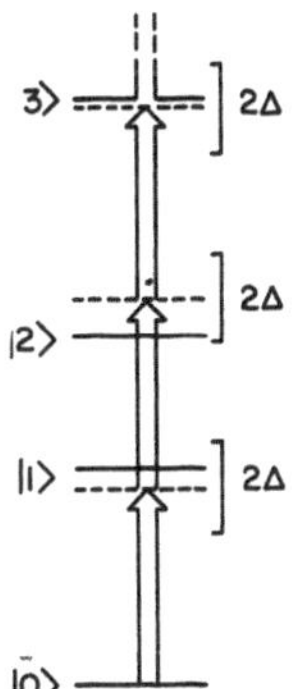

Fig.5.6. Diagram of levels for which the requirement $\gamma \cong \Delta$ is sufficient for compensation of the detuning due to power broadening

5.3 Interaction of Nonequidistant Multilevel Systems with a Quasi-Resonant Field

In this section we consider a more complicated question than in the previous section: the effect of a monochromatic light field on multilevel systems with nonequidistant energy levels. Such systems are of great interest for understanding why rather moderate laser intensities are sufficient for the excitation of very high vibrational levels (up to dissociation) in polyatomic molecules. Even in the simplest case of a single nondegenerate vibrational mode (i.e., a quasi-diatomic molecule) vibrational anharmonicity and rotation of the molecular framework cause the frequencies of successive vibrational transitions to differ. As a result, a study of nonequidistant multilevel systems must be the starting point of any theory of laser excitation of polyatomic molecules. The effects of vibrational degeneracy and molecular rotation will be considered in a later section; here we consider some examples of nondegenerate, nonequidistant multilevel systems.

Of course, the excitation of any nearly equidistant multilevel system can be described in the same term as an equidistant system if the field is sufficiently strong, as indicated by the criteria derived in Sect.5.2.3. However, the results for an equidistant system are not generally applicable to nonequidistant systems under low to moderate laser field strengths. In Sects.5.3.1 and 2 we shall concentrate on situations in which nonequidistant behavior becomes apparent, that is, for moderate laser intensities. In this case multiphoton resonances seem to be by far the most important route for excitation. We consider the dynamics of multiphoton transitions in detail in Sect.5.3.1. The main features of multiphoton resonances, such as the width and the excitation rate, are illustrated by numerical results in Sect.5.3.2.

For several reasons the most interesting multilevel systems, polyatomic molecules, cannot be considered as quasi-diatomic. First, the presence of many vibrational degrees of freedom with equal or nearly equal frequencies (degenerate or nearly degenerate vibrations) is of importance, because their interaction through vibrational anharmonicity allows the splitting of degenerate or nearly degenerate levels, thus affecting both the number and the detuning of allowed optical transitions. Second, as the energy above the ground state increases, the density of vibrational states increases very rapidly in a polyatomic molecule. This results in many closely spaced vibrational transitions starting from a given level, though each one may have a relatively small dipole moment. Such reasoning leads us to a class of multilevel models which we denote for short as "oscillator-type systems". These models employ two subsystems: a lower, nonequidistant one (subsystem B) with a few levels, and an upper, infinite one (subsystem A) with equidistant levels (see Figs.5.7,8). The lower subsystems is obviously intended to model low molecular vibration-rotation levels, for the excitation of which multiphoton transitions play an essential role. The upper subsystem models high-lying vibrational levels, where a sequence of exactly resonant transitions can be selected from the many weakly allowed ones. In this context, models in which the dipole moments of the upper transitions are considerably smaller than the dipole moments of the lower transitions are of special interest. We consider the dynamics of excitation of such systems in Sects.5.3.3,4. In this case some additional effects appear which are different from simple power broadening (see Sect.5.2.3). The dynamics of excitation of multilevel oscillator-type systems has been investigated in [5.8,32-34].

For calculations of the excitation of molecular vibrations for high levels of excitation, where the density of energy levels is very high, practical considerations may necessitate considering each step of the vibrational ladder separately. It has naturally been suggested [5.35,36] that the situation need not be described by the Schrödinger equation, but may instead be described by an ordinary rate equation, using rates calculated in accordance with the well-known "Fermi golden rule" for transitions into the continuous spectrum. We give a discussion of the

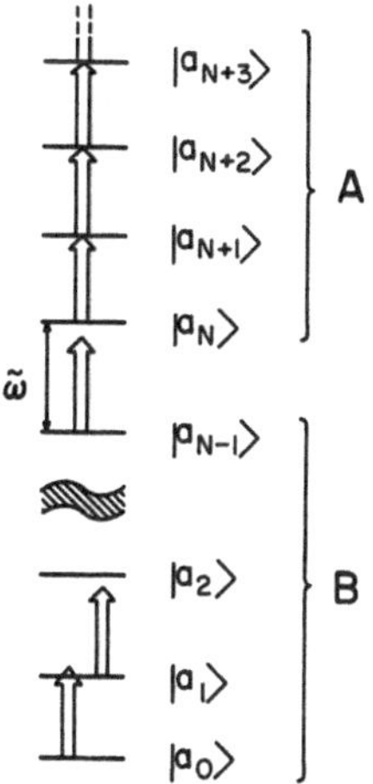

Fig.5.7. Diagram of levels for oscillator-type systems

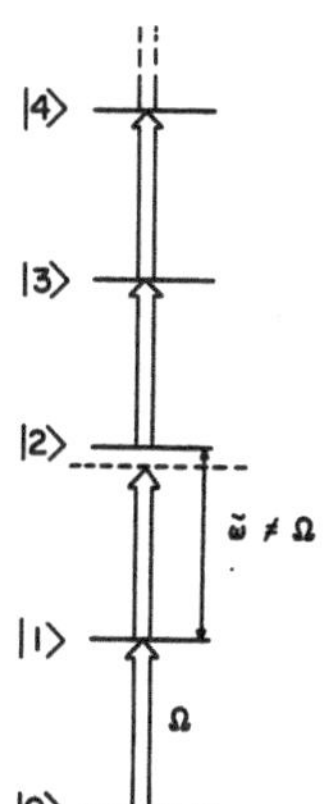

Fig.5.8. Diagram of levels for detuning of transition 1 → 2 with respect to exact resonance

validity of this hypothesis in Sect.5.3.5 through an analytical solution of the Schrödinger equation for the simplest multiplet systems with a quasi-continuous spectrum of transitions.

5.3.1 Multiphoton Resonances

A characteristic feature of the dynamics of multilevel quantum systems driven by a coherent optical-frequency electric field is the occurrence of sharp resonances when the laser frequency is equal to 1/N of the energy difference between the ground state and an excited state with N vibrational quanta, when none of the intermediate transitions is resonant. Such an N-photon resonance may be described to lowest order of approximation by equations similar to those describing resonant (one-photon) excitation of a two-level system, as we shall show below. When none of the intermediate steps is resonant, the N-photon Rabi frequency is typically much smaller than the one-photon Rabi frequency (if we assume that all the intermediate dipole moments are of the same order of magnitude). The time required to build up significant population in a state $|Ns_N\rangle$ is of the order of magnitude of the reciprocal $(\gamma^{(N)})^{-1}$ of the N-photon Rabi frequency $\gamma^{(N)}$, and is therefore typically much longer than the time required to build up significant population in a single-photon resonance. As a function of the optical electric field amplitude $\mathcal{E}$, the population of an N-quantum state $|Ns_N\rangle$ in N-photon resonance is proportional to $|\mathcal{E}|^{2N}$ for times short compared to the N-photon Rabi frequency. Multiphoton resonances with N ~ 7 to 10 have been observed in atomic ionization, where return of population from the excited state to the ground state is prevented by irreversible decay of the excited state. Because the experimentally observed number of laser quanta absorbed per molecule does not depend upon a very high power of the laser intensity, the possibility that multiphoton resonances may play an important role in the multiple-photon excitation of SF_6 and other polyatomic molecules has been discounted by some. However, it appears probable, in view of calculations which will be reviewed in Sect.5.4, that multiphoton resonances play the most important role in the excitation of levels with 2, 3, or 4 vibrational quanta. It also appears to be the case that enhancement of the probability of N-photon excitation by near-resonance of intermediate transitions is very common in multiplet systems, and has much to do with determining the dependence of excitation probability on laser frequency.

Rabi Frequency for Multiphoton Transitions

We shall first derive the usual expression [5.6] for the Rabi frequency $\gamma^{(N)}$ for an N-photon transition, and shall then comment on an appropriate interpretation of this result using the example of a two-photon resonance.

To derive the usual expression for $\gamma^{(N)}$, we shall employ (5.1.7). Let the detuning of the transition $j \to j+1$ be

$$\delta_j = (\hbar)^{-1}(j\Omega - \varepsilon_j) \quad . \tag{5.3.1}$$

We assume that none of the intermediate steps is resonant, i.e.,

$$|\delta_j| \gg \gamma_j \quad \text{for} \quad j = 1,\ldots,N-1 \tag{5.3.2}$$

where γ_j, the resonant Rabi frequency for the transition $j \to j+1$, is given in (5.2.2). On the other hand, we assume also that $|\delta_{N-1}| \lesssim \gamma_{N-1}$, i.e., that the last step is resonant or nearly resonant. The equations of motion governing the first step are

$$\frac{da_0}{dt} = i\gamma_1 a_1 \tag{5.3.3}$$

$$\frac{da_1}{dt} = i\delta_1 a_1 + i\gamma_1 a_0 + i\gamma_2 a_2 \quad . \tag{5.3.4}$$

An approximate solution for $a_1(t)$ is, in view of (5.3.2),

$$a_1(t) \cong \frac{\gamma_1}{\delta_1}\,[1 - \exp(-i\delta_1 t)]a_0(t) \quad , \tag{5.3.5}$$

provided we assume that $a_0(t)$ is slowly varying on the time scale of $(\delta_1)^{-1}$ and neglect a_2. Physically, (5.3.5) expresses $a_1(t)$ as the sum of a slowly varying and a rapidly varying part. Since one might hope that the rapidly varying second term in (5.3.5) will not contribute to the secular time development of a_j for $j > 1$, we discard it, leaving

$$a_1(t) \cong \frac{\gamma_1}{\delta_1}\,a_0(t) \quad . \tag{5.3.6}$$

Note that $|a_1| \ll |a_0|$, by virtue of (5.3.2). If we continue this process up the ladder, we find that

$$\frac{da_N}{dt} = i\delta_N a_N + i\gamma_N \frac{\gamma_{N-1}}{\delta_{N-1}} \cdots \frac{\gamma_1}{\delta_1} a_0 \tag{5.3.7}$$

which is of the same form as the equation for the amplitude of the upper state in a two-level system. The analog of the two-level Rabi frequency is [5.6]

$$\gamma^{(N)} = \gamma_N \left|\frac{\gamma_{N-1}}{\delta_{N-1}}\right| \cdots \left|\frac{\gamma_1}{\delta_1}\right| \quad . \tag{5.3.8}$$

We shall provisionally call $\gamma^{(N)}$ the N-photon Rabi frequency. If the detuning is small compared to the N-photon Rabi frequency, i.e., if

$$|\delta_N| \ll |\gamma^{(N)}| \quad , \tag{5.3.9}$$

then the amplitude a_N can be comparable in magnitude to a_0, and much larger than any of the intermediate amplitudes. Equivalently, from (5.3.7), the time scale on which $|a_N|$ becomes large is $[\gamma^{(N)}]^{-1}$; by (5.3.2) and (5.3.8),

$$[\gamma^{(N)}]^{-1} \gg [\gamma_N]^{-1} \quad . \tag{5.3.10}$$

This means that the N-photon Rabi period is much longer than a single-photon Rabi period. Equation (5.3.7) therefore describes a very narrow resonance which grows correspondingly slowly in time, and in which population cycles periodically between levels $|0\rangle$ and $|N\rangle$ at approximately the frequency $2\gamma^{(N)}$.

Our derivation for $\gamma^{(N)}$ is completely equivalent to the lowest nonvanishing order of time-dependent perturbation theory. A formally different, but physically equivalent, derivation of (5.3.8) has been given by LARSEN and BLOEMBERGEN [5.6]. Our derivation of $\gamma^{(N)}$, and others which are physically equivalent, become invalid when $\gamma_j \lesssim |\delta_j|$ for any one level $|j\rangle$, since in that case the amplitude a_j can also have appreciable magnitude in addition to a_0 and a_N. In this case the magnitude of $\gamma^{(N)}$ is also much larger than when no intermediate transitions are resonant. The cycling of population between the levels $|0\rangle$ and $|N\rangle$ typically becomes aperiodic. Nevertheless the qualitative statement remains true that $|0\rangle$ and $|N\rangle$ are connected by a narrow resonance in which $|a_N|$ becomes appreciable over times on the order of $[\gamma^{(N)}]^{-1}$.

Evidently one of the conditions which must be satisfied as long as (5.3.7) is valid, is

$$|a_{N-1}| \ll |a_{N-2}| \ll \ldots \ll |a_1| \ll |a_0| \quad . \tag{5.3.11}$$

Initially this condition may be satisfied. However, the coupling of adjacent levels through the Schrödinger equation makes it very difficult for $|a_{N-1}|$ to remain the smallest amplitude when $|a_N|$ becomes comparable in magnitude to $|a_0|$, as must occur in the course of an N-photon Rabi cycle. Hence one is led to suspect that (5.3.8) is not the correct N-photon Rabi frequency at all, but is simply the initial rate of change of a_N, when $a_0 \cong 1$. We shall now show for a simple example that this is indeed the case. The dynamics of a three-level system, including the two-photon transition from level 1 to level 3, may be discussed exactly (within the framework of the RWA) using the method of dressed states. The quasi-energy spectrum is discrete, and the eigenvalue equation for the quasi-energy λ derived by setting the secular determinant of (5.1.11) equal to zero reduces to

$$\lambda^3 - (\delta_2 + \delta_3)\lambda - (\gamma_2^2 + \gamma_3^2 - \delta_2\delta_3)\lambda + \delta_3\gamma_2^2 = 0 \quad . \tag{5.3.12}$$

Under a condition of two-photon resonance, the laser frequency Ω is such that $2\Omega = (E_3 - E_1)/\hbar$; this means that $\delta_3 = 0$. For a two-photon resonance, then, the eigenvalues λ are

$$\begin{aligned} \lambda_1 &= 0 \\ \lambda_{2(3)} &= \frac{1}{2}\,\delta_2 \pm \left(\frac{1}{4}\,\delta_2^2 + \gamma_2^2 + \gamma_3^2\right)^{\frac{1}{2}} \quad . \end{aligned} \tag{5.3.13}$$

When the single-photon Rabi frequencies are small compared to the detuning δ_2 of the intermediate level, that is, when (5.3.2) is satisfied, then the eigenvalues λ_2 and λ_3 may be written approximately as follows

$$\lambda_2 \cong |\delta_2| \tag{5.3.14}$$

$$\lambda_3 \cong -\frac{\gamma_2^2 + \gamma_3^2}{|\delta_2|} \quad . \tag{5.3.15}$$

The eigenvectors may be determined straightforwardly and substituted into (5.3.37), which is derived below, to give the time dependence of the upper-state amplitude a_2

$$a_2(t) \cong \frac{\gamma_2\gamma_3}{\gamma_2^2 + \gamma_3^2}\,[\exp(-i\lambda_3 t) - 1] \quad . \tag{5.3.16}$$

It is evident from (5.3.16) that the two-photon Rabi frequency is (5.3.15), and not the result predicted by the usual formula (5.3.8),

$$\lambda_3' = \frac{\gamma_2\gamma_3}{|\delta_2|} \quad . \tag{5.3.17}$$

However, expansion of the correct result (5.3.16) for short times reveals that

$$\frac{da_2}{dt} \cong i\lambda_3' \quad \text{for} \quad t \ll |\lambda_3|^{-1} \quad . \tag{5.3.18}$$

Thus the usual formula for the two-photon Rabi frequency [5.6] does not give the correct frequency of oscillation of the upper-state amplitude, but instead gives the correct initial rate of change of the upper-state amplitude. It should be noted that the correct two-photon Rabi frequency (5.3.15) is also proportional to the square of the field amplitude, as expected.

5.3.2 Numerical Calculations for Multilevel Systems

The analytical methods described in Sect.5.2 are difficult to apply in a number of cases of theoretical and practical interest: 1) when multiple ladders of energy levels originate at a single ground state, i.e., when the system is degenerate; 2) when the transition frequencies $\omega_j - \omega_{j-1}$ of a nondegenerate system, or the laser frequency, are such that many or all levels are detuned from exact resonance; 3) when the single-photon Rabi frequencies γ_j are such that the special analytically solvable cases discussed in Sect.5.2 do not apply. However, in all these cases the quasi-energy or dressed-states approach still allows one to make major simplifications in numerical calculations, as long as the electric-field amplitude remains constant in time. Specifically, the dressed-states formalism allows one to calculate almost trivially the populations $a^*_{js}(t)a_{js}(t)$, either as functions of time or averaged over a time long compared to any single-photon or multiphoton Rabi frequencies. (In what follows we shall use the notation for a multiplet system; the simplification to a nondegenerate system should be obvious). If $|a_{js}(t)|^2$ is desired as a function of time, then a single diagonalization of an $N \times N$ matrix (where N is the number of energy levels) allows one to calculate $|a_{ns}(t)|^2$ for arbitrary initial conditions and for arbitrarily long times. This flexibility contrasts very favorably with step-by-step numerical integration of the Schrödinger equation, which is expensive to carry out for long times or for many different sets of initial conditions. Step-by-step integration is still necessary, however, for arbitrary (general) pulse shapes $\mathcal{E}(t)$, when detuning is present.

In order to have a general framework for discussing the time dependence of the amplitudes $a_{ns}(t)$ and populations $|a_{ns}(t)|^2$, it is useful to introduce the unitary time-development operator $\hat{U}^S(t_2,t_1)$. This operator may be used to express the time dependence of the state vector at time t_2 in terms of the state vector at time t_1

$$|\Psi^S(t_2)\rangle = \hat{U}^S(t_2,t_1)|\Psi^S(t_1)\rangle \quad . \tag{5.3.19}$$

The superscript S indicates that the operator and state vectors are in the Schrödinger picture, where all time dependence is carried by the probability amplitudes. If $\hat{U}^S$ is known, then merely specifying the initial conditions (at $t_1 = 0$) allows us to write out the populations immediately

$$|\langle n|\Psi^S(t)\rangle|^2 = |\langle n|\hat{U}^S(t,0)|\Psi^S(0)\rangle|^2 \quad . \tag{5.3.20}$$

For example, if it is assumed that the system is initially in the ground state, i.e., that

$$\langle ms|\Psi^S(0)\rangle = \delta_{m0} \quad , \tag{5.3.21}$$

then

$$|\langle ns|\Psi^S(t)\rangle|^2 = |\langle ns|\hat{U}^S(t,0)|mq\rangle|^2 \quad . \tag{5.3.22}$$

Thus the square of the absolute value of each matrix element of $\hat{U}^S$ is the probability $P(ns,t_2|mq,t_1)$ that the system will be found in a specific final state at time t_2, given that it was in a specific initial state at time t_1. This idea may be expressed formally as

$$P(mq,t_2|ns,t_1) = |\langle ns|\hat{U}^S(t_2,t_1)|mq\rangle|^2 \quad . \tag{5.3.23}$$

It is clear from this definition that if the system is initially in the state $|mq\rangle$, then $P(ns,t_2|mq,t_1)$ is the fraction of the total population found in the state $|ns\rangle$ at time t_2. For pulses which have a duration long compared to the longest single-photon or multiphoton Rabi period of the system, it is convenient to define the time-averaged probability

$$\bar{P}(ns|mq) = \lim_{T\to\infty} T^{-1}\int_0^T P(ns,t'|mq,0)dt' \quad . \tag{5.3.24}$$

The computational advantage of the dressed-states approach rests on the fact that, for a constant field amplitude $\mathcal{E}$, the matrix elements of the time-development operator $\hat{U}^S(t_2,t_1)$ may be expressed directly in terms of the dressed-state eigenvalues and eigenvectors, which may be determined by a single matrix diagonalization of the effective Hamiltonian. We shall now show how to calculate $\hat{U}^S(t_2,t_1)$.

The RWA equation of motion (5.1.7) may be derived from the equation of motion in the Schrödinger picture,

$$i\hbar\frac{\partial}{\partial t}|\Psi^S(t)\rangle = \hat{H}^S|\Psi^S(t)\rangle \tag{5.3.25}$$

where H^S has the following nonzero matrix elements

$$\langle ns|\hat{H}^S|ns\rangle = \hbar\varepsilon_{ns} \tag{5.3.26}$$

$$\langle ns|\hat{H}^S|n-1,q\rangle = -\langle ns|\hat{d}|n-1,q\rangle\mathcal{E}\cos\Omega t \quad . \tag{5.3.27}$$

The state vector $|\Psi^S(t)\rangle$ may be expanded as follows

$$|\Psi^S(t)\rangle = \sum_{ns} a^S_{ns}(t)|n\rangle \quad . \tag{5.3.28}$$

The transformation from $|\Psi^S(t)\rangle$, the state vector in the Schrödinger picture, to

$$|\Psi^I(t)\rangle \equiv |\Psi(t)\rangle \quad , \tag{5.3.29}$$

the state vector given in (5.1.4), is essentially a transformation to an interaction picture. This transformation may be accomplished by the unitary operator $V(t)$

$$|\Psi^I(t)\rangle = \hat{V}(t)|\Psi(t)\rangle \tag{5.3.30}$$

where

$$\hat{V}(t) = \sum_{mq} |mq\rangle e^{im\Omega t}\langle mq| \quad (5.3.31)$$

$$= \exp(\frac{i}{\hbar}\hat{H}_0 t) \quad (5.3.32)$$

and

$$\hat{H}_0 = \sum_{mq} \hbar|mq\rangle m\Omega\langle mq| \quad . \quad (5.3.33)$$

The use of (5.3.23) transforms the Schrödinger equation (5.3.25) into the following equation for $|\Psi^I(t)\rangle$

$$i\hbar \frac{\partial}{\partial t}|\Psi^I(t)\rangle = \hat{H}^{eff}|\Psi^I(t)\rangle \quad (5.3.34)$$

where the effective Hamiltonian in the interaction (dressed-states) picture is

$$\hat{H}^{eff} = \hat{V}(t)\hat{H}^S\hat{V}(t)^{\dagger} - \hat{H}_0 \quad . \quad (5.3.35)$$

The diagonal matrix elements of $\hat{H}^{eff}$ are

$$ns|\hat{H}^{eff}|ns = \hbar(\varepsilon_{ns} - n\Omega) \quad (5.3.36)$$

and in the RWA the nonzero off-diagonal matrix elements

$$ns|\hat{H}^{eff}|n\text{-}1q = -\hbar\gamma_{ns,(n-1)q} \quad (5.3.37)$$

are independent of time. The H^{eff} defined in this way is identical with the RWA effective Hamiltonian used by other authors.

In the interaction picture defined by (5.3.30), the unitary time-development operator $\hat{U}^I(t_2,t_1)$, such that

$$|\Psi^I(t_2)\rangle = \hat{U}^I(t_2,t_1)|\Psi^I(t_1)\rangle \quad , \quad (5.3.38)$$

satisfies the equation

$$i\hbar\frac{\partial}{\partial t_2}\hat{U}^I(t_2,t_1) = H^{eff}\hat{U}^I(t_2,t_1) \quad , \quad (5.3.39)$$

the solution of which is

$$\hat{U}^I(t_2,t_1) = \exp\left[\frac{i}{\hbar}\hat{H}^{eff}(t_2 - t_1)\right] \quad . \quad (5.3.40)$$

The unitary time-development operator in the Schrödinger picture may be shown from (5.3.19,30,39) to be

$$\hat{U}^S(t_2,t_1) = \hat{V}(t_2)^{\dagger}\hat{U}^I(t_2,t_1)\hat{V}(t_1) \quad . \quad (5.3.41)$$

Using (5.3.32) and the secular equation for $\hat{H}^{eff}$,

$$\sum_{mq}\langle ns|\hat{H}^{eff}|mq\rangle p_{mq}(\lambda) = \lambda p_{ns}(\lambda) \quad , \quad (5.3.42)$$

where $p_{mq}(\lambda)$ are the components of the dressed state $|\lambda\rangle$ with respect to the basis $\{|mq\rangle\}$, we can easily express $\hat{U}^I(t_2,t_1)$ as follows

$$\hat{U}^I(t_2,t_1) = \sum_\lambda |\lambda\rangle \exp[-i\lambda(t_2,t_1)]\langle\lambda| \quad . \tag{5.3.43}$$

From (5.3.41,43), with the use of the completeness relation for $\{|ns\rangle\}$, we find that

$$\langle ns|\{U^S(t_2,t_1)|mq\rangle\} = \sum_\lambda p_{ns}(\lambda)p_{mq}(\lambda)^* \cdot \exp\{i[\lambda(t_1 - t_2) + \Omega(mt_1 - nt_2)]\} \quad . \tag{5.3.44}$$

From (5.1.7,38,43), the time evolution of the probability amplitudes $a_{ns}(t)$ is given by

$$\begin{aligned} a_{ns}(t_2) &= \langle n|\Psi^I(t_2)\rangle \\ &= \sum_\lambda p_{ns}(\lambda)e^{-i\lambda(t_2-t_1)} \langle\lambda|\Psi^I(t_1)\rangle \\ &= \sum_{\lambda mq} p_{ns}(\lambda)e^{-i\lambda(t_2-t_1)} p_{mq}(\lambda)a_{mq}(t) \quad , \end{aligned} \tag{5.3.45}$$

which is consistent with (5.3.44), in view of (5.3.30,31).

The probability that the system will be found in the state $|n\rangle$ at time t_2 if it was in the state $|m\rangle$ at time t_1 is therefore

$$P(ns,t_2|mq,t_1) = \left|\sum_\lambda p_{ns}(\lambda)p_{mq}(\lambda)\exp[-i\lambda(t_2 - t_1)]\right|^2 \quad , \tag{5.3.46}$$

and the time-averaged probability $\bar{P}(ns|mq;\Omega)$ is, from (5.3.24),

$$\bar{P}(ns|mq;\Omega) = \sum_\lambda |p_{ns}(\lambda)p_{mq}(\lambda)|^2 \quad . \tag{5.3.47}$$

The formulas (5.3.46,47) are very simple and are well suited to computer evaluation for an arbitrary time difference $t_2 - t_1$. To provide a simple example, we have evaluated (5.3.46) as a function of the laser frequency Ω, and (5.3.47) as a function of $t \equiv t_2 - t_1$ (assuming $m = 0$), for a simple model of a nondegenerate anharmonic oscillator. In this model, the vibrational energy levels are

$$\varepsilon_n = n\omega + n(n-1)X \tag{5.3.48}$$

and the single-photon Rabi frequencies are

$$\gamma_n = \frac{\mu\sqrt{n}}{2\hbar} \quad . \tag{5.3.49}$$

The frequency of an N-photon resonance from $n = 0$ to $n = N$ is [5.37]

$$\Omega^{(N)} = \omega + (N-1)X \quad , \tag{5.3.50}$$

so that one expects successively higher multiphoton resonances to occur at successively larger displacements from the one-photon resonant frequency ω, the displacement being proportional to $(N-1)$. The results obtained by numerical diagonalization of $\hat{H}^{eff}$ for a basis $\{|m\rangle; a \leq m \leq 9\}$ and evaluation of (5.3.46,47) are shown in Figs.5.9 and 10. In these calculations we used parameters appropriate for SF_6: $\omega = 948\ cm^{-1}$, $X = -2.54\ cm^{-1}$, even though our model anharmonic oscillator is nondegenerate and bears little resemblance to SF_6.

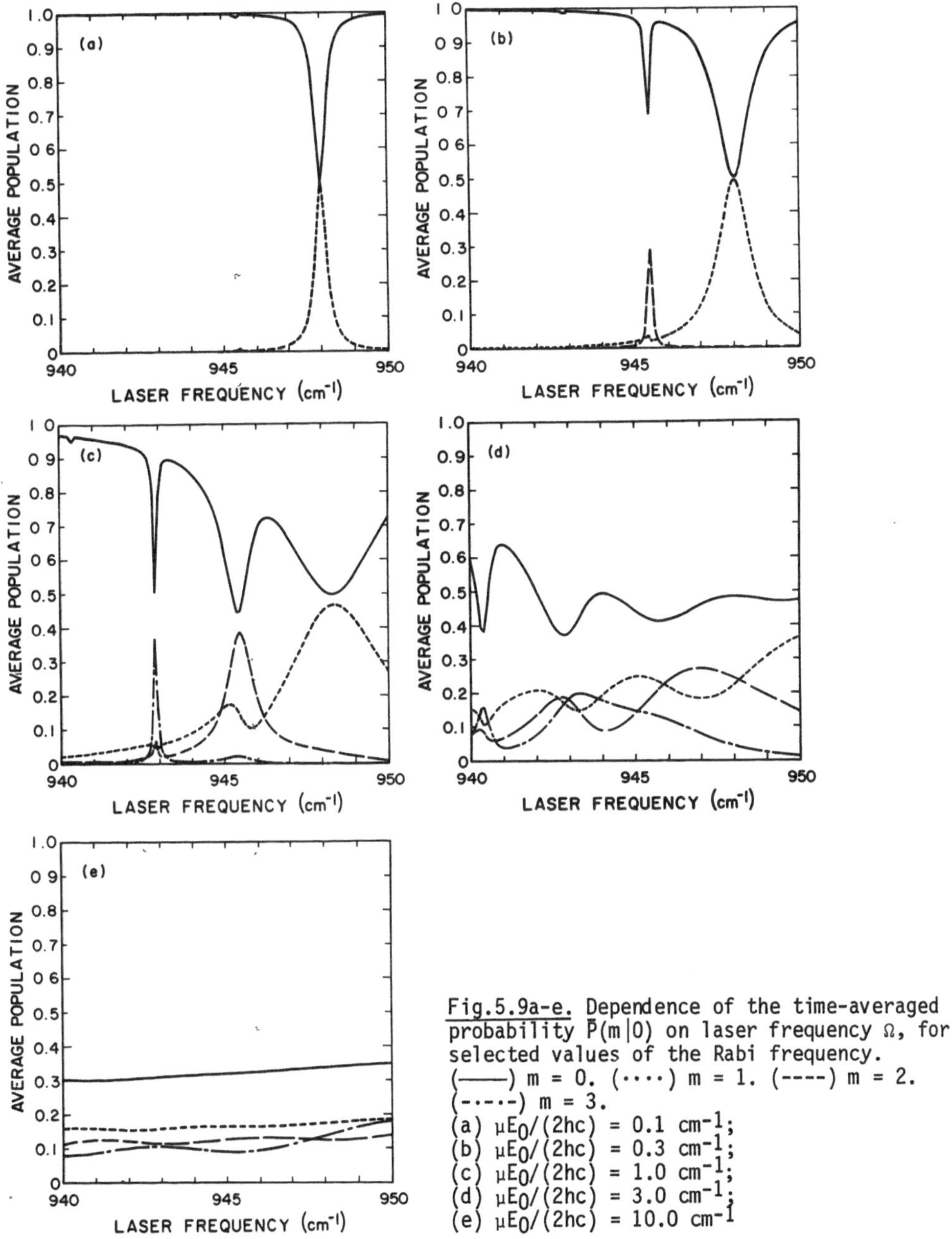

Fig.5.9a-e. Dependence of the time-averaged probability $\bar{P}(m|0)$ on laser frequency Ω, for selected values of the Rabi frequency. (——) m = 0. (····) m = 1. (----) m = 2. (-·-·-) m = 3.
(a) $\mu E_0/(2hc) = 0.1\ cm^{-1}$;
(b) $\mu E_0/(2hc) = 0.3\ cm^{-1}$;
(c) $\mu E_0/(2hc) = 1.0\ cm^{-1}$;
(d) $\mu E_0/(2hc) = 3.0\ cm^{-1}$;
(e) $\mu E_0/(2hc) = 10.0\ cm^{-1}$

In Fig.5.9 we show the time-averaged probabilities $\bar{P}(m|0)$ of the levels with zero to three vibrational quanta ($m = 0,\ldots, 3$) as a functions of the laser frequency, for selected values of the optical field strength E_0. Since $\bar{P}(m|0)$ is the time-averaged probability that the system will be found in level m, given that it was initially in level 0, the quantity $\bar{P}(m|0)$ is the time-averaged fraction of the total population found in level m, for a system which started in level 0. In Fig. 5.9a, the field strength was chosen such that the Rabi frequency $\mu E_0/(2hc)$ was $0.1\ cm^{-1}$. The ground and first excited vibrational states ($m = 0$ and 1) behave essentially as a two-level system, with $\bar{P}(1|0) = \bar{P}(0|0) = \frac{1}{2}$ at resonance ($\Omega/(2\pi c) = 948\ cm^{-1}$). The small peak evident at $\Omega/(2\pi c) = 942.92\ cm^{-1}$ is due to two-photon excitation of $m = 2$. In Fig.5.9b, for $\mu E_0/(2hc) = 0.3\ cm^{-1}$, the one-photon resonance ($m = 0$ to $m = 1$) has been perceptibly broadened, and the two-photon resonance ($m = 0$ to $m = 2$) is evident as a narrow spike. In Fig.5.9c, for $\mu E_0/(2hc) = 1.0\ cm^{-1}$, the one-, two-, and three-photon resonances ($m = 0$ to $m = 1, 2, 3$, respectively) are evident as a series of progressively narrower peaks. The N-photon resonance occurs at a frequency $\Omega^{(N)} = \omega - (N-1)X$ according to (5.3.50), and this is clearly evident in Fig.5.9c. In Figs.5.9d,e, for Rabi frequencies $\mu E_0/(2hc) = 3.0\ cm^{-1}$ and $10.0\ cm^{-1}$, respectively, the N-photon resonances with $N \leq 3$ have become so greatly power broadened as to be almost unrecognizable as resonances.

In Fig.5.10 we show the time dependence of the probabilities $P(m,t|0,0)$ for $m = 0,\ldots, 3$, for selected values of the laser frequency. In Figs.5.10a-e we show the time dependence of $m = 0,\ldots, 3$ for various values of the Rabi frequency $\mu E_0/(2hc)$, at the one-photon and two-photon resonance frequencies $\Omega/(2\pi c) = 948.0\ cm^{-1}$ and $942.92\ cm^{-1}$, respectively. At the lower field strengths the levels $m = 0$ and $m = 1$ behave as a two-level system, undergoing periodic oscillations at the one-photon Rabi frequency. At the higher field strengths the oscillation is aperiodic, as expected from the analytical results discussed elsewhere in this section. However, a quasiperiod is evident in the time dependent of the $m = 0$ level [$P(0,t\ 0,0)$] in Figs.5.10b,c,e. This quasiperiod is the time required for

Fig.5.10a-f. Dependence of the probability $P(m,t|0,0)$ on time t, for selected values of the laser frequency Ω and selected Rabi frequencies. In (a)-(e): Curve beginning at upper left corner: $m = 0$. Curves beginning at lower left corner, from left, respectively: $m = 1, 2, 3$. Width of horizontal (time) axis = one Rabi period = $2h/\mu E_0$
(a) $\Omega/(2\pi c) = 948\ cm^{-1}$, $\mu E_0/(2hc) = 1.0\ cm^{-1}$.
(b) $\Omega/(2\pi c) = 948\ cm^{-1}$, $\mu E_0/(2hc) = 3.0\ cm^{-1}$.
(c) $\Omega/(2\pi c) = 948\ cm^{-1}$, $\mu E_0/(2hc) = 10.0\ cm^{-1}$.
(d) $\Omega/(2\pi c) = 942.92\ cm^{-1}$, $\mu E_0/(2hc) = 1.0\ cm^{-1}$.
(e) $\Omega/(2\pi c) = 942.92\ cm^{-1}$, $\mu E_0/(2hc) = 3.0\ cm^{-1}$.
In (f): Curve beginning at upper left corner: $m = 0$. Curve beginning at lower left corner: $m = 3$. Width of horizontal (time) axis: 8 Rabi periods = $16\ h/\mu E_0$. Rabi frequency $\mu E_0/(2hc) = 0.7\ cm^{-1}$

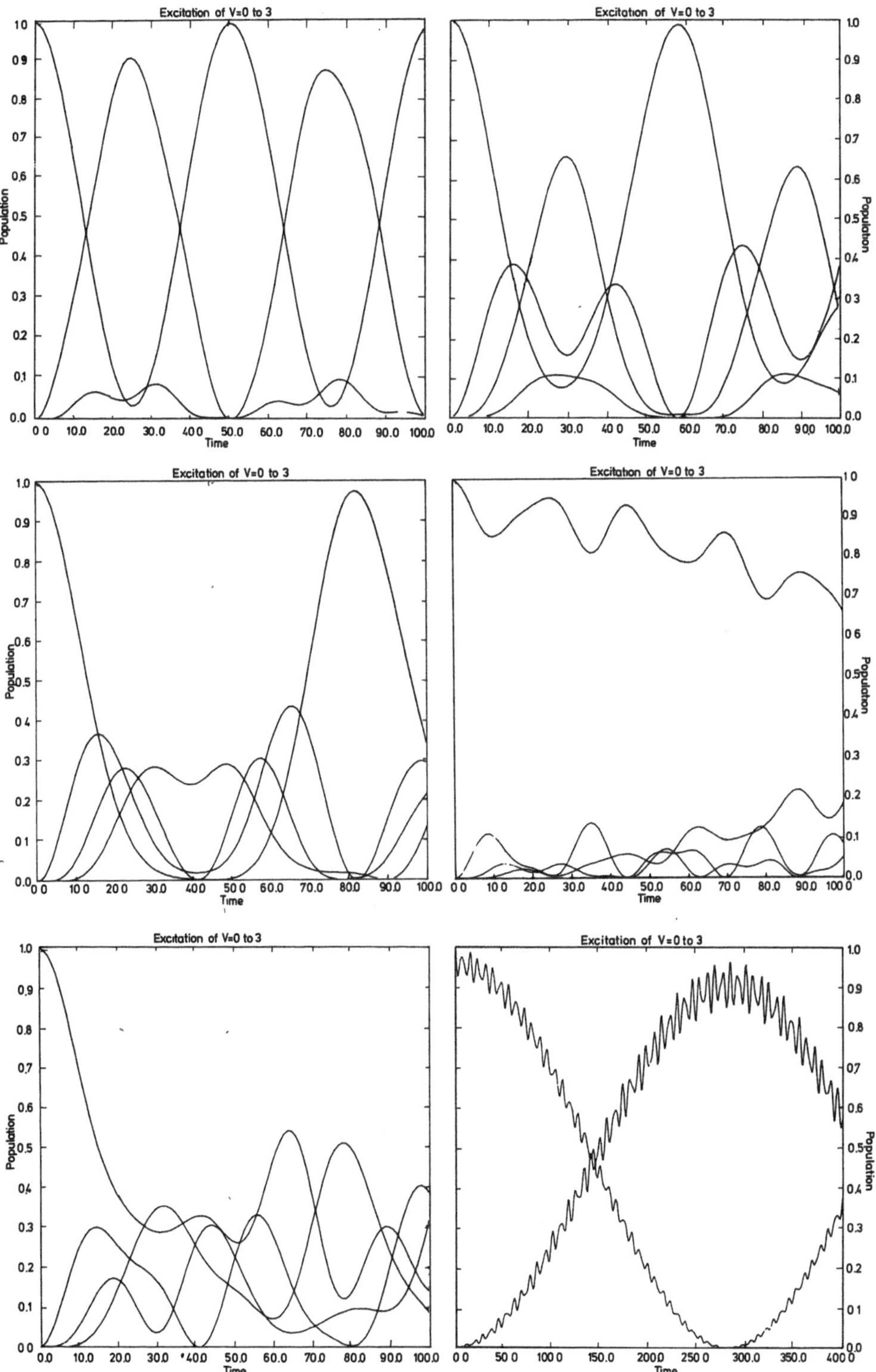

Fig.5.10a-f. Figure caption see opposite page

a "wave" of excitation to travel from $m = 0$ up the ladder of vibrational states to the highest level with significant excitation (for the given value of E_0), "reflect" from this "anharmonicity barrier", and return to $m = 0$. It is important to stress that "reflection" in this case is not a result of truncation of the basis set, but is a consequence of progressively larger detunings for higher and higher excitation, as a result of the physical phenomenon of vibrational anharmonicity.

Finally, in Fig.5.10f we show the time dependence of the probabilities $P(m,t|0,0)$ for $m = 0$ and 2 at the two-photon resonance frequency $\Omega/(2\pi c) = 942.92\ \mathrm{cm}^{-1}$, for $\mu E_0/(2hc) = 0.7\ \mathrm{cm}^{-1}$. As is evident from the change of time scale in this figure, the two-photon Rabi frequency is substantially lower than the one-photon Rabi frequency. Further, in Fig.5.10f we see a small-amplitude oscillation superimposed upon the slow secular variation of the $m = 0$ and $m = 2$ populations, as was supposed in deriving (5.3.7).

5.3.3 Dynamic Stark Effect and Frequency Shifts

In Sect.5.2.2 we have considered a system with a jump in the dependence of the dipole moment on the transition number ($\mu_{01} \gg \mu_{n,n+1}$ for $n \geqq 1$). We note that in this case even the existence of resonance for all successive transitions does not result in effective excitation. The natural question arises whether such specificity with regard to dipole moments should be an insuperable difficulty for excitation, or whether the situation can be improved by changing the value of the $1 \to 2$ transition frequency. The answer seems to be qualitatively clear from Fig.5.5. To provide effective excitation, the upper subsystem of relatively weak transitions must be tuned into resonance with either of the two dressed states which result from the interaction of the field with the lower two-level system.

An Analytically Solvable Example

To demonstrate this, let us consider the solution of (5.1.7) in the case of resonance for all transitions, except the transition $1 \to 2$, the frequency $\tilde{\omega}$ of which is arbitrary (see Fig.5.8). The equations corresponding to (5.1.11) which determine the quasi-energy eigenvalues and the structure of the eigenvectors have the following form

$$-\lambda p_0 = \gamma\beta p_1$$

$$-\lambda p_1 = \gamma\beta p_0 + \gamma p_2$$

$$(\tilde{\omega} - \Omega - \lambda)p_n = \gamma(p_{n-1} + p_{n+1}) \quad (n \geqq 2) \quad . \tag{5.3.51}$$

Here the notation of Sects.5.1.3 and 5.2.2 is used. We recall that $\beta = \gamma_1/\gamma$ and $\gamma_n = \gamma = \text{constant for } n \geqq 2$.

To obtain the solution of (5.3.51), it is possible to use the procedure given in Sect.5.2.2 for (5.2.32). As a result, for any given eigenvalue λ the coefficients $p_n (n \geqq 2)$ may be represented as a linear combination of Chebyshev polynomials. In the case under consideration this linear combination may conveniently be written in the following form

$$p_n = yU_{n-3}(z) + [\beta(y^2 - 1) + \beta^{-1}]U_{n-2}(z) + 2yU_{n-1}(z) + \beta^{-1}U_n(z) \quad \text{for} \quad n \geqq 2 \ , \tag{5.3.52}$$

where $z = (\tilde{\omega} - \Omega - \lambda)/2\gamma$, $y = (\Omega - \omega)/\beta\gamma$. The coefficients p_0 and p_1 are given by the following relations

$$p_0 = 1 \ , \quad p_1 = y + 2z/\beta \quad . \tag{5.3.53}$$

From the representation of the solution in the form of Chebyshev polynomials one concludes that the spectrum of eigenvalues contains the continuous segment

$$\tilde{\omega} - \Omega - 2\gamma \leqq \lambda \leqq \tilde{\omega} - \Omega + 2\gamma \quad \text{or} \quad -1 \leqq z \leqq 1 \quad . \tag{5.3.54}$$

In general case there also exist two discrete eigenvalues which may be found from (5.2.36).

To estimate the effectiveness of excitation, it is not necessary to find an analytical form for the discrete eigenvalues. It suffices to find the weight function $u(z)$ which effects the expansion of the solution with initial conditions $a_n(0) = \delta_{n0}$ in terms of the continuous-spectrum eigenvectors, and then to find the probability W_c (5.2.40) for escape of the system to an unbound state.

In accordance with the general methods which have been formulated in Sect.5.1.2, the equation for the determination of the weight function $u(z)$ has the following form

$$\sum_{n=0}^{\infty} p_n(z)p_n(z') = \frac{1}{u(z)}\,\delta(z - z') \quad , \tag{5.3.55}$$

where the $p_n(z)$ are given by (5.3.52,53) as functions of n. Multiplying (5.3.55) by $(1 - z'^2)^{\frac{1}{2}}$, integrating with respect to z' in the segment $-1 \leqq z \leqq 1$, and using an explicit form for the Chebyshev polynomials we arrive at the following expression for the weight function $u(z)$

$$u(z) = 2\pi^{-1}(1 - z^2)^{\frac{1}{2}}\{8y\beta^{-1}z^3 + 4(3y^2 - 1 + \beta^{-2})z^2 + [6\beta y(y^2 - 1) + 4y\beta^{-1}]z + y^2 + \beta^2(1 - y^2)^2\}^{-1} \quad . \tag{5.3.56}$$

From this the probability W_c can be calculated by the following formula

$$W_c = \sum_{n=0}^{\infty} \left| \int_{-1}^{1} u(z)p_n(z)dz \right|^2 \quad . \tag{5.3.57}$$

We already know the solution for a particular value of the parameter y, namely, y = 0, that is, for resonance at all transitions. For $\beta \gg 1$ the contribution of the continuous spectrum to the solution is very small, see (5.2.44). From (5.3.52, 53,56,57) one can see that under the condition

$$|1 - y^2| \gg \beta^{-1} \tag{5.3.58}$$

the solution differs only slightly from the solution when y = 0. It is easy to show that under the condition (5.3.58) the estimate $W_c \sim \beta^{-2} \ll 1$ also holds, as in the exactly resonant case.

The solution differs significantly from the solution when all transitions are resonant only when the condition

$$|1 - y^2| \lesssim \beta^{-1} \tag{5.3.59}$$

is met. This condition corresponds to resonance of the second excited level with the one of the dressed states formed by the interaction of the field with the lower pair of levels. Under this condition there exists only a single discrete eigenvalue, and the other one goes into the continuous spectrum.

We shall consider only the case of exact equality, $y^2 = 1$. If in (5.3.52,53, 56) we neglect the terms $\sim\beta^{-1}$ and $\sim\beta^{-2}$, then the integrals in (5.3.57) reduce to tabulated ones. As a result we find that $W_c \cong \frac{1}{2}$ for $y^2 = 1$. Thus with probability $\sim\frac{1}{2}$ the system goes into the continuous spectrum, resulting in its unlimited excitation, and with probability $\sim\frac{1}{2}$ the system remains in a single discrete dressed state. To excite the system from the ground state with probability ~ 1, two upper subsystems are needed.

Figure 5.11 illustrates the different cases considered. The collection of solutions we have obtained allows us to formulate the principle of "quasi-energy resonance" that is, the condition of resonance taking into account the shifts and splittings of levels because of the dynamic Stark effect. The principle of quasi-energy resonance may also be deduced in a general form for more complicated oscillator-type systems.

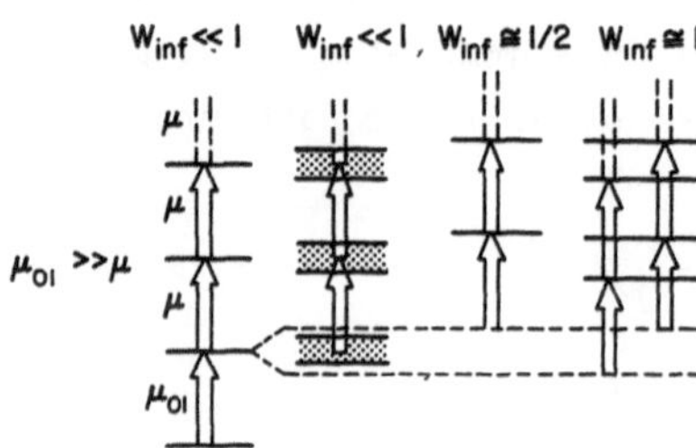

Fig.5.11. Illustration of the principle of quasi-energy resonance

General Approach to an Approximate Description of the Dynamics of Oscillator-Type Systems

The approximate method which will help us to obtain the solution of (5.1.7) in the case of oscillator-type systems is based on simple physical considerations. Since (by assumption) the dipole moments in the lower subsystem B are much greater than those in the upper subsystem A, the probability amplitudes of the lower subsystem change more quickly, and the upper subsystem A can be treated mathematically as a perturbation.

Let the lower subsystem B consist of N levels. We rewrite (5.1.7) in a form which suggests resonance for transition in the upper subsystem, that is, beginning from the Nth level:

$$\begin{aligned}
&\frac{da_0}{dt} = i\gamma_1 a_1 \\
&\frac{da_n}{dt} + i(\varepsilon_n - n\Omega)a_n = i\gamma_n a_{n-1} + i\gamma_{n+1}a_{n+1} \quad \text{for} \quad 1 \le n \le N-2 \\
&\frac{da_{N-1}}{dt} + i[\varepsilon_{N-1} - (N-1)\Omega]a_{N-1} = i\gamma_{N-1}a_{N-2} + i\gamma_N a_N \\
&\frac{da_N}{dt} + i(\varepsilon_{N-1} + \tilde{\omega} - N\Omega)a_N = i\gamma_N a_{N-1} + i\gamma_{N+1}a_{N+1} \\
&\frac{da_n}{dt} + i(\varepsilon_{N-1} + \tilde{\omega} - N\Omega)a_n = i\gamma_n a_{n-1} + i\gamma_{n+1}a_{n+1} \quad \text{for} \quad n \ge N+1 \ .
\end{aligned} \tag{5.3.60}$$

Since we assume that

$$\gamma_{n \le N-1} \gg \gamma_{n \ge N} \quad , \tag{5.3.61}$$

we consider the lower subsystem B separately in order to obtain a zero-order approximation for the solution of (5.3.60). [That is, in (5.3.60) we assume $\gamma_N = 0$]. The general solution of such a truncated set of equations can be represented as a linear combination of the following functions:

$$a_n^{(j)}(t) = p_n^{(j)} \exp(-i\lambda_j t) \quad , \quad n \le N-1 \ , \quad 1 \le j \le N \quad , \tag{5.3.62}$$

which correspond to N quasi-energy eigenvalues λ. The dressed-state eigenvectors $||p_n^{(j)}||$ obey the normalization condition (5.1.17).

We now follow perturbation theory, that is, we substitute the zero-order approximation $a_{N-1}(t)$ into (5.3.60) for the upper probability amplitudes a_n ($n \ge N$). Since the equations are linear, the general solution for the upper levels has the form of a linear superposition of responses to each term $a_{N-1}^{(j)}(t)$ given by (5.3.62). Therefore all further information can be obtained from the following equations, where the index j is, for simplicity, omitted

$$\frac{da_N}{dt} + i(\varepsilon_{N-1} + \tilde{\omega} - N\Omega)a_N = i\gamma_N p_{N-1} \exp(-i\lambda t) + i\gamma_{N+1}a_{N+1}$$

$$\frac{da_n}{dt} + i(\varepsilon_{N-1} + \tilde{\omega} - N\Omega)a_n = i\gamma_n a_{n-1} + i\gamma_{n+1}a_{n+1} \quad \text{for} \quad n \geq N+1 \quad . \tag{5.3.63}$$

We renumber all the upper levels involved in (5.3.63) beginning from the lowest level of subsystem A, and we shift the energy by an amount which is common to all levels. Such a shift is equivalent to multiplying all probability amplitudes by a common phase factor. This procedure amounts to the transformation

$$a_n(t) = c_{n-N}(t) \exp[i(N\Omega - \varepsilon_{N-1} - \tilde{\omega})t] \quad . \tag{5.3.64}$$

We finally arrive at the following equations:

$$\frac{dc_0}{dt} = i\tilde{\gamma}_1 c_1 + ir \exp(iGt)$$

$$\frac{dc_n}{dt} = i\tilde{\gamma}_n c_{n-1} + i\tilde{\gamma}_{n+1}c_{n+1} \quad \text{for} \quad n \geq 1 \quad , \tag{5.3.65}$$

where $r = p_{N-1}\gamma_N$, $\tilde{\gamma}_n = \gamma_{n+N}$, and the frequency G is determined by the relation

$$G = \varepsilon_{N-1} + \tilde{\omega} - N\Omega - \lambda \quad . \tag{5.3.66}$$

Figure 5.12 shows that G denotes the detuning from exact resonance of the transition between the dressed state under consideration and the lowest level of A.

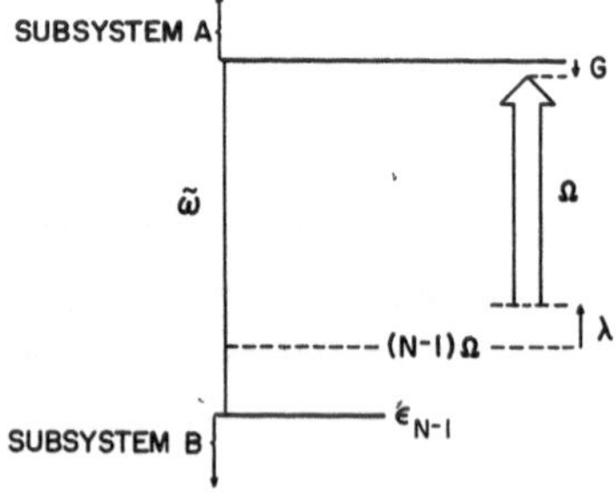

Fig.5.12. Relation between the frequencies λ, G and $\tilde{\omega}$

We are now able to find the solution of the inhomogeneous set of equations (5.3.65) complying with the initial conditions $c_n(0) = 0$. Assume that we know the solution $\tilde{c}_n(t)$ of the appropriate homogeneous set complying with the initial conditions $\tilde{c}_n(0) = \delta_{n0}$. Then the solution of (5.3.65) can be easily written as

$$c_n(t) = r \exp(iGt) \int_0^t \tilde{c}_n(\tau)\exp(-iG\tau)d\tau \quad . \tag{5.3.67}$$

We consider two particular cases for which the solutions $\tilde{c}_n(t)$ are known from Sect.5.2, where the dynamics of some equidistant systems has been analyzed.

Before turning to this discussion, let us make a general remark concerning the effectiveness of excitation for oscillator-type systems. It is obvious that the criterion for effective excitation from a given dressed state is

$$M_0 = \sum_{n=0}^{\infty} |c_n(t)|^2 \sim 1 \tag{5.3.68}$$

for the total population M_0 of the levels of the upper subsystem.

We shall not discuss below the details of the procedure for obtaining formulas for the probability amplitudes and the value of M_0. These formulas are compiled in Table 5.4. To obtain them, one has to combine (5.3.67) with (5.2.7,12). Asymptotic estimates for $\tilde{\gamma}t \gg 1$ are straightforward. Below in the text we merely comment on the final results.

Table 5.4. Summary of solutions describing the dynamics of oscillator-type systems

Rabi frequencies in the upper subsystem	$C_n(t)$	Condition on the detuning G	Asymptotic behavior of $M_0(t)$ for $\gamma t \gg 1$	Remarks
$\tilde{\gamma}_n = \tilde{\gamma}_n^{\frac{1}{2}}$	$ir(i\tilde{\gamma})^n(n!)^{-\frac{1}{2}} \exp(iGt)$ $\int_0^t \tau^n \exp(-iG\tau-\tilde{\gamma}^2\tau^2/2)d\tau$	Arbitrary	$(2\pi)^{\frac{1}{2}}|r|^2\tilde{\gamma}^{-1}t \exp(-G^2/2\tilde{\gamma}^2)$ $+2|r|^2\tilde{\gamma}^{-2}\phi(1,\tfrac{1}{2};-G^2/2\tilde{\gamma}^2)$	ϕ is the degenerate hypergeometric function
$\tilde{\gamma}_n = \tilde{\gamma} = \text{const}$	$i^{n+1}(n+1)r\tilde{\gamma}^{-1} \exp(iGt)$ $\int_0^t \tau^{-1} \exp(-iG\tau)J_{n+1}(2\tilde{\gamma}\tau)d\tau$	$|G| < 2\tilde{\gamma}$ and $\frac{2\tilde{\gamma}-|G|}{|G|} \gtrsim 1$	$2\left(1-\frac{G^2}{4\tilde{\gamma}^2}\right)^{\frac{1}{2}} \frac{|r|^2 t}{\tilde{\gamma}}$	J_{n+1} are Bessel functions
		$|G| > 2\tilde{\gamma}$ and $\frac{|G|-2\tilde{\gamma}}{2\tilde{\gamma}} \gtrsim 1$	$\frac{4|r|^2}{(G^2-4\tilde{\gamma}^2)^{\frac{1}{2}}[|G|+(G^2-4\tilde{\gamma}^2)^{\frac{1}{2}}]}$	
		$|G| = 2\tilde{\gamma}$	$\frac{2|r|^2}{\tilde{\gamma}^2}\left(\frac{2}{\pi}\tilde{\gamma}t\right)^{\frac{1}{2}}$	

Upper Subsystem—Harmonic Oscillator ($\tilde{\gamma}_n = \tilde{\gamma}n^{\frac{1}{2}}$)

In this case the expression for $M_0(t)$ consists of two terms (see Table 5.4). The first term describes the increase in the population of the levels of the upper subsystem, the total population being proportional to time. However, when the absolute value of the detuning G is great, this term is exponentially small. The second stationary term is also small when $|G| \gg \tilde{\gamma}$ (it is $\cong 2|r|^2/G^2$ or $\sim |p_{N-1}|^2\tilde{\gamma}^2/G^2$ since $\gamma_N \sim \tilde{\gamma}$). Thus, the excitation of the system is effective only when the detuning G of the transition between the upper quasi-level of the lower "rapid" subsystem B and the lowest level of the upper "slow" subsystem A is no larger than the Rabi frequency of the upper subsystem, that is, when

$$|G| \lesssim \tilde{\gamma} \quad . \tag{5.3.69}$$

This requirement is in strict agreement with the principle of quasi-energy resonance which has been formulated above.

Upper Subsystem with Equal Dipole Moments ($\tilde{\gamma}_n = \tilde{\gamma}$ = constant)

In this case the principle of quasi-energy resonance appears to be even more clearly operative than in the previous case. A contribution to the total upper-level population which is proportional to time appears only if the requirement (5.3.69) is met. In a more precise form this requirement is given in Table 5.4, where one can observe a sharp change of the features of the solution with a change of G. In accordance with the analytical solution considered above, the condition $|G| = 2\tilde{\gamma}$ corresponds to a situation in which a discrete dressed state enters the continuous spectrum of the upper subsystem, and therefore effective excitation can be realized.

The principle of quasi-energy resonance allows us to formulate a requirement on the frequency $\tilde{\omega}$ which must be satisfied for effective excitation of any oscillator-type system. Below we assume for simplicity that the optimum situation is realized with G = 0. In the next section we shall be interested in estimates of the excitation rate, and also in resonant properties which originate in the lower subsystem.

5.3.4 "Leakage" from the Lower Quantum States into the Upper Levels

Let us turn to the formulas describing the total population M_0 of the levels of the upper subsystem (see Table 5.4). Strictly speaking, these formulas are valid under the assumption that M_0 is considerably less than unity. Since M_0 is proportional to time, this assumption is violated for long times. We should therefore derive more exact formulas which may be used for the full time interval. To obtain such equations, we note that the rate of increase of M_0 is proportional to the population of the upper level of the lower subsystem B. Since transitions occur more rapidly in the lower subsystem than in the upper one, the population $a^*_{N-1}a_{N-1}$ can be represented as the product of $|p_{N-1}|^2$ and a correction factor $(1 - M_0)$, which was not taken into account above. This correction gives the following result

$$\frac{dM_0}{dt} \cong (1 - M_0) \quad \text{or} \quad M_0 \cong 1 - \exp(\mathscr{K}t) \quad . \tag{5.3.70}$$

Here, for G = 0, $\mathscr{K} = 2|p_{N-1}|^2\gamma_N^2/\tilde{\gamma}$ (for $\tilde{\gamma}_n = \tilde{\gamma}$ = constant) and $\mathscr{K} = (2\pi)^{\frac{1}{2}} \cdot |p_{N-1}|^2\gamma^2/\tilde{\gamma}$ (for $\tilde{\gamma}_n = \tilde{\gamma}n^{\frac{1}{2}}$). Since $\gamma_N \sim \tilde{\gamma}$ and since the exact value of the numerical coefficients is unimportant, we have used the following simple estimate in (5.3.70)

$$\mathscr{K} \sim |p_{N-1}|^2\tilde{\gamma} \quad . \tag{5.3.71}$$

It follows from these results that the decay of a given dressed state of the lower subsystem into the levels of the upper subsystem does not require a large broadening. The population of the $(N-1)^{th}$ level may be very small, but, nevertheless, the system "leaks" into the upper levels if the observation time is sufficiently long. For an illustration, we consider an example of a lower two-level subsystem in which the difference between the transition frequency and the field frequency is $\sim 10^{12}$ s^{-1} (this value is ~ 5 cm^{-1} in the units which are usually used by spectroscopists). If the Rabi frequency for the lower two-level subsystem is $\sim 10^{11}$ s^{-1} (for a typical allowed molecular vibration-rotation transition dipole moment of approximately 0.1 Debye, this Rabi frequency corresponds to an intensity of the exciting field of about 5×10^8 W/cm^2) the dressed state which is adiabatically connected with the ground one gives $|p_1|^2 \sim 10^{-2}$ for the upper population. If, for example, the dipole moment of the transitions in the upper subsystem is two orders of magnitude smaller than that in the lower one, then $\tilde{\gamma} \sim 10^9$ s^{-1}. As a result the expression (5.3.71) which gives the "leakage" rate leads to the estimate $\mathcal{K} \sim 10^7$ s^{-1}.

The physical interpretation of leakage is clear. This phenomenon is analogous to some extent to the well-known phenomenon of quantum tunneling through a potential barrier. Leakage into the levels of the upper subsystem may be treated as occurring by stages. At first, a small (but nonzero) population of the $(N-1)^{th}$ level decays irreversibly due to transitions into the states of the upper subsystem. Then the population of the $(N-1)^{th}$ level is reestablished due to pumping by the field from lower states, and again decays, and so on. That is, a stationary upwards flow of population is established. In time this flow embraces more and more of the levels of the upper subsystem.

This physical picture can be illustrated through a detailed analysis of the cases given in Table 5.4. We limit ourselves to the case when $\tilde{\gamma}_n = \tilde{\gamma} =$ constant. We know from the analysis made in Sect.5.2.1 that in this case the integrand changes substantially only during a time $t_n \sim n/2\tilde{\gamma}$. For $\tilde{\gamma}t \gg n$ the integration can be extended to the whole semi-infinite axis. As a result we obtain the following expressions for the stationary populations

$$c_n^{(st)} = i^{n+1} r/\tilde{\gamma} \quad \text{or} \quad |c_n^{(st)}|^2 = (|r|/\tilde{\gamma})^2 \sim |p_{N-1}|^2 \quad . \tag{5.3.72}$$

Thus, if one considers the behavior in time of the population distribution, this behavior appears to consist of a stationary upwards flow of population through the boundary between the lower and upper subsystems. Of course, the location of the upper end of the distribution depends on time and can be estimated as $n_{end} \sim 2\tilde{\gamma}t$. (This estimate follows from the corresponding result obtained in Sect.5.2.1).

In the discussion above, we rather schematically assumed the lower subsystem to be in a given dressed state. Such a schematic picture describes well the regime of adiabatically slow switching on of the laser field, since in that case a single dressed state $|0\rangle$ participates, which is adiabatically connected with the zero-field ground state. When the laser field is switched on instantaneously, the lower subsystem proceeds to make transitions to a superposition of its dressed states. In general, then, leakage with a probability equal to unity occurs only if a complete set of states of the upper subsystem are tuned in resonance with all the dressed states of the lower subsystem (compare with Fig.5.11). However, if the power broadening is small in comparison with a typical detuning, then the dressed state $|0\rangle$ plays the main role, and therefore only one upper subsystem is needed for effective leakage.

It can also happen that an (N-1)-photon resonance (see Sect.5.3.1) mixes two stationary states. Such a mixing leads to the participation of two dressed states, the splitting of which is equal to the doubled Rabi frequency $\gamma^{(N-1)}$ of the (N-1)-photon resonance. Under the condition

$$\gamma^{(N-1)} \gg \tilde{\gamma} \tag{5.3.73}$$

leakage with probability ~ 1 requires two upper subsystems, as in the case of one-photon resonance in the lower two-level subsystem (see Fig.5.11).

If the power broadening is small and the inequality (5.3.73) is satisfied, then the (N-1)-photon resonance frequency is optimum for leakage from the lower N-level subsystem. In this case $|p_{N-1}|^2 \sim \frac{1}{2}$. On the other hand, if the laser frequency is detuned with respect to the (N-1)-photon resonance, then $|p_{N-1}|^2 \ll 1$. Multiphoton resonances of order lower than (N-1) (see Sect.5.3.2) appear as peaks in the frequency dependence of $|p_{N-1}|^2$. It follows that these resonances also appear in the dependence of the leakage probability on the laser frequency. A simple numerical example which deals with a three-level subsystem B is given in Fig.5.13.

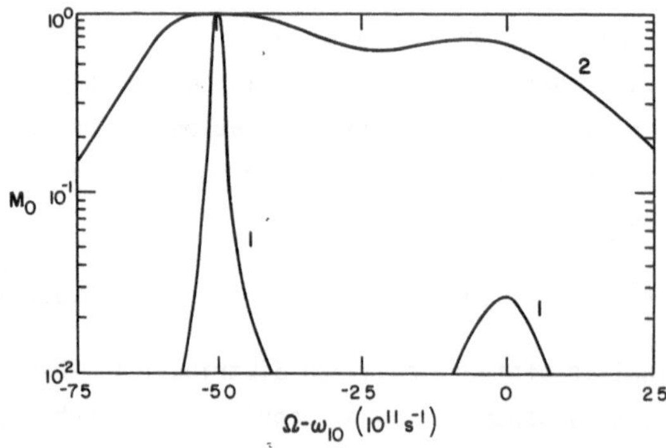

Fig.5.13. Dependence of the leakage probability M_0 on the detuning $(\Omega - \omega_{10})$ for a three-level lower subsystem B. The following values of the parameters were used: $\omega_{10} - \omega_{21} = 10^{12}$ s-1; $\gamma_2 = \sqrt{2}\gamma_1$; $\tilde{\gamma} = 10^{-2}\gamma_1$; the pulse duration is 10^{-7}s; curve (1) corresponds to $\gamma_1 = 3\times 10^{10}\,s^{-1}$, curve (2) corresponds to $\gamma_1 = 10^{11}$ s-1

However, even if the lower transition dipole moments are considerably greater than the upper, the above estimates are fully realistic as long as the inequality

$$\gamma^{(N-1)} \ll \tilde{\gamma} \qquad (5.3.74)$$

is obeyed. In that case a multiphoton resonance cannot appear as a sharp spike in the laser-frequency dependence of the leakage probability. This fact is a result of the phenomenon of broadening of the resonance by a relatively fast decay into the upper levels (see [5.32]). We can estimate the leakage rate in this case. Since the periodic cycling of populations under a multiphoton resonance is analogous to that under a one-photon resonance, an estimate for the leakage rate may be obtained by use of the solution given in Sect.5.2.2 for a relatively weak one-photon resonance in the lower two-level subsystem. The corresponding expression (5.2.47) can be easily modified, with the following result for the leakage rate

$$\mathcal{K} \cong 2\left(\gamma^{(N-1)}\right)^2/\tilde{\gamma} \ . \qquad (5.3.75)$$

In the particular case of two-photon resonance, analytical solutions have been obtained in [5.32]; this solution coincides with (5.3.75). It should be noted that the leakage rate is less than the multiphoton Rabi frequency, the ratio being $2\gamma^{(N-1)}/\tilde{\gamma}$. Thus, the broadening of multiphoton resonances must be taken into account in order to make true estimates of the efficiency of excitation and of the width of the interval in laser frequency for which efficient excitation occurs.

We have completed our analysis of the effects of a quasi-resonant field on non-degenerate multilevel systems. We now turn to a consideration of some aspects of the excitation of more complicated multiplet systems.

5.3.5 Excitation of Multiplet Systems with a Quasi-Continuous Structure of Transitions

The following well-known formula gives the transition rate from a discrete level into a continuous spectrum under the influence of monochromatic radiation

$$W = 2\pi |\langle\phi_\omega|\hat{d}\mathcal{E}/2\hbar|0\rangle|^2 \ . \qquad (5.3.76)$$

A schematic diagram of the process is shown in Fig.5.14a. The transitions obey the law of conservation of energy. This formula is known as the Fermi "golden rule" and, strictly speaking, is valid only within the framework of perturbation theory [5.38]. In this section we formulate a criterion for use of the golden rule for transitions from a discrete level into a quasi-continuous spectrum, that is, into a band consisting of discrete but sufficiently close levels. Of course, by this procedure we shall find the requirement which must be imposed on the ratio of the Rabi frequency to the distance between adjacent levels of the band.

Let the multilevel system consist of a lower discrete state $|0\rangle$ and an array of upper states $|1,q\rangle$ (see Fig.5.14b). We denote the Rabi frequencies for the transitions $|0\rangle \to |1,q\rangle$ as γ_q, and the corresponding transition frequencies as ω_q. (Of

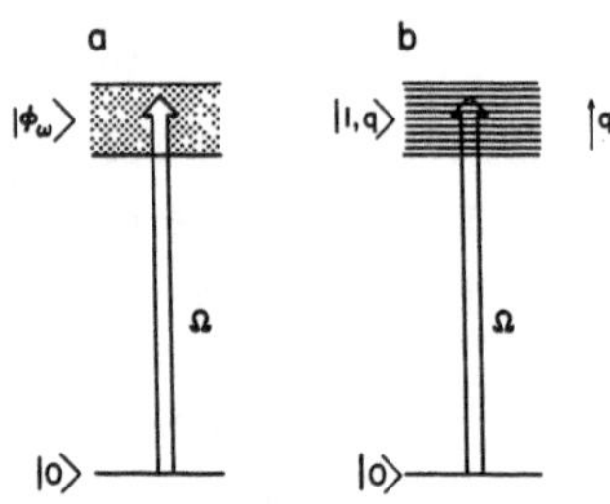

Fig.5.14a,b. Transitions from a discrete state into a continuous spectrum (a) and a quasi-continuous spectrum (b)

course, one can regard all γ_q as positive.) Then (5.1.8) for probability amplitudes has the following form

$$\frac{da_0}{dt} = i\sum_q \gamma_q a_{1q}$$

$$\frac{da_{1q}}{dt} + i(\omega_q - \Omega)a_{1q} = i\gamma_q a_0 \quad . \tag{5.3.77}$$

In order to obtain an analytical solution we consider the case [5.39] when the levels $|1,q>$ are equidistant and the dependence of γ_q^2 on q is Lorentzian

$$\gamma_q^2 = \frac{\sigma\delta}{\pi}\frac{\gamma^2}{q^2\gamma^2 + \sigma^2}\tanh\left(\frac{\pi\sigma}{\delta}\right) \quad , \quad \omega_q = \omega_0 + q\delta \ (-\infty < q < +\infty) \quad . \tag{5.3.78}$$

Here σ is the halfwidth at half maximum of the curve of γ_q^2 versus $(q\delta)$; δ is the distance between the adjacent levels of the band; and the normalization is chosen so that

$$\sum_{q=-\infty}^{+\infty} \gamma_q^2 = \gamma^2 \quad .$$

Of course, we shall assume below that the condition

$$\sigma > \delta \tag{5.3.79}$$

is met.

The solution of (5.3.77) with parameters given by (5.3.78) can be obtained more easily by use of a Laplace transform than by use of the quasi-energy approach. The intermediate calculations are rather cumbersome, and we give the final result in Table 5.5 [for the initial conditions $a_0(0) = 1$ and $a_{1q}(0) = 0$].

Let us analyze the behavior of the probability amplitude a_0 within the time interval $0 < t < 2\pi/\delta$. Here $a_0(t) = A_0(t)$ (see Table 5.5). Using the binomial expansion and neglecting terms which are small by the parameter κ we arrive at the following expression

$$a_0(t) \cong \frac{r_1 + \sigma - i\Delta}{r_1 - r_2}e^{r_1 t} + \frac{r_2 + \sigma - i\Delta}{r_2 - r_1}e^{r_2 t} \quad , \tag{5.3.80}$$

Table 5.5. Notation and formulas for excitation from a discrete level into a band of equidistant sublevels with a Lorentzian distribution of Rabi frequencies

The defined values	Formulas	Remarks
Probability amplitude $a_0(t)$	$a_0(t) = \sum_{m=0}^{\infty} A_m(t)$	
Function $A_0(t)$	$A_0(t) = \frac{(p_1-i\Delta)^2-\sigma^2}{(p_1-p_2)(p_1-p_3)} e^{p_1t} + \frac{(p_2-i\Delta)^2-\sigma^2}{(p_2-p_1)(p_2-p_3)} e^{p_2t} + \frac{(p_3-i\Delta)^2-\sigma^2}{(p_3-p_1)(p_3-p_2)} e^{p_3t}$	1. $\Delta = \Omega - \omega_0$ is the detuning of the transition of the center of the band with respect to resonance
Functions $A_m(t)$	0 for $t < 2\pi m/\delta$	2. $\kappa = 1 - \tanh(\pi\sigma/\delta)$. Under condition (5.3.79) $\kappa \ll 1$
	$A_m(t) = 2\sigma\gamma^2 \tanh(\pi\sigma/\delta)\exp(2\pi i m\Delta/\delta) \sum_{j=1}^{3} \underset{p=p_j}{\mathrm{res}} \left\{[(p-i\Delta)^2-\sigma^2] \prod_{k=1}^{3} \left[\frac{(p+p_k^*)^{m-1}}{(p-p_k)^{m+1}}\right] \exp[(t-2\pi m/\delta)p]\right\}$ for $t > 2\pi m/\gamma$	
Roots $p_k (k=1,2,3)$	$p^3 - 2i\Delta p^2 + (\gamma^2-\sigma^2-\Delta^2)p - i\gamma^2\Delta - \sigma\gamma^2 \tanh(\pi\sigma/\delta) = 0$	
Approximate expressions for roots p_k under condition (5.3.79)	$p_1 \approx -(\sigma-i\Delta)/2 + [(\sigma-i\Delta)^2/4-\gamma^2]^{1/2}$; $p_2 \approx -(\sigma-i\Delta)/2 - [(\sigma-i\Delta)^2/4-\gamma^2]^{1/2}$; $p_3 \approx \sigma + i\Delta - \kappa\sigma\gamma^2/(2\sigma^2+2i\sigma\Delta+\gamma^2)$	

where $r_{1,2} = -(\sigma - i\Delta)/2 \pm [(\sigma - i\Delta)^2/4 - \gamma^2]^{1/2}$. In the case of a strong field, that is, when the condition

$$\gamma \gg |\sigma - i\Delta| \tag{5.3.81}$$

is met, the probability amplitude a_0 is described by a damped oscillation

$$a_0(t) \sim \exp\left(-\frac{\sigma - i\Delta}{2}t\right)\cos\gamma t \quad . \tag{5.3.82}$$

In the opposite case of a weak field, when the condition

$$\gamma \ll |\sigma - i\Delta| \tag{5.3.83}$$

is met, the probability amplitude a_0 decays exponentially without oscillation

$$a_0(t) \sim \exp\left(-i\frac{\gamma^2\Delta}{\sigma^2+\Delta^2}t\right)\exp\left(-\frac{\gamma^2\sigma}{\sigma^2+\Delta^2}t\right) \quad . \tag{5.3.84}$$

Only the last case, when condition (5.3.83) is met, is equivalent to the golden rule. If we substitute into (5.3.84) a typical value γ_r of the Rabi frequency near resonance we find the following result:

$$|a_0(t)|^2 \sim \exp(-Wt) \ , \ W = 2\pi\gamma_r^2/\delta \tag{5.3.85}$$

the excitation rate W coinciding with that in (5.3.76).

So far we have confined our solution to the time interval $0 < t < 2\pi/\delta$. This limitation is not essential if the inequality

$$|a_0(2\pi/\delta)|^2 \ll 1 \tag{5.3.86}$$

is also obeyed. Combining (5.3.86) with (5.3.85) we conclude that the natural condition

$$\gamma_r > \delta \tag{5.3.87}$$

serves together with (5.3.83) as the second criterion for validity of the golden rule. Condition (5.3.87) means that the "resonant" Rabi frequency has to exceed the distance between the adjacent levels of the band. However, inequality (5.3.87) does not appear to be very strict. Even when $\gamma_r \cong \delta$ the golden rule describes well the dynamics of excitation.

When $\gamma_r < \delta$ the lower level interacts mainly with the single upper level which is closest to resonance. This results in a return of coherent features in the dynamics of excitation. Finally, for $\gamma_r \ll \delta$ the oscillations which are typical of a two-level system take place, as illustrated by numerical results (see Fig.5.15) obtained by use of the exact formulas given in Table 5.5.

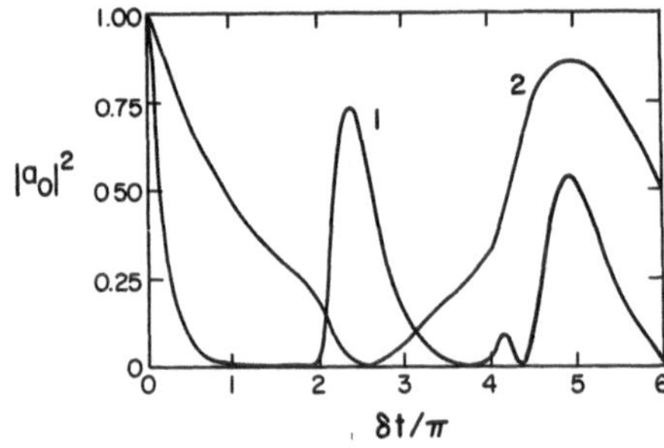

Fig.5.15. Time dependence of $|a_0|^2$ under excitation into a band with a Lorentzian distribution of Rabi frequencies. The following values of the parameters were used: $\Delta/\delta = 0$; $\sigma/\delta = 5$; in curve (1), $\gamma_0/\delta = 0.5$ (or $\gamma/\delta = 1.98$); in curve (2), $\gamma_0/\delta = 0.2$ (or $\gamma/\delta = 0.79$)

The two criteria (5.3.83,87) make it possible to use the golden rule for the describing the dynamics of excitation from a discrete state into a quasi-continuous spectrum. Of course, these criteria are also required for use of the golden rule in more complicated multiplet systems. A detailed analysis has been carried out in [5.40], where it has been shown by approximate methods, the discussion of which is beyond the scope of the present article, that the dynamics of excitation of multiplet systems (see Fig.5.1b) may be described by rate equations if conditions which are analogous to (5.3.83,87) are met. These rate equations involve the total populations of successive bands, and have the following form

$$\frac{d\rho_n}{dt} = W(\rho_{n-1} + \rho_{n+1} - 2\rho_n) \quad , \tag{5.3.88}$$

where for example, $\rho_n = \sum_s |a_{ns}|^2$. Equations (5.3.88) serve in fact as a generalization of the law of exponential decay (5.3.87). Another natural conclusion has been drawn in [5.40] that only the narrow vicinity of each resonance location is populated in each band.

Equations (5.3.88) have been formulated using the assumptions that the densities P_n of states are equal for all bands, and that the average transition rates are equal for all steps. For a more general situation the rate equations have been formulated in [5.36,41]

$$\frac{d\rho_n}{dt} = W_{n-1,n}\left(\rho_{n-1} - \frac{P_{n-1}}{P_n}\rho_n\right) - W_{n,n+1}\left(\rho_n - \frac{P_n}{P_{n+1}}\rho_{n+1}\right) . \quad (5.3.89)$$

Of course, the use of rate equations simplifies considerably the analysis of the dynamics of excitation of multiplet systems. However, when the spectrum of transitions cannot be treated as quasi-continuous, then it is necessary to turn again to the initial equations (5.1.8). A detailed analysis of this situation using computer methods is the subject of Sect.5.4.

5.4 Excitation of Triply-Degenerate Vibrational Modes of Spherical-Top Molecules

As was mentioned in Sect.5.1, part of our motivation for studying the excitation of triply-degenerate vibrational modes is that spherical-top molecules, particularly SF_6, have played a major role as subjects of experimental studies of multiple-photon excitation [5.1-3,7,9]. Another reason for our interest in spherical-top molecules is the fact that certain mathematical and computational simplifications result from the high degree of symmetry of these molecules, as we shall discuss below. Another strong reason for studying SF_6 is the ready availability of spectroscopic data and analyses on at least the low-lying vibration-rotation states [5.4, 42-51]. Other spherical-top molecules besides SF_6 in which multiphoton laser excitation has been studied include SiF_4 [5.52], CCl_4 [5.53], OsO_4 [5.54] and UF_6 [5.55].

The complexity of even the fundamental bands of the triply-degenerate modes of spherical-top molecules is the result of many physical effects: a) the splitting of levels with two or more vibrational quanta (of degenerate modes) by vibrational anharmonic effects (anharmonic splitting); b) Teller Coriolis splitting (and Coriolis interaction between different vibrational states) due to interactions between the vibrational and rotational angular momenta; c) the splitting of each rotational level into as many states as are allowed by the molecular point-group symmetry, by tensor vibration-rotation interactions; d) nuclear hyperfine splitting. All of these effects are significant in understanding and assigning the experimentally observed high-resolution spectra of the SF_6 ν_3 fundamental band. Even the nuclear hyperfine splitting has been resolved and assigned in recent studies of the SF_6 Q branch by saturation spectroscopy [5.47]. Naturally, it is not feasible to take all of these interactions into account, even in numerical studies employing large computers. It seems quite probable that the vibrational, rotational and Teller Coriolis structures are the most significant for multiphoton excitation, in the sense that these interactions probably determine the gross structure and basic properties of multiphoton excitation in SF_6 and other spherical-top molecules. Accordingly we begin our discussion of the theoretical foundations for understanding the multiphoton excitation of spherical-top molecules with a review of the vibrational basis states and energy levels, before discussing vibration-rotation bases, energy levels, and transition moments. We then give a systematic procedure for reducing the number of quantum numbers by neglecting the energy-level splittings identified by the other, neglected quantum numbers. Since real spherical-top molecules are often studied at temperatures where many vibrational and rotational levels are thermally excited, we show how to take these initially excited levels into account (in an approximate way) in a multiphoton calculation. Finally, we illustrate the physical ideas developed in this section with selected numerical results.

5.4.1 Vibrational States and Vibrational Hamiltonian

In this section a derivation of the anharmonic Hamiltonian for a triply-degenerate vibrational mode is presented. This derivation supplements the original derivation by HECHT [5.56] in that the Cartesian vibrational basis is used in this article, while HECHT used the spherical vibrational basis, which we discuss below. In addition, the present derivation is more elementary and appears to give more physical insight than the original derivation, which was based on a canonical transformation of the vibrational potential energy. For future reference, let us note that the triply-degenerate modes of tetrahedral molecules (such as CH_4) are ν_3 and ν_4, while the triply-degenerate modes of octahedral molecules (such as SF_6) are ν_3, ν_4, ν_5, and ν_6.

The Hamiltonian for a single triply-degenerate vibrational mode may be expressed as the sum of the three-dimensional harmonic-oscillator Hamiltonian, and terms of third and higher order in the vibrational normal coordinates arising from a power-series expansion of the vibrational potential energy. Those terms in the vibrational potential energy which are of third or higher order in the normal coordinates together compose the anharmonic potential energy. Let $\hat{q}_j$ (j = 1,2,3) be the normal-coordinate operators of the triply-degenerate mode ν_i. A direct consideration of the transformations of the $\hat{q}_j$ by the elements of the proper octahedral group (as tabulated, for example, in [5.57]) shows that the only important contributions to the anharmonic potential energy $\hat{V}^{(i)}_{anh}$ through fourth order in the normal coordinates $\hat{q}_i$ are the following

$$\hat{V}^{(i)}_{anh} = C^{(i)}_0 \hat{V}^{(i)}_0 + C^{(i)}_1 \hat{V}^{(i)}_1 + C^{(i)}_2 \hat{V}^{(i)}_2 + C^{(i)}_3 \hat{O}^{(i)}_2 \tag{5.4.1}$$

$$\hat{V}^{(i)}_0 = (\hat{q}^2_1 + \hat{q}^2_2 + \hat{q}^2_3)^2 \tag{5.4.2}$$

$$\hat{V}^{(i)}_1 = \hat{q}^4_1 + \hat{q}^4_2 + \hat{q}^4_3 \tag{5.4.3}$$

$$\hat{V}^{(i)}_2 = \hat{q}^2_1\hat{q}^2_2 + \hat{q}^2_1\hat{q}^2_3 + \hat{q}^2_2\hat{q}^2_3 \quad . \tag{5.4.4}$$

The operator $\hat{O}^{(i)}_2$ will be defined below, in (5.4.25).

Since $\hat{V}^{(i)}_0 = \hat{V}^{(i)}_1 + 2\hat{V}^{(i)}_2$, we need not consider $\hat{V}^{(i)}_0$ further. Other contributions of third or fourth order may be eliminated as in the following examples: The potential energy

$$\hat{V}^{(i)}_3 = \hat{q}^3_1 + \hat{q}^3_2 + \hat{q}^3_3 \tag{5.4.5}$$

is not invariant under rotations by π about the main coordinate axes in Fig.5.16, which induce the following transformations of the normal coordinates

$$C_2(x): \hat{q}_1 \rightarrow \hat{q}_1, \ \hat{q}_2 \rightarrow -\hat{q}_2, \ \hat{q}_3 \rightarrow -\hat{q}_3 \tag{5.4.6}$$

$$C_2(y): \hat{q}_1 \rightarrow -\hat{q}_1, \ \hat{q}_2 \rightarrow \hat{q}_2, \ \hat{q}_3 \rightarrow -\hat{q}_3 \tag{5.4.7}$$

$$C_2(z): \hat{q}_1 \rightarrow -\hat{q}_1, \ \hat{q}_2 \rightarrow -\hat{q}_2, \ \hat{q}_3 \rightarrow \hat{q}_3 \tag{5.4.8}$$

(which are correct if ν_i has either F_1 or F_2 symmetry) [5.57]. Terms such as $\hat{q}^2_1\hat{q}_2$ + permutations may be eliminated similarly. The term $\hat{q}_1\hat{q}_2\hat{q}_3$ is allowed if the mode ν_i is of F_2 symmetry (in a tetrahedral molecule) or of F_{2g} symmetry (in an octahedral molecule) but will be eliminated by order of magnitude arguments. Apart from this there are no terms in the anharmonic potential energy which are cubic in the normal coordinates $\hat{q}_1,\hat{q}_2,\hat{q}_3$ of the mode ν_i. Among the quartic terms, $\hat{q}^3_1\hat{q}_2$ + permutations is also not invariant under the transformations (5.4.6-8). The potential energy $\hat{q}^2_1\hat{q}_2\hat{q}_3$ + permutations may be eliminated similarly. The invariance of the remaining quartic terms, (5.4.3,4) may be checked under all the matrices of the generators of the F_1 and F_2 representations of the octahedral group O as given in [5.57]. As an aid to the imagination, we show the nuclear displacements corresponding to the normal coordinate q_1 in the ν_3 and ν_4 modes of SF_6, as calculated by McDOWELL [5.58], in Fig.5.17.

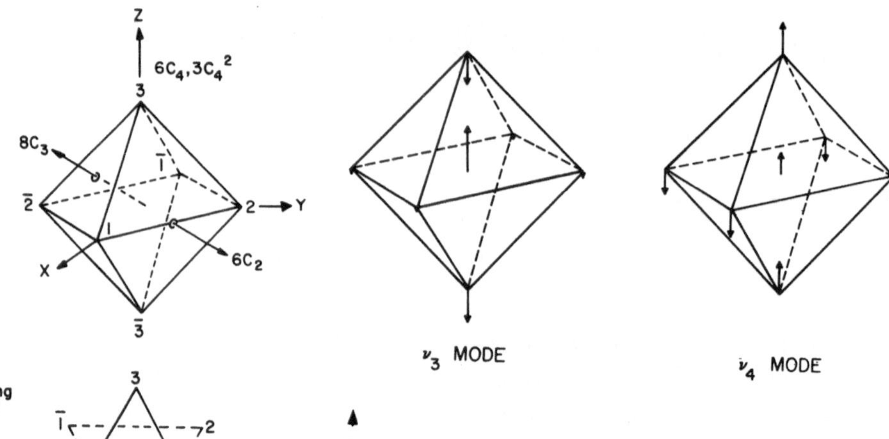

View along C_3 Axis:

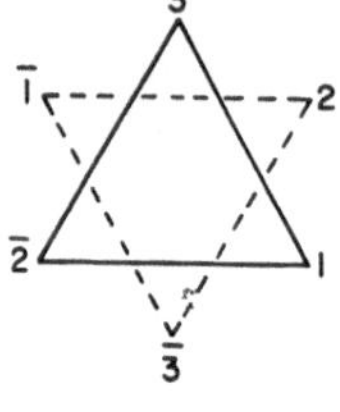

Symmetries of a Regular Octahedron

Fig.5.17. Relative nuclear displacements in the ν_3 and ν_4 vibrational modes of SF_6, calculated from the vibrational force field of [5.58]

Fig.5.16. Rotational symmetry operations of an octahedral molecule

Expression of the Hamiltonian in Terms of Cartesian Creation and Annihilation Operators

The normal-coordinate operators $\hat{q}_j$ ($j = 1,2,3$) may be expressed in terms of the creation ($\hat{a}_j^\dagger$) and annihilation ($\hat{a}_j$) operators. The resulting expressions for $\hat{V}_1^{(i)}$ and $\hat{V}_2^{(i)}$ in terms of $\hat{a}_j$ and $\hat{a}_j^\dagger$ lead trivially to expressions for the matrix elements of the anharmonic potential energy in the Cartesian basis. These expressions also illustrate in simple terms the approximations involved in the derivation of the HECHT Hamiltonian [5.56].

The creation and annihilation operators may be defined as follows [5.46]:

$$\hat{a}_j^\dagger = \frac{1}{\sqrt{2}}(\hat{q}_j - i\hat{p}_j) \tag{5.4.9}$$

$$\hat{a}_j = \frac{1}{\sqrt{2}}(\hat{q}_j + i\hat{p}_j) \tag{5.4.10}$$

where $\hat{p}_j$ is the canonical momentum conjugate to $\hat{q}_j$. The Cartesian vibrational basis functions are

$$|n_1, n_2, n_3\rangle = \sum_{j=1}^{3} \frac{(\hat{a}_j^\dagger)^{n_j}}{\sqrt{n_j!}} |0,0,0\rangle \tag{5.4.11}$$

in terms of the creation operator $\hat{a}_j^\dagger$ and the ground-state wave function

$$|0,0,0\rangle = (2\pi)^{-3/2} \exp\left[-\frac{1}{2}(q_1^2 + q_2^2 + q_3^2)\right] \quad . \tag{5.4.12}$$

Physically, (5.4.11) represents a state in which there are n_1 quanta in the first of the three modes which make up ν_i, n_2 in the second, and n_3 in the third. The total number of vibrational quanta is

$$n = n_1 + n_2 + n_3 \quad . \tag{5.4.13}$$

Hence the basis states for the manifold $n\nu_i$ (i = 3,4,5,6) are all those functions (5.4.11) which satisfy (5.4.13) in an approximation in which anharmonic mixing of states with different values of n is neglected.

The expression

$$\hat{q}_j = \frac{1}{\sqrt{2}} (\hat{a}_j + \hat{a}_j^\dagger) \tag{5.4.14}$$

which follows from (5.4.9,10), may be used to express any integer power $(\hat{q}_j)^n$ in terms of $\hat{a}_j$ and $\hat{a}_j^\dagger$. To extract the matrix elements of $(\hat{q}_j)^n$ in the Cartesian basis, it is convenient to rewrite the expression for $(\hat{q}_j)^n$ with every $\hat{a}_j^\dagger$ to the left of every $\hat{a}_j$, i.e., in normal order [5.59]. This may be accomplished straightforwardly although tediously by use of the commutation relation

$$\left[\hat{a}_j, \hat{a}_k^\dagger\right] = \delta_{jk} \tag{5.4.15}$$

where δ_{jk} is the Kronecker delta: $\delta_{jk} = 1$ if j = k, and $\delta_{jk} = 0$ if $j \neq k$. However, an explicit expression for $(\hat{q}_j)^n$ in normal order may be derived using the techniques developed by LOUISELL [5.59]. Glauber's theorem [5.60] for the exponentiation of two operators $\hat{A}$, $\hat{B}$ which both commute with the commutator $[\hat{A}, \hat{B}]$, may be applied to $\hat{a}_j$ and $\hat{a}_j^\dagger$, with the result

$$\exp(\eta\hat{a}_j^\dagger) \exp(\zeta\hat{a}_j) \exp\left(\frac{1}{2}\eta\zeta\right) = \exp(\eta\hat{a}_j^\dagger + \zeta\hat{a}_j) \quad . \tag{5.4.16}$$

When $\xi = \eta = \zeta$, the r.h.s. of (5.4.16) becomes $\exp(\sqrt{2}\xi q)$. A straightforward power-series expansion of both sides of (5.4.16) leads to the normally ordered expression

$$(\hat{q}_j)^n = \sum_{k,\ell,m}\sum\sum \frac{n!}{2^{m+n/2}k!\ell!m!} (\hat{a}_j^\dagger)^k (\hat{a}_j)^\ell \tag{5.4.17}$$

where the summation on the nonnegative integers k,ℓ, m is restricted by the condition

$$k + \ell + 2m = n \quad . \tag{5.4.18}$$

For n = 1, (5.4.17) is equivalent to (5.4.14); for n = 2,3 and 4, (5.4.17) becomes

$$(\hat{q}_j)^2 = \frac{1}{2} (\hat{a}_j^{\dagger 2} + \hat{a}_j^2 + 2\hat{a}_j^\dagger\hat{a}_j + 1) \tag{5.4.19}$$

$$(\hat{q}_j)^3 = \frac{1}{2\sqrt{2}} (\hat{a}_j^{\dagger 3} + \hat{a}_j^3 + 3\hat{a}_j^{\dagger 2}\hat{a}_j + 3\hat{a}_j^\dagger\hat{a}_j^2 + 3\hat{a}_j^\dagger + 3\hat{a}_j) \tag{5.4.20}$$

$$(\hat{q}_j)^4 = \frac{1}{4}\,(\hat{a}_j^{\dagger 4} + \hat{a}_j^4 + 4\hat{a}_j^{\dagger 3}\hat{a}_j + 4\hat{a}_j^{\dagger}\hat{a}_j^3 + 6\hat{a}_j^{\dagger 2}\hat{a}_j^2 + 6\hat{a}_j{}^2 + 6\hat{a}_j^{\dagger 2} + 12\hat{a}_j^{\dagger}\hat{a}_j + 3) \quad . \tag{5.4.21}$$

Equations (5.4.19-21) may now be substituted into (5.4.2-4) to give the anharmonic potential-energy operators for any triply-degenerate mode ν_i of an octahedral XF_6 molecule, through fourth order in the normal coordinates of ν_i

$$\hat{V}_1^{(i)} = \frac{1}{4}\sum_{j=1}^{3}\left[(\hat{a}_j^{\dagger})^4 + \hat{a}_j^4 + (4\hat{n}_j + 6)\hat{a}_j^2 + (\hat{a}_j^{\dagger})^2(4\hat{n}_j + 6) + (6\hat{n}_j^2 + 6\hat{n}_j + 3)\right] \tag{5.4.22}$$

$$\hat{V}_2^{(i)} = \hat{O}_1^{(i)} + \hat{O}_2^{(i)} + (\hat{n}_1\hat{n}_2 + \hat{n}_1\hat{n}_3 + \hat{n}_2\hat{n}_3) + (\hat{n}_1 + \hat{n}_2 + \hat{n}_3) \tag{5.4.23}$$

$$\hat{O}_1^{(i)} = \frac{1}{2}\sum_{j=1}^{3}\left[(\hat{a}_j^{\dagger})^2 + \hat{a}_j^2\right] + \frac{1}{2}\sum_{j=1}^{3}\sum_{k\neq j}\hat{n}_j\left[(\hat{a}_k)^2 + \hat{a}_k^2\right]$$
$$+ \frac{1}{4}\sum_{j=1}^{3}\sum_{k>j}\left[(\hat{a}_j^{\dagger})^2(\hat{a}_k^{\dagger})^2 + \hat{a}_j^2\hat{a}_k^2\right] \tag{5.4.24}$$

$$\hat{O}_2^{(i)} = \frac{1}{4}\sum_{j=1}^{3}\sum_{k>j}\left[(\hat{a}_j^{\dagger})^2\hat{a}_k^2 + \hat{a}_j^2(\hat{a}_k^{\dagger})^2\right] \quad . \tag{5.4.25}$$

In (5.4.21,22), the number operators

$$\hat{a}_j^{\dagger}\hat{a}_j = \hat{n}_j \tag{5.4.26}$$

which give the number of quanta in the jth component of ν_i, have been introduced by the use of (5.4.15) in order to give the diagonal matrix elements of $\hat{V}_1^{(i)}$ and $\hat{V}_2^{(i)}$ by inspection.

Orders of Magnitude of Anharmonic Operators

The off-diagonal operators in (5.4.22,23) are of two kinds: I) those which change the total number of vibrational quanta n, (5.4.13), such as $(\hat{a}_j^{\dagger})^4$; and II) those which do not change n, but which move quanta from one of the components of ν_i to another, such as $(\hat{a}_1^{\dagger})^2\hat{a}_2^2$. Physically these two types of operators are very different: Operators of type I) cause anharmonic mixing of the states $n\nu_i$ with $(n \pm 2)\nu_i$ and $(n \pm 4)\nu_i$. Operators of type II) transfer excitation from one of the components of ν_i to another, with no change of n. On physical grounds one might expect operators of type II) to be much more important than those of type I). We shall now estimate the effect of terms of type I) on the energy eigenvalues.

One of the major advantages of expressing the vibrational potential energy in terms of the creation ($\hat{a}_j^{\dagger}$) and annihilation ($\hat{a}_j^{\dagger}$) operators is that the simple matrix elements

$$\langle n_1', n_2', n_3' | (a_j^{\dagger} | n_1, n_2, n_3 \rangle) = \sqrt{n_j'}\,\delta_{n_j', n_j+1} \prod_{k\neq j} \delta_{n_k', n_k} \tag{5.4.27}$$

and

$$\langle n_1',n_2',n_3'|(\hat{a}_j|n_1,n_2,n_3\rangle) = \sqrt{n_j}\delta_{n_j',n_j-1}\prod_{k\neq j}\delta_{n_k',n_k} \quad (5.4.28)$$

permit straightforward calculation of the dependence of various terms in the anharmonic potential energy upon the vibrational quantum numbers. In order to make worst-case estimates of the order of magnitude of a particular type of term it is evidently adequate to consider only states for which n_1' is as large as possible, i.e., $|n_1',n_2',n_3'\rangle = |n,0,0\rangle$.

The effect of the term $(\hat{a}_1^\dagger)^2\hat{n}_1 + \hat{n}_1\hat{a}_1^2$ in (5.4.22) on the energy eigenvalue of the state $|n,0,0\rangle$ may be written, using second-order perturbation theory, as

$$\Delta E = C_1^{(i)}\sum_{n''}\frac{|\langle n'',0,0|(\hat{a}_1^\dagger)^2\hat{n}_1 + \hat{n}_1\hat{a}_1^2|n,0,0\rangle|^2}{E(n'',0,0) - E(n,0,0)} \quad (5.4.29)$$

where $C_1^{(i)}$ is the coefficient of $\hat{V}_1^{(i)}$ in the vibrational Hamiltonian (5.4.1). The matrix element in (5.4.29) vanishes unless $n'' = n \pm 2$.

The energy denominator in (5.4.29) will be approximately $2\nu_i$ as long as the anharmonic splitting is small compared to ν_i. Then

$$\Delta E = \frac{C_1^{(i)}}{2\nu_i}\left[n\sqrt{(n+1)(n+2)} - (n-2)\sqrt{n(n-1)}\right] \quad . \quad (5.4.30)$$

For $n = 5$, the quantity in braces is equal to 18.99. Since $|C_1^{(i)}| \cong 1\ \mathrm{cm}^{-1}$ for the parameter sets currently in use for the ν_3 mode of SF_6 we see that $\Delta E \sim 10^{-2}\ \mathrm{cm}^{-1}$ for $n = 5$ in the ν_3 mode. On the other hand, we shall see below that the off-diagonal operator $\hat{O}_2$ of type II) (5.4.25) causes corrections which are greater than $1\ \mathrm{cm}^{-1}$ for $n = 5$ for SF_6 and for reasonable values of the parameters.

The remaining type I) off-diagonal operator is $\hat{q}_1\hat{q}_2\hat{q}_3$, which will contribute terms of the form $\hat{a}_1^\dagger\hat{a}_2\hat{a}_3^\dagger$, etc. The resulting correction to the energy eigenvalue of $|n,0,0\rangle$ is

$$\Delta E \cong \frac{c^2}{32\nu_i}n(n+4), \quad (5.4.31)$$

where c is the coefficient of $\hat{q}_1\hat{q}_2\hat{q}_3$. For $n = 5$, and for the ν_5 mode of SF_6, (5.4.31) predicts $\Delta E \cong 2.5\times 10^{-2}\ c^2\ \mathrm{cm}^{-1}$. Since $|c|$ may be as small as X_{55}, i.e., $0.5\ \mathrm{cm}^{-1}$ [5.58], the resulting correction E is probably negligible in comparison with the uncertainties induced by our lack of knowledge of the anharmonicity constants.

Final Form of the Vibrational Hamiltonian

Let us turn now to a more fundamental point affecting the magnitude of various contributions to the anharmonic potential energy. The scaling of the diagonal matrix elements of the anharmonic potential energy $\hat{V}_{anh}$ (5.4.1) as a function of the vibrational quantum numbers n_1, n_2 and n_3 is determined by the highest power m of the creation and annihilation operators occurring in (5.4.17), which indicates that all powers of $\hat{q}_j$ equal to or greater than m will contribute terms of order m in the quantum numbers n_j to the diagonal matrix element in question. As a result, to obtain the correct coefficient of the kth power of the vibrational-quantum-number operator $\hat{n}_j$ (or a product of powers of the $\hat{n}_j$) it is necessary in principle to sum the contributions of all orders ($\geq 2k$) of the power-series expansion of the anharmonic potential energy in terms of the normal coordinates $\hat{q}_j$. The orders higher than 2k will alter the coefficient of $(\hat{n}_j)^k$, but will probably not change its order of magnitude.

Let us see how the preceding discussion applies to the anharmonic potential energy developed above. The complete Hamiltonian, including the three-dimensional harmonic-oscillator Hamiltonian plus anharmonic terms through fourth order in the normal coordinates, but after elimination of the relatively unimportant type I) off-diagonal terms, is

$$\hat{H}_i = \left(\omega_i + \frac{3}{2} C_1^{(i)} + C_2^{(i)}\right)\sum_{j=1}^{3} \hat{n}_j + \frac{3}{2} C_1^{(i)} \sum_{j=1}^{3} \hat{n}_j^2 + C_2^{(i)} \sum_{j=1}^{3} \sum_{k>j} \hat{n}_j\hat{n}_k + C_2^{(i)}\hat{O}_2^{(i)} \quad . \tag{5.4.32}$$

In (5.4.32), ω_i is the frequency of the ν_i mode prior to anharmonic corrections. In (5.4.32), the zero-point energy has been discarded.

The equality of the coefficients of $\sum_{j<k} \hat{n}_j\hat{n}_k$ and $\hat{O}_2^{(i)}$ (5.4.32) is the direct result of going only to fourth order in the normal coordinates in expanding the anharmonic potential energy in a power series. Terms of sixth order in the normal coordinates can, according to (5.4.17), contribute terms of fourth (or lower) order in the creation and annihilation operators. For example, the sixth-order term

$$\hat{q}_1^2\hat{q}_2^4 + \hat{q}_1^2\hat{q}_3^4 + \hat{q}_2^2\hat{q}_1^4 + \hat{q}_2^2\hat{q}_3^4 + \hat{q}_3^2\hat{q}_1^4 + \hat{q}_3^2\hat{q}_2^4$$

will contribute terms which, when collected together, are proportional to $\hat{O}_2^{(i)}$, (5.4.25). Consequently the coefficients of $\hat{O}_2^{(i)}$ and $\sum \hat{n}_j\hat{n}_k$ in (5.4.32) need not be equal, and the anharmonic Hamiltonian for the ν_i mode should be amended to read

$$\hat{H}_i = \nu_i \sum_{j=1}^{3} \hat{n}_j + \frac{3}{2} C_1^{(i)} \sum_{j=1}^{3} \hat{n}_j(\hat{n}_j - 1) + C_2^{(i)} \sum_{j=1}^{3} \sum_{k>j} \hat{n}_j\hat{n}_k + C_3^{(i)}\hat{O}_2^{(i)} \tag{5.4.33}$$

correct to fourth order in the creation and annihilation operators. In (5.4.33), ν_i has been set equal to the mode frequency including anharmonicity corrections.

Comparison with Hecht's Hamiltonian

HECHT [5.56] derived the following form for the purely vibrational Hamiltonian of a single triply-degenerate mode, designated ν_i

$$\hat{H}_i^{(H)} = \hat{H}^{(0)}(i) + X_{ii}[\hat{H}^{(0)}(i)/\omega_i]^2 + G_{ii}\hat{\underline{\ell}}_i^2 + T_{ii}\hat{O}_{ii} \qquad (5.4.34)$$

where $H^{(0)}(i)$ is the three-dimensional harmonic-oscillator Hamiltonian,

$$\hat{H}^{(0)}(i) = \frac{1}{2}\omega_i \sum_{j=1}^{3} (\hat{p}_j^2 + \hat{q}_j^2) \quad , \qquad (5.4.35)$$

$\hat{\underline{\ell}}_i$ is the vibrational-angular-momentum operator

$$\hat{\underline{\ell}}_i = (\hat{q}_2\hat{p}_3 - \hat{q}_3\hat{p}_2, \hat{q}_3\hat{p}_1 - \hat{q}_1\hat{p}_3, \hat{q}_1\hat{p}_2 - \hat{q}_2\hat{p}_1) \qquad (5.4.36)$$

and $\hat{O}_{ii}$ is an octahedral splitting operator

$$\hat{O}_{ii} = \sum_{j=1}^{3} (\hat{p}_j^2 + \hat{q}_j^2)^2 - 3\sum_{j=1}^{3}\sum_{k>j} (\hat{p}_j^2 + \hat{q}_j^2)(\hat{p}_k^2 + \hat{q}_k^2)$$
$$+ 2\hat{\underline{\ell}}_i^2 + 6 \quad . \qquad (5.4.37)$$

A reordering of Hecht's Hamiltonian (5.4.34) in terms of the number operators $\hat{n}_j$ and the square of the vibrational angular momentum, $\hat{\underline{\ell}}_i^2$, has been described in [5.61]. The vibrational Hamiltonian (5.4.33) has two advantages over the reordered Hecht Hamiltonian presented in [5.61]: a) The only operator in (5.4.33) which is not diagonal in the Cartesian basis (5.4.11) is $\hat{O}_2^{(i)}$, which is completely off-diagonal in that basis. Consequently if, as suggested in [5.10], the coefficient of $\hat{O}_2^{(i)}$ is small, the vibrational energy levels can be written down instantly once the quantum numbers n_1, n_2, and n_3 are specified. In the representation of Hecht's Hamiltonian employed in [5.61], the partially off-diagonal operator $\hat{\ell}_3^2$ appears instead of $O_2^{(i)}$; thus neglecting the off-diagonal terms in that case also entails the more serious error of neglecting some diagonal contributions. b) The second and more important advantage of the Hamiltonian (5.4.33) is that the derivation of (5.4.33) makes clear the origin of the various terms in the Hamiltonian from the vibrational force field, thereby enabling one to see clearly the implications of an approximate force field for the vibrational energy levels. In principle this is also possible with Hecht's Hamiltonian [5.56]; in practice the contact transformation used to derive Hecht's Hamiltonian obscures the physical origin of the terms in (5.4.34).

Let us now cast Hecht's Hamiltonian into the same form as (5.4.33), in order to effect a comparison of the two different sets of parameters.

In terms of (5.4.9,10) the vibrational angular momentum is

$$\hat{\underline{\ell}}_i = i(\hat{a}_2\hat{a}_3^\dagger - \hat{a}_3\hat{a}_2^\dagger, \hat{a}_3\hat{a}_1^\dagger - \hat{a}_1\hat{a}_3^\dagger, \hat{a}_1\hat{a}_2^\dagger - \hat{a}_1\hat{a}_2^\dagger) \qquad (5.4.38)$$

(where $i = \sqrt{-1}$ on the r.h.s.), and its square is

$$\hat{\underline{\ell}}_i^2 = 2 \sum_{j=1}^{3} \hat{n}_j + 2 \sum_{j=1}^{3} \sum_{k>j} \hat{n}_j \hat{n}_k - 4\hat{0}_2^{(i)} \quad . \tag{5.4.39}$$

It is obvious from a comparison of (5.4.39) and (5.4.23) that the presence of $\hat{\underline{\ell}}_i^2$ in Hecht's Hamiltonian has little to do with rotation, but is instead a consequence of the term $\hat{V}_2^{(i)}$ in the anharmonic potential energy. Using (5.4.39) and the relation

$$\frac{1}{2}\left[\hat{p}_j^2 + \hat{q}_j^2\right] = \hat{n}_j + \frac{1}{2} \tag{5.4.40}$$

one may show that

$$\hat{0}_{ii} = \frac{20}{3}\left[\hat{V}_1^{(i)} - \frac{3}{5}\hat{V}_0^{(i)}\right] = \frac{8}{3}\hat{V}_1^{(i)} - 3\hat{V}_2^{(i)} \tag{5.4.41}$$

apart from operators which change the total number of vibrational quanta n. One may then reorder (5.4.34) into the same form as (5.4.33), with the following relations among the coefficients

$$c_1^{(i)} = \frac{2}{3} X_{ii} + \frac{8}{3} T_{ii}$$

$$c_2^{(i)} = 2\, X_{ii} + 2\, G_{ii} - 8\, T_{ii}$$

$$c_3^{(i)} = -4\, G_{ii} - 8\, T_{ii} \quad . \tag{5.4.42}$$

A different reordering of Hecht's Hamiltonian has been employed by AKULIN et al. [5.34]

$$\hat{H}_i = \hat{H}_0(i) + A\hat{V}_i + B\hat{V}_2 + C\hat{\underline{\ell}}_i^2 \quad . \tag{5.4.43}$$

Like the reordering presented in [5.61], this form for $\hat{H}_i$ suffers from the disadvantage that $\hat{\underline{\ell}}_i^2$ has nonzero matrix elements both on- and off-diagonal in the Cartesian basis. The relationships between the coefficients A, B, C; $c_1^{(i)}$, $c_2^{(i)}$, $c_3^{(i)}$; and X_{ii}, G_{ii}, T_{ii} are given in Table 5.6.

In view of the confusion which can arise due to the widespread use of different sets of parameters for the same vibrational Hamiltonian, it is suggested that published spectroscopic results use no parameter sets other than that introduced by HECHT [5.56].

When the coefficient of the operator $\hat{0}_2^{(i)}$ is negligibly small, i.e., when $G_{ii} \cong -2T_{ii}$ according to (5.4.42), (5.4.33) gives the vibrational energy eigenvalues explicitly in terms of the Cartesian quantum numbers n_1, n_2, n_3

$$E_i(n_1 n_2 n_3) = \nu_i \sum_{j=1}^{3} n_j + \frac{3}{2} c_1^{(i)} \sum_{j=1}^{3} n_j(n_j - 1) + c^{(i)} \sum_{j=1}^{3} \sum_{k>j} n_j n_k \quad . \tag{5.4.44}$$

Table 5.6. Relations between parameter sets for the vibrational Hamiltonian of a triply-degenerate mode

A) Hecht parameters in terms of other parameters

The parameters X_{ii}, G_{ii}, T_{ii} were introduced by HECHT [5.56]. The parameters $c_1^{(i)}$, $c_2^{(i)}$, $c_3^{(i)}$ are introduced in this chapter; the parameters A, B, C were introduced by AKULIN et al. [5.34].

$$X_{ii} = \frac{9}{10} c_1^{(i)} + \frac{1}{5} c_2^{(i)} + \frac{1}{10} c_3^{(i)}$$

$$= \frac{9}{10} A + \frac{3}{10} B$$

$$G_{ii} = - \frac{3}{10} c_1^{(i)} + \frac{1}{10} c_2^{(i)} - \frac{1}{5} c_3^{(i)}$$

$$= - \frac{3}{10} A - \frac{1}{10} B + C$$

$$T_{ii} = \frac{3}{20} c_1^{(i)} - \frac{1}{20} c_2^{(i)} - \frac{1}{40} c_3^{(i)}$$

$$= \frac{3}{20} A - \frac{3}{40} B$$

B) AKULIN et al. [5.34]. Parameters in terms of other parameters

$$A = c_1^{(i)}$$

$$= \frac{2}{3} X_{ii} + \frac{8}{3} T_{ii}$$

$$B = \frac{2}{3} c_2^{(i)} + \frac{1}{3} c_3^{(i)}$$

$$= \frac{4}{3} X_{ii} - 8 T_{ii}$$

$$C = \frac{1}{6} c_2^{(i)} - \frac{1}{6} c_3^{(i)}$$

$$= \frac{1}{3} X_{ii} + G_{ii}$$

C) The Parameters of (5.4.33) in terms of other parameters

$$c_1^{(i)} = \frac{2}{3} X_{ii} + \frac{8}{3} T_{ii}$$

$$= A$$

$$c_1^{(i)} = 2 X_{ii} + 2 G_{ii} - 8 T_{ii}$$

$$= B + 2 C$$

$$c_3^{(i)} = - 4 G_{ii} - 8 T_{ii}$$

$$= B - 4 C$$

The levels $E_i(n_1n_2n_3)$ are 1-, 3-, or 6-fold degenerate, according to whether three, two, or none of the n_j are equal to one another. The total number of states for a given n is (n+1)(n+2)/2.

We shall turn now to a discussion of the vibrational energy levels when the coefficient of $\hat{O}_2^{(i)}$ is not small. First, it may happen that $C_3^{(i)}$ is not small, but that T_{ii} is small; in that case the vibrational Hamiltonian is approximately diagonal in a different basis, which we shall discuss immediately below. Second, if neither of these alternatives gives a satisfactory approximation, numerical diagonalization may be employed to calculate the anharmonic energies, as we shall discuss later.

The Spherical Vibrational Basis

As we have shown above, the Cartesian basis (5.4.11) approximately diagonalizes the vibrational Hamiltonian $\hat{H}_i$ when the octahedrally or tetrahedrally symmetric anharmonic potential energy $C_1^{(i)}\hat{V}_1^{(i)} + C_2^{(i)}\hat{V}_2^{(i)}$ of (5.4.3,4) is large compared to the contribution $C_3^{(i)}\hat{O}_2^{(i)}$ [see (5.4.25)] which is off-diagonal in the Cartesian basis. It may happen, however, that the coefficients are such that $T_{ii} \cong 0$. From the fifth line of Table 5.6, this will occur when $6C_1^{(i)} \cong 2C_2^{(i)} + C_3^{(i)}$. It is known from analysis of experimental spectra that this situation probably arises in the ν_4 mode of CH_4. In this case the basis which approximately diagonalizes $\hat{H}_i$ is one adapted to spherical symmetry, since all the terms in (5.4.34) other than $T_{ii}\hat{O}_{ii}$ are evidently invariant under arbitrary rotations and not merely under the rotations which belong to the tetrahedral or octahedral point groups. The spherical vibrational basis functions $|n\ell m\rangle$ are eigenfunctions of the vibrational angular momentum operators $\hat{\underline{\ell}}_i^2$ and $\hat{\ell}_{3i}$ [see (5.4.36,39)], and of the operator $\hat{n}$ giving the total number of vibrational quanta in the mode ν_i,

$$\hat{n} = \hat{n}_1 + \hat{n}_2 + \hat{n}_3 \tag{5.4.45}$$

where n_j is defined in (5.4.26) . The eigenfunctions $|n\ell m\rangle$ are shown by BARGMANN and MOSHINSKY [5.62] using general group theoretic arguments to be

$$|n\ell m\rangle = A_{n\ell}[\hat{a}^\dagger]^{2n}\mathcal{Y}_{\ell m}(\hat{a}^\dagger)|0,0,0\rangle \tag{5.4.46}$$

where the normalization constant is

$$A_{n\ell} = (-1)^n(4\pi)^{\frac{1}{2}}[(2n + 2\ell + 1)!!(2n)!!]^{-\frac{1}{2}} \tag{5.4.47}$$

and $(\hat{\underline{a}}^\dagger)^2$ is

$$(\hat{\underline{a}}^\dagger)^2 = \sum_{j=1}^{3} (\hat{a}_j^\dagger)^2 \quad . \tag{5.4.48}$$

The function $\mathcal{Y}_{\ell m}$ appearing in (5.4.46) is a solid harmonic, which is given as a function of the coordinate vector $\underline{r}$ by

$$\mathcal{Y}_{kq}(\underline{r}) = r^k Y_{kq}(\theta,\phi) \tag{5.4.49}$$

Y_{kq} being a spherical harmonic.

A different expression for $|n\ell m\rangle$ may be derived [5.63] by noting that the Schrödinger equation for a triply-degenerate harmonic oscillator in spherical polar coordinates may be reduced by separation of variables to an angular equation which is solved by the spherical harmonics, and a radial equation which is solved by certain associated Laguerre polynomials. Since the derivation is not of general interest for the theory of multiple-photon excitation we omit the details and give only the result

$$|n\ell m\rangle = N_{n\ell} Y_{\ell m}(\underline{q}/|\underline{q}|)e^{-q^2/2}(q^2)^{\ell/2} L^{(\ell+\frac{1}{2})}_{\frac{1}{2}(n-\ell)}(q^2) \tag{5.4.50}$$

where

$$q^2 = \sum_{j=1}^{3} q_j^2 \equiv |\underline{q}|^2 \tag{5.4.51}$$

$$\underline{q} = (q_1,q_2,q_3) \tag{5.4.52}$$

$$\ell = n,n-2,\ \ldots,\ 1 \text{ or } 0 \tag{5.4.53}$$

$$N_{n\ell} = \{2[\tfrac{1}{2}(n-\ell)]!\}^{\frac{1}{2}}\{[\tfrac{1}{2}(n+\ell+3)]!\}^{-\frac{1}{2}} \quad . \tag{5.4.54}$$

By (5.4.49), the states $|n\ell m\rangle$ are eigenfunctions of $\hat{\underline{\ell}}_i^2$ and $\hat{\ell}_{3i}$. Hence if $T_{ii} = 0$ in the Hecht Hamiltonian (5.4.34), the vibrational energies may be written down by inspection

$$E_i(n\ell m) = \left(n + \frac{3}{2}\right)\omega_i + X_{ii}\left(n + \frac{3}{2}\right)^2 + G_{ii}\ell(\ell+1) \tag{5.4.55}$$

$$= n\nu_i + X_{ii}n(n-1) + G_{ii}[\ell(\ell+1) - 2n] \quad . \tag{5.4.56}$$

For a given value of n, there are n/2 distinct eigenvalues $E_i(n\ell m)$, one for each value of ℓ given by (5.4.53). Each eigenvalue is $(2\ell + 1)$-fold degenerate, since $m = -\ell, -\ell + 1, \ldots, \ell$, and $E_i(n\ell m)$ is independent of m. It is easy to show by summing the arithmetic progression $(2n + 1) + (2n - 1) + \ldots + 3$ or 1 that the total number of levels for a given value of n is $(n + 1)(n + 2)$. This is, of course, the same as found for the Cartesian basis.

The matrix elements of the operator $\hat{O}_{ii}$ in (5.4.37), which is completely off-diagonal in the spherical vibrational basis, may be written down at once by use of the Wigner-Eckart theorem [5.64], once it is recognized that $\hat{O}_{ii}$ is a particular tetrahedrally or octahedrally invariant linear combination of some of the spherical components of a tensor operator of rank four. The Wigner-Eckart theorem [5.64] implies that

$$\langle n\ell m|T_q^{(k)}|n'\ell'm'\rangle = \langle \ell m|\ell' kmq\rangle\langle n\ell||T^{(k)}||n'\ell'\rangle \tag{5.4.57}$$

for the spherical component $T_q^{(k)}$ of any tensor operator $T^{(k)}$ of rank k. The symbol $\langle \ell m|\ell' k m' q\rangle$ is a Clebsch-Gordan coefficient; the symbol $\langle n\ell||T^{(k)}||n'\ell'\rangle$ is the reduced matrix element [5.64] of the tensor operator $T^{(k)}$.

What linear combination $\hat{O}_{ii}$ is of spherical tensor components, is suggested by considering the explicit form of the solid harmonics (5.4.49) for k = 4. It is straightforward to show from standard formulas [5.64] for the spherical harmonics that

$$\sqrt{70}\,\mathcal{Y}_{4,0} + 5\left[\mathcal{Y}_{4,4} + \mathcal{Y}_{4,-4}\right] = 3\left(\frac{70}{4\pi}\right)^{\frac{1}{2}} [x^4 + y^4 + z^4 - 3(x^2y^2 + x^2z^2 + y^2z^2)] \quad . \tag{5.4.58}$$

The quantity in square brackets in (5.4.58) is of exactly the same form as the operator $\hat{V}_1^{(i)} - 3\hat{V}_2^{(i)}$ in (5.4.41). In fact, a comparison of (5.4.41) with (5.4.58) shows directly that

$$\hat{O}_{ii} = \frac{4}{9}\left(\frac{8\pi}{35}\right)^{\frac{1}{2}}\left\{\sqrt{70}\,\mathcal{Y}_{4,0}(\hat{q}) + 5\left[\mathcal{Y}_{4,4}(\hat{q}) + \mathcal{Y}_{4,-4}(\hat{q})\right]\right\} \quad . \tag{5.4.59}$$

Since the spherical harmonics and solid harmonics are spherical tensor operators [5.64], (5.4.59) and the Wigner-Eckart theorem (5.4.57) allow us to write down the matrix elements of $\hat{O}_{ii}$

$$\begin{aligned}\langle n\ell m|\hat{O}_{ii}|n'\ell' m'\rangle = \frac{4}{9}\left(\frac{8\pi}{35}\right)^{\frac{1}{2}} \langle n\ell||O_{ii}||n'\ell'\rangle \\ \cdot\, [\sqrt{70}\langle \ell,m|\ell',4,m',0\rangle + 5(\langle \ell,m|\ell',4,m',-4\rangle \\ + \langle \ell,m|\ell',4,m',4\rangle)] \quad .\end{aligned} \tag{5.4.60}$$

In fact, it is well known [5.56,65] that the linear combination $\sqrt{70}\;T_0^{(4)} + 5\left(T_{-4}^{(4)} + T_4^{(4)}\right)$ is tetrahedrally or octahedrally invariant, for any fourth-rank spherical tensor operator $T^{(4)}$. The reduced matrix element $\langle v\ell||\hat{O}_{ii}||v\ell'\rangle$ is given in Table 5.7 [5.56]. The resulting vibrational Hamiltonian (5.4.34), the matrix elements of which in the spherical basis are given by (5.456,60), must be diagonalized numerically in general to find the vibrational eigenvalues and eigenvectors. HECHT [5.56] has, however, given analytical results for $n \leq 4$, where no more than 2×2 matrices need be considered.

Eigenvalues of the Vibrational Hamiltonian

The calculation of the energy levels into which the overtone "state" $n\nu_i$ is split by the Hamiltonian H_i (5.4.3) involves the solution of a secular equation of an order equal to the greatest number of levels of the same symmetry type occurring in $n\nu_i$. For n = 2, the secular equation is linear and may be solved by inspection.

Table 5.7. Reduced matrix elements of $\hat{O}_{ii}$ [a]

$$\langle n\ell||\hat{O}_{ii}||n\ell\rangle = \frac{[3(2n+3)^2-(2\ell-1)(2\ell+3)][(2\ell+4)(2\ell+2)(2\ell+1)2\ell(2\ell-2)]^{\frac{1}{2}}}{8[70(2\ell+5)(2\ell+3)(2\ell-1)(2\ell-3)]^{\frac{1}{2}}}$$

$$\langle n,\ell-2||\hat{O}_{ii}||n\ell\rangle = \frac{(2n+3)[(n+\ell+1)(n-\ell+2)(2\ell+2)\,2\ell\,(2\ell-2)(2\ell-4)]^{\frac{1}{2}}}{4[7(2\ell-5)(2\ell-1)(2\ell+3)]^{\frac{1}{2}}}$$

$$\langle n,\ell-4||O_{ii}||n\ell\rangle = \frac{[(n-\ell+4)(n-\ell+2)(n+\ell-1)(n+\ell+1)(2\ell-6)(2\ell-4)(2\ell-2)2\ell]^{\frac{1}{2}}}{4[(2\ell-5)(2\ell-3)(2\ell-1)]^{\frac{1}{2}}}$$

[a] [5.56]

For n = 3 and 4, the secular equation is quadratic; the solutions were given long ago by HECHT [5.56]. For n = 5 and 6, the secular equation is of fourth order and can in principle be solved analytically, although the solution is tedious. For $n \geq 7$, the secular equation is of sixth or higher order, and a numerical solution is necessary.

The experimental determination of the anharmonic parameters of SF_6, which is currently a subject of active discussion in the scientific community, will be summarized (as of the time of writing this review) in Sect.5.4.7.

In Fig.5.18 are shown the calculated vibrational energy levels of SF_6, in three different approximations. The arrows indicate the position of the levels when the anharmonic splitting is zero, i.e., when the vibrational levels undergo a simple anharmonic shift as in a diatomic molecule, with $X_{ii} = -2.13\ \mathrm{cm}^{-1}$. According to (5.4.33) this will occur if the energy eigenvalues depend only on n; that is, if

$$c_2^{(i)} = 3\,c_1^{(i)},\ c_3^{(i)} = 0\quad , \tag{5.4.61}$$

i.e.,

$$G_{ii} = T_{ii} = 0\quad . \tag{5.4.62}$$

The levels labeled with three integers, e.g., [210], are the energy levels obtained analytically by setting $c_3^{(i)} = 0$, i.e., neglecting the off-diagonal operator $\hat{O}_2^{(i)}$ in (5.4.33). The integers in square brackets are the values of n_1, n_2, and n_3 (or permutations of n_1, n_2, and n_3) for the level in question. Each level in this approximation is degenerate, with a degeneracy of 6, 3, or 1. The reason for this degeneracy is clear from (5.4.33) (with $\hat{O}_2^{(i)}$ neglected): the symmetry of the remaining terms in (5.4.33) under permutations of the n_j (the eigenvalues of the $\hat{n}_j$) implies that the number of degenerate levels will be equal to the number of distinguishable permutations of n_1, n_2, n_3. This is six when all three n_j are different; three when two are equal; and one when all are equal. Previous discussions of the approximate analytical solution of (5.4.33) and the degeneracy of the levels [5.61]

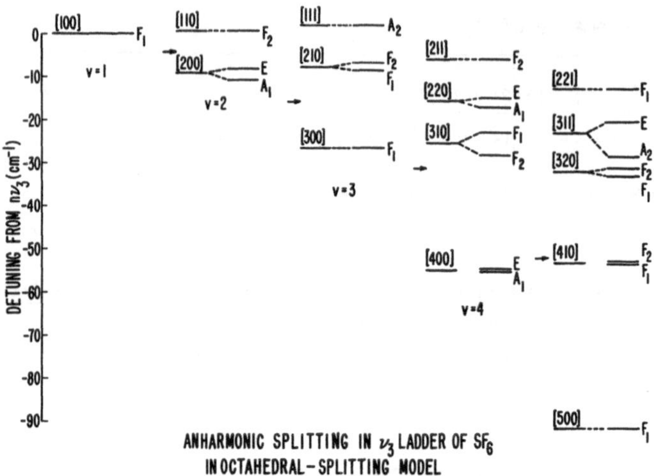

Fig.5.18. Overtone vibrational energy levels of a triply-degenerate vibrational mode in three different approximations (see text)

have omitted consideration of the cross terms $\hat{n}_1\hat{n}_2$ + permutations, in (5.4.33). For illustrative purposes, we have chosen to pass from the first set of energy levels (shown by arrows in Fig.5.18) to the second set of energy levels by making the center of gravity of the second set of levels equal to the position of the first set of levels. Since we ultimately wish to show the energy levels determined by the parameters X_{33}, G_{33}, and T_{33} calculated from a model force field [5.66] for the ν_3 mode of SF_6, we have forced the centers of gravity of the first and second (approximate) sets of energy levels to coincide with the center of gravity of the levels determined by the parameters of [5.66]. We have therefore: a) determined $C_1^{(i)}$, $C_2^{(i)}$ and $C_3^{(i)}$ from Table 5.6 and the parameters X_{33}, G_{33} and T_{33} given in [5.66]; b) set $C_3^{(i)} = 0$ to determine the second set of energy levels analytically, using (5.4.33); c) chosen X_{ii}^{eff} for the first set of energy levels (shown by arrows in Fig.5.18) such that the these energy levels fall at the center of gravity of the second set of energy levels. This may be accomplished by choosing $X_{ii}^{\mathrm{eff}} = (3C_1^{(i)} + C_2^{(i)})/4$.

The third set of levels shown in Fig.5.18 are those obtained by exact (numerical) diagonalization of $\hat{H}_i$ (5.4.33), using the parameters $X_{33} = -2.8\ \mathrm{cm}^{-1}$, $G_{33} = 1.34\ \mathrm{cm}^{-1}$, and $T_{33} = -0.44\ \mathrm{cm}^{-1}$ of [5.66]. The assignment of the symmetry types of the levels (A_1, A_2, E, F_1 or F_2) was performed by calculating the transformation matrix of each set of degenerate eigenfunctions under a rotation by π about an axis in the direction $(\underline{x} + \underline{y})\sqrt{2}$ (for the octahedral group). As a glance at the character table [5.57] of the octahedral group O will show, the trace of this matrix and the degeneracy of the level suffice to determine the symmetry type. These assignments are consistent with the results obtained by explicit analytical construction of the vibrational eigenfunctions, neglecting $\hat{O}_2^{(i)}$ [5.61].

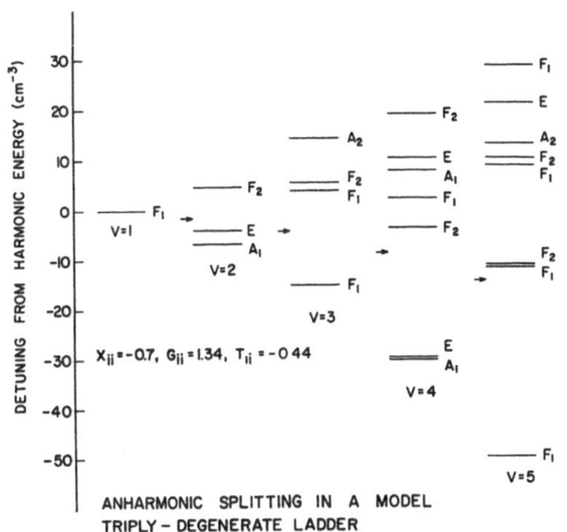

Fig.5.19. Overtone vibrational energy levels of a model triply-degenerate vibrational mode (see text)

For comparison, in Fig.5.19 are shown the energy levels of $n\nu_i$ for a set of parameters such that the anharmonically split manifolds always overlap the position the energy levels of a harmonic oscillator with frequency ν_i would have. From the point of view of multiple-photon excitation, such a set of levels should show qualitatively different behavior than those shown in Fig.5.18. The levels shown in Fig.5.18 become increasingly detuned from the position of the levels of a harmonic oscillator as n increases, and the transition frequencies are increasingly shifted towards the red, away from ν_i. Accordingly an "anharmonicity barrier" will develop, which can be overcome only by an intramolecular transition to the "quasicontinuum". The set of levels shown in Fig.5.19 should display no "anharmonicity barrier"; in this case, a transition with frequency near ν_i is always possible. A similar discussion has been given by AKULIN [5.34], using the analytical approximation of neglecting $\hat{V}_2$. In the following section we shall give an extended discussion of the physical implications of different sets of anharmonicity parameters.

5.4.2 Physical Significance of Vibrational Anharmonic Parameters

In all cases, the relative magnitude of the anharmonic shift and the anharmonic splitting is determined by the anharmonic parameters X_{ii}, G_{ii} and T_{ii}, which can be determined experimentally only by analysis of spectra involving transitions to, from, or among states $n\nu_i$ with $n \geqq 2$.

As we can see from (5.4.42,44), the anharmonic shift of energy levels (if the Cartesian basis is approximately correct) is determined by the combination $3C_1^{(i)}/2 = X_{ii} + 4T_{ii}$ of anharmonic parameters, while the anharmonic splitting is determined by T_{ii} and G_{ii}. According to (5.4.56), the anharmonic shift (if the spherical vibrational basis is approximately correct) is determined solely by X_{ii}, while G_{ii} and T_{ii} determine the anharmonic splitting. Given the rotational parameters B_0 and ζ_i, which can be determined by analysis of the fundamental band

ν_i, the values of the anharmonic parameters T_{ii} and G_{ii} also determine the relative importance of rotational and anharmonic splitting. It is evident from (5.4.44,54) that the total range of energy of the states with a given number of vibrational quanta, n, scales approximately as n^2. Thus, depending on the relative magnitudes of X_{ii}, G_{ii} and T_{ii}, the anharmonic splitting may partially or totally cancel the anharmonic shift; that is, the manifold of anharmonically split states may, for all n, overlap the positions $n\nu_i$ the vibrational energy levels would have in the harmonic approximation. This possibility was pointed out in [5.10], where the potential importance for multiple-photon excitation of the cancellation of the anharmonic shift by anharmonic splitting was also suggested.

As we have shown, the basic physics of excitation of the first few vibrational states of a polyatomic molecule is multiphoton resonance, possibly enhanced by nearby levels which are connected by nonzero dipole matrix elements to the levels which are directly involved in the multiphoton resonance. These processes are strongly affected by molecular rotation, as we shall discuss below. However, the vibrational anharmonic parameters determine the regime in which multiple-photon excitation occurs. The excitation of a molecular vibration-rotation band in which the anharmonic splitting equals or exceeds the detuning from resonance for each n will be qualitatively different from the excitation of a vibration-rotation band in which the anharmonic splitting becomes much less than the detuning from resonance as n increases. In the former case (anharmonic splitting equal to or greater than detuning) the many rotational pathways of excitation to an excited state will provide abundant excitation through many near-resonant multiphoton transitions. In this case, anharmonic splitting obviates any necessity for postulating a quasicontinuum. More precisely, in this case anharmonic splitting plus rotations constitute the quasicontinuum, without any necessity to postulate resonant interactions or transfer of energy between the laser-pumped mode and other vibrational modes. In the latter case (anharmonic splitting less than detuning for $n > n_0$) an anharmonicity barrier will develop: when an excited state can be reached only through an N-photon resonance, the Rabi frequency $\gamma^{(N)}$ of which is so small that

$$\left(\gamma^{(N)}\right)^{-1} >> \tau_p \qquad (5.4.63)$$

(where τ_p is the length of the laser pulse), then only a meager excitation can be achieved in the duration of a pulse. In this case we may use (5.3.8) for $\gamma^{(N)}$, which gives the magnitude of the initial rate of change of the upper-state probability amplitude. Since we expect $N \sim n$, the quantity $\gamma^{(N')}$ for $N' > N$ will be even smaller than $\gamma^{(N)}$ according to (5.3.8), because $|\delta_N| \propto N^2$ when the anharmonic shift exceeds the anharmonic and rotational splittings. In other words, once (5.4.63) is satisfied for all states with a given number of vibrational quanta n(=N), we expect it to be satisfied for all $N' > N$ as well, so that effective excitation ceases for $n < N$. Although this statement may not be strictly valid for all possible states with $n' = N' > N$, owing to some unusual combination of detunings and transition moments, it will generally be true.

We note that rotational energy cannot compensate anharmonic energy shifts for transitions from n to n+1 for arbitrarily large n unless the anharmonic splitting is comparable to the anharmonic shift. For, if the anharmonic shift dominates all anharmonic and rotational splittings, the detuning from resonance is quadratic in n, while the maximum change in rotational energy achievable in transitions from 0 to n is linear in n, owing to the $\Delta J = \pm 1$ or 0 selection rule for each dipole-allowed step, which implies $|\Delta J_{max}| = n$ for a transition from 0 to n. Thus if an anharmonicity barrier develops owing to the positions of the purely vibrational energy levels, the inclusion of rotation will not remove the barrier.

In mathematical form these arguments can be summarized as follows. From (5.4.33,56), we can write the diagonal elements of the multilevel dressed-states Hamiltonian analogous to (5.3.36,37) as

$$(\hat{H}^{eff})_{ns,ns} = n(\nu_i - \Omega) + X_{ii}\, n(n-1) + E_{ahs}(n,s) \qquad (5.4.64)$$

where s specifies the quantum numbers other than n. The term $X_{ii}\, n(n-1)$ gives the anharmonic shift. For the spherical basis the anharmonic splitting energy E_{ahs} is, from (5.4.56),

$$E_{ahs}(n,\ell) = G_{ii}[\ell(\ell+1) - 2n] \qquad (5.4.65)$$

while for the Cartesian basis, from (5.4.33),

$$E_{ahs}(n_1,n_2,n_3) = 4T_{ii} \sum_{j=1}^{3} n_j^2 - 4T_{ii}\, n + 2(G_{ii} - 4T_{ii}) \sum_{j=1}^{3} \sum_{k>j} n_j n_k \quad . \qquad (5.4.66)$$

There are two possibilities: either I) E_{ahs} is such that the total anharmonic energy

$$E_{anh} = X_{ii}\, n(n-1) + E_{ahs}(n,s) \qquad (5.4.67)$$

has one sign for some levels (s) and the opposite sign for other levels (s') for a fixed n; or II) this is not the case. Sufficient conditions for case I) to hold in the spherical basis are

$$X_{ii} + G_{ii} > 0 \ , \quad X_{ii} < 0 \qquad (5.4.68)$$

(or the opposite inequalities, with < replaced by > and vice versa). For the Cartesian basis, there are two extrema of E_{ahs}. One extremum occurs when $n_1 = n$, $n_2 = n_3 = 0$, and in this case

$$E_{ahs}(n_1,n_2,n_3) = 4T_{ii}\, n(n-1) \quad . \qquad (5.4.69)$$

The other extremum of E_{ahs} in the Cartesian basis occurs when $n_1 \cong n_2 \cong n_3 \cong [n/3]$ (where the square brackets denote the integer part), and in this case

$$E_{ahs}(n_1,n_2,n_3) \cong \frac{2}{3}(G_{ii} - 2T_{ii})n^2 - 4T_{ii}\, n \quad . \tag{5.4.70}$$

Hence sufficient conditions for case I) to hold are

$$X_{ii} + \frac{2}{3}(G_{ii} - 2T_{ii}) > 0 \ , \quad X_{ii} + 4T_{ii} < 0 \tag{5.4.71}$$

or the reversed inequalities. In case I), nearly-resonant one-photon transitions will be possible for all n, since the diagonal elements of H^{eff} are all nearly zero for $\Omega \cong \nu_i$.

In the other case, II), where E_{anh} has one sign for all $n > n_0$, the main process of excitation is (nearly-resonant) multiphoton excitation. An N-photon resonance to a level $(n = N,s)$ occurs when

$$\Omega = \nu_i + (N - 1)X_{ii} + \frac{1}{N} E_{ahs}(N,s) \tag{5.4.72}$$

which should be compared with (5.3.50). The detunings δ_{nq} of intermediate states with $n < N$ are

$$\delta_{nq} = (n - N)(n - 1)X_{ii} + E_{ahs}(n,q) - \frac{n}{N} E_{ahs}(N,s) \tag{5.4.73}$$

which is quadratic in n and (approximately) linear in N. As N increases, one qualitatively expects the intermediate detunings $|\delta_{nq}|$ to increase, or at least not to decrease. If all the $|\delta_{nq}|$ are small compared to the one-photon Rabi frequencies, and if only a single ladder of most nearly resonant intermediate levels need be considered, then (5.3.8) implies that the initial rate of change $\gamma^{(N)}$ of $n = N$ will become smaller and smaller as N increases, apart from exceptional cases. For some N_0, one will find

$$\gamma^{(N_0)} \sim (\tau_p)^{-1} \tag{5.4.74}$$

where τ_p is the length of the laser pulse. For $N > N_0$, one expects very little excitation even at the resonant laser frequencies (5.4.72), because the laser pulse will be too short to pump the narrow N-photon resonances effectively.

We may summarize the arguments of this section by observing that the anharmonic paramters X_{ii}, G_{ii} and T_{ii} determine whether nearly-resonant one-photon transitions, or multiphoton transitions, dominate the multiple-photon excitation of a spherical-top molecule. Nonresonant multiphoton transitions can dominate the excitation only when the anharmonic shift (quadratic in n) is the dominant contribution to the detuning for highly excited vibrational levels. In that case, an anharmonicity barrier develops owing to the fact that a laser pulse of finite duration cannot effectively pump a very narrow multiphoton resonance.

5.4.3 Rotational States and Vibration-Rotation Bases

In this section we shall briefly discuss the eigenstates for a rigid rotator, and shall then summarize some pertinent results from the voluminous literature on vibration-rotation bases for spherical-top and symmetric-top molecules. Within the framework of the Born-Oppenheimer approximation, the vibration-rotation basis states for a spherical-top molecule may be taken as products of vibrational and rigid-rotor eigenfunctions. However, the presence of a large vibration-rotation interaction (the Coriolis interaction, which arises because of the common and useful practice of performing calculations in a frame of reference rotating with the molecule) complicates the choice of a vibration-rotation basis. If the vibrational anharmonic splitting is large compared to the Coriolis energy, then the vibration-rotation basis functions may be conveniently taken to be product of purely vibrational eigenfunctions with rigid-rotor eigenfunctions. In the opposite limit, where the Coriolis energy is large compared to the vibrational anharmonic splitting, it is convenient to employ a coupled angular momentum basis in which the Coriolis energy is diagonal. Vibration-rotation interactions of higher order than the Coriolis energy generally split the vibration-rotation energy levels due to the Coriolis energy into as many levels as are allowed by the symmetry of the molecule. To investigate this structure MORET-BAILLY [5.67] introduced another basis, the elements of which belong to specific rows of irreducible representations of the molecular symmetry group. The Moret-Bailly basis functions approximately diagonalize the higher-order vibration-rotation interactions as well [5.67]. In this section, we shall summarize some important results concerning these different types of basis function.

Rigid-Rotor Wave Functions

The rotational wave functions of a polyatomic molecule may conveniently be considered as functions of the real orthogonal 3×3 matrix which rotates the components of a vector in the laboratory frame into the components of the same vector in the molecule-fixed frame

$$\underline{x}'_{mol} = \Omega \underline{x}_{lab} \quad . \tag{5.4.75}$$

The Euler angles, in terms of which one may specify the orientation of the molecular axes relative to axes fixed in the laboratory, are a parametrization of Ω^{-1}. The rotational wave function ψ^J_M of a rigid polyatomic molecule may always be expressed as a linear combination of the rotation matrix elements $D^J_{KM}(\Omega)$. If the equilibrium moment of inertia tensor possesses an axis of symmetry, i.e., if the molecule is a symmetric top or a spherical top, then the linear combination reduces to a single term

$$\psi^{J}_{KM}(\Omega) = \left(\frac{2J+1}{8\pi^2}\right)^{\frac{1}{2}} D^{J}_{KM}(\Omega^{-1}) \quad . \tag{5.4.76}$$

In all cases the projection M of the total angular momentum $\underline{J}$ along the laboratory z axis is a good quantum number. For symmetric tops or spherical tops the projection of $\underline{J}$ along the molecule-fixed z axis is also a good quantum number, so long as the molecule may be considered a rigid rotor. We shall see below that, because of the Coriolis interaction, K is not a good quantum number in general. However, if the vibrational energy levels are separated by an amount which is large compared to the Coriolis energy, then the vibration-rotation wave function which most nearly diagonalizes the vibrational anharmonic Hamiltonian and the vibration-rotation Hamiltonian is the product

$$\phi^{nsJ}_{KM} = \psi_{ns}(\{q\})D^{J}_{KM}(\Omega^{-1}) \tag{5.4.77}$$

where n is the number of vibrational quanta and s represents the other quantum numbers needed to specify the vibrational state. In the spherical vibrational basis, $s = \{\ell m\}$.

Coupled Vibration-Rotation Basis

The Coriolis contribution to the vibration-rotation Hamiltonian of a spherical-top molecule for states in which quanta of only one triply-degenerate mode are excited is [5.38]

$$\hat{H}_{Cor} = -2B\zeta_i \hat{\underline{J}} \cdot \hat{\underline{\ell}}_i \tag{5.4.78}$$

where ζ_i is a function of the vibrational (quadratic) force constants and the molecular geometry [5.63], $\hat{\underline{J}}$ is the total angular momentum operator, and $\hat{\underline{\ell}}_i$ is the vibrational angular momentum operator. By analog with the spinorbit interaction in atomic physics, one sees that $\hat{H}_{Cor}$ will be diagonal in a basis in which $\hat{\underline{J}}$ and $\hat{\underline{\ell}}_i$ are coupled to form a new angular momentum

$$\hat{\underline{R}} = \hat{\underline{J}} - \hat{\underline{\ell}}_i \quad , \tag{5.4.79}$$

which has the physical significance of being the purely rotational angular momentum of the molecular framework. The coupled vibration-rotation basis functions are

$$\phi^{n\ell JR}_{K_R M} = \sum_m \langle \ell J;mK|RK_R\rangle[\phi^{n\ell}_m]^* \psi^{J}_{KM} \quad . \tag{5.4.80}$$

The presence of the complex conjugate in (5.4.80) is the result of the subtraction of $\hat{\underline{\ell}}$ from $\hat{\underline{J}}$ to form $\hat{\underline{R}}$. The good quantum numbers in the coupled spherical basis (5.4.80) are n, ℓ, J, R, the projection K_R of $\hat{\underline{K}}$ along the molecule-fixed z axis, and M. Thus the coupled spherical basis vectors span the representation $D^{(R)}$ of the proper orthogonal group $O_3^{(+)}$.

Symmetry-Adapted Vibration-Rotation Basis

The symmetry-adapted basis functions are obtained from the coupled vibration-rotation basis by transformation with a unitary matrix ${}^{(R)}G$ [5.66]

$$\psi_{pM}^{n\ell JR} = \sum_{K} {}^{(R)}G_p^K \phi_{KM}^{n\ell JR} \quad . \tag{5.4.81}$$

In (5.4.81),

$$p = C^{(n)} \tag{5.4.82}$$

where

$$C = (\Gamma\gamma) \quad ; \tag{5.4.83}$$

γ is the row of the irreducible representation Γ of the molecular symmetry group; and n is an index which distinguishes the different state vectors with the same C which arise from the reduction of $D^{(R)}$ into irreducible representations of the molecular symmetry group. The elements of ${}^{(R)}G$ have been calculated for R up to 130 [5.68] and are tabulated [5.67] for $R \leq 20$. Since the sum in (5.4.81) runs only over the projection K_R, the Coriolis Hamiltonian is diagonal in the symmetry-adapted basis, in which the good quantum numbers are n, ℓ, J, R, p, and M. However, since the vibrational anharmonic operator (5.4.37) evidently mixes states of different ℓ [see (5.4.60)], the full anharmonic Hamiltonian (5.4.33) is not diagonal in either of the bases (5.4.80) or (5.4.81). Since the spherically symmetric part $\hat{H}_i^{(H)} - T_{ii}\,\hat{O}_{ii}$ (5.4.34) of the anharmonic Hamiltonian is diagonal in both bases (5.4.80,81), however, these bases are appropriate when the coefficient of the non-spherically-symmetric operator $\hat{O}_{ii}$ (5.4.37) is small compared to the coefficient G_{ii}.

5.4.4 Vibration-Rotation Hamiltonian

In our discussion of the vibrational anharmonic Hamiltonian, we employed primarily the Cartesian vibrational basis, because it permits calculation of the matrix elements of the anharmonic Hamiltonian in a completely elementary manner. However, the introduction of molecular rotation, and the complications of calculating matrix elements of vibration-rotation operators in the bases (5.4.80,81), make more powerful methods imperative. If the vibration-rotation Hamiltonian is expressed in terms of irreducible tensor operators, as was done earlier for the purely vibrational Hamiltonian, then the Racah-Wigner angular momentum calculus allows us to calculate the matrix elements of each tensor operator in a completely systematic manner. This fact makes it possible to construct a family of models of an infrared-active triply-degenerate vibrational mode, including more or less detail of the energy-level structure as deemed necessary. Such models will form the subject of Sect.5.4.7.

A vibration-rotation tensor operator may be formed from a purely vibrational tensor operator of rank k_1 and a purely rotational tensor operator of rank k_2 in the standard manner [5.64]

$$\hat{T}_q^{(k_1k_2k)} = \sum_{q'} \hat{T}_{q'}^{(k_1)}(\mathrm{vib})\hat{T}_{q-q'}^{(k_2)}(\mathrm{rot})\langle k_1k_2;q',q-q'|kq\rangle \quad . \tag{5.4.84}$$

In this equation, $T_q^{(k)}$ belongs to the qth row of the representation $D^{(k)}$ of $O_3^{(+)}$. For example, the Coriolis operator (5.4.78) may be expressed in terms of the tensors [5.69]

$$\hat{T}_0^{(1)}(\mathrm{rot}) = -\hat{J}_z \tag{5.4.85}$$

$$\hat{T}_{\pm1}^{(1)}(\mathrm{rot}) = \pm\frac{1}{\sqrt{2}}(\hat{J}_x \pm i\hat{J}_y) \tag{5.4.86}$$

$$\hat{T}_0^{(1)}(\mathrm{vib}) = \hat{\ell}_z \tag{5.4.87}$$

$$\hat{T}_{\pm1}^{(1)}(\mathrm{vib}) = \mp\frac{1}{\sqrt{2}}(\hat{\ell}_x \pm i\hat{\ell}_y) \tag{5.4.88}$$

(where components are expressed in the molecule-fixed axes) as

$$\hat{H}_{\mathrm{Cor}} = 2\sqrt{3}\, B\zeta\hat{T}_0^{(110)} \quad . \tag{5.4.89}$$

The matrix elements of (5.4.84) in the coupled spherical basis (5.4.80) are

$$\left(\phi_{K_R'M'}^{n'\ell'JR'}\ ,\ \hat{T}_q^{(k_1k_2k)}\ \phi_{K_RM}^{n\ell JR}\right)$$

$$= \delta_{M'M}(-1)^{R+K_R'}\begin{pmatrix} R & K & R' \\ K_R & q & -K_R' \end{pmatrix}[(2k+1)(2R+1)(2R'+1)]^{\frac{1}{2}}$$

$$\cdot \begin{Bmatrix} \ell' & \ell & k_1 \\ J & J & k_2 \\ R' & R & k \end{Bmatrix} \langle n'\ell'||\hat{T}^{(k_1)}(\mathrm{vib})||n\ell\rangle\langle J||\hat{T}^{(k_2)}(\mathrm{rot})||J\rangle \quad . \tag{5.4.90}$$

Most of the reduced matrix elements, which must be evaluated using the explicit forms of the operators $T^{(k)}(\mathrm{vib})$ and $T^{(k)}(\mathrm{rot})$, were tabulated by HECHT [5.56] and by ROBIETTE et al. [5.69]. For example,

$$\langle n\ell||T^{(1)}(\mathrm{vib})||n\ell\rangle = [\ell(\ell+1)(2\ell+1)]^{\frac{1}{2}} \tag{5.4.91}$$

$$\langle J||\hat{T}^{(1)}(\mathrm{rot})||J\rangle = [J(J+1)(2J+1)]^{\frac{1}{2}} \quad . \tag{5.4.92}$$

To construct a vibration-rotation Hamiltonian which is invariant under the operations of the molecular symmetry group, it is necessary to form appropriate linear combinations of the spherical tensor components

$$\hat{T}_q^{(k_1k_2k)} \quad .$$

The invariant linear combinations are

$$\hat{T}_{A_1}^{(k_1k_2k)} = \sum_q {}^{(k)}G_{A_1}^q \; \hat{T}_q^{(k_1k_2k)} \equiv T_{k_1k_2k} \quad , \tag{5.4.93}$$

where A_1 is the totally symmetric representation of the molecular symmetry group. In view of (5.4.90,93) the matrix elements of all tensor operators of rank k are proportional to the linear combination

$$\sum_q {}^{(k)}G_{A_1}^q (-1)^{K_R'} \begin{pmatrix} R & k & R' \\ K_R & q & -K_R' \end{pmatrix} \quad . \tag{5.4.94}$$

This fact has been employed by KROHN [5.68] to calculate the matrix elements of ${}^{(R)}G$, by diagonalization of a matrix with elements proportional to (5.4.94) for k = 4.

The matrix elements of (5.4.93) take on a particularly simple form when expressed in the symmetry-adapted basis (5.4.81)

$$\begin{aligned} &\left(\Psi_{p'M'}^{n'\ell'JR'} \;,\; \hat{T}_{A_1}^{(k_1k_2k)} \; \Psi_{pM}^{n\ell JR}\right) \\ &= \delta_{M'M} \begin{Bmatrix} \ell' & \ell & k_1 \\ J & J & k_2 \\ R' & R & k \end{Bmatrix} [(2k+1)(2R+1)(2R'+1)]^{\frac{1}{2}} \\ &\quad \cdot \langle n'\ell' || T^{(k_1)}(\mathrm{vib}) || n\ell \rangle \langle J || \hat{T}^{(k_2)}(\mathrm{rot}) || J \rangle (-1)^R F_{A_1p'p}^{(kR'R)} \quad . \end{aligned} \tag{5.4.95}$$

The Moret-Bailly $F^{(k)}$ coefficient, which is defined as

$$F_{A_1p'p}^{(kR'R)} = \sum_{m_1} \sum_{m_2} \sum_{m_3} {}^{(k)}G_{A_1}^{m_1} {}^{(R')}G_{p'}^{m_2} {}^{(R)}G_p^{m_3} \begin{pmatrix} R' & K & R \\ m_2 & m_1 & m_3 \end{pmatrix} \quad , \tag{5.4.96}$$

may be calculated numerically as a function of its indices.

An approximate vibration-rotation Hamiltonian for a triply-degenerate mode of tetrahedral or octahedral, expressed in terms of the operators (5.4.81) and the operators

$$T_{k_1k_2k}^{mn} = (\hat{n})^m (\underline{J}^2)^n T_{k_1k_2k_3} \tag{5.4.97}$$

is shown in Table 5.8 [5.69]. In the calculations of multiple-photon excitation discussed below, only the first row in Table 5.8 will be important, since the operators of the first row are responsible for the anharmonic and Coriolis splitting. Of these, all are diagonal in the bases (5.4.80,81) except T_{404}, which is

Table 5.8. Vibration-rotation Hamiltonian for a triply-degenerate mode [5.56,69]

$$H = \omega_s T^{10}_{000} + X_{ss}\, T^{20}_{000} + G_{ss}\ell_s^2 + T_{ss}\, T_{404} + B_0 T^{01}_{000} + 2\sqrt{3}B\zeta_s T_{110}$$
$$- \alpha T^{11}_{000} + \alpha_{220}T_{220} + \alpha_{224}T_{224}$$
$$- D_0 T^{02}_{000} - D_{t0}T_{044} + F_{110}T^{01}_{110} + F_{134}T_{134}$$
$$- (D - D_0)T^{12}_{000} - (D_t - D_{t0})T^{10}_{044} + G_{220}T^{01}_{224}$$
$$+ G_{244}T_{244} + G_{246}T_{246}$$
$$+ HT^{03}_{000} - H_{4t}T^{01}_{044} + H_{6t}T_{066}$$

equal to $\hat{O}_{ii}$ [(5.4.37)]. It is evident from the 9-J symbols in (5.4.90,95) that the operators (5.4.84,93) will possess nonzero off-diagonal matrix elements in R as long as

$$|R - k| \leq R' \leq R + k \tag{5.4.98}$$

except in the ground vibrational state, where R = J = R' according to (5.4.53,80). Thus in general R will not be a good quantum number in vibrationally excited states, owing to the presence of tensor operators of rank k > 0. T_{404}, which is the most likely of the tensor operators with k > 0 in Table 5.8 to have a large coefficient, contributes to the energy only in states with $n \geq 2$, since the vibrational reduced matrix element vanishes for $\ell = 0$ and $\ell = 1$. Thus the mixture of states with different R may always be expected to be small when n = 1, regardless of the anharmonic Hamiltonian, while the admixture may be large for more highly excited vibrational states, depending on the anharmonic force field. It is known [5.44-47] that in SF_6, the mixing of states with different R in n = 1 is primarily due to the operator T_{224}.

The energy eigenvalues of the following approximate vibration-rotation Hamiltonian consisting of the first row and the first term of the second row of Table 5.8 (with $T_{ii} = 0$)

$$\hat{H}_{approx} = \omega_i T^{10}_{000} + X_{ii}T^{20}_{000} + G_{ii}\hat{\ell}_i^2 + B_0 T^{01}_{000} + 2\sqrt{3}B_0\zeta_i T_{110}$$
$$- \alpha T^{11}_{000} \tag{5.4.99}$$

are

$$E(n\ell JR) = n\nu_i + n(n-1)X_{ii} + [\ell(\ell+1) - 2n]G_{ii}$$
$$+ B_0 J(J+1) + B_0\zeta_i[R(R+1) - J(J+1) - \ell(\ell+1) + 2n]$$
$$- \alpha nJ(J+1) \quad . \tag{5.4.100}$$

This expression for the energy levels, without the term $nJ(J+1)$, was presented in [5.14], and is a minor modification of the expression given in [5.10]. The bases (5.480,81) are eigenstates of (5.4.99). It is trivial to add the operators $-\alpha_{220}T_{220}$ and $F_{110}T^{01}_{110}$ if desired, since these are also diagonal in either basis. The approximate eigenvalues of the full Hamiltonian of Table 5.8, neglecting the perturbations which are off-diagonal in R, may be calculated in the symmetry-adapted basis (5.4.81) using (5.4.95). Although the latter procedure has been widely employed in the fitting of experimental spectra [5.44-47,70], it has not yet been applied to calculations of multiple-photon excitation.

In Fig.5.20 we show a portion of the Doppler-limited spectrum of SF_6 [5.45]. The figure shows portions of several manifolds near the P(16)CO_2 laser line; the splitting is primarily due to the operators T_{224}. The lines show the frequencies calculated by diagonalization of the vibration-rotation Hamiltonian. The excellent agreement between theory and experiment indicates that the theoretical approach described in this section can lead to a highly accurate fit of observed to calculated spectra for a fundamental band (n = 0 to n = 1) in a spherical-top molecule. A detailed discussion of methods for analysis of fundamental bands in spherical-top molecules is forthcoming [5.70].

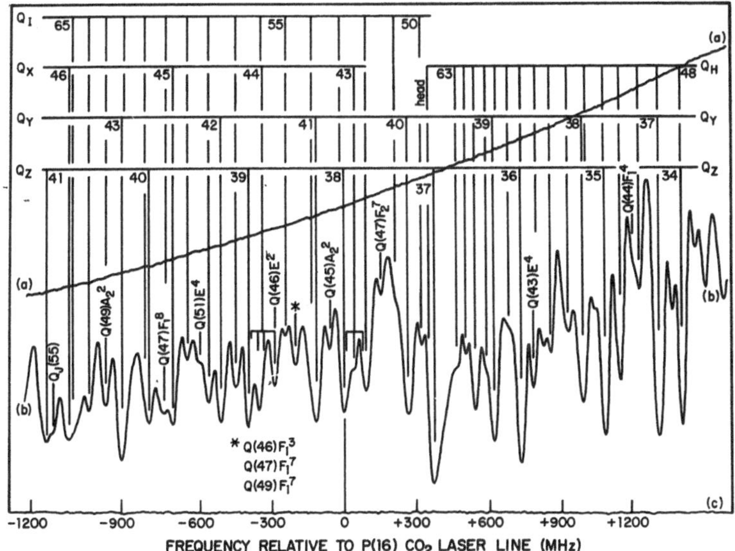

Fig.5.20. The spectrum of a portion of the Q branch of the ν_3 fundamental band of SF_6 near the P(16)CO_2 laser line [5.45]

5.4.5 Dipole Transition Moments in Spherical-Top Molecules

The dipole-moment operator for a polyatomic molecule restricted to the ground electronic state, may be expressed in the molecule-fixed frame approximately as the sum of the permanent dipole moment, and the dipole derivatives times the normal coordinates of the infrared-active vibrational modes. Physically the quantities which mediate the interaction between the molecule and an external electromagnetic field are the components of the dipole moment in the laboratory frame, which are related to the components in the molecule-fixed frame by a rotation matrix (the direction-cosine matrix). Because of this rotation, the permanent dipole moment can cause transitions in which the rotational (but not the vibrational) quantum numbers change. Although tetrahedral molecules can have a weak rotational spectrum, a rotational spectrum is totally lacking in octahedral molecules. Vibration-rotation transitions are the result of the part of the dipole-moment operator which depends nontrivially on the normal-coordinate operators $\{\hat{q}\}$. To lowest order for a spherical-top molecule this may be written (in the molecule-fixed frame)

$$\hat{d}_\tau = A\hat{q}_\tau \quad , \tag{5.4.101}$$

where A is a constant (the dipole derivative) characteristic of the molecule, and $\hat{d}_\tau$ is any of the three spherical components of the dipole-moment operator $\hat{\underline{d}}$. In the laboratory frame the spherical components of $\hat{\underline{d}}$ are

$$\hat{d}_\sigma = A \sum_\tau D^1_{\tau\sigma}(\Omega^{-1})\hat{q}_\tau \quad , \tag{5.4.102}$$

where D^1 is the rotation matrix for J = 1.

The matrix elements of $\hat{d}_\sigma$ may be evaluated in either of the bases (5.4.80) or (5.4.81) by a simple calculation: express the states $\phi^{n\ell JR}_{KM}$ using (5.4.80); use the Wigner-Eckart theorem to evaluate $(\phi^{v'\Omega'}_{m'}, \hat{q}_\tau\phi^{v\ell}_{m})$; evaluate the rotational matrix elements using the well-known integral over a product of three D^J's; and resum all the remaining 3-J symbols to 6-J or 9-J symbols. The result is [5.71-74]

$$\left(\phi^{n'\ell'J'R'}_{p'M'} , \hat{d}_\sigma\phi^{n\ell JR}_{pM}\right)$$

$$= A\delta_{R'R}\delta_{p'p}(-1)^{J+M'+1}\begin{pmatrix} J & 1 & J' \\ M & \sigma & -M' \end{pmatrix}$$

$$\cdot[3(2R + 1)(2J' +1)(2J + 1)]^{\frac{1}{2}}\langle n\ell||\hat{q}||n'\ell'\rangle$$

$$\cdot\begin{Bmatrix} \ell' & \ell & 1 \\ J' & J & 1 \\ R' & R & 0 \end{Bmatrix} \tag{5.4.103}$$

where the reduced matrix element is

$$\langle n\ell||\hat{q}||n'\ell'\rangle = \begin{cases} [(n+\ell+3)(\ell+1)/2]^{\frac{1}{2}} \text{ when } n' = n+1\ ,\ \ell' = \ell+1 \\ [(n-\ell+2)\ell/2]^{\frac{1}{2}} \text{ when } n' = n+1\ ,\ \ell' = \ell-1\ . \end{cases} \quad (5.4.104)$$

The selection rules implied by (5.4.102) are

$$R' = R \quad (5.4.105)$$

$$p' = p \quad (5.4.106)$$

$$n' = n \pm 1 \quad (5.4.107)$$

$$\ell' = \ell \pm 1 \quad (5.4.108)$$

$$J = J \ \text{ or } \ J \pm 1 \quad (5.4.109)$$

$$M' = M + \sigma\ . \quad (5.4.110)$$

Selection rules (5.4.107-109), which imply physically that one vibrational quantum is changed by the absorption of one photon, hold for all dipole-allowed transitions. Selection rules (5.4.105,106) are satisfied when the states $\phi_{pM}^{n\ell JR}$ are eigenstates of the exact Hamiltonian, but are violated if the exact eigenstates are a superposition of states with different rotational angular momenta R or symmetry types p. We have already remarked that the possibly important vibrational anharmonic operator T_{404} mixes states with different R, and that this admixture may be large, depending on the magnitude of the coefficient T_{ii} compared to G_{ii}. Other vibration-rotation operators also mix states with different R, but are probably not as important as T_{404} except in unusual cases; for SF_6 the coefficient t_{224}, which is the largest of the coefficients of the tensor operators of nonzero rank, is of the order of 10^{-5} cm^{-1} [5.44-47].

While the admixture of states with different R is important primarily for vibrational states above $n = 1$, the admixture of states with different p is expected to be important primarily in the vibrational ground state. An interaction which mixes states of different molecular symmetry types cannot be symmetric under the (discrete) molecular symmetry group. In the ground electronic state, the most important example of such an interaction is the nuclear hyperfine interaction, which has been shown experimentally in SF_6 to produce significant mixing of states with different symmetry types in the ground vibrational state [5.47], where the splitting due to the tensor centrifugal distortion operators T_{044} and T_{066} may (depending on the states involved) easily be less than the splitting due to the spin-rotation hyperfine interaction. The admixture of states with different symmetry type p is expected to be far smaller in states with $n \geq 1$, since the vibration-rotation splitting operator with the largest coefficient, T_{224}, is nonzero for states with $n \neq 0$, and thus causes a vibration-rotation splitting for $n \geq 1$ which in most cases far exceeds the nuclear hyperfine splitting. Since the initial state defines (through the selection rules) the pathways available for excitation, the mixture of rotational states within the same cluster or (for small J) nearby clusters, in the vibrational ground state, may relax the selection rule (5.4.106).

Other effects which we have not discussed, but which may break some of the selection rules (5.4.105-110), include the perturbation of the state $n\nu_i$ by other vibrational states; and the terms of quadratic or higher degree in the normal co-cordinates $\{\hat{q}\}$ in the expansion of the dipole-moment operator $\hat{\underline{d}}$.

The importance of "weak" transitions, i.e., transitions which do not satisfy the selection rules (5.4.105-110), has been suggested by KNYAZEV et al. [5.75] to lie in the occurrence of many multiphoton resonances, some steps of which are nearly in single-photon resonance, as a result of the great number of pathways of excitation involving some "weak" transitions. As we have already indicated, some of the transitions which do not satisfy (5.4.105-110) may not be weak at all; in that case the basic physics of excitation is similar to that discussed in Sect. 5.3. A quantitative discussion of multiphoton excitation in the presence of many transitions which do not satisfy (5.4.105-110) and in the presence of some transitions which do satisfy these selection rules has not yet been given.

5.4.6 Experimental Determination of the Anharmonic Parameters of the ν_3 Mode of SF_6

It is evident from (5.4.100) that the vibrational anharmonic parameters G_{33} and T_{33} do not affect the ν_3 state with n = 1, which is by all odds the best-analysed state of SF_6 [5.42-47]. These parameters must be determined from an analysis of transitions involving the overtone states $n\nu_i$, with $n \geqq 2$. Analyses of $2\nu_3$ and $2\nu_4$ have been made for some tetrahedral molecules [5.76] but no $3\nu_i$ state of a triply-degenerate mode had been made until very recently, when analyses of the n = 0 to n = 3 spectra of CH_4 [5.77] and SF_6 [5.4,49] have been published. In octahedral spherical-top molecules, a single-photon transition from n = 0 to n = 2q cannot occur; these levels are visible only in Raman spectra. Hence the lowest overtone level of the ν_3 (or ν_4) mode of SF_6 or UF_6 for which an absorption spectrum exists is $3\nu_3$ (or $3\nu_4$). The overtone bands of spherical-top molecules possess a very weak absorption, making it difficult in SF_6 to obtain the Doppler-limited spectrum which is an essential prerequisite to a full assignment of vibration-rotation quantum numbers. That such a full analysis is an essential prerequisite to obtaining believable anharmonic parameters is well illustrated by the history of attempts to obtain anharmonic constants for SF_6 from the unresolved contour of the $3\nu_3$ spectrum, which we shall summarize briefly below.

The vibration-rotation analysis of an overtone band in a spherical-top molecule is similar in many respects to the analysis of a fundamental band, which has been described elsewhere [5.70]. The selection rules for overtone transitions, which are quite different from those for transitions in which only one vibrational quantum changes, may be calculated by standard methods [5.76]. The unique difficulty in analyzing the n = 0 to n = 3 overtone spectrum is due to the fact that there are two vibrational levels (ℓ = 1 and 3 in the spherical basis, or two F_{1u} states in the Cartesian basis) which are connected to the ground vibrational state by single-photon transitions. Accordingly there are two rotational series in the $3\nu_3$ overtone spectrum which may perturb one another. Parts of such a band may, in favorable cases, be analyzed as if the vibrational states were isolated, and preliminary constants may be determined in this way. Once the essential and difficult step of finding an approximately correct set of constants has been taken, the perturbed parts of the band may be analyzed by diagonalization of the vibration-rotation Hamiltonian to obtain a preliminary predicted spectrum; comparison of the calculated and experimental spectra to assign more lines; and iteration of this procedure to obtain successively more accurate sets of spectroscopic constants, as for a fundamental band [5.44-47,70]. The anharmonic parameters for the ν_3 mode of SF_6 determined by a partial analysis of a Doppler-limited $3\nu_3$ spectrum are [5.49]

$$X_{33} \cong -1.56\ \mathrm{cm}^{-1}\ ,\ G_{33} \cong 0.9\ \mathrm{cm}^{-1}\ ,\ T_{33} \cong -0.25\ \mathrm{cm}^{-1}\quad . \tag{5.4.111}$$

Previous attempts to analyze the $3\nu_3$ band of SF_6 depended upon spectra of SF_6 in cryogenic solvents [5.78] or unresolved band contours of gas-phase SF_6 [5.79]. The first cryogenic spectrum [5.78] showed only a single main peak, which was interpreted [5.66] as one of the F_{1u} components in the Cartesian basis. The absence of the other F_{1u} component was explained [5.80] by a calculation of the transition moments, which showed that one may expect very unequal transition strengths from n = 0 to the two F_{1u} components of n = 3, for certain classes of model vibrational force fields which are dominated by octahedrally (rather than spherically) symmetric anharmonic forces. The anharmonic constants predicted [5.66] by such a force field appeared to be in qualitative agreement with an earlier set of parameters [5.10,11]. A gas-phase band contour of $3\nu_3$ [5.79] revealed two intense features, which were interpreted by FOX [5.81] as corresponding to the ℓ = 1 and ℓ = 3 components of n = 3. A calculation of the transition moments from n = 0 to n = 3 in the spherical basis [5.81] showed that the ratio of intensities of transitions from n = 0 to the ℓ = 1 and ℓ = 3 components of n = 3 expected in the spherical basis was very nearly the intensity ratio observed for the two main components of the band contour, even at low temperature [5.82]. The anharmonic parameters derived from this interpretation of the $3\nu_3$ spectrum are [5.14]

$$X_{33} = -2.54\ \mathrm{cm}^{-1}\ ,\ G_{33} = +0.303\ \mathrm{cm}^{-1}\ ,\ T_{33} = 0\quad . \tag{5.4.112}$$

Several authors [5.14,15] explored the implications of the new set of anharmonic parameters for multiple-photon excitation. At roughly the same time, a different interpretation of the gas-phase band contour of the $3\nu_3$ spectrum of SF_6 was made, in which one of the two strong features observed in [5.79] was assigned as belonging to one of the F_{1u} components of n = 3 in the Cartesian basis, while another weak feature was assigned to the other F_{1u} component [5.83]. Finally, an apparently reliable analysis [5.4,49] of the Doppler-limited spectrum of $3\nu_3$ has removed the ambiguities arising from the many interpretations to which the band contour was susceptible.

It has been pointed out in [5.4] that the anharmonic parameters obtained there for the ν_3 mode of SF_6 are reasonably close to the parameters of [5.10,11,66], but are not in agreement with the anharmonic parameters of [5.14,15,80,81]. The same is true of the slightly more accurate parameters of [5.49] (5.4.111). The main difference between the anharmonic parameters propsed in [5.10,11,66] appears to be in the value of X_{33}, which is appreciably smaller in the actual parameters (5.4.11) [5.49] than in the proposed parameters. This means that the spacings of the actual anharmonicity split vibrational energy levels of the ν_3 mode are close to those calculated in [5.11], but that the anharmonic shift is substantially smaller than in [5.11]. Although this means that the compensation of the anharmonic shift by anharmonic splitting is even greater than envisaged in [5.11], a total compensation of the anharmonic shift by anharmonic splitting does not occur in the ν_3 mode of SF_6 because the anharmonic parameters do not satisfy the criterion (5.4.71).

5.4.7 Effective-State Models for Molecular Multiphoton Calculations

In view of the complexity of the vibration-rotation wave functions, energy levels, and dipole transition moments discussed in Sects.5.4.3 and 5, practical considerations make it imperative to employ simplified models of such a molecular system even in numerical calculations with large computers. From a physical point of view the most natural course is to discard some of the interactions which are responsible for certain energy-level splittings or spectral fine structure which one wishes to disregard in an approximate calculation. This leads to assuming that some states are degenerate, constituting a single "effective state"; i.e., one assumes certain energy-level splittings to be zero which are actually nonzero. In such a model it is essential to choose dipole transition moments between different effective states by demanding that the total absorption strength between pairs of effective states be equal to the summed absorption strength of the real physical spectral lines which are forced to coincide in the effective-state model. The calculation of appropriate transition dipole matrix elements is therefore of central importance in multiple-photon calculations based on effective-state models of the molecular energy levels.

The relation between the absorption strength of a group of spectral lines, the dipole transition matrix elements in the molecule-fixed frame of reference, and the transition dipole moments appropriate for use in effective-state models, is both subtle and unconventional, and has not previously been presented in detail in the published literature. In this section we give a detailed explanation and classification of effective-state models of molecular energy levels and the appropriate transition dipole moments.

Absorption Strength of an Ensemble of Two-Level Systems

Since the calculation of the absorption strength between pairs of effective states is a central step in our calculation of the effective-state transition dipole moments, we begin by calculating the absorption strength of an ensemble of two-level systems in which the transition frequency $\nu_{fi} = \omega_{fi}/c$ is distributed over a range of frequencies due to inhomogeneous broadening. The dipole interaction Hamiltonian between the molecular system and the field is

$$\hat{H}_{int} = -\hat{\underline{d}} \cdot \mathcal{E}_0 \cos\Omega t \tag{5.4.113}$$

where Ω is the laser frequency, and $\hat{\underline{d}}$ is the molecular dipole operator in the laboratory-fixed frame, the matrix elements of which were calculated in Sect.5.4.5. For fully polarized light, with a polarization specified by the spherical component σ, the interaction Hamiltonian is

$$\hat{H}_{int} = -\hat{d}_\sigma \mathcal{E}_0 \cos\Omega t \quad . \tag{5.4.114}$$

The solution of the Schrödinger equation in the RWA with the interaction (5.4.114) is well known, and will not be repeated here. An external optical field of constant amplitude $\mathcal{E}_0$ generates a superposition of the lower and upper states $|i\rangle$ and $|f\rangle$, with the result that the expectation value of the dipole operator has the time dependence

$$\langle i(t)|\hat{d}_\sigma|f(t)\rangle = -\frac{2\gamma_0\mu_{fi}(\sigma)}{\beta}\left[\sin\beta t+\frac{2i\delta}{\beta}\sin^2(\tfrac{1}{2}\beta t)\right] \tag{5.4.115}$$

for a molecule with detuning $\delta = \Omega - \omega_{fi}$. In (5.4.115),

$$\gamma_0 = \frac{\mu_{fi}(\sigma)\mathcal{E}_0}{2\hbar} \tag{5.4.116}$$

$$\mu_{fi}(\sigma) = \langle i(0)|\hat{d}_\sigma|f(0)\rangle \tag{5.4.117}$$

$$\beta = [\delta^2 + 4\gamma_0^2]^{\frac{1}{2}} \quad . \tag{5.4.118}$$

The macroscopic oscillating dipole moment induced by the optical field $\mathcal{E}_0$ is the integral of the induced dipole moments, over all molecules and all transition frequencies ν_{fi}. If the normalized probability density (per Hz) of finding a system with detuning δ is $g(\delta/2\pi)$, and if the total molecular number density in the initial state $|i\rangle$ at time $t = 0$ is N_i, then the macroscopic oscillating dipole moment induced by the optical electric field $\underline{\mathcal{E}}_0 \cos\Omega t$ is $\mathrm{Re}\{-\mathscr{P}_i \exp(-i\Omega t)\}$, where

$$\mathscr{P}_i = N_i \int_{-\infty}^{\infty} g\left(\frac{\delta}{2\pi}\right)\langle i(t)|\hat{d}_\sigma|f(t)\rangle \; d\left(\frac{\delta}{2\pi}\right) \tag{5.4.119}$$

$$\cong -N_i\mu_{fi}(\sigma)\gamma_0 g(0)J_0(2\gamma_0 t) \tag{5.4.120}$$

and J_0 is the zero-order Bessel function. In passing from (5.4.119) to (5.4.120), we have assumed that $g(\delta/2\pi)$ is symmetric about $\delta = 0$, and that the inhomogeneous distribution function g varies much less rapidly in δ than the function $\beta^{-1}\sin\beta t$. The rate at which energy is absorbed per unit volume from the incident optical field by the molecular systems is

$$\frac{dW}{dt} = \frac{\Omega}{4}\left(\mathcal{E}_0^*\mathcal{P}_i + \mathcal{E}_0\mathcal{P}_i^*\right) \tag{5.4.121}$$

$$= \frac{8\pi^3}{hc}\left[\mu_{fi}(\sigma)\right]^2 \nu_{fi} N_i I(\nu_{fi}) \tag{5.4.122}$$

where

$$I(\nu_{fi}) = \frac{c|\mathcal{E}_0|^2}{8\pi}\, g(0) \tag{5.4.123}$$

is the intensity distribution of the incident optical field at the resonant frequency ν_{fi} about which the distribution of transition frequencies is symmetric. If initially (at time t = 0) number density of systems in the final state $|f\rangle$ is N_f, then N_i in (5.4.122) must be replaced by $N_i - N_f$.

It is often true that the initial or final energy levels are degenerate, but only a single final state $|f_i\rangle$ is connected to a given initial state $|i\rangle$. This is the case, for example, when a set of degenerate levels with different spatial quantum numbers M are excited by fully polarized light, since it is true in general that the matrix element of $\hat{d}_\sigma$ between $|i,M\rangle$ and $|f,M'\rangle$ vanishes unless $M' = M + \sigma$ [see, for example, (5.4.103)]. Since the total polarization amplitude $\mathcal{P}$ is the sum of amplitudes $\mathcal{P}_i$ (5.4.119), we have the result that the rate at which energy is absorbed per unit volume from the incident field is

$$\frac{dW}{dt} = SI(\nu_{fi}) \tag{5.4.124}$$

where the absorption strength S is

$$S = \frac{8\pi^3}{hc}\nu_{fi}(N_i - N_f)\sum_{i,f}|\mu_{fi}(\sigma)|^2 \quad . \tag{5.4.125}$$

We have introduced the sum of final states $|f\rangle$ in view of the fact that $\mu_{fi}(\sigma)$ vanishes unless $f = f_i$. In (5.4.125), N_i is the number densit- in one single non-degenerate initial state $|i\rangle$.

If the sum on $|i\rangle$ and $|f\rangle$ in (5.4.125) includes a sum over all initial and final magnetic quantum numbers (M and M') corresponding to initial and final total angular momenta J and J', then the entire sum is independent of the polarization σ. In that case

$$|\mu_{fi}(\sigma)|^2 = \frac{1}{3}\sum_{\sigma,i,f}|\mu_{fi}(\sigma)|^2 = \frac{1}{3}\sum_{i,f}|\langle i|\underline{\hat{d}}|f\rangle|^2 \tag{5.4.126}$$

and the equation for S becomes

$$S = \frac{8\pi^3}{3hc} \nu_{fi}(N_i - N_f) \sum_{i,f,\sigma}\sum\sum |\mu_{fi}(\sigma)|^2 \quad , \tag{5.4.127}$$

the expression which is most often encountered in the literature. It is worth emphasizing, however, that the matrix elements which are relevant to laser excitation are in general $\mu_{fi}(\sigma)$ and not $|\langle i|\hat{\underline{d}}|f\rangle|$ since laser light is ordinarily fully polarized.

Effective-State Equations of Motion

We shall now derive the analog of the equations of motion (5.1.8) in an effective-state model, using a generalization of the method of GOODMAN et al. [5.84]. Let

$$|nq\rangle = |ers\rangle \tag{5.4.128}$$

where the set of quantum numbers {e} constitutes the effective quantum numbers, i.e., the quantum numbers which distinguish different effective states; and {s} is the set of quantum numbers (such as M) for which, given the initial quantum numbers {ers} and the partial final quantum numbers {e'r'}, the remaining final quantum numbers {s'} are uniquely determined by selection rules. The set {r} includes all the remaining quantum numbers not included in {e} or {s}. Then the Schrödinger equation (5.1.8) becomes

$$\frac{da_{ers}}{dt} = -i\delta_{ers}a_{ers} + \frac{i\mathscr{E}_0}{2\hbar} \sum_{e'r'} \mu_{ers,e'r's'}(\sigma)a_{e'r's'} \quad , \tag{5.4.129}$$

where {s'} is uniquely determined by {s}, {e'} and {r'}. Physically an effective-state model assumes that all states with the same quantum numbers {e} are degenerate, so that

$$\delta_{ers} = \delta_e \quad . \tag{5.4.130}$$

To derive the effective-state equation of motion, we must also assume (in the effective-state model) that all the transition moments between states with a given {e} and {e'} are equal to a "typical" transition moment

$$\mu_{ers,e'r's'}(\sigma) = \mu_{ee'}^{typ}(\sigma) \quad . \tag{5.4.131}$$

We now define the degeneracies

$$g_e = \sum_{r,s} \delta_{ee'} \quad , \quad h_e = \sum_r \delta_{ee'} \tag{5.4.132}$$

where the notation indicates by use of the Kronecker delta that {e} is held fixed during the summation. Then (5.4.129) becomes independent of the indices {rs} and {r's'}, so that all amplitudes a_{ers} with the same {e} are equal to one another. If we now define the effective-state amplitude

$$c_e = \sqrt{g_e}\, a_{ers} \tag{5.4.133}$$

(where a_{ers} is any of the g_e equal amplitudes), so that the effective-state population is

$$|c_e|^2 = \sum_{r,s} |a_{ers}|^2 \quad , \tag{5.4.134}$$

then the Schrödinger equation for c_e is

$$\frac{dc_e}{dt} = -i\delta_e c_e + \frac{i\mathcal{E}_0}{2\hbar} \sum_{e'} \mu_{ee'}^{typ}(\sigma) h_e (g_e/g_{e'})^{\frac{1}{2}} c_{e'} \quad . \tag{5.4.135}$$

We turn now to a calculation of the "typical" transition moments $\mu_{ee'}^{typ}(\sigma)$.

Dipole Transition Moments Between Effective States

Since (with GOODMAN et al. [5.84]) we insist that the transition strength be the same for the real physical states and the corresponding effective states, we must have

$$\sum_{rr'}\sum_{s} [\mu_{ee'}^{typ}(\sigma)]^2 = T_{ee'}(\sigma) \quad , \tag{5.4.136}$$

where $T_{ee'}(\sigma)$ is defined as

$$T_{ee'}(\sigma) = \sum_{rr'}\sum_{s} [\mu_{ers,e'r's'}(\sigma)]^2 \quad . \tag{5.4.137}$$

GOODMAN et al. [5.84] used $\sum_\sigma T_{ee'}(\sigma)$; however, $T_{ee'}(\sigma)$ is the appropriate transition strength for polarized light, while $(1/3)\sum_\sigma T_{ee'}(\sigma)$ is the appropriate transition strength for unpolarized, isotropic light. If (5.4.126) holds, then the latter two transition strengths are equal, but differ by a factor of 3 from $\sum_\sigma T_{ee'}(\sigma)$. Then the typical transition moment between e and e' is

$$\mu_{ee'}^{typ}(\sigma) = [g_e h_{e'}]^{-\frac{1}{2}} [T_{ee'}(\sigma)]^{\frac{1}{2}} \quad . \tag{5.4.138}$$

The transition moment between the effective states {e} and {e'} is

$$\mu_{ee'}^{eff}(\sigma) = (h_{e'}/g_{e'})^{\frac{1}{2}} [T_{ee'}(\sigma)]^{\frac{1}{2}} \quad . \tag{5.4.139}$$

For conservation of probability we must have

$$\mu_{e'e}^{eff}(\sigma) = \mu_{e'e}^{eff}(\sigma) \quad . \tag{5.4.140}$$

In most of the cases discussed below this is true. However, if (5.4.140) is not satisfied by the prescription (5.4.139), then (5.4.139) must simply be regarded as a heuristic guide, and a set of symmetric transition moments must be chosen arbitrarily.

In the models developed by GOODMAN et al. [5.84] it was assumed that

$$h_e = g_e \quad ; \tag{5.4.141}$$

that is, that no selection rules hold. In that case

$$\mu^{eff}_{ee'}(\sigma) = \mu^{typ}_{ee'}(\sigma)(g_e g_{e'})^{\frac{1}{2}} \quad . \tag{5.4.142}$$

However, this form is not appropriate when selection rules must be satisfied, i.e., when {s} is not the empty set.

Explicit formulas for $\mu^{eff}_{ee'}(\sigma)$ for several models, i.e., for several choices of effective quantum numbers {e}, are given in Table 5.9. We turn now to an illustration of the technique of calculating the degeneracies g_e and h_e, and the sums $T_{ee'}$; subsequently we discuss the numerical magnitude of the transition moments, which is determined by the constant A appearing in Table 5.9.

Calculation of Degeneracies and Transition Strengths

We calculate the degeneracies g_e starting from two bases of molecular eigenstates: $|n\ell JR;K_RM>$, in which nuclear spin, and the restriction of possible quantum numbers due to the application of the Pauli principle, are both disregarded; and $|n\ell JR;pM>$, in which the symmetry type p is uniquely correlated [5.38,85,86] with a nuclear spin state, and in which only some symmetry types p are permitted by the Pauli principle.

In the basis $|n\ell JR;K_RM>$, we have

$$g_{nR} = \sum_{\ell=0 \text{ or } 1}^{n} \sum_{J=R-\ell}^{R+\ell} \sum_{M=-J}^{J} \sum_{K=-R}^{R} \tag{5.4.143}$$

$$= \sum_{\ell=0 \text{ or } 1}^{n} \sum_{J=R-\ell}^{R+\ell} (2J+1)(2R+1) \tag{5.4.144}$$

$$= \sum_{\ell=0 \text{ or } 1}^{n} (2\ell+1)(2R+1)^2 \tag{5.4.145}$$

$$= d_n^{(v)}(2R+1)^2 \tag{5.4.146}$$

where

$$d_n^{(v)} = \frac{1}{2}(n+1)(n+2) \tag{5.4.147}$$

is the number of all states with n quanta in a triply-degenerate oscillator. Other degeneracies may be read off directly from (5.4.144-146): $g_{n\ell JR} = (2J+1)(2R+1)$, and $g_{n\ell R} = (2\ell+1)(2R+1)^2$.

If we include the nuclear-spin quantum numbers I, M_I in the list of molecular quantum numbers, and then form the linear combinations of the nuclear-spin wave functions $|IM_I>$ which transform according to irreducible representations of the

Table 5.9. Effective-State Transition Moments

{e}	{r}	{s}	g_e	h_e	$T_{ee'}(\sigma)$	$\mu^{eff}_{ee'}(\sigma)$
n,R	ℓ,J	K_R,M	$d_n^{(v)}(2R+1)^2$	$d_n^{(v)}$	$\frac{1}{6}A^2(n+3)d_n^{(v)}(2R+1)^2$	$\frac{1}{\sqrt{6}}A[\frac{1}{2}(n+1)(n+2)(n+3)]^{\frac{1}{2}}$
n,R	ℓ,J	p,M	$d_n^{(v)}(2R+1)d_R^{(r)}$	$d_n^{(v)}$	$\frac{1}{6}A^2(n+3)d_n^{(v)}(2R+1)d_R^{(r)}$	$\frac{1}{\sqrt{6}}A[\frac{1}{2}(n+1)(n+2)(n+3)]^{\frac{1}{2}}$
n,ℓ,R	J	K_R,M	$(2\ell+1)(2R+1)^2$	2ℓ+1	$\frac{1}{3}A^2\langle n\ell\|\|\hat{q}\|\|n'\ell'\rangle^2(2R+1)^2$	$\frac{1}{\sqrt{3}}A\langle n\ell\|\|\hat{q}\|\|n'\ell'\rangle$
n,ℓ,R	J	p,M	$(2\ell+1)(2R+1)d_R^{(r)}$	2ℓ+ 1	$\frac{1}{3}A^2\langle n\ell\|\|\hat{q}\|\|n'\ell'\rangle^2(2R+1)d_R^{(r)}$	$\frac{1}{\sqrt{3}}A\langle n\ell\|\|\hat{q}\|\|n'\ell'\rangle$
n,ℓ,J,R	-	K_R,M	(2J+1)(2R+1)	1	$\frac{1}{3}A^2\langle n\ell\|\|\hat{q}\|\|n'\ell'\rangle^2(2J+1)$ $\cdot W(\ell'\ell J'J;1R)^2(2J'+1)(2R+1)$	$\frac{1}{\sqrt{3}}A\langle n\ell\|\|\hat{q}\|\|n'\ell'\rangle$ $\cdot W(\ell'\ell J'J;1R)(2J+1)$
n,ℓ,J,R	-	p,M	$(2J+1)d_R^{(r)}$	1	$\frac{1}{3}A^2\langle n\ell\|\|\hat{q}\|\|n'\ell'\rangle^2(2J+1)$ $\cdot W(\ell'\ell J'J;1R)^2(2J'+1)d_R^{(r)}$	$\frac{1}{\sqrt{3}}A\langle n\ell\|\|\hat{q}\|\|n'\ell'\rangle$ $\cdot W(\ell'\ell J'J;1R)(2J+1)$
n,ℓ,J,R,M	-	K_R	2R+1	1	$A^2\begin{pmatrix}J & 1 & J'\\ M & \sigma & -M'\end{pmatrix}^2\langle n\ell\|\|\hat{q}\|\|n'\ell'\rangle^2$ $\cdot W(\ell'\ell J'J;1R)^2(2J'+1)(2J+1)(2R+1)$	$A\begin{pmatrix}J & 1 & J'\\ M & \sigma & -M\end{pmatrix}\langle n\ell\|\|\hat{q}\|\|n'\ell'\rangle$ $\cdot W(\ell'\ell J'J;1R)[(2J+1)(2J'+1)]^{\frac{1}{2}}$
n,ℓ,J,R,M	-	p	$d_R^{(r)}$	1	$A^2\begin{pmatrix}J & 1 & J'\\ M & \sigma & -M'\end{pmatrix}^2\langle n\ell\|\|\hat{q}\|\|n'\ell'\rangle^2$ $\cdot W(\ell'\ell J'J;1R)^2(2J+1)(2J'+1)d_R^{(r)}$	$A\begin{pmatrix}J & 1 & J'\\ M & \sigma & -M'\end{pmatrix}\langle n\ell\|\|\hat{q}\|\|n'\ell'\rangle$ $\cdot W(\ell'\ell J'J;1R)[(2J+1)(2J'+1)]^{\frac{1}{2}}$
n,ℓ,J,R,p	-	M	2J+1	1	$\frac{1}{3}A^2\langle n\ell\|\|\hat{q}\|\|n'\ell'\rangle^2 W(\ell'\ell J'J;1R)^2$ $\cdot(2J+1)(2J'+1)$	$\frac{1}{\sqrt{3}}A\langle n\ell\|\|\hat{q}\|\|n'\ell'\rangle$ $\cdot W(\ell'\ell J'J;1R)(2J+1)^{\frac{1}{2}}$

molecular symmetry group [5.87,88], we discover that the Pauli principle implies that only one nuclear-spin symmetry type can be associated with each vibration-rotation symmetry type p (in the vibration-rotation basis $|n\ell JR;pM\rangle$). The allowed combinations were determined by WILSON for CH_4 and CD_4 [5.89]. Wilson's method may also be proven to be mathematically correct for centrally symmetric molecules [5.90] starting from the group of feasible operations [5.91]. The statistical weights for XF_6 calculated by Wilson's method [5.85] have been confirmed subsequently [5.86] by a calculation based on the group of feasible operations.

Since the coefficients ${}^{(R)}G_p^K$ which reduce $|n\ell JR;KM\rangle$ to $|n\ell JR;pM\rangle$ depend only on R and p, and since the nuclear-spin state is determined by p alone, R and p determine the degeneracy $d_{Rp}^{(r)}$ of the family of vibration-rotation-spin states which all have the quantum numbers n,ℓ,J,R,p and M. The degeneracies $d_{Rp}^{(r)}$ and

$$d_R^{(r)} = \sum_p d_{Rp}^{(r)} \tag{5.4.148}$$

are given in Table 5.10. In this case, the degeneracy g_{nR} may be calculated as follows

$$g_{nR} = \sum_{\ell=0 \text{ or } 1}^{n} \sum_{J=R-\ell}^{R+\ell} \sum_{M=-J}^{J} \sum_{p} d_{pR}^{(r)} \tag{5.4.149}$$

$$= \sum_{\ell=0 \text{ or } 1}^{n} \sum_{J=R-\ell}^{R+\ell} (2J+1) d_R^{(r)} \tag{5.4.150}$$

$$= \sum_{\ell=0 \text{ or } 1}^{n} (2\ell+1)(2R+1) d_R^{(r)} \tag{5.4.151}$$

$$= d_n^{(v)}(2R+1) d_R^{(r)} \quad . \tag{5.4.152}$$

To calculate $T_{nR,n'R}(\sigma)$ expeditiously, we note that the orthogonality relations for the 3-J symbols and Racah coefficients [5.64] imply that

$$\sum_{M=-J}^{J} \begin{pmatrix} J & 1 & J' \\ M & \sigma & -M' \end{pmatrix}^2 = \frac{1}{3} \tag{5.4.153}$$

and

$$\sum_{J'} (2J'+1)[W(\ell'\ell J'J;1R)]^2 = \sum_{J'} (2J'+1)[W(\ell'R1J;J'\ell)]^2$$

$$= (2\ell+1)^{-1} \quad . \tag{5.4.154}$$

Also, from (5.4.104),

$$\sum_{\ell'=\ell-1}^{\ell+1} \langle n\ell||\hat{q}||n'\ell'\rangle^2 = \frac{1}{2}(n+3)(2\ell+1) \quad . \tag{5.4.155}$$

Then

$$T_{nR,n'R}(\sigma) = \sum_{\ell\ell'} \sum_{JJ'} \sum_{M} \sum_{K \text{ or } p} [\mu_{n'\ell'J'R'RKM',n\ell JRKm}(\sigma)]^2$$

Table 5.10. Nuclear spin degeneracies $d_{pR}^{(r)}$ for XF_6 [a]

R [b]	d_{pR} for symmetry type p [c]					$d_{pR}^{(r)}$
	A_1	A_2	E	F_1	F_2	
12r	2(r+1)	10r	8(2r)	6(3r)	6(3r)	64r+2
12r+1	2r	10r	8(2r)	6(3r+1)	6(3r)	64r+6
12r+2	2r	10r	8(2r+1)	6(3r)	6(3r+1)	64r+14
12r+3	2r	10(r+1)	8(2r)	6(3r+1)	6(3r+1)	64r+22
12r+4	2(r+1)	10r	8(2r+1)	6(3r+1)	6(3r+1)	64r+22
12r+5	2r	10r	8(2r+1)	6(3r+2)	6(3r+1)	64r+26
12r+6	2(r+1)	10(r+1)	8(2r+1)	6(3r+1)	6(3r+2)	64r+38
12r+7	2r	10(r+1)	8(2r+1)	6(3r+2)	6(3r+2)	64r+42
12r+8	2(r+1)	10r	8(2r+2)	6(3r+2)	6(3r+2)	64r+42
12r+9	2(r+1)	10(r+1)	8(2r+1)	6(3r+3)	6(3r+2)	64r+50
12r+10	2(r+1)	10(r+1)	8(2r+2)	6(3r+2)	6(3r+3)	64r+58
12r+11	2r	10(r+1)	8(2r+2)	6(3r+3)	6(3r+3)	64r+62

[a] From [Ref.5.85, Tables I and VIII]
[b] The integer r = 0,1,2,...
[c] Although p includes more quantum numbers than the irreducible representation label Γ, only Γ (A_1, A_2, E, F_1, F_2) is shown here, since d_{pR} depends only on R and Γ

$$= \frac{1}{6} A^2(n + 3)d_n^{(v)}(2R + 1)\left\{(2R + 1) \quad \text{or} \quad d_R^{(r)}\right\} \quad , \tag{5.4.156}$$

where the first (second) entry in curly brackets here and in subsequent equations up to (5.4.159) indicates the vibration-rotation bases $|n\ell JR;K_RM\rangle$ ($|n\ell JR;pM\rangle$), respectively.

We now perform the humble but extremely important task of relating the constant A in Table 5.9 (the derivative of the dipole moment with respect to the normal coordinates) to the effective-state transition moments $\mu_{ee'}^{eff}(\sigma)$. FOX and PERSON [5.92] define the quantity $\langle\mu_{01}\rangle^2$ as the square of the nonvanishing (n = 0, n = 1) purely vibrational matrix element of the molecule-fixed z component of the dipole operator

$$\langle\mu_{01}\rangle^2 = |([\phi_0^{11}]^* \ , \ A\hat{q}_0[\phi_0^{00}]^*)|^2 = \frac{1}{2} A^2 \quad . \tag{5.4.157}$$

Values of $\langle\mu_{01}\rangle$ for many spherical-top molecules are tabulated in [5.92].

From (5.4.153,154) we see that the orders of magnitude of the 3-J and Racah coefficients there are $\lesssim[3(2J + 1)]^{-\frac{1}{2}}$ and $\sim[(2\ell + 1)(2R + 1)]^{-\frac{1}{2}}$, respectively, so that in all cases $\mu_{ee'}^{eff}(\sigma)$ in Table 5.9 is nearly independent of J and J' when J and J' are large. Also, in all cases where {e} does not include M, the effective transition moments from {e} to {e'} are all of the same order of magnitude.

Finally, we must relate $T_{ee'}(\sigma)$ to observable quantities such as the absorption strength of a group of spectral lines. The population density N_i of a single (non-degenerate) initial state in (5.4.125) is $N \exp(-E_i/kT)/Z$, where N is the total molecular number density and Z is the partition function. From (5.4.125) we find that the absorption strength of the group of lines originating from the M sublevels with quantum numbers n,ℓ,J,R, and {K or p} and ending on n',ℓ',J',R, and {K or p} is

$$S = \frac{8\pi^3}{hc} \nu_{fi} \frac{N}{Z} \left(e^{-E_i/kT} - e^{-E_f/kT}\right) T_{n\ell JR\{K \text{ or } p\},\ n'\ell'J'R\{K \text{ or } p\}}(\sigma) \quad . \tag{5.4.158}$$

If we define

$$\{1 \text{ or } d_{Rp}\} \langle \mu_{if} \rangle^2 = T_{n\ell JR\{K \text{ or } p\},\ n'\ell'J'R\{K \text{ or } p\}}(\sigma) \tag{5.4.159}$$

(which corrects the published definition [5.92] in a minor way), we find that

$$\langle \mu_{if} \rangle^2 = A^2 \langle n\ell || \hat{q} || n'\ell' \rangle^2 [W(\ell'\ell J'J;1R)]^2 (2J + 1)(2J' + 1) \quad . \tag{5.4.160}$$

Explicit evaluation of the Racah coefficient and reduced matrix element for $n = \ell = 0$, $n' = \ell' = 1$ shows that

$$\langle \mu_{if} \rangle^2 = \langle \mu_{01} \rangle^2 (2J' + 1) \tag{5.4.161}$$

in that case. Use of (5.4.126) and (5.4.158-161) allows one to determine $\langle \mu_{01} \rangle$ directly from measurements of the absorption strength of single "lines" originating from states with single symmetry types p in the ground vibrational state [5.93].

In this section we have defined and classified effective-state models for molecular multiphoton excitation, and have shown how to relate the effective-state transition moments to experiment. In the next section we shall consider how to take a thermal distribution of initial states into account in a calculation of multiphoton excitation and in Sect.5.4.10 we shall give numerical results for the excitation of certain multilevel systems, based on the results of all the preceding sections.

5.4.8 Calculation of Multiphoton Excitation Including a Thermal Distribution of Initial State

Up to this point in our discussion of probabilities for multiphoton excitation, we have considered only the probabilities for finding a system in a particular final state, given that it was in a particular initial state when the optical field was switched on. In reality, at finite temperatures a distribution of initial states is populated, and formulas such as (5.3.47) must be modified accordingly. It has been suggested by some that the initial thermal distribution may play a qualitatively important role in multiphoton excitation, by reducing the importance of coherent processes. We shall see that, at least in the cases studied numerically in the next section, this is not so: one of the most characteristic features of multiphoton excitation, multiphoton resonances (Sect.5.3.2), are still evident in the frequency dependence of the excited-state populations, even when a full thermal distribution of vibrational and rotational states is included in the calculation. In this section, we give a general formalism for including a thermal distribution of initial states in a multiphoton calculation, as part of the foundation for our discussion of numerical results in Sect.5.4.9.

We shall begin with a brief discussion of a simple model [5.14] for taking into account the presence, and initial thermal population, of vibrational states other than the ladder of states directly pumped by the laser. In SF_6 there are five other vibrational modes ($\nu_1, \nu_2, \nu_4, \nu_5$, and ν_6), of which ν_1 is nondegenerate, ν_2 is doubly degenerate, and the rest are triply degenerate. Although the excited states of doubly- and triply-degenerate modes, and combination states in which more than one degenerate mode is excited, are certainly subject to anharmonic splitting and to complex Coriolis splittings, we shall ignore all of these complications in our model. Specifically, we assume that the excited states of vibrational modes other than the one (ν_3 in SF_6) pumped by the laser are degenerate, and that the anharmonic splitting and rotational structure of combination states which include n quanta of ν_3 are the same for a given n regardless of the excitation of other modes. Then the total energy of an effective state with the ν_3 quantum numbers n,ℓ,J,R, and the quantum numbers $\{v\} = \{v_1,v_2,v_4,v_5,v_6\}$ giving the total number of quanta of other modes excited, is [5.94]

$$E_{tot}(n\ell JR,\{v\}) = E(n\ell JR) + \sum_{k\neq 3} X_{3k}\left(n+\frac{3}{2}\right)\left(v_k+\frac{d_k}{2}\right) + \sum_{k\neq 3} v_k\left(\omega_k+\frac{d_k}{2}\right) \quad (5.4.162)$$

where $E(n\ell JR)$ is given in (5.4.100); X_{3k} is the anharmonicity coefficient between the ν_3 and ν_k modes; and d_k is the degeneracy of the ν_k mode. If we introduce the (corrected) frequencies ν_k, then (5.4.162) becomes, apart from terms which are independent of n and {v},

$$E_{tot}(n\ell JR,\{v\}) = E(n\ell JR) + E_{0th}(\{v\}) + nS(\{v\}) \quad (5.4.163)$$

where

$$E_{0th}(\{v\}) = \sum_{k\neq 3} \nu_k v_k \quad (5.4.164)$$

$$S(\{v\}) = \sum_{k\ 3} X_{3k} v_k \quad . \quad (5.4.165)$$

Since the anharmonic interaction energy between the ν_3 mode and the other modes is linear in n [in the approximation (5.4.162)], the shift $S(\{v\})$ can be considered as a shift of the ν_3 mode frequency due to excitation of other modes. Since $X_{3k} < 0$ in general, one expects $S(\{v\}) < 0$, so that the effective ν_3 frequency is shifted to lower frequencies. However, since the diagonal elements of the effective

Hamiltonian (5.3.36) are proportional to $E(n\ell JR;\{v\}) - n\nu_\ell$ (where $\nu_\ell = \Omega/2\pi$ is the laser frequency), it is mathematically equivalent to shift the laser frequency ν_ℓ to $\nu_\ell - S(\{v\})$. That is, the time-averaged conditional probability (5.3.47) that a system will be found in the effective state $|n\ell JR,\{v\}\rangle$, given that when the laser was switched on it was in the effective state $|n'\ell'J'R',\{v\}\rangle$, is, in our model,

$$\bar{P}(n\ell JR,\{v\}|n'\ell'J'R',\{v\};\nu_\ell)$$
$$= \bar{P}[n\ell JR\ n'\ell'J'R';\nu_\ell - S(\{v\})] \quad . \tag{5.4.166}$$

In (5.4.166), the left-hand side is the conditional probability (5.3.47) calculated with the energy levels (5.4.162); the right-hand side is the conditional probability calculated with the energy levels (5.4.100). This simple model for taking the interaction energy between the ν_3 mode and other modes into account was used in [5.14].

We now proceed to give a simple model for taking the thermal distribution of initial states into account. In its most general form this problem can be solved only by use of the density matrix [5.95]. However, if all off-diagonal elements of the density matrix are zero at the time when the optical field is switched on, and if there are no relaxation processes during the time when the field is turned on, then the process of excitation may be described using a wave function. Specifically, if $P_{th}(n'\ell'J'R',\{v\})$ is the (thermal) total probability of finding the system in the effective state $|n'\ell'J'R';\{v\}\rangle$ at time $t = 0$, then the total time-averaged probability for finding the system in all states $|n\ell JR,\{v\}\rangle$ with the quantum numbers n,ℓ,J and R is

$$\langle\bar{P}_{tot}(n\ell JR;\nu_\ell)\rangle = \sum_{\{v\}} \sum_{n'}\sum_{\ell'}\sum_{J'}\sum_{R'} P(n\ell JR,\{v\}|n'\ell'J'R',\{v\};\nu_\ell) \cdot P_{th}(n'\ell'J'R',\{v\}) \tag{5.4.167}$$

$$= \sum_{\{v\}} \sum_{n'}\sum_{\ell'}\sum_{J'}\sum_{R'} \bar{P}[n\ell JR\ n'\ell'J'R';\nu_\ell - S(\{v\})] \, P_{th}(n'\ell'J'R',\{v\}) \quad . \tag{5.4.168}$$

The probability $P'_{th}(n'\ell'J'R',\{v\})$ is, according to statistical mechanics,

$$P_{th}(n'\ell'J'R',\{v\}) = Z^{-1}\, d_{\{v\}} g_{n'\ell'J'R'}\, e^{-\beta E(n'\ell'J'R',\{v\})} \tag{5.4.169}$$

where $\beta = h/kT$ (since all energies are measured in frequency units); the vibrational degeneracy of $\{v\}$ is

$$d_{\{v\}} = \frac{1}{8}(v_2+1)(v_4+1)(v_4+2)(v_5+1)(v_5+2)(v_6+1)(v_6+2) \quad ; \tag{5.4.170}$$

and the partition function is

$$Z = \sum_{\{v\}} \sum_{n'}\sum_{\ell'}\sum_{J'}\sum_{R'} d_{\{v\}} g_{n'\ell'J'R'}\, e^{-\beta E(n'\ell'J'R',\{v\})} \quad . \tag{5.4.171}$$

For the purpose of evaluating (5.4.169,171) only, we neglect the vibration-rotation interaction energy and the anharmonic splitting in (5.4.100), so that

$$E(n'\ell'J'R',\{v\}) \cong E_v(n') + E_r(J') + E_{0th}(\{v\}) \tag{5.4.172}$$

where $E_v(n') = n'\nu_3$ and $E_r(J') = BJ'(J'+1)$. Physically this approximation is justified because $E_r(J')$ is very large compared to the vibration-rotation interaction energy [proportional to $B\zeta$ in (5.4.100)], the anharmonic splitting, and the anharmonic energy $nS(\{v\})$, for values of J' which are near the maximum of the population distribution in SF_6. This is true even at temperatures as low as 140 K.

The probability for finding the system in all states $|n\ell JR,\{v\}\rangle$ with the ν_3 quantum number n is, from (5.4.168-172) and Table 5.9,

$$\begin{aligned}\langle\bar{P}_{tot}(n;\nu_\ell)\rangle &= \sum_{\ell JR} \langle\bar{P}_{tot}(n\ell JR;\nu_\ell)\rangle \\ &= Z^{-1} \sum_{\{v\}} d_{\{v\}}\, e^{-\beta E_{0th}(\{v\})} \sum_{n'} e^{-\beta n'\nu_3} \sum_{RR'} d_{R'}^{(r)} \\ &\cdot \sum_{\ell\ell'} \sum_{JJ'} (2J'+1)\, e^{-\beta E_r(J')} P[n\ell JR|n'\ell'J'R;\nu_\ell - S(\{v\})] \ .\end{aligned} \tag{5.4.173}$$

This equation, which is the principal result of this section, gives a straightforward prescription for taking into account a thermal distribution of initial states, while preserving the coherent multiphoton features of the process of excitation.

For numerical evaluation it is necessary to limit the number of terms in the sums in (5.4.173), and it is convenient to make certain additional approximations as well. First, we use a model for the transition moments in which R' = R (see Sect.5.4.5). Second, we cut off the sums on n', {v}, and R after the total statistical weight (represented by the states which are included in the sum) reaches a predetermined value. Third, we limit n to states which are significant populated at laser intensities of interest. Fourth, we sum over a subset I of values of R, chosen to give a good approximation to the results obtained by summing on all values of R (up to the maximum). Finally, we employ the approximation

$$\begin{aligned}&Z^{-1} \sum_R d_R^{(r)} \sum_{\ell'} \sum_{J'} (2J'+1) e^{-\beta E_r(J')} \bar{P}[n\ell JR|n'\ell'J'R;\nu_\ell - S(\{v\})] \\ &\cong Z_v^{-1} d_n^{(v)} [W(n')]^{-1} \sum_{R\varepsilon I} \sum_{\ell'} d_R^{(r)} \sum_{J'=R-\ell'}^{R+\ell'} (2J'+1) e^{-\beta E_r(J')} \\ &\quad \bar{P}[n\ell JR|n'\ell'J'R;\nu_\ell - S(\{v\})]\end{aligned} \tag{5.4.174}$$

where

$$Z_v = \sum_n \sum_{\{v\}} d_d^{(v)} d_{\{v\}}\, e^{-\beta[n'\nu_3 + E_{0th}(\{v\})]} \tag{5.4.175}$$

$$= \prod_{k=1}^{6} \{[1 - e^{-\beta\nu_k}]^{d_k}\}^{-1} \tag{5.4.176}$$

is the vibrational partition function in the harmonic approximation, and

$$W(n') = \sum_{R\varepsilon I} \sum_{\ell'} \sum_{J'=R-\ell'}^{R+\ell'} (2J' + 1)d_R^{(r)} e^{-\beta E_r(J')} \tag{5.4.177}$$

$$\cong d_{n'}^{(v)} \sum_{R\varepsilon I} (2R + 1)d_R^{(r)} e^{-\beta E_r(R)} \tag{5.4.178}$$

is the statistical weight of the rotational states actually used in numerical computation. The approximate form (5.4.178) for W(n'), although not actually used in the computation reported in Sect.5.4.9, is useful for checking the plausibility of the basic approximation (5.4.174) used in evaluating

We turn now to a discussion of some numerical studies of the conditional (single-initial-state) probability $\bar{P}(n\ell JR|n'\ell'J'R;\nu_\ell)$ and the total probability $\langle\bar{P}_{tot}(n;\nu_\ell)\rangle$.

5.4.9 Numerical Calculations of Multiphoton Excitation of SF_6

Previous sections have, for the most part, dealt with analytical studies of multiphoton excitation of nondegenerate and multiplet systems, making use of special models and approximations. In this section we concentrate on numerical results for a model of SF_6 which contains much, though not all, of the complexity actually present in the SF_6 molecule. Our purpose is not to model a specific system with spectroscopic accuracy, but rather to explore some aspects of the physical processes associated with multiphoton excitation of a complex system which are difficult to study by analytical methods. Exact analytical solutions, while very useful, are generally available only for special combinations of Rabi frequencies and detunings, and are probably impossible for the complex multiplet systems considered here. Even for a study of the qualitative physics of excitation, a numerical model is necessary in such complex cases.

The objects of our study in this section are the role which compensation of anharmonicity (by vibration or rotation), and complexity of the energy-level and transition-moment structure, play in excitation of levels in which a few quanta of the laser-pumped vibrational mode are excited. It is our point of view that the process of excitation of these relatively low-lying levels is fundamentally coherent, and must be described by quantum mechanics and not by rate equations. We shall see here that the compensation of the vibrational anharmonic shift by anharmonic splitting or rotations is important not because of the possibility of successive exactly resonant transitions, which is realized only for special initial vibration-rotation states, but because the presence of intermediate states with small detuning has an effect on multiphoton processes, as we shall show below. Molecular complexity, i.e., the splitting of energy levels by anharmonic effects and by vibration-rotation interactions, produces many multiphoton resonance frequencies even when a few vibrational quanta are excited; produces many intermediate levels which can assist multiphoton transitions; and, perhaps most important, provides many pathways of transitions to a given excited vibrational state, as shown in Fig.5.21. The existence of many pathways to a given excited state increases the probability of one or more nearly resonant intermediate states.

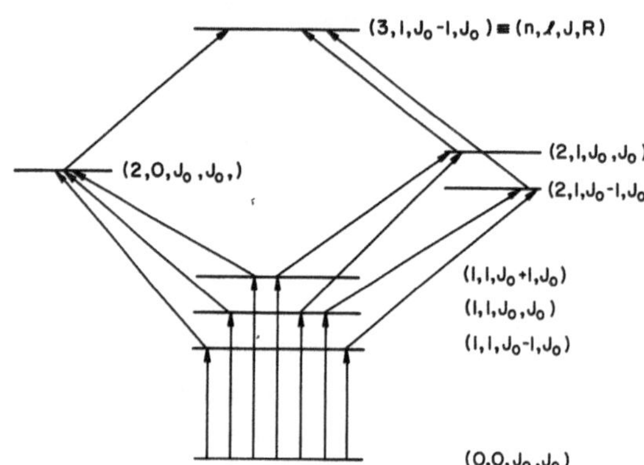

Fig.5.21. Pathways of excitation from one ground effective state to one effective state with n = 3

In the numerical calculations reported in this section we shall use the effective states $|n\ell JR\rangle$ adapted to spherical symmetry (Table 5.9), despite the fact that the most accurate sets of anharmonic parameters available indicate that the spherical basis functions are not eigenstates of the vibration-rotation Hamiltonian for the ν_3 mode of SF_6. However, the approximate energy levels (5.4.100) in the basis $|n\ell JR\rangle$ include both rotations and anharmonic splitting, and permit the relative importance of these effects to be adjusted for model calculations by choice of X_{33} and G_{33}. The use of a Hamiltonian which is diagonal in the effective states $|n\ell JR\rangle$ and transition moments which satisfy the selection rule $\Delta R = 0$ (5.4.105) permits a major computational simplification. The number of *vibration-rotation* states which are radiatively coupled to a fixed ground-state angular momentum $J_0(=R)$ is equal to the sum of the number of values of J for each value of n and ℓ used in the calculation. By (5.4.143-147), this is equal to $\sum_n d_n^{(v)}$, which is the same number of effective states used in the purely *vibrational* calculation reported in [5.11]. The spherically adapted effective states permit an extremely economial computation involving most of the major features of real spherical-top molecules and enable us to pursue effectively our primary interest, the theoretical foundations for an understanding of the first few steps of multiphoton excitation of polyatomic molecules.

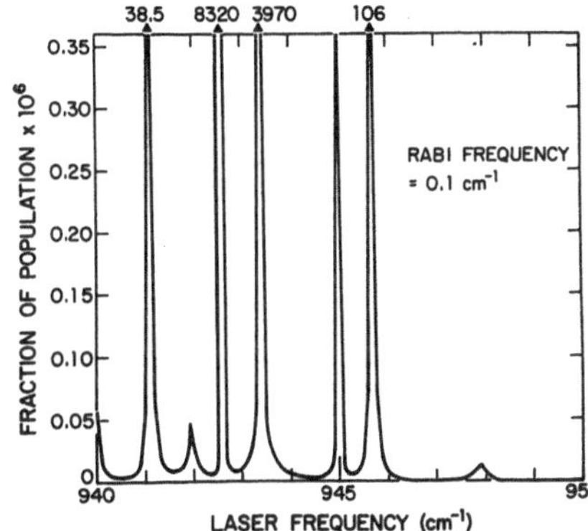

Fig.5.22. Dependence of vibrational populations on laser frequency for the effective states {e} = nℓR and the energy levels (5.4.46), for a Rabi frequency (5.4.182) equal to 0.1 cm^{-1}

To demonstrate multiphoton resonances for a relatively sparse level structure, in the absence of rotation, we show in Fig.5.22 the probability

$$\bar{P}(n|0;\nu_\ell) = \sum_\ell \bar{P}(n\ell|00;\nu_\ell) \tag{5.4.179}$$

for n = 3, as a function of laser frequency ν_ℓ, calculated using the transition moments for the effective state {e} = nℓR from Table 5.9 and the energy levels (5.4.46). The narrow, isolated multiphoton resonances expected at low laser intensity are seen at the frequencies predicted by (5.4.72,65) (for example, the 3-photon resonance to n = 3, ℓ = 3 occurs at ν_ℓ = 943.4 cm^{-1}, and the 3-photon resonance to n = 3, ℓ = 1 occurs at ν_ℓ = 942.6 cm^{-1}). However, it is evident that these are not the only resonances. For example, the two peaks near ν_ℓ = 945.0 cm^{-1} and ν_ℓ = 945.9 cm^{-1} are near the frequencies expected for the 2-photon resonances to n = 2 and ℓ = 2 or 0, respectively. In other words, the two peaks near ν_ℓ = 945 cm^{-1} arise from resonant two-photon transitions, followed by nonresonant one-photon transitions. Similarly, the remaining peaks in Fig.5.22 are correlated with multiphoton resonances to states other than n = 3.

Analytical studies of multiphoton resonances are frequently conducted assuming that the laser field is in exact multiphoton resonance, for the good reason that even for two-photon resonance (a three-level system) the secular equation is an irreducible cubic off resonance [if $\delta_3 \neq 0$ in (5.3.12)]. Although a general analytical analysis is possible in that case, it is not particularly informative. The analytical results obtained by the methods of Sect.5.3 for a three-level system in exact two-photon resonance show that the time-averaged population of the n = 2 state is

$$\bar{P}(2|0) = 2\,\frac{(\gamma_2\gamma_3)^2}{(\gamma_2^2 + \gamma_3^2)^2} \tag{5.4.180}$$

for the nondegenerate system considered there. In this case the population averaged over a time long compared to the two-photon Rabi frequency is independent of the laser intensity, while the time-averaged population of the intermediate (n = 1) state is proportional to the intensity

$$\bar{P}(1|0) = 2\gamma_2^2/\delta_2^2 \quad . \tag{5.4.181}$$

However, exact or nearly exact multiphoton resonance will occur only fortuitously. For most laser frequencies, all multiphoton resonances will be slightly (or substantially) detuned and (5.4.180,181) will not apply. In Fig.5.23 we show the time-averaged populations of one effective state $|n\ell JR\rangle$ as a function of the incident optical electric field $\mathcal{E}_0$, for fixed laser frequency. The calculation employed the effective transition moments for $\{e\} = \{n,\ell,J,R\}$ from Table 5.9 (which were forced to be symmetric), and the energy levels (5.4.100). The Rabi frequency for the purpose of this calculation is taken to be the Rabi frequency of the transition $|00RR\rangle \rightarrow |11RR\rangle$, i.e., the Rabi frequency of a single rotational line in the n = 0 to n = 1 Q branch. From Table 5.9, this implies that the Rabi frequency is

$$\gamma_0 = \frac{\mathcal{E}_0 A}{2\sqrt{3}\hbar} \langle 00||\hat{q}||11\rangle W(10RR;1R)(2R+1)$$

$$= \frac{\langle\mu_{01}\rangle\mathcal{E}_0}{2\sqrt{3}\hbar} \quad . \tag{5.4.182}$$

The detunings of the final and closest intermediate states are 0.47 and 1.28 cm^{-1}. At very low intensities ($\gamma_0 \lesssim 0.03$ cm^{-1}), the n = 2 population approaches proportionality to $(\mathcal{E}_0)^4$. For $\gamma_0 \sim 0.2$ cm^{-1}, the n = 2 population is more nearly proportional to $(\mathcal{E}_0)^2$, as is the n = 1 population. At higher intensities the populations display a complex saturation behavior. In general, the probability of excitation of this particular n = 2 state by a nonresonant two-photon transition (in which the detuning of the intermediate n = 1 state is of the same order of magnitude as the detuning of the final n = 2 state) is *not* proportional to the square of the intensity over the range of $\mathcal{E}_0$ where the most significant excitation occurs. Substantial excitation and even saturation occur for values of γ_0 which are of the same order of magnitude as the detunings of the n = 2 or n = 1 states. It appears quite probable that these properties are true of excitation to higher vibrational states as well, and that as a result the excitation probability for small values of n need not be proportional to the nth power of the laser intensity.

We turn now to a discussion of the effect of multiphoton resonances on the frequency dependence of excitation to n = 2 and 3 in this model of SF_6. The N-photon resonances to $|n\ell JR\rangle$ with n = N, including the effects of rotations, according to (5.4.100) for the energy levels, are

$$\Omega_{\ell JR}^{(N)} = \nu_3 + (N-1)X_{33} + \left(\frac{\ell(\ell+1)}{N} - 2\right)(G_{33} - B\zeta_3)$$
$$+ \frac{B(1-\zeta_3)\Delta J}{N}(2R + \Delta J + 1) - \alpha[R(R+1) + \Delta J(2R+1)] \tag{5.4.183}$$

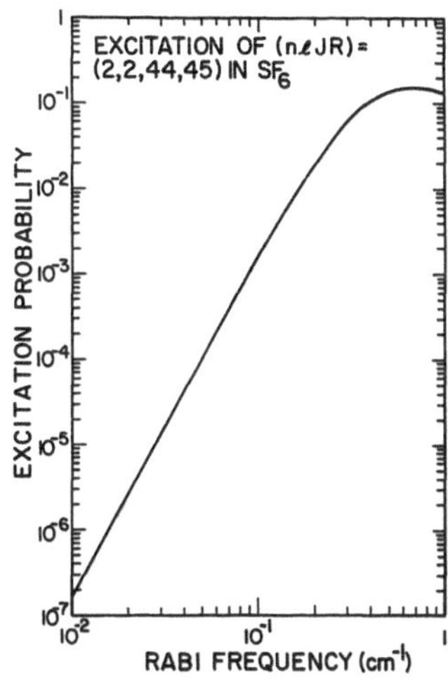

Fig.5.23. Dependence of population in the effective state (nℓJR) = (2,2,44,45) upon the Rabi frequency (5.4.182)

where

$$\Delta J = J - R \qquad (5.4.184)$$

is the change in total angular momentum from n = 0 to n = N. The first line of (5.4.183) is the purely vibrational contribution to the multiphoton resonance frequency from (5.4.72,65), evident in Fig.5.22. The anharmonic shift (proportional to X_{33}) is the dominant vibrational contribution to the difference $\Omega^{(N)}_{JR} - \nu_3$, for the parameters (5.4.122) [5.14]. The second line of (5.4.183) is the rotational contribution, the principal part of which is the term proportional to B. This term arises from the purely rotational energy BJ(J + 1) and the Coriolis energy (proportional to $B\zeta_3$) in (5.4.100). The quantity $2B(1 - \zeta_3)$ is twice the effective rotational constant, i.e., $2B(1 - \zeta_3)$ is approximately equal to the spacing between successive P-branch or R-branch lines in the n = 1 to n = 1 spectrum. The magnitude of the term proportional to B in (5.4.184) for $\Delta J \cong N/2$, R = 39 and N = 4 is about 2.3 cm^{-1} for SF_6, which is substantially greater than the contribution of 0.72 cm^{-1} from vibrational anharmonic splitting (the term proportional to $G_{33} - B\zeta_3$) for N = 4, ℓ = 4. Thus, in this model [5.14] and in subsequent, similar studies [5.15] rotations are the dominant contribution to the multiphoton resonance frequencies except for the anharmonic shift.

According to the selection rule (5.4.109),

$$|\Delta J| \leqq N \qquad (5.4.185)$$

as was pointed out on qualitative grounds in [5.37]. According to (5.4.98), R must form a triangle with J and ℓ. Thus

$$|\Delta J| \leqq \ell \quad . \qquad (5.4.186)$$

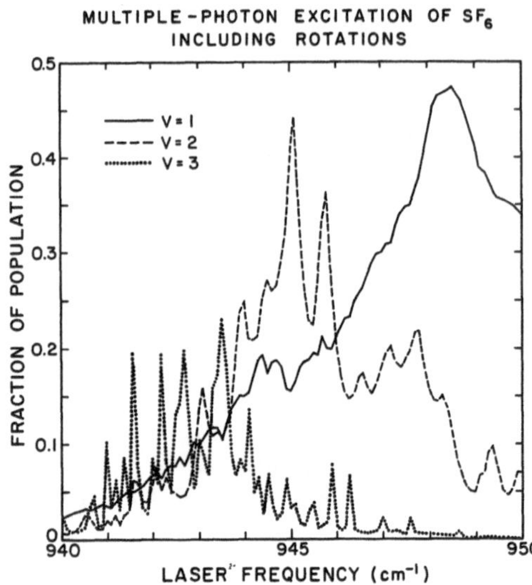

Fig.5.24. Dependence of populations in the vibrational levels $n=1,2$, and 3 upon laser frequency, for a Rabi frequency (5.4.182) of 0.5 cm^{-1}, for molecules which are initially in the ground vibrational state. A distribution of populations in the initial rotational levels corresponding to a temperature T = 300 K was assumed

This selection rule, and the rule $\Delta R = 0$ (5.4.105), hold in our model but not necessarily in general, while the rule (5.4.185) holds in general. A schematic diagram of successive nonzero one-photon transition moments leading to excited states which satisfy the selection rules (5.4.185,186) is shown in Fig.5.24 [5.14].

The spacing between two N-photon resonances with the same R and ℓ but successive values of J, is

$$\Omega^{(N)}_{\ell,J+1,R} - \Omega^{(N)}_{\ell JR} = \frac{2B(1-\zeta_3)}{N}(R+\Delta J+1) - \alpha(2R+1) \quad . \tag{5.4.187}$$

For R = 39 and $\Delta J = 0$, this spacing is about 0.4/N cm^{-1} for SF_6. The spacing between two resonances with the same ℓ and ΔJ, but successive values of ground-state angular momentum R, is

$$\Omega^{(N)}_{\ell,J+1,R+1} - \Omega^{(N)}_{\ell JR} = \frac{2B(1-\zeta_3)\Delta J}{N} - 2\alpha(R+\Delta J+1) \tag{5.4.188}$$

which is smaller than (5.4.187) for $R \gg N$. For R = 39 and $\Delta J = N/2$, the spacing given by (5.4.188) is about 0.04 cm^{-1} for SF_6. Thus the N-photon resonances originating from the same and from different ground-state angular momenta R form an overlapping network of resonance frequencies, with a spacing which is of the same order of magnitude as the spacing between successive rotational lines in the n = 0 to n = 1 spectrum (approximately 0.11 cm^{-1} in SF_6).

The terms proportional to α in (5.4.100,183,187,188) arise physically from a change in the molecular rotation constant due to vibrational excitation, and mathematically from the term $-\alpha T^{11}_{000}$ in Table 5.8. The principal effect of this term in the n = 0 to n = 1 spectrum is to separate the Q branch (which would otherwise be a single "line" consisting of the superposition of all the $\Delta J = 0$ transitions from n = 0 to n = 1) into distinct lines, one for each value of the ground-state angular momentum. Although this is the usual Q-branch structure in linear molecules, it is known that in spherical-top molecules in general and in SF_6 in particular the structure of the Q branch is actually primarily due to the tensor splitting, primarily the term $\alpha_{224}T_{224}$ in Table 5.8, which is physically a tensor change of rotation constant with vibrational excitation [5.46,96,97]. Our analytical analysis above of the N-photon resonance frequencies and their spacing indicates that the effect represented by the term $-\alpha T^{11}_{000}$ in the Hamiltonian is of minor, not major, importance, in SF_6, in view of the value of α (1.31×10^{-4} cm^{-1}) [5.47] and the magnitudes of N and R of interest in our calculations. The major role in our case is played by the rotational energy $BJ(J+1)$ and the Coriolis energy (proportional to $B\zeta_3$).

From Fig.5.23 we see that substantial 2-photon excitation occurs at Rabi frequencies which are of the same order of magnitude as the spacing between multiphoton resonances. The same is true of 3-photon excitation. This fact, and the occurrence of several overlapping families of multiphoton resonances over a broad spectral range, lead one to expect that the frequency dependence of $\bar{P}(n|0;\nu_\ell)$ at modest Rabi frequencies (i.e., $\gamma_0 \cong 0.5$ cm^{-1}) will show broad peaks centered at the N-photon resonance frequencies determined by X_{33} alone (5.3.50) [5.37] in the effective-states model $\{e\} = n\ell JR$, with energy levels (5.4.100). This is in fact the case, as seen in Fig.5.24. The peaks of the successive 1-, 2-, and 3-photon regions in Fig.5.24 are equally spaced, as expected from (5.3.50). The excitation spectrum shows few isolated multiphoton peaks, in contrast to Fig.5.23.

In Fig.5.25 we show the thermally averaged multiphoton excitation probability $\langle\bar{P}(n|0;\nu_\ell)\rangle$ for n = 1, 2, and 3, calculated using (5.4.173,174) for a temperature T = 300 K. This calculation included enough initial vibrational states to encompass 99.72% of the molecular population at 300 K. It can be seen that the major effect of the thermally populated excited vibrational states is to smooth the distribution (as one would expect), and to shift the peaks of excitation towards lower laser frequencies as expected from the shift predicted by (5.4.166) . We find no evidence in this calculation for a dramatic increase of multiphoton excitation probability as a result of thermally excited initial states [5.14].

In this model, with the parameters (5.4.112), there is little excitation of $n > 3$ at laser frequencies ν_ℓ near 944 cm^{-1}, at Rabi frequencies $\gamma_0 < 0.5$ cm^{-1}. At lower laser frequencies there are narrow multiphoton resonances to n = 4 and 5. This model, with the parameters (5.4.112), is therefore unable to explain the very large excitation of SF_6 observed for laser frequencies near 944 cm^{-1}, at modest laser intensities.

In conclusion, we have established by numerical calculation that in the effective-state model $\{e\} = n\ell JR$ of SF_6, with the energy levels (5.4.100), rotations play a key role in the following aspects of multiphoton excitation: 1) Rotations

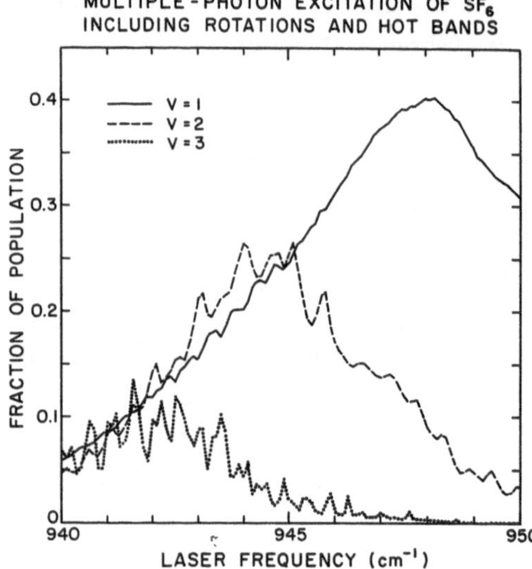

Fig.5.25. Dependence of populations in the vibrational levels n = 1,2, and 3 upon laser frequency, for a Rabi frequency (5.4.182) of 0.5 cm^{-1}, calculated for a thermal (T = 300 K) distribution of initial vibrational and rotational states using (5.4.173)

provide many slightly detuned intermediate states, which lead to a different dependence of multiphoton excitation upon laser intensity than simple analytical multiphoton calculations have suggested. 2) Rotations give rise to a dense network of multiphoton resonances, and to broad bands of N-photon transitions centered at approximately the frequency of a purely vibrational N-photon resonance. These two features are probably common to all multiphoton models of molecular excitation which include rotation, and are in that sense fundamental. In particular, point 1) represents a generalization of the idea of compensation of anharmonicity by rotations [5.6,12]. The final conclusion, that no significant excitation of n > 3 occurred at ν_ℓ = 944.2 cm^{-1} for modest Rabi frequencies ($\gamma_0 \sim$ 0.5 cm^{-1}), is probably the result of the specific set of vibrational anharmonic parameters used in the calculation, and may be subject to change.

References

5.1 R.V. Ambartzumian, V.S. Letokhov, E.A. Ryabov, N.V. Chekalin: ZhETF Pis. Red. *20*, 597 (1974)

5.2 R.V. Ambartzumian, Yu.A. Gorokhov, V.S. Letokhov, G.N. Makarov: ZhETF Pis. Red. *21*, 375 (1975) [English transl.: JETP Lett. *21*, 171 (1975)]
J.L. Lyman, R.J. Jensen, J. Rink, C.P. Robinson, S.D. Rockwood: Appl. Phys. Lett. *27*, 87 (1975)

5.3 M.J. Coggiola, P.A. Schulz, Y.T. Lee, Y.R. Shen: Phys. Rev. Lett. *38*, 17 (1977)
P. Kolodner, C. Winterfeld, E. Yablonovitch: Opt. Commun. *20*, 119 (1977)

5.4 A.S. Pine, A.G. Robiette: J. Mol. Spectrosc. (in press)

5.5 S. Mukamel, J. Jortner: Chem. Phys. Lett. *40*, 150 (1976)

5.6 D.M. Larsen, N. Bloembergen: Opt. Commun. *17*, 254 (1976)

5.7 R.V. Ambartzumian, Yu. A. Gorokhov, V.S. Letokhov, G.N. Makarov: Zh. Eksp. Teor. Fiz. *69*, 1956 (1975) [English transl.: Sov. Phys. JETP *42*, 993 (1976)]

5.8 V.S. Letokhov, A.A. Makarov: "Coherent Excitation of Multilevel Molecular Systems in Intense Quasi-Resonant Laser IR Field", preprint (1976)
5.9 N.R. Isenor, V. Merchant, R.S. Hallsworth, M.C. Richardson: Can. J. Phys. *51*, 1281 (1973)
5.10 C.D. Cantrell, H.W. Galbraith: Opt. Commun. *18*, 513 (1976)
5.11 C.D. Cantrell, H.W. Galbraith: Opt. Commun. *21*, 374 (1977)
5.12 R.V. Ambartzumian, Yu. A. Gorokhov, V.S. Letokhov, G.N. Makarov, A.A. Puretzky: ZhETF Pis. Red. *23*, 26 (1976) [English transl.: JETP Lett. *23*, 22 (1976)]
5.13 D.M. Larsen: Opt. Commun. *19*, 404 (1976)
5.14 C.D. Cantrell, K. Fox: Opt. Lett. *2*, 151 (1978)
5.15 J.R. Ackerhalt, H.W. Galbraith: J. Chem. Phys. *69*, 1200 (1978)
5.16 R.P. Feynman, F.L. Vernon, Jr., R.W. Hellwarth: J. Appl. Phys. *28*, 49 (1957)
5.17 F. Bloch: Phys. Rev. *70*, 460 (1946)
5.18 L. Allen, J.H. Eberly: *Optical Resonance and Two-Level Atoms* (Wiley, New York 1975)
5.19 N.N. Bogolyubov, Yu. A. Mitropolskii: *Asimptoticheskie metody v teorii nelineinykh kolebanii (Asymptotic methods in the theory of nonlinear vibrations)* (Fismatgis, Moscow 1963)
5.20 Ya.B. Zel'dovich: Usp. Fiz. Nauk *110*, 139 (1973) [English transl.: Sov. Phys. Usp. *16*, 427 (1973)]
5.21 I.I. Rabi: Phys. Rev. *51*, 652 (1937)
5.22 G. Szego: *Orthogonal Polynomials* (Am. Math. Society, 1939)
5.23 H. Bateman, A. Erdelyi: *Higher Transcendental Functions*, Vol.2 (McGraw-Hill, New York 1953)
5.24 R.P. Feynman: Phys. Rev. *84*, 108 (1951)
5.25 J. Schwinger: Phys. Rev. *91*, 758 (1953)
5.26 C.J. Elliott, B.J. Feldman: Bull. Am. Phys. Soc. *20*, 1282 (1975)
5.27 V.M. Akulin, S.S. Alimpiev, N.V. Karlov, B.G. Sartakov, L.A. Shelepin: Zh. Eksp. Teor. Fiz. *71*, 454 (1976) [English transl.: Sov. Phys. JETP *44*, 239 (1976)]
5.28 A.A. Makarov: Zh. Eksp. Teor. Fiz. *72*, 1749 (1977) [English transl.: Sov. Phys. JETP *45*, 918 (1977)]
5.29 J.H. Eberly, B.W. Shore, Z. Bialynicka-Birula, I. Bialynicki-Birula: Phys. Rev. A*16*, 2038 (1977)
5.30 V.A. Fock: Dokl. Akad. Nauk USSR *1*, 97 (1934)
5.31 A.I. Baz', Ya.B. Zel'dovich, A.M. Perelomov: *Rasseyanie, reaktsii i raspady v nerelyativistskoi kvantovoi mekhanike (Scattering, Reactions, and Decays in Nonrelativistic Quantum Mechanics)* (Nauka, Moscow 1966)
5.32 V.S. Letokhov, A.A. Makarov: Appl. Phys. *16*, 47 (1978)
5.33 V.S. Letokhov, A.A. Makarov: Opt. Commun. *17*, 250 (1976)
5.34 V.M. Akulin, S.S. Alimpiev, N.V. Karlov, B.G. Sartakov: Zh. Eksp. Teor. Fiz. *72*, 88 (1977)
5.35 N. Bloembergen: Opt. Commun. *15*, 416 (1975)
5.36 E.R. Grant, P.A. Schulz, Aa.S.Sudbo, Y.R. Shen, Y.T. Lee: Phys. Rev. Lett. *40*, 115 (1978)
5.37 N. Bloembergen, C.D. Cantrell, D.M. Larsen: In *Tunable Lasers and Applications*, ed. by A. Mooradian, T. Jaeger, P. Stokseth, Springer Series in Optical Sciences, Vol.3 (Springer, Berlin, Heidelberg, New York 1976) pp.162-176
5.38 L.D. Landau, E.M. Lifshitz: *Kvantovaya mekhanika (Quantum Mechanics)* (Fismatgiz, Moscow 1963) [English transl.: Pergamon Press, Oxford 1965)]
5.39 A.A. Makarov, V.T. Platonenko, V.V. Tyakht: Zh. Eksp. Teor. Fiz. *75*, 2075 (1978) [English transl.: Sov. Phys. JETP *48*, 1044 (1978)]
5.40 V.M. Akulin, A.M. Dykhne: Zh. Eksp. Teor. Fiz. *73*, 2098 (1977)
5.41 V.T. Platonenko: Kvantovaya Elektron. *5*, 1783 (1978)
5.42 J.P. Aldridge, H. Filip, H. Flicker, R.F. Holland, R.S. McDowell, N.G. Nereson: J. Mol. Spectrosc. *58*, 165 (1975)
5.43 C.D. Cantrell, H.W. Galbraith: J. Mol. Spectrosc. *58*, 158 (1975)
5.44 R.S. McDowell, H.W. Galbraith, B.J. Krohn, C.D. Cantrell, E.D. Hinkley: Opt. Commun. *17*, 178 (1976)
5.45 R.S. McDowell, H.W. Galbraith, C.D. Cantrell, N.G. Nereson, E.D. Hinkley: J. Mol. Spectrosc. *68*, 288 (1977)

5.46 R.S. McDowell, H.W. Galbraith, C.D. Cantrell, N.G. Nereson, P.F. Moulton, E.D. Hinkley: Opt. Lett. *2*, 97 (1978)
5.47 Ch.J. Bordé, M. Ouhayoun, A. Van Lerberghe, C. Salomon, S. Avrillier, C.D. Cantrell, J. Bordé: In *Laser Spectroscopy IV*, ed. by H. Walther, K.W. Rothe, Springer Series in Optical Sciences, Vol.21 (Springer, Berlin, Heidelberg, New York 1979) pp.142-153
5.48 K.C. Kim, W.B. Person, D. Seitz, B.J. Krohn: J. Mol. Spectrosc. *76*, 322 (1979)
5.49 C.W. Patterson, B.J. Krohn: To be published
5.50 P.F. Moulton, D.M. Larsen, J.N. Walpole, A. Mooradian: Opt. Lett. *1*, 51 (1977)
5.51 C.C. Jensen, T.G. Anderson, C. Reiser, J.I. Steinfeld: J. Chem. Phys. *71*, 3648 (1979)
5.52 N.R. Isenor, M.C. Richardson: Appl. Phys. Lett. *18*, 224 (1971); Opt. Commun. *3*, 360 (1971); Proc. Tenth Intern. Gaseous Electronics Conference (D. Parsons, Oxford 1971)
N.R. Isenor, V. Merchant, R.S. Hallsworth, M.C. Richardson: Can. J. Phys. *51*, 1281 (1973)
R.S. Hallsworth, N.R. Isenor: Chem. Phys. Lett. *22*, 283 (1973)
5.53 R.V. Ambartzumian, Yu.A. Gorokhov, V.S. Letokhov, G.N. Makarov, A.A. Makarov, A.A. Puretzky: Phys. Lett. *56*A, 183 (1976)
5.54 R.V. Ambartzumian, Yu.A. Gorokhov, V.S. Letokhov, G.N. Makarov: ZhETF Pis. Red. *22*, 96 (1975)
R.V. Ambartzumian, Yu.A. Gorokhov, G.N. Makarov, A.A. Puretzky, N.P. Furzikov: Kvantovaya Elektron. *4*, 1590 (1977)
5.55 J.J. Tiee, C. Wittig: Opt. Commun. *27*, 377 (1978)
P. Rabinowitz, A. Stein, A. Kaldor: Opt. Commun. *27*, 381 (1978)
5.56 K.T. Hecht: J. Mol. Spectrosc. *5*, 355 (1960)
5.57 M.I. Petrashen, E.D. Trifonov: *Applications of Group Theory in Quantum Mechanics*, translated by S. Chomet, ed. and revised by J.L. Martin (MIT Press, Cambridge 1969)
5.58 R.S. McDowell, J.P. Aldridge, R.F. Holland: J. Phys. Chem. *80*, 1203 (1976)
5.59 W.H. Louisell: *Quantum Statistical Properties of Radiation* (Wiley, New York 1973)
5.60 A. Messiah: *Quantum Mechanics*, Vol.I (North-Holland, Amsterdam 1961)
5.61 H.W. Galbraith, C.D. Cantrell: In *The Significance of Nonlinearity in the Natural Sciences*, ed. by B. Kursunoglu, A. Perlmutter, L.F. Scott (Plenum, New York 1977) pp.227-264
5.62 V. Bargmann, M. Moshinsky: Nuclear Phys. *18*, 697 (1960)
5.63 H.H. Nielsen: In *Encyclopedia of Physics*, Vol.37/1, ed. by S. Flügge (Springer, Berlin, Göttingen, Heidelberg 1959) pp.173-313
5.64 D.M. Brink, G.R. Satchler: *Angular Momentum*, 2nd ed. (University Press, Oxford 1968)
5.65 H.A. Jahn: Proc. Roy. Soc. A*168*, 469 (1938)
5.66 C.C. Jensen, W.B. Person, B.J. Krohn, J. Overend: Opt. Commun. *20*, 275 (1977)
5.67 J. Moret-Bailly, L. Gautier, J. Montagutelli: J. Mol. Spectrosc. *15*, 355 (1965)
5.68 B.J. Krohn: J. Mol. Spectrosc. *68*, 497 (1977); and private communications
5.69 A.G. Robiette, D.L. Gray, F.W. Birss: Mol. Phys. *32*, 1591 (1976)
5.70 C.D. Cantrell, S. Avrillier, Ch. Salomon, Ch.J. Bordé: To be published
5.71 E. Pascaud: J. Phys. (Paris) *37*, 1287 (1976)
5.72 H.W. Galbraith: Opt. Lett. *3*, 154 (1978)
5.73 C.D. Cantrell: Unpublished
5.74 J.S. Briggs: Unpublished
5.75 I.N. Knyazev, V.S. Letokhov, V.V. Lobko: Opt. Commun. *25*, 337 (1978)
5.76 K. Fox: J. Mol. Spectrosc. *9*, 381 (1962)
5.77 G. Pierre, J.-C. Hilico, C. de Berghe, J.-P. Maillard: J. Mol. Spectrosc. (to be published)
5.78 V.V. Bertsev, T.D. Kolomiitseva, N.M. Tsyganenko: Opt. Spectrosc. *37*, 263 (1974)
5.79 H. Kildal: J. Chem. Phys. *67*, 1287 (1977)
5.80 C. Marcott, W.G. Golden, J. Overend: Spectrochim. Acta *34*A, 661 (1978)
5.81 K. Fox: Opt. Lett. *1*, 214 (1977)
5.82 J.R. Ackerhalt, H. Flicker, H.W. Galbraith, J. King, W.B. Person: J. Chem. Phys. *69*, 1461 (1978)

5.83 V.M. Akulin, S.S. Alimpiev, N.V. Karlov, A.M. Prokhorov, B.G. Sartakov, E.M. Khokhlov: J. Opt. Soc. Am. *68*, 694 (1978)
5.84 M.F. Goodman, J. Stone, D.A. Dows: J. Chem. Phys. *65*, 5062 (1976)
5.85 C.D. Cantrell: Los Alamos Scientific Laboratory Informal Report LA-5464-MS (1973)
5.86 C.D. Cantrell, H.W. Galbraith: J. Mol. Spectrosc. *58*, 158 (1975)
5.87 J. Bordé: J. de Phys. Lett. *39*, L175 (1978)
5.88 F. Michelot, B. Bobin, J. Moret-Bailly: J. Mol. Spectrosc. *76*, 374 (1979)
5.89 E.B. Wilson: J. Chem. Phys. *3*, 276 (1935)
5.90 C.D. Cantrell: Unpublished
5.91 J.D. Louck, H.W. Galbraith: Rev. Mod. Phys. *48*, 69 (1976)
5.92 K. Fox, W.B. Person: J. Chem. Phys. *64*, 5218 (1976)
5.93 K. Fox: Opt. Commun. *19*, 397 (1976)
5.94 G. Herzberg: *Molecular Spectra and Molecular Structure, II. Infrared and Raman Spectra of Polyatomic Molecules* (Van Nostrand, New York 1945) p.210
5.95 C.D. Cantrell, S.M. Freund, J.L. Lyman: In *Laser Handbook III*, ed. by M.L. Stitch (North-Holland, Amsterdam 1979) pp.485-576
5.96 F. Michelot, J. Moret-Bailly: J. de Phys. Lett. *39*, L275 (1978)
5.97 E.G. Brock, B.J. Krohn, R.S. McDowell, C.W. Patterson, D.F. Smith: J. Mol. Spectrosc. *76*, 301 (1979)

6. Coherent Picosecond Interactions

A. Laubereau and W. Kaiser

With 10 Figures

The advent of a coherent light source, the laser, marks the beginning of coherent experiments in the optical part of the electromagnetic spectrum. Numerous coherent phenomena have been elucidated in recent years, some of which are discussed in various chapters of this volume. Most of these experiments were performed with metal vapors or molecular gases at low pressure where the dephasing times are long, of the order of 10^{-9} s or more. In these cases, commercial electronic detection systems may be used even for the study of transient processes. In the condensed phases at room temperature fast interactions occur between the closely packed molecules. It is not surprising that coherently excited systems lose the phase relation quite rapidly at 300 K. In fact, new experimental methods working in the time domain of picoseconds (10^{-12} s) were required to observe directly the vibrational dephasing time T_2 of molecules in liquids and solids.

6.1 Overview

For several decades conventional infrared and Raman spectroscopy have been the main source of information on vibrational dynamics. The analysis of the spectroscopic data faces several basic difficulties. We mention briefly the finite instrumental resolution which limits the determination of the true band contour (in particular for Raman transitions). Even more important for polyatomic molecules is the problem of overlapping neighboring bands. Contributions of overtones and hot bands of other vibrational modes - although not accurately known - have to be taken into account in many practical cases; they represent a nonnegligible source of experimental uncertainty [6.1]. Once a band contour is determined the question arises about its interpretation. In general, there are several physical processes which contribute to an observed band shape. Line broadening factors are: Rotational motion, vibrational dephasing processes, energy redistribution and transfer, and a (inhomogeneous) distribution of vibrational transition frequencies. At the present time, the different contributions to the band contour cannot be separated unambiguously in conventional spectroscopy.

In this chapter, new experimental techniques are presented and results are discussed on vibrational dephasing phenomena. Picosecond laser pulses are used for time-resolved coherent anti-Stokes (CARS) and Stokes Raman spectroscopy yielding - for the first time - direct information on homogeneous and inhomogeneous broadening factors. Several recent experiments will be presented to illustrate the state of the art. We note very briefly that different experimental methods allowed us to measure as well the population lifetime T_1 of vibrational transitions. For a discussion of these results we refer to a recent review by LAUBEREAU and KAISER [6.2].

6.2 Theory of Investigations

In our investigations, a vibrational mode is first excited by transient stimulated Raman scattering of an intense pumping pulse and subsequently monitored via coherent Raman scattering of a weak probing pulse. The probe scattering process is sensitive to the phase relation between the vibrating molecules and requires a carefully adjusted geometry of the relevant wave vectors. It will be discussed below that the dephasing of a molecular subgroup selected out of an inhomogeneously broadened vibrational band may be studied by a suitable k-vector geometry.

6.2.1 Excitation Process

Some theoretical remarks on transient stimulated Raman scattering should be given here [6.2-4,7]. It is convenient to treat the vibrational system in terms of a two-level model [6.5]. Transitions involving higher excited states are negligible on account of the anharmonic frequency shift. One vibrational mode is coherently excited by a nonresonant light pulse via stimulated Raman scattering. We present here a semiclassical theory where the vibrational system is treated quantum mechanically and the light fields classically. The large number of photons involved in the stimulated excitation process justifies such a treatment. The Hamiltonian of the molecular system interacting with the light field has the form

$$H = H^0 - \frac{1}{2} q \sum_{h,i} \left(\frac{\partial\alpha}{\partial q}\right)_{h,i} E_h E_i \quad . \tag{6.1}$$

In the Raman process, the interaction between the electromagnetic field and the normal vibrational mode is described by a polarizability tensor (α_{hi}) which depends on the relevant vibrational coordinate q

$$\alpha_{hi} = \alpha^0_{hi} + \left(\frac{\partial\alpha}{\partial q}\right)_{h,i} q \quad . \tag{6.2}$$

The subscripts h,i refer to a coordinate system (x,y,z) which is fixed to the symmetry axes of the individual molecules; q denotes the operator of the vibrational coordinate; E represents the amplitude of the total electromagnetic field which consists of the incident laser field and the generated Stokes field. The propagation of the light pulses and the interaction with the vibrating molecules is described by Maxwell's equations which lead to the wave equation

$$\Delta \underline{E} - \frac{1}{c^2}\frac{\partial^2}{\partial t^2}(\mu^2 \underline{E}) = \frac{4\pi}{c^2}\frac{\partial^2}{\partial t^2}\underline{P}^{NL} \quad , \tag{6.3}$$

μ denotes the refractive index of the medium; the nonlinear polarization P^{NL} represents the coupling between the light fields and the vibrational mode of interest. The molecular motion under the effect of the electromagnetic fields is described by two quantities: The expectation value of the displacement operator $\langle q\rangle$ and the probability n of finding the molecule in the upper vibrational state. It can be shown for molecules approximated as two-level systems and for the Raman interaction of (6.1) that the equations of motion have the form [6.6]

$$\frac{\partial^2}{\partial t^2}\langle q\rangle + \frac{2}{T_2}\frac{\partial}{\partial t}\langle q\rangle + \omega_0^2\langle q\rangle = \frac{1}{m}F(t)[1-2(n+\bar{n})] \tag{6.4}$$

$$\frac{\partial}{\partial t}n + \frac{1}{T_1}n = \frac{1}{\hbar\omega_0}F(t)\frac{\partial}{\partial t}\langle q\rangle \quad , \tag{6.5}$$

n is the occupation number in excess of the thermal equilibrium value $\bar{n}$. The reduced mass and the frequency of the vibrational mode are denoted by m and ω_0, respectively. F represents an effective force exerted by the electromagnetic field on the vibrating molecules

$$F = \frac{1}{2}\sum_{h,i}\left(\frac{\partial\alpha}{\partial q}\right)_{h,i}E_h E_i \quad . \tag{6.6}$$

Of special interest for this chapter are the two time constants T_1 and T_2 introduced in (6.5.6). T_2 denotes the dephasing time of the vibrational amplitude of a homogeneously broadened system (single transition frequency ω_0) and T_1 represents the population lifetime (an energy relaxation time) of the first excited vibrational state. We point to the related time constants in the Bloch equations of a two-level spin system.

The induced dipole moment $(\partial\alpha/\partial q)\langle q\rangle E$ of an individual molecule is directly proportional to the excited vibrational amplitude $\langle q\rangle$. Summing over the molecular ensemble with number density N yields the nonlinear polarization

$$P_i^{NL} = N\langle q\rangle \sum_h \left(\frac{\partial\alpha}{\partial q}\right)_{i,h}E_h \quad . \tag{6.7}$$

Equations (6.4,7) refer to a system of spatially isolated molecules.

$\hbar = h/2\pi$ (normalized Planck's constant)

The following points should be noted:

1) According to (6.4,5), the coherent vibrational excitation and the excited state population decay exponentially after the excitation process of the pump pulse has terminated. The exponential time depndence does not hold for times shorter than τ_c, where $\tau_c \sim 10^{-13}$ s is a correlation time connected to intermolecular interactions (e.g., rapid translational motion in the liquid). This limitation is not critical in our experiments since we work with light pulses $t_p \gg \tau_c$ and we observe time constants T_1 and $T_2 \gg \tau_c$. A large number of individual physical events leads to the experimentally observed exponential decay with time constant T_2.

2) The physical processes which determine the time constant T_1 and T_2 are different. In fact, in several experiments where T_1 and T_2 were determined for the same mode and molecule, a considerable difference was observed. For instance, a ratio of $T_1/T_2 = 10^{12}$ was found for the fundamental vibrational mode of liquid N_2 [6.8,9], and $T_1/T_2 = 2.6$ [6.10] was observed for a CH_3-stretching mode of CH_3CCl_3 at 300 K.

3) In general, the decay rate of $\langle q \rangle$ is affected by the loss of phase correlation and by the loss of vibrational energy (occupation density). For our experimental dephasing time T_2 we write

$$2/T_2 \simeq 1/\tau_{ph} + 1/T_1 \quad . \tag{6.8}$$

The time constant τ_{ph} represents the "pure dephasing" processes which were recently studied theoretically (see Sect.6.5). In a semiclassical collision model the molecular interactions are described by quasi-elastic collisions. In the quantum-mechanical picture the molecules undergo virtual excitation processes ending up in the initial energy state. The calculated pure dephasing times are in reasonable agreement with experimental observations.

4) A comparison was made between our dynamic measurements under nonequilibrium conditions and conventional Raman spectroscopy at thermal equilibrium [6.2]. It can be shown that for a homogeneous vibrational system the linewidth (FWHM in units of cm^{-1}) of the (approximately) Lorentzian vibrational band shape may be written as follows:

$$\delta\tilde{\nu}_{hom} = (\pi c \; T_2)^{-1} \quad , \tag{6.9}$$

i.e., the linewidth is determined by the dephasing time T_2.

For the following discussion it is advantageous to consider the more general case of an ensemble of two-level systems with an inhomogeneous distribution of vibrational transition frequencies. Molecules in a small frequency interval of this distribution are grouped together to vibrational components j with coherent amplitude $\langle q_j \rangle$ and with number density Nf_j, where N denotes the total number density and

$\sum f_j = 1$. Equation (6.4) is replaced by a corresponding set of differential equations for the amplitudes $\langle q_j \rangle$ of the molecular subroups with transition frequencies ω_j. Introducing the absolute values Q_j, the phase factors ϕ_j of the amplitudes $\langle q_j \rangle$, and the field amplitudes E_L and E_S of the laser and Stokes scattered light, (6.1-7) yield the following set of equations [6.2,11]

$$\left(\frac{\partial}{\partial x} + \frac{1}{v_S}\frac{\partial}{\partial t}\right) E_S = \kappa_1 E_L \sum_j f_j Q_j \cos(\Delta\omega_j t + \phi_j) \tag{6.10}$$

$$\left(\frac{\partial}{\partial t} + \frac{1}{T_2}\right) Q_j = \kappa_2 E_L E_S \cos(\Delta\omega_j t + \phi_j) \tag{6.11}$$

$$\frac{\partial}{\partial t}(\Delta\omega_j t + \phi_j) + \frac{\kappa_2 E_L E_S}{Q_j}\sin(\Delta\omega_j t + \phi_j) = \Delta\omega_j \tag{6.12}$$

where

$$\kappa_1 = \frac{\pi\omega_S^2 N}{c^2 k_S}\frac{\partial\alpha}{\partial q}\ ; \quad \kappa_2 = \frac{1}{4m(\omega_L - \omega_S)}\frac{\partial\alpha}{\partial q} \tag{6.13}$$

and

$$\Delta\omega_j = \omega_L - \omega_S - \omega_j \quad . \tag{6.14}$$

The following approximations are used for the derivation of (6.10-14) which are well justified for the experiments discussed below: Plane waves propagating in the x direction, slowly varying amplitudes (neglect of second order derivatives), small excitation with negligible population changes $n \ll 1$ and negligible depletion of the pump pulse.

The coupling coefficients κ_1 and κ_2 in (6.13) collect various material parameters. We assume for all components j equal values of the reduced molecular mass m, the Raman polarizability $\partial\alpha/\partial q$, and the dephasing time T_2. The group velocity of the Stokes pulse is v_S, and $\Delta\omega_j$ denotes the frequency distance of the molecular subgroup j from the band center. Equation (6.5) for the time evolution of the excited state population is not important in the subsequent discussion and it is not repeated here. We note that the rotational motion of the molecules does not explicitly enter (6.10-14) when the vibrational transition has an isotropic Raman tensor, i.e., when $\partial\alpha/\partial q$ is a scalar. This approximation holds for the experiments to be discussed here. A careful analysis including the anisotropy of the Raman scattering tensor shows that the molecular rotation has a negligible effect on the coherent probing technique, even for a moderately large depolarization factor $\rho_S \lesssim 0.2$; as a result, a deconvolution of a rotational contribution is not required. This point is an important advantage as compared to conventional Raman spectroscopy [6.2].

At this point we wish to note the analogy to a two-level spin system which is governed by the well-known Bloch equations. For optical dipole transitions a pseudo-spin vector has been introduced [6.38] describing the instantaneous state of the two-level system. The z component of this vector represents the population of the system, while the x and y components determine the induced dipole moment. An analogous pseudo-spin vector may be defined for our Raman transition. The coherent amplitude $\langle q \rangle$ is proportional to the induced dipole moment. In the terminology of this model the decay of $\langle q \rangle$ represents the free induction decay process.

We have evaluated numerical solutions of (6.10-14). The excitation and subsequent free decay of the vibrational system is illustrated by theoretical data depicted in Figs.6.1-3 for three different physical situations. Figure 6.1 considers the simple case of a homogeneous line. The exponential decay of the coherent excitation, Q^2, with time constant $T_2/2$ should be noted. The dotted curve represents the excitation pulse of a Gaussian shape assumed in the computations; t_p denotes the pulse duration (FWHM of laser intensity). In Fig.6.2, a vibrational system with isotopic substructure consisting of three equidistant lines is considered. The total excitation $|Q_{tot}|^2 = |\sum f_j \langle q_j \rangle|^2$ is plotted. The free decay of the total system shows a beating effect which is a direct consequence of the (isotopic) line splitting and the phase relation between the (isotopic) species established in the excitation process. The beating time reflects the constant frequency spacing $\Delta\omega$ of the vibrational components. The envelope curve decays exponentially with time constant $T_2/2$.

Quite different is the result shown in Fig.6.3 for a broad (quasi-continuous) distribution of transition frequencies of width $\delta\omega_{inh} T_2 = 45$. The total coherent excitation $|Q_{tot}|^2 = |\sum f_j \langle q_j \rangle|^2$ disappears very rapidly in the figure on account of the broad frequency distribution. The individual components, on the other hand, decay more slowly with time constant $T_2/2$ (broken curve in Fig.6.3).

The vibrational excitation illustrated by the numerical results of Figs.6.1-3 may be achieved in two different experimental systems:

I) *The Raman generator*. A single intense excitation pulse at ω_L generates the Stokes field E_S at the frequency ω_S by a high gain process (starting from quantum noise). An amplification of approximately $\exp(25) \simeq 10^{11}$ is common.

II) *The Raman amplifier*. This approach necessitates two light pulses, a laser field E_L and an input Stokes field E_S of appropriate frequency difference $\omega_L - \omega_S$. Compared with the generator setup, the gain requirements are strongly reduced due to the presence of the input Stokes pulse [6.36,37].

Both methods have been successfully applied for the coherent excitation of molecular systems.

6.2.2 Coherent Probing

The following experimental situation is considered here. The vibrational mode of interest is first coherently excited by transient stimulated scattering using the Raman generator or amplifier setup mentioned above. Then, a second weak light pulse

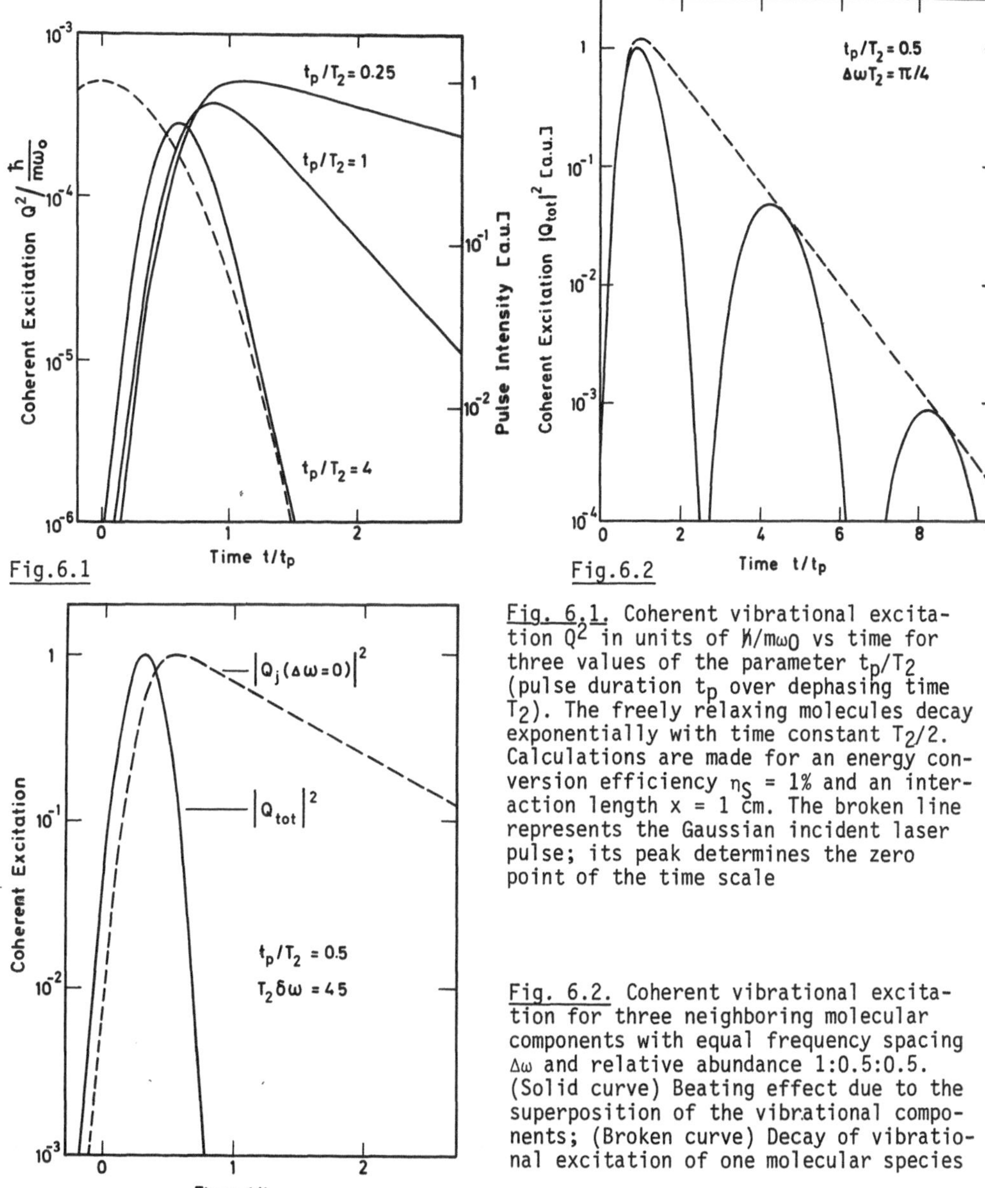

Fig. 6.1. Coherent vibrational excitation Q^2 in units of $\hbar/m\omega_0$ vs time for three values of the parameter t_p/T_2 (pulse duration t_p over dephasing time T_2). The freely relaxing molecules decay exponentially with time constant $T_2/2$. Calculations are made for an energy conversion efficiency η_S = 1% and an interaction length x = 1 cm. The broken line represents the Gaussian incident laser pulse; its peak determines the zero point of the time scale

Fig. 6.2. Coherent vibrational excitation for three neighboring molecular components with equal frequency spacing $\Delta\omega$ and relative abundance 1:0.5:0.5. (Solid curve) Beating effect due to the superposition of the vibrational components; (Broken curve) Decay of vibrational excitation of one molecular species

Fig. 6.3. Calculated coherent excitation of a vibrational system with a (Gaussian) distribution of transition frequencies. The solid curve represents the total vibrational excitation with rapid nonexponential decay due to destructive interference of the various components. The broken line shows the exponential time decay of a molecular subensemble with time constant $T_2/2$.

with variable time delay with respect to the pumping pulse(s) traverses the sample and probes the instantaneous decay of the vibrational excitation as a function of delay time t_D. The probe wave interacts with the vibrational amplitudes $\langle q_j \rangle$ giving rise to scattered radiation via coherent Raman scattering. This experimental method was first used by VON DER LINDE et al. [6.12] for the determination of the dephasing time of molecular vibrations in liquids. A similar technique was independently applied to $CaCO_3$ crystals by ALFANO and SHAPIRO [6.33].

There exists some analogy of the coherent Raman scattering to the interaction of light with coherent sound waves. The molecules in the excited volume vibrate with a definite phase relation and produce a macroscopic modulation of the optical refractive index via the coupling parameter $\partial\alpha/\partial q$. The system behaves like an oscillating three-dimensional phase grating. Scattering off this phase grating produces side bands shifted by the vibrational frequency to higher (anti-Stokes) and lower (Stokes) frequencies. Comparing coherent Raman scattering with conventional spontaneous Raman scattering we note drastic differences: The latter process is known to scatter an exceedingly small fraction of the incident light beam ($\sim 10^{-6}$/cm) into the whole solid angle of 4π. Coherent scattering, on the other hand, is more intense by many orders of magnitude and generates a scattering signal which is highly collimated close to the direction of the pump beam. Most important, the scattering efficiency depends critically on the scattering geometry according to the k-matching condition.

In CARS experiments under stationary conditions the excitation and probing processes involving four light fields are often treated as a four-photon interaction. This approach is not convenient for the nonstationary situation investigated here, where excitation and probe scattering are clearly separated by a delay time t_D which may be longer than the light pulses by a factor of up to 100.

The theoretical description of coherent Raman probe scattering is closely related to the stimulated excitation process discussed above. Both processes are produced by the same nonlinear polarization. The experimental conditions, however, are significantly different for coherent scattering and stimulated excitation. The latter process generates the vibrational excitation by a sufficiently intensive light pulse. Probe scattering is done with a weak light pulse which does not noticeably disturb the vibrational excitation.

Using picosecond pulses we detect scattering signals of several 10^{-12} s duration which is two to three orders of magnitude below the time resolution of conventional photo detectors. As a result, time-integrated signals are observed, $S^{coh} \propto \int_{-\infty}^{\infty} |E_{A,S}|^2 dt$ at Stokes (S) or anti-Stokes (A) frequency position. For a probe pulse of amplitude $E_{L2}(t - x/v)$ and for a homogeneously broadened vibrational transition we find a scattering signal $S^{coh}(t_D)$ at the end of the sample, $x = \ell$

$$S^{coh}(t_D) \propto \frac{\kappa_{AS}^2 \Delta\ell^2}{1+(2\Delta k_{AS}\Delta\ell)^2} \int_{-\infty}^{\infty} dt |E_{L2}(\ell, t - t_D)|^2 \times |Q(\ell, t)|^2 \; ; \tag{6.15}$$

$\Delta\ell$ represents the effective interaction length at the end of the sample where the maximum excitation occurs. The coupling coefficient $\kappa_{AS} = \pi\omega_{AS}^2(\partial\alpha/\partial q)N/(c^2 k_{AS})$ contains several material parameters analogous to (6.13); Δk_{AS} represents the k-vector mismatch for the Stokes and anti-Stokes process, respectively,

$$\Delta k_A = k_A - k_{L2} - k_L + k_{\bar{S}} \;;\quad \Delta k_S = k_{L2} - k_S - k_L - k_{\bar{S}} \quad . \tag{6.16}$$

The quantities k_L and $k_{\bar{S}}$ denote the wave vectors of the excitation pulse and of the (primary) stimulated Stokes emission, respectively; k_{L2}, k_S, and k_A represent the wave vectors of the probe pulse and of the scattered Stokes and anti-Stokes waves; Δk_A and Δk_S are determined by the wavelength dispersion of the optical refractive index of the medium and have different magnitudes; i.e., k-matching $\Delta k_{AS} = 0$ may be achieved either for the anti-Stokes or for the Stokes scattering process. Equation (6.15) represents a convolution of the coherent vibrational excitation $|Q|^2$ (in the homogeneous case) and the probing light pulse. It is important to note that the magnitude of the scattering signal strongly depends on the wave vector mismatch. $\Delta k_{AS} = (2\Delta\ell)^{-1}$ reduces the signal by a factor of two. Equation (6.15) indicates that the time dependence of $S^{coh}(t_D)$ is identic for Stokes or anti-Stokes probe scattering.

We have extended our theoretical investigations to the coherent excitation of inhomogeneously broadened vibrational systems. Here, the k-matching situation requires a more detailed discussion. During the excitation process the molecules of the various subgroups j vibrate with a certain spatial phase relation described by a wave vector $\underline{k}_0 = \underline{k}_L - \underline{k}_{\bar{S}}$, where $\underline{k}_0$ is prepared by the laser ($\underline{k}_L$) and stimulated Stokes pulse ($\underline{k}_{\bar{S}}$). In condensed systems the material k-vector $\underline{k}_0$ is conserved when the excitation process terminates, while the molecular subgroups oscillate freely with their individual transition frequencies ω_j. As a result, the mismatch Δk_j is different for different subensembles. For Stokes scattering we have

$$\Delta k_j = |\underline{k}_{L2} - \underline{k}_{Sj} - \underline{k}_0| \quad , \tag{6.17a}$$

while a similar expression holds for the anti-Stokes component

$$\Delta k_j = |\underline{k}_{Aj} - \underline{k}_{L2} - \underline{k}_0| \quad . \tag{6.17b}$$

The quantities $\underline{k}_{Sj}$ and $\underline{k}_{Aj}$, respectively, represent the wave vectors relevant for the Stokes and anti-Stokes components of the probe scattering generated by the molecular subgroup j; i.e., $k_{AS,j} = (\omega_{L2} \pm \omega_j)n/c$. Molecular subgroups with approximate k-matching $\Delta k_j \Delta\ell \lesssim 0.5$ determine the coherent scattering signal, while components with large mismatches $\Delta k_j \Delta\ell \gg 0.5$ give negligible contributions to the probe scattering. The frequency interval of allowed transition frequencies depends on the interaction length $\Delta\ell$. We wish to distinguish two limiting cases [6.11,13]:

I) *Selective k-matching.* For large values of $\Delta\ell(\gtrsim 1$ cm) a highly selctive k-vector geometry may be adjusted in order to observe probe scattering of a molecular subensemble with a negligible spread of transition frequencies. This subgroup is expected to decay with a characteristic (homogeneous) dephasing time.

II) *Nonselective k-matching.* For small interaction length ($\Delta\ell \ll 1$ cm), the k-matching is not sensitive to the distribution of transition frequencies and coherent superposition of the various molecular components may be seen. In this case, interesting beating effects of fast dephasing of the molecules by the distribution of transition frequencies was observed; i.e., the beats of the coherent excitation in Fig. 6.2 and $|Q_{tot}|^2$ of Fig.6.3 may be studied.

The different values of $\Delta\ell$ are experimentally adjusted by proper choice of the scattering geometry.

6.3 Experimental

In this section, several remarks concerning the quality of ultrashort laser pulses should be made. Experimental systems are presented which were successfully applied to the investigation of dynamical processes in liquids and solids. In these experiments transient stimulated Raman scattering was used as the excitation process. The molecular system was interrogated by coherent Raman probe scattering.

6.3.1 Generation of Ultrashort Laser Pulses

The coherent excitation and probing of molecular vibrations requires high quality laser pulses. For reliable experimental data it is necessary to work with single picosecond laser pulses with the pulse properties under careful control. One has to measure the pulse duration and one should work with bandwidth limited pulses. It is desirable to know the peak intensity and the pulse shape of the pulses used. The application of the whole mode-locked pulse train of a high-gain solid-state should be avoided since the important pulse parameters, pulse duration, bandwidth, peak intensity, and pulse shape, vary from the beginning to the end of the pulse train [6.14-16]. The selection of chirp-free pulses from the pulse train is discussed by ZINTH et al. [6.17].

In our experimental systems we work with mode-locked Nd-glass lasers operating at a frequency $\tilde{\nu}_L = 9455$ cm^{-1}. Satellite pulses on a picosecond time scale are eliminated by contacting the dye cell to one mirror of the laser oscillator [6.18]. A single picosecond pulse is cut from the leading part of the pulse train using an electro-optic switch. This switch consists of a high-pressure spark gap in conjunction with an optical Kerr cell [6.19]. When one pulse of the mode-locked train reaches a certain power level the spark gap operates and the Kerr shutter transmits the subsequent pulse. The pulse duration was measured by a quantitative application of the two-photon fluorescence (TPF) technique of GIORDMAINE et al. [6.20,21], by the use of a fast streak camera of ~3 ps time resolution and by a nonlinear optical technique using a rapidly rising and decaying material excitation for probing the laser pulse [6.22]. The latter system has the advantage that the wings of the laser pulse may be measured several orders of ten below the peak of the pulse.

The various parameters of our ultrashort light pulses are as follows: Pulse duration $t_p \simeq 6$ ps; frequency bandwidth $\Delta\tilde{\nu}_L = 3\ \mathrm{cm}^{-1}$ leading to $t_p \times \Delta\nu_L = 0.5$, i.e., our pulses are essentially bandwidth limited; peak power $P_L = 10^8$ W; pulse shape approximately Gaussian; peak to background ratio 10^4; mode pattern TEM_{00}.

6.3.2 Coherent Excitation and Probing Techniques

Stimulated Raman scattering has the advantage of being able to coherently excite known vibrations of different molecules with a single laser frequency. In the high gain situation, where the Stokes scattered field builds up from quantum noise (Raman generator), no input Stokes pulse is required.

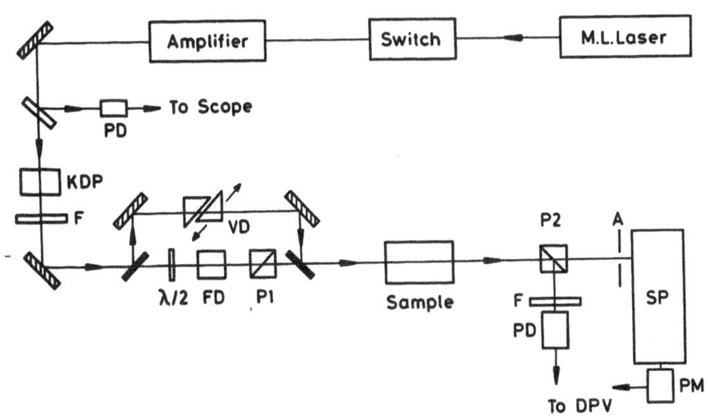

Fig. 6.4. Experimental setup to measure coherent probe scattering in a collinear geometry. A beam splitter generates the probe pulse which is properly delayed before travelling collinearly to the exciting pulse. The Stokes radiation of the exciting pulse is suppressed by the Polarizer P2

The experimental system for excitation by the stimulated Raman process and for coherent probe scattering with collinear wave-vector geometry is depicted schematically in Fig.6.4. The mode-locked laser oscillator is followed by an electro-optic switch. The selected pulse passes through an optical amplifier with gain of approximately 100; it is subsequently converted (in most experiments) to the second harmonic at 18,995 cm^{-1} (0.53 μm) in a potassium dihydrogen phosphate crystal. The powerful light pulse traverses the sample, a cell containing the liquid. Stimulated Raman scattering is effectively generated in samples of several millimeters to several centimeters in length. In our measurements, the energy conversion of laser to Stokes emission is kept at a few percent in order to avoid depletion of the input pulse and thus stay within the validity of our calculations. The stimulated Stokes pulse is simply measured by a fast photodiode and suitable filters (Fig.6.4). More detailed information was obtained using a high-resolution spectrometer in conjunction with an optical multichannel analyzer system.

A beam splitter in the path of the input pulse provides a second pulse of smaller intensity, approximately 10^{-2} of the exciting pulse. This weak pulse serves as an interrogating pulse with variable delay time t_D. Our optical delay system of high precision consists of two movable prisms. The delay times were determined taking into account the group velocities of various optical components. The accuracy of the t_D scale is better than 2%.

A second beam splitter is carefully adjusted to make the probe pulse travel collinearly to the exciting pulse through the medium. In a number of experiments, noncollinear geometries (not shown here) have been used. Due to the k-matching requirements, the coherent Raman scattering signal is highly collimated and observed in forward direction. Measurements are made as a function of time delay between excitation and probing pulse. In Fig.6.4, the radiation of the excitation and probing processes are distinguished using perpendicular planes of polarization. The pump pulse passes a $\lambda/2$ plate and polarizer P1 defining the plane of polarization of the excitation process. The corresponding stimulated Stokes radiation is effectively blocked by a factor of $\sim 10^5$ with the help of the second polarizer P2. The probe pulse and the corresponding coherent scattering signal, on the other hand, pass P2 without significant attenuation. It is then possible to observe the scattered Stokes or anti-Stokes signals of the interrogating pulse and to study their magnitude as a function of delay time over a measuring range of three to four order of ten.

The experimental system for coherent Raman probe scattering using excitation by a Raman amplifier setup is depicted in Fig.6.5. A single picosecond pulse from a Nd: glass laser system is frequency-doubled in a KDP crystal to $\tilde{\nu}_L$ = 18,995 cm^{-1}. The laser pulse is then directed into a Raman cell of suitable Stokes shift to generate the input Stokes pulse of frequency ν_S required in the experiment. The laser pulse and the generated Stokes pulse traverse a polarizer P1 and are imaged into the sample cell ($\ell = 1$ cm). Both pulses produce a coherent material excitation in the sample via stimulated Raman amplification. The probing light pulse E_{L2} of smaller intensity is generated by the beam splitter (BS) and is properly delayed by the variable delay (VD). The probe field is polarized perpendicular to the excitation fields; it crosses the excitation beam in the sample under a small angle of 7 mrad. The scattered Stokes light passes the aperture D, a second polarizer P2, and is detected by the photomultiplier (PM). The generator system of Fig.6.5 has the advantage that diluted systems and vibrational modes of smaller Raman cross sections may be studied in the separate sample cell.

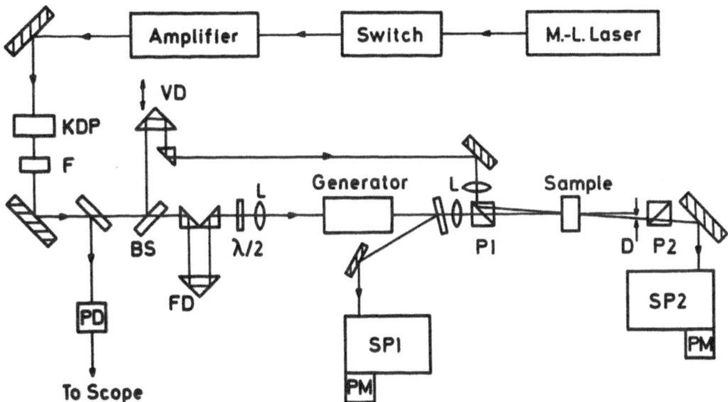

Fig. 6.5. Schematic of experimental setup for coherent probe experiments. The coherent material excitation is generated in the sample by two light pulses, a laser pulse and a Stokes shifted pulse. The latter is produced in the preceding generator cell

6.4 Experimental Results and Discussion

The following four Subsections are concerned with the study of the time evolution of the coherent material excitation of a variety of vibrational modes. We begin with the simplest case of a single molecular vibration with a homogeneously broadened Raman line. In the following section, we investigate vibrational modes with discrete but closely spaced vibrational components. In particular, we present a collective beating phenomenon originating from neighboring isotope species. Next, we discuss the more complex situation of inhomogeneously broadened Raman lines, where a range of vibrational transitions is excited in the pumping process. The last section is concerned with optical phonons in solids.

6.4.1 Modes with Homogeneous Line Broadening

As an example of homogeneous line broadening, we discuss the fundamental vibrational mode of liquid N_2 at 2326 cm^{-1}. Figure 6.6 presents the experimental data [6.8]. From the exponential decay of the signal curve we deduce a dephasing time of $T_2/2$ = 75 ps, which corresponds to a dynamic line broadening of 0.070 cm^{-1}. This number is in excellent agreement with the linewidth of the isotropic scattering component observed in spontaneous Raman spectroscopy [6.23,24]; i.e., the dephasing time T_2 fully accounts for the spectrosocpic line broadening. It is interesting to note that an exceptionally long population lifetime T_1 of the fundamental mode of approximately one minute was reported by several authors [6.8,9]. As a result, population decay does not affect the dephasing process of liquid N_2.

The data of Fig.6.6 were subsequently extended to mixtures of N_2 with liquid Ar. A comparatively small variation of T_2 with concentration and temperature was reported

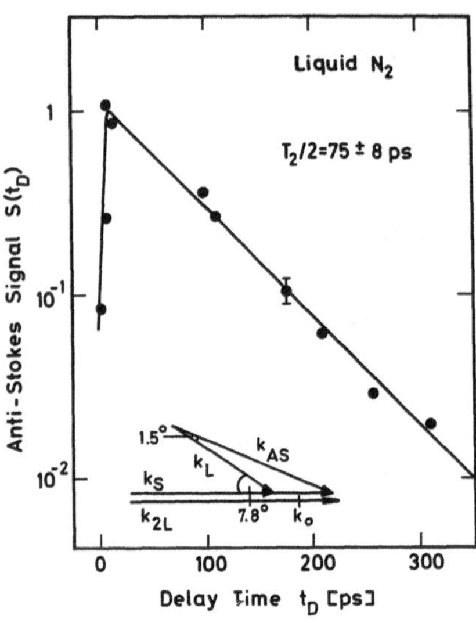

Fig. 6.6. Coherent anti-Stokes probe signal vs delay time t_D. The dephasing time of the fundamental mode of liquid N_2 at $\tilde{\nu}$ = 2326 cm cm $^{-1}$ is found to be $T_2/2$ = 75 ps. (Inset) Schematic of k-matching

[6.25]. The picosecond data agree well with spectrosopic findings on mixtures of N_2 : Ar.

Quite different numbers were found on the symmetric CH_3-stretching mode of CH_3CCl_3 at 2939 cm^{-1}. Exponential decay of the coherent signal was observed with a very short time constant of $T_2/2$ = 1.2 ps. This number is in good agreement with the spontaneous Raman linewidth of 4.3 cm^{-1}. We emphasize that the same value of T_2 was measured for different k-matching geometries. The results on CH_3CCl_3 strongly suggest that the investigated mode is predominantly homogeneously broadened. A list of measured T_2 times is shown in Table 6.1; the values vary over approximately two orders of magnitude. The agreement with time constants deduced from the Raman linewidth is good for several modes and molecules. In these cases, we may state that the linewidth is predominantly determined by fast dephasing processes. For the three remaining molecules $SnBr_4$, CH_3OH, and (CH_2OH) a distribution of frequencies determines the spontaneous Raman line profile. A discussion of the dephasing mechanism determining T_2 will be presented in Section 6.5.

In a modified experimental system, HERITAGE [6.26] investigated the dephasing of the mode at 656 cm^{-1} of liquid CS_2. The molecules were excited by pairs of pulses having a frequency difference equal to the mode frequency. The gain (loss) of delayed pairs of the same pulses was measured. Pulse trains of two synchronized picosecond dye lasers allowed working with low peak powers ($P_L \sim 100$ W). Since gain measurements depend on phase differences very stable optical conditions are required. The observed exponential decay time of T_2 = 21 ps agrees with the value deduced from the spontaneous Raman linewidth.

Table 6.1. Measured dephasing times of several modes and molecules. Comparison with spontaneous Raman linewidths

	$\tilde{\nu}_0$ [cm^{-1}]	$T_2/2$ [10^{-12} s]	
		Measured	From linewidth
N_2(77K)	2326	75 ± 8	79 ± 8
CCl_4	459	3.5 ± 0.4	3.8 ± 0.5
$SiCl_4$	425	3.0 ± 0.5	2.8 ± 0.5
$SnCl_4$	368	2.8 ± 0.3	2.5 ± 0.5
CH_3CCl_3	2939	1.2 ± 0.2	1.2 ± 0.1
$SnBr_4$	221	3.0 ± 0.3	>1.6
CH_3OH	2835	2.3 ± 0.5	>0.25
$(CH_2OH)_2$	2935	3.0 ± 0.5	>0.1

6.4.2 Modes with Discrete Substructure

As example of vibrational bands with discrete substructure we consider lines with isotopic multiplicity. Many liquids are composed of several isotopic species. The resulting line splitting may or may not be resolved by spontaneous Raman (or infrared) spectroscopy. It will be shown here that coherent excitation and coherent probing allow us to determine the dephasing time of one isotope component. In addition, a new beating phenomenon is observed which gives values of the frequency difference of the vibrational isotope components.

CCl_4 is discussed here as an example. The two isotopes ^{35}Cl and ^{37}Cl give rise to vibrational multiplicity of the totally symmetric tetrahedron vibration around 460 cm^{-1}. The isotopic structure is readily seen in the spontaneous Raman spectrum with a frequency spacing of approximately 3 cm^{-1} of the various lines (see inset of Fig.6.7a). The results for coherent probe scattering with a selective k-geometry [6.1] are depicted in Fig.6.7a. The decaying part of the signal curve represents loss of phase correlation of a single isotope component with a time constant of $T_2/2$ = 3.6 ps. The measured dephasing time fully accounts for the Raman linewidth of one isotope species of 1.4 cm^{-1}. Additional information is obtained by coherent probing with nonselective k-matching. In these experiments, a coherent superposition of the different isotopic species is observed (Fig.6.7b). The various isotope components are first excited with approximately equal phases (rising part of the signal curves). The excitation process then terminates and free relaxation of the collective excitation is observed. The frequency differences of the individual species lead to a striking interference phenomenon with a beating period of ~12 ps, in good agreement with the theoretical results of Fig.6.2. We recall that the coherent beat

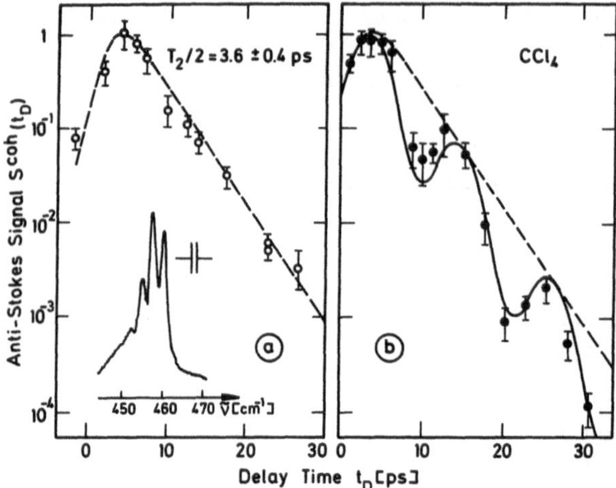

Fig. 6.7. (a) Coherent Stokes signal vs delay time of one isotope component of CCl_4; a selective collinear k-matching geometry was applied; (b) Coherent anti-Stokes signal observed in a nonselective off-axis geometry. The beating of the different isotope components should be noted . (Inset) Spontaneous Raman-Stokes spectrum of CCl_4

signal represents the convolution of the coherent excitation with a probe pulse of finite pulse duration [see (7.15)].

6.4.3 Modes with Inhomogeneous Line Broadening

There is evidence that numerous vibrational bands consist of a distribution of transition frequencies. Spontaneous spectroscopic techniques provide the total Raman or infrared band, but do not allow a separation between homogeneous and inhomogeneous broadening factors. As a result, it is impossible, in general, to decide to what extent a band contour contains inhomogeneous contributions. Our coherent Raman probing technique allows one to study this problem.

As an example of a small inhomogeneity we discuss the totally symmetric ring vibration at 945 cm^{-1} of deuterated benzene C_6D_6 [6.27]. This mode gives rise to a strong and narrow Raman line of width 1.5 cm^{-1} which has been measured repeatedly by conventional Raman equipment. We have investigated this mode with our picosecond Raman technique and found interesting results which are presented in Fig.6.8. With the experimental setup of Fig.6.4 we have measured the coherent probe scattering under a selective k-vector geometry (Fig.6.8a). Monitoring the coherent decay over a factor of 200 we see approximately exponential relaxation with a time constant of $T_2/2 = 5.0$ ps. The broken line in Fig.6.8a is a theoretical curve for $T_2 = 10$ ps taking into account the finite k-resolution of the experimental system (interaction length $\Delta\ell = 0.4$ cm). For nonselective k-matching conditions, Fig.6.8b shows a notably accelerated signal decay suggesting a distribution of vibrational frequencies. For

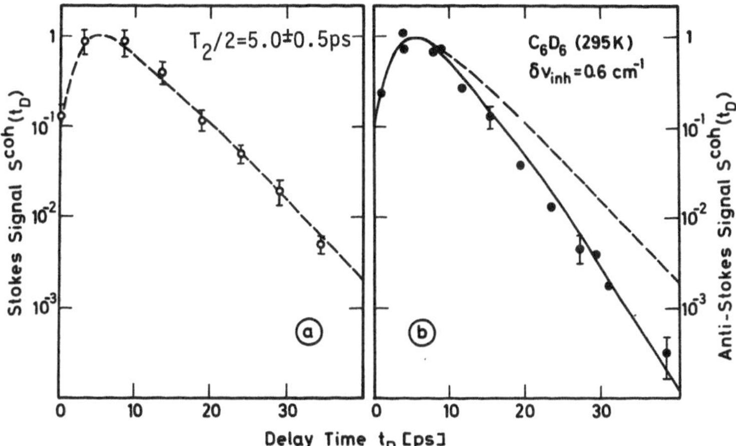

Fig. 6.8. (a) Coherent Stokes probe scattering vs delay time using a selctive k-vector geometry; (b) Coherent anti-Stokes signal under nonselective k-matching conditions. Data suggest a distribution of vibrational frequencies. The broken curves are calculated for $T_2/2$ = 5 ps

an assumed Gaussian frequency distribution, the theoretical curve in Fig.6.8b (solid line) was fitted to the picosecond data using the value T_2 = 10 ps of Fig.6.8a. This fitting procedure suggests an inhomogeneous broadening of $\delta\tilde{\nu}_{inh} \simeq 0.6$ cm^{-1}.

Additional support for our picosecond results is obtained by a comparison with the spontaneous Raman band. A careful analysis shows that the isotropic scattering component has a slightly non-Lorentzian band contour. Our value of T_2 = 10 ps corresponds to a homogeneous broadening of Lorentzian shape and width (FWHH) $\delta\tilde{\nu}_{hom} \simeq$ 1.0 cm^{-1}. Convolution of the homogeneous part (1.0 cm^{-1}) with the inhomogeneous contribution of 0.6 cm^{-1} leads to a Voigt profile which fully agrees with the measured band contour of this mode. The data of Fig.6.8 raise the interesting question how a vibrational mode in a weakly associated liquid can at the same time be homogeneously and inhomogeneously broadened. An answer will be given below.

Significant inhomogeneous broadening may be expected for the extended Raman bands of molecules with strong hydrogen bonding. We discuss here the CH-stretching mode of pure $(CH_2OH)_2$ at 2935 cm^{-1}. The corresponding spontaneous Raman band is broad with a linewidth of ~60 cm^{-1}. In Fig.6.9 we present experimental data of two widely different experimental situations [6.28]. In Fig.6.9a, a highly selective k-matching situation is used with a sample length of 10 cm and with a small divergence of the Stokes beam of 3 mrad. In Fig.6.9b, on the other hand, we devised a less selective k-vector geometry by using a shorter cell of 1 cm and a larger Stokes divergence of 10 mrad. The decaying part of the signal curve in Fig.6.9a indicates exponential decay with a time constant of $T_2/2$ = 3.0 ps. This time constant is believed to represent the homogeneous dephasing of a subgroup of molecules, the frequency of which is close to the center frequency of the broad Raman band at 2935 cm^{-1}

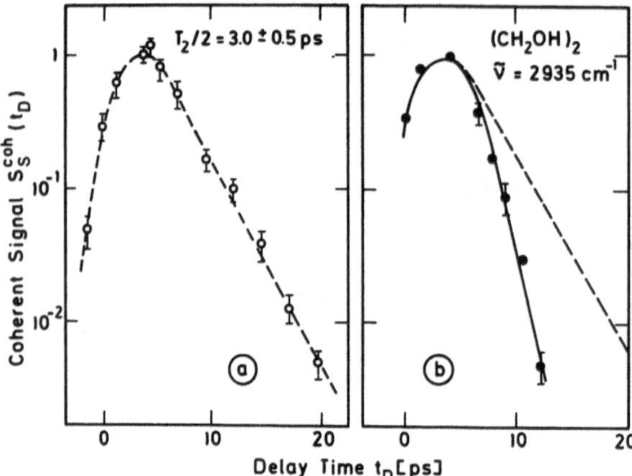

Fig. 6.9a,b. Coherent Stokes probe scattering vs time for a selective (a) and a nonselective (b) k-matching situation. (a) shows the dephasing of a molecular subgroup and (b) represents destructive interference of molecules with a distribution of frequencies

The time dependence of $S^{coh}(t_D)$ is completely different for less selective k-matching of Fig.6.9b. Under these experimental conditions the coherent scattering signal disappears rapidly. The measured time dependence obviously represents the destructive interference of molecules which vibrate with a wide distribution of frequencies. In fact, the value of T_2 = 6 ps of Fig.6.9a corresponds to a homogeneous line of width 1.8 cm^{-1}, which is smaller than the total band by a factor of approximately 30.

6.4.4 Vibrational Modes in Solids

In a number of crystals, optical phonons were investigated after coherent excitation by the stimulated Raman process. The interesting optical phonons are located very near to the center of the Brillouin zone on account of the long optical wavelengths. We note briefly a basic difference between liquids and solids. In liquids we treat the excited molecules as two-level systems. A moderate excess population n of the first vibrational state is generated. In solids, the optical phonons represent a collective lattice excitation with harmonic oscillator levels. Within a collimated optical beam a small number of lattice modes with very large phonon occupation numbers is excited.

In Fig.6.10a, we present the coherent anti-Stokes probe signal measured on diamond at 295 and 77 K. The buildup and decay of the coherent excitation is readily seen for both temperatures. The fundamental transverse optical (TO) phonon at $\tilde{\nu}_p$ = 1332 cm^{-1} was excited [6.29]. A comparison with time constants deduced from the linewidths of spontaneous Raman measurements gives good agreement with our directly measured lifetimes. The phonon dispersion depicted in Fig.6.10b should serve to illustrate the

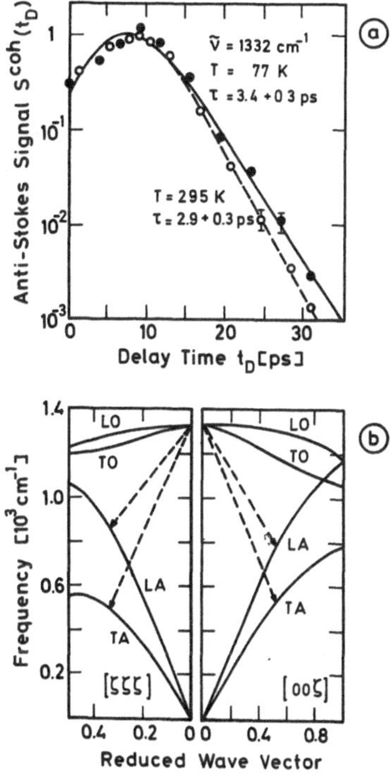

Fig. 6.10. (a) Coherent anti-Stokes probe signal vs delay time. The TO phonon in diamond is excited; (b) Phonon dispersion curves of diamond. The dotted lines indicate two delay processes into one TA and one LA phonon each

decay process. The excited TO phonon decays - according to the selection rules - into one TA and one LA phonon as indicated in the figure. Energy and crystal momentum are conserved in this process. Calculations of the lattice dynamics of diamond suggest that the observed relaxation times correspond to the energy decay of the TO phonon into two acoustic phonons [6.30]. The coherent experiment provides here information of a T_1- process.

As a second example, experimental investigations of $CaCO_3$ crystals should be discussed. The stimulated Raman process excites the internal A_{1g} mode of the CO_3^{--} islands with $\tilde{\nu}$ = 1086 cm^{-1} and the coherent anti-Stokes probe signal allows one to determine decay times of 4.4 ps and 8.7 ps for 295 K and 90 K, respectively [6.31]. These numbers compare favorably with time constants calculated from the line profile of the spontaneous Raman process [6.32]. Different data of the decay time [6.33] and of the spontaneous Raman linewidth [6.34] appear to be due to experimental errors [6.35]. The decay process of the internal CO_3^{--} mode in $CaCO_3$ is not yet determined. The excited phonon is far above the acoustic branches ruling out a decay process similar to the case of diamond. The independent observation of the incoherent probe signal would be of great interest for the understanding of the decay mechanism.

6.5 Interaction Processes

The picosecond experiments clearly show that ultrafast dephasing processes occur in the investigated samples. These processes lead to the dephasing time T_2. In order to explain the exponential decay observed over several orders of magnitude the limiting case of fast modulation has to be considered, i.e., homogeneous broadening with motional narrowing is suggested by the experimental data. We note that a time scale of ~1 ps or less is required for the individual interaction event in the fast modulation limit. The small amplitude rattling motion of the molecules in their liquid cages (time scale 10^{-13} s) and/or the fast molecular rotation are important factors for an exponential dephasing time T_2.

The following processes have been suggested to contribute to the measured T_2 values of polyatomic molecules:

I) "Direct" dephasing by quasi-elastic collisions of nearest neighbors via the repulsive part of the intermolecular potential [6.39-42]. Numerical estimates of the process agree with experimental results within a factor of three. Improved agreement is reported when the coupling with rotational motion and vibrational anharmonicities are included [6.43].

II) Indirect dephasing by anharmonic coupling to low frequency vibrational modes [6.44,45].

III) Resonant exchange of vibrational quanta via the repulsive part of the intermolecular potential of transition dipole-dipole interaction [6.46,47].

IV) Energy transfer to neighboring vibrational levels converting small energy mismatches to rotational and translational motion [6.48].

Processes I) to III) have been termed "pure dephasing". Processes IV) may be distinguished from the mechanisms I) to III) measuring the time evolution of the excited state population (T_1-techniques) not discussed in the present paper.

For the symmetric CH-stretching mode of CH_3CCl_3, theoretical estimates suggest process I) as dominant T_2 mechanism while process II) appears to be unimportant [6.45]. The small concentration dependence of the linewidth rules out a significant contribution of III). Approximately 20% of the total line broadening is due to population decay IV). For many vibrational modes the dominant mechanism of T_2 cannot be stated definitely at the present time.

Turning to the investigations of inhomogeneously broadened lines the following preliminary physical picture is suggested.

The picosecond data present clear evidence for a distribution of vibrational frequencies in several strongly and weakly associated liquids. The observed frequency distribution appears to be static in the picosecond time scale; it does not change notably for several 10^{-11} s or even longer. The different frequencies of the individual molecules may be interpreted as variations of the solvent shift exerted by the molecular environment. Existing static theories successfully explain the well-known frequency shifts of the band centers in liquid systems [6.49]. More elaborate stochastic theories also reveal the relationship with the frequency distribution [6.50]. In the slow modulation limit, the inhomogeneous broadening and the solvent shift $\Delta\tilde{\nu}$ of the band center are related by (6.18)

$$\delta\tilde{\nu}_{inh} = \Delta\tilde{\nu}/\sqrt{Z} \quad , \tag{6.18}$$

where Z denotes the effective number of interacting neighboring molecules. Equation (6.18) gives smaller values of $\delta\tilde{\nu}_{inh}$ as compared with the solvent shift. It is suggested that the long-range attractive part of the intermolecular potential determines the inhomogeneous broadening. It may be expected that this interaction is governed by the diffusional motion on the time scale of 10^{-10} s.

In conclusion, it is pointed out that new techniques have been developed for the investigation of the coherent material excitation. These experiments benefit from the improved understanding of the excitation process and the high time resolution available with picosecond light pulses. The coherent Raman excitation and probing of molecules in a macroscopic volume offers the possibility of selecting a molecular subgroup by k-vector spectrosopy. In this way, new and detailed information on relaxation phenomena is obtained. The continued application of the picosecond techniques will give deeper insight in the liquid and solid state and will advance our understanding of the ultrafast molecular dynamics.

References

6.1 R.T. Bailey: In *Molecular Spectroscopy*, Vol. 2 (The Chemical Society, London 1974)
6.2 A. Laubereau, W. Kaiser: In *Reviews of Modern Physics*, Vol. 50 (American Phys. Society, New York 1978)
6.3 S.A. Akhmanov, K.N. Drabovich, A.P. Sukhorukov, A.K. Shchednova: Sov. Phys. JETP *35*, 279 (1972)
6.4 R.L. Carman, F. Shimizu, C.S. Wang, N. Bloembergen: Phys. Rev. A*2*, 60 (1970)
6.5 J.A. Giordmaine, W. Kaiser: Phys. Rev. *144*, 676 (1966)
6.6 M. Maier, W. Kaiser, J.A. Giordmaine: Phys. REv. *177*, 580 (1969)
6.7 D. von der Linde: In *Ultrashort Light Pulses*, Topics in Applied Physics, Vol. 18, ed. by S.L. Shapiro (Springer, Berlin, Heidelberg, New York 1977)
6.8 A. Laubereau: Chem. Phys. Lett. *27*, 600 (1974)
6.9 W.F. Calaway, G.E. Ewing: Chem. Phys. Lett. *30*, 485 (1975)
S.R. Brueck, R.M. Osgood, Jr.: Chem. Phys. Lett. *39*, 568 (1976)
6.10 A. Laubereau, D. von der Linde, W. Kaiser: Phys. Rev. Lett. *28*, 1162 (1972)
6.11 A. Laubereau, G. Wochner, W. Kaiser: Phys. Rev. A*13*, 2212 (1976)
6.12 D. von der Linde, A. Laubereau, W. Kaiser: Phys. Rev. Lett. *26*, 954 (1971)
6.13 A. Laubereau, G. Wochner, W. Kaiser: Chem. Phys. *28*, 363 (1978)
6.14 M.A. Duguay, J.W. Hansen, S.L. Shapiro: IEEE J. QE-*6*, 725 (1970)
6.15 R.C. Eckardt, C.H. Lee, J.N. Bradford: Appl. Phys. Lett. *19*, 420 (1971)
6.16 D. von der Linde: IEEE J. QE-*8*, 328 (1972)
6.17 W. Zinth, A. Laubereau, W. Kaiser: Opt. Commun. *22*, 161 (1977)
6.18 D.J. Bradley, G.H. New, S.J. Caughey: Phys. Lett. A*30*, 78 (1969)

6.19 D. von der Linde, O. Bernecker, A. Laubereau: Opt. Commun. *2*, 215 (1970)
6.20 J.A. Giordmaine, P.M. Rentzepis, S.L. Shapiro, K.W. Wecht: Appl. Phys. Lett. *11*, 216 (1967)
6.21 D. von der Linde, O. Bernecker, W. Kaiser: Opt. Commun. *2*, 149 (1970)
6.22 D. von der Linde, A. Laubereau: Opt. Commun. *3*, 279 (1971)
6.23 W.R.L. Clements, B.P. Stoicheff: Appl. Phys. Lett. *12*, 246 (1968)
6.24 M. Scotto: J. Chem. Phys. *49*, 5362 (1968)
6.25 H.M.M. Hesp, J. Langelaar, B. Bebelaar, J.D.W. van Voorst: Phys. Rev. Lett. *39*, 1376 (1977)
S.A. Akhmanov, N.I. Koroteev, R.Yu. Orlov, I.L. Shumay: JETP Lett. *23*, 276 (1976)
6.26 J.P. Heritage: Appl. Phys. Lett. *34*, 470 (1979)
6.27 A. Laubereau et al.: To be published
6.28 A. Laubereau, G. Wochner, W. Kaiser: Chem. Phys. *28*, 363 (1978)
6.29 A. Laubereau, D. von der Linde, W. Kaiser: Phys. Rev. Lett. *27*, 802 (1971)
6.30 H. Bilz: To be published
6.31 A. Laubereau, G. Wochner, W. Kaiser: Opt. Commun. *14*, 75 (1975)
6.32 K. Park: Phys. Lett. *22*, 39 (1966); A*25*, 490 (1967)
6.33 R.R. Alfano, S.L. Shapiro: Phys. Rev. Lett. *26*, 1247 (1971)
6.34 C.H. Lee, D. Ricard: Appl. Phys. Lett. *32*, 168 (1978)
6.35 W. Kiefer, A. Laubereau: To be published
6.36 A. Laubereau, D. von der Linde, W. Kaiser: Opt. Commun. *7*, 173 (1973)
6.37 W. Zinth, A. Laubereau, W. Kaiser: Opt. Commun. *26*, 457 (1978)
6.38 R.P. Feynman, F.C. Vernon, R.W. Hellwarth: J. Appl. Phys. *28*, 49 (1957)
6.39 S.F. Fischer, A. Laubereau: Chem. Phys. Lett. *35*, 6 (1975)
6.40 P.A. Madden, R.M. Lynden-Bell: Chem. Phys. Lett. *38*, 163 (1976)
6.41 W.G. Rothschild: J. Chem. Phys. *65*, 2958 (1976)
6.42 D.W. Oxtoby, S.A. Rice: Chem. Phys. Lett. *42*, 1 (1976)
6.43 R.K. Wertheimer: Mol. Phys. *35*, 257 (1978)
6.44 S. Bratos: J. Chem. Phys. *63*, 3499 (1975)
6.45 S.F. Fischer, A. Laubereau: Chem. Phys. Lett. *55*, 189 (1978)
6.46 D. Döge, R. Arndt, A. Khuen: Chem. Phys. *21*, 53 (1977)
6.47 M. Gilbert, M. Drifford: J. Chem. Phys. *66*, 3205 (1977)
6.48 A. Miklavc, S.F. Fischer: Chem. Phys. Lett. *44*, 209 (1976)
6.49 A.D. Buckingham: Proc. R. Soc. London A*248*, 169 (1958)
6.50 S. Bratos, J. Rios, Y. Guissany: J. Chem. Phys. *52*, 439 (1970)

7. Coherent Raman Spectroscopy

M. D. Levenson and J. J. Song

With 32 Figures

The theory, practice and applications of Coherent Raman Spectroscopy are comprehensively reviewed. In addition to the now standard Coherent Anti-Stokes Raman Spectroscopy (CARS) technique, related methods such as Raman Induced Kerr Effect Spectroscopy (RIKES), Stimulated Raman Spectroscopy (SRS) and four-wave mixing are presented in detail. An extended two-level model is employed to calculate the nonlinear polarization and third-order nonlinear susceptibility in the cases where one-photon resonances are unimportant, and the results from a complete four-level density matrix treatment are quoted for the resonant case. Also included are discussions of symmetry and focusing geometries applicable to isotropic samples and to both centrosymmetric and accentric crystals. A general formulation of signal-to-noise considerations allows a comparison of the various techniques and configurations. Applications for which the coherent Raman techniques are superior to spontaneous scattering are emphasized with citations to the pre-1979 literature.

7.1 Historical Background

7.1.1 Prehistory

When a transition occurs between states of opposite parity, the nonvanishing dipole matrix element produces a line in the absorption or emission spectrum. Transitions between levels of the same parity, however, do not generally result in such readily identifiable features in optical spectra. Thus RAMAN's discovery in 1928 of the inelastic photon scattering mechanism which bears his name opened a vast new realm to spectroscopy [7.1]. In Raman scattering, a material simultaneously absorbs one photon while emitting another, the energy of the second photon differing from the first by the splitting between two levels of the same parity. In molecules and solids, such levels have different rotational or vibrational quantum numbers, and the energy differences correspond to frequencies in the infrared region. The incident and Raman scattering light, however, can be in the more easily managed visible or ultraviolet region. In the 50 years since its discovery, spontaneous Raman scattering has proved an invaluable tool in physics and chemistry [7.2-6].

Spontaneous Raman scattering is, however, a fearsomely weak effect; the scattering cross sections are typically less than 10^{-30} cm^2. The practitioners of spontaneous scattering must illuminate their samples as strongly as possible and collect a substantial fraction of the randomly scattered quanta. In the days before the CW laser, Raman spectroscopists nearly surrounded their samples with powerful mercury arc lamps and recorded the spectra of the scattered radiation in hours-long photographic exposures [7.3]. In this way many of the molecular vibrational and rotational constants tabulated by HERTZBERG, for example, were determined [7.7].

The invention of the laser revolutionized the practice of Raman spectroscopy in the early 1960s. A variety of new techniques were developed which employed the intense and directional sources in place of the conventional discharge lamps. At the same time, low noise photoelectric detectors became available along with tandem monochromators which could separate the elastic and inelastically scattered photons more efficiently [7.4,5].

Major difficulties, however, remained. Because the Raman effect was so weak, even faint luminescence could obscure the desired spectra. It was difficult or impossible to study flames, plasmas, crystals with fluorescent impurities and many biological molecules. The time required to obtain a Raman spectrum was reduced when a CW laser was employed, but not to the point where short-lived phenomena could be probed. While laser sources soon gave linewidths of a few megacycles, the resolution of a practical Raman spectrometer was limited to 0.3 cm^{-1} or so by the need to collect as many photons as possible.

Soon after the invention of the laser, however, WOODBURY and NG discovered a radically different aspect of the Raman effect [7.8,9]. At the high intensities characteristic of focused pulsed lasers, they found stimulated emission occurred at the Raman shifted frequencies. This effect, now called stimulated Raman oscillation, produced intense laser-like beams from the Raman active sample. Stimulated Raman oscillation occurs only when the incident intensity exceeds a threshold value, but once that value was exceeded, fully half of the incident power could be Raman shifted.

It soon developed, however, that stimulated Raman oscillations was nearly inextricably linked with other phenomena. The threshold intensity, for example, was often reached as the result of catastrophic self-focusing of the incident laser beam to be a diameter of a few microns due to the intensity dependence of the index of refraction of the sample. Self-focusing also led to self-phase modulation which broadened the spectrum of the transmitted and Raman shifted radiation [7.10]. At the highest intensities, dielectric breakdown occurred, destroying the sample [7.11]. While the simplest theory predicted that oscillation should occur only at Stokes shifted frequencies, with anti-Stokes shifted light being absorbed, multiple bright cones of anti-Stokes radiation were observed [7.12].

It is not our purpose here to review the development of the theory of the stimulated Raman effect. That task has been performed admirably by BLOEMBERGEN [7.13] in his 1967 review of the subject and also more recently by SHEN [7.14]. Rather we intend to mention only the most essential milestones in the development of the three viable coherent Raman spectroscopic techniques: Coherent Anti-Stokes Raman Spectroscopy, Stimulated Raman Gain and Loss Spectroscopy, and Raman Induced Kerr Effect Spectroscopy. Let it then suffice to say that even when all of the complexities of stimulated Raman oscillation were understood, that effect still proved to have no spectroscopic utility.

Stimulated Raman oscillation did, however, prove useful as a means of producing intense frequency shifted laser beams. Using such Raman laser sources, various investigators confirmed each aspect of the theory of the stimulated Raman effect and developed the understanding necessary to contrive useful spectroscopic tools. The amplification of a Stokes-shifted probe wave was soon observed without oscillation, and the dependence upon polarization, pump intensity and sample length was verified [7.5]. LALLEMAND et al. even observed the reduction of the Doppler linewidth and Dicke narrowing using stimulated Raman amplification of co-propagating beams in H_2 [7.15,16]. Attenuation of incident radiation at the anti-Stokes frequency was also soon demonstrated and termed the inverse Raman effect [7.17]. The nonlinear mixing process actually responsible for the generation of the anti-Stokes cones was analyzed by BLOEMBERGEN and SHEN [7.18] and demonstrated directly by YAJIMA [7.19] and by MAKER and TERHUNE [7.27].

Other nonlinear optical effects had by this time been observed in media where the more familiar second-order nonlinear susceptibility could not exist. Among these effects were third harmonic generation (1962) [7.20], electric field induced second harmonic generation (1963) [7.21], the optical Kerr effect (1964) [7.22], nonlinear ellipse rotation (1964) [7.23], and two-photon absorption (1961) [7.24]. All these phenomena can be described in a general framework utilizing the third-order susceptibility tensor, first introduced by ARMSTRONG et al. (1962) [7.25]. This concept was soon thereafter generalized to complex tensor elements by BLOEMBERGEN (1963) [7.26].

MAKER and TERHUNE (1965) [7.27] used this formalism to give a systematic description of all these phenomena. They also made the first comprehensive and systematic experimental investigation of the degenerate four-wave mixing process in which input light beams at frequencies ω_1 and ω_2 generate an output at $2\omega_1 - \omega_2$. The importance of the Raman contributions to the optical Kerr effect predicted in their paper went unrecognized for many years. Nevertheless, the physical phenomena now being exploited in coherent Raman spectroscopy had been observed and correctly understood by 1965.

The accepted picture of these coherent Raman interactions was one in which waves at the Stokes and laser frequencies produced a force that acted upon the generalized coordinate that described the Raman active mode of the sample [7.14]. At each point in the sample the phase of the driven vibration depended upon the phase of the two input waves. GIORDAMAINE and KAISER showed that the driven mode could efficiently scatter a probe wave if the wave vectors of the probe and vibration added to give the wave vector of the scattered wave [7.28]. In stimulated Raman gain and loss, the phase matching condition is automatically fulfilled, and the "scattered wave" is responsible for the amplification and attenuation. In wave mixing spectroscopy where the output frequency is $\omega_s = \omega_0 + \omega_1 - \omega_2$, the wave vector matching conditions $\underline{k}_s = \underline{k}_0 + \underline{k}_1 - \underline{k}_2$ leads to geometrical constraints. Further experiments demonstrated, however, that nonresonant processes in wave mixing interfered with the Raman resonances altering the line shape and reducing the sensitivity [7.29,30].

7.1.2 The Tunable Laser Era

The availability of reliable tunable lasers triggered rapid development of coherent Raman spectroscopy. STANSFIELD et al. were the first to use these sources in an experiment in which the beams of a ruby laser at frequency ω_0 and dye lasers at frequency ω_1 and ω_2 were crossed in a plasma [7.31]. They found an increase in the four-wave mixing signal at $\omega_0 + \omega_1 - \omega_2$ when $\omega_1 - \omega_2$ and $\underline{k}_1 - \underline{k}_2$ were equal to the plasma frequency and wave vector. Very soon thereafter came the first demonstration of phonon resonance in degenerate four-wave mixing in centrosymmetric and accentric crystals [7.32,33]. The first high resolution wave mixing spectrum of a gas was obtained by DeMARTINI et al. who also observed Dicke narrowing of the $Q(1) v = 0 \to 1$ line of H_2 [7.34]. REGNIER and TARAN measured hydrogen concentration in a methane flame using spatial filtering to suppress the strong background luminescence [7.35]. Thus the advantages of degenerate four-wave mixing and many potential applications were well known by the time that BEGLEY et al. renamed the technique "Coherent Anti-Stokes Raman Spectroscopy" and coined the acronym CARS [7.36]. A deluge of papers followed.

Meanwhile the development of CW dye lasers with relatively stable outputs increased the sensitivity of stimulated Raman spectroscopy (SRS) techniques. OWYOUNG accurately measured the Raman contribution to the nonlinear susceptibility of benzene in 1976 and proceeded to push the sensitivity of his technique to the quantum limit [7.37]. Working intracavity, WERNCKE et al. demonstrated the excellent sensitivity of the inverse Raman effect [7.38].

The Raman contributions to the optical Kerr effect continued to be ignored until HELLWARTH pointed out the possibility of a truly background-free Raman induced Kerr effect spectroscopy (RIKES). HEIMAN et al. proceeded to demonstrate that a circularly polarized pump wave could mix with and depolarize a linearly polarized probe only if the difference of the two frequencies equalled the frequency of a Raman mode [7.39]. Later, optical heterodyne detection was added to RIKES to yield a technique with greater sensitivity than conventional spontaneous Raman scattering [7.40].

While spontaneous Raman scattering will continue to play a role in spectroscopy, the coherent Raman techniques have demonstrated significant advantages in many areas. Because the resolution of these techniques depends only on the laser linewidth and because the Doppler effect is largely cancelled for copropagating beams, Raman spectra of gases can now be obtained with sub-megacycle precision. Since diffraction limited beams are employed, the dispersion curves of excitations in solids can be plotted more accurately. Spatial filtering and heterodyne detection completely eliminate the background signals from luminescence and elastic scattering which have previously obscured the Raman spectra of many materials. Using multiplex techniques, entire Raman spectra can be taken in a few nanoseconds [7.41]. The strengths of lines at different frequency shifts can be readily compared by double resonant four-wave mixing [7.42]. Certainly other advantages and applications will appear as the scientific community becomes more familiar with the coherent Raman techniques.

The purpose of this chapter is to review the theory and practice of coherent Raman spectroscopy. The next section develops the necessary theory while the third section catalogs the demonstrated experimental techniques along with their various merits and weaknesses. The fourth section describes the present day applications of coherent Raman spectroscopy along with key results. We conclude with predictions of the role that we expect coherent Raman spectroscopy to play in the future of spectroscopy and nonlinear optics.

7.2 Theory

The optical mixing processes employed in coherent Raman spectroscopy all have the structure diagrammed in Fig.7.1 [7.43,44]. Laser fields with two or three different frequency components excite a sample, producing a coherently oscillating polarization at a frequency corresponding to a threefold sum of the incident frequencies. That polarization then radiates a signal amplitude E_s at the output frequency, which is then separated from some of the incident waves and detected photoelectrically or photographically. (In photoacoustic Raman spectroscopy, the energy deposited in the sample is detected directly [7.45].) When the difference of two input frequencies equals the frequency of a Raman transition, a resonance occurs in the detected signal.

The quantum mechanics of the coherent Raman interactions is generally described in terms of the third-order nonlinear susceptibility tensor $\chi^{(3)}_{\alpha\beta\gamma\delta}(-\omega_s,\omega_0,\omega_1,-\omega_2)$

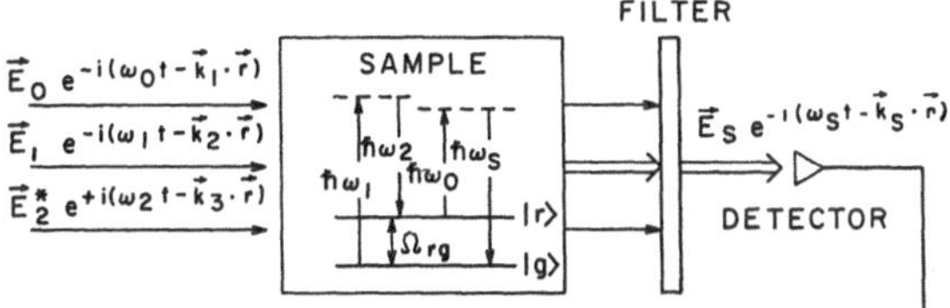

Fig. 7.1. Schematic of a general coherent Raman process. The input waves mix together in the sample to produce a signal amplitude that is separated from the inputs and detected. Two of the input frequencies are often equal: in CARS $\omega_0=\omega_1$ while in RIKES and SRS $\omega_0=-\omega_1$

[7.26,27]. The incident beams interact by way of this tensor to produce the oscillating dielectric polarization vector

$$P_{\alpha}^{NL}(\omega_s)e^{-i\omega_s t} = 6\chi^{(3)}_{\alpha\beta\gamma\delta}(-\omega_s,\omega_0,\omega_1,-\omega_2)E_\beta(\omega_0)E_\gamma(\omega_1)E^*_\delta(\omega_2)e^{-i\omega_s t} \tag{7.1}$$

where $\omega_s = \omega_0 + \omega_1 - \omega_2$ and the subscripts refer to the components of the vectors and tensors. The factor of six results from the indistinguishability of the various permutations of three distinct incident field components. That polarization then radiates the signal amplitude according to Maxwell's wave equation

$$\nabla \times \nabla \times \underline{E}_s(r,t) - \frac{\underline{\underline{\varepsilon}}}{c^2} \cdot \ddot{\underline{E}}_s(r,t) = -\frac{4\pi\omega_s^2}{c^2} \underline{P}(\omega_s)e^{-i(\omega_s t - \underline{k}_p \cdot \underline{r})} \quad , \tag{7.2}$$

where $\underline{\underline{\varepsilon}}$ is the dielectric constant tensor and $\underline{k}_p = \underline{k}_0 + \underline{k}_1 - \underline{k}_2$ is the wave vector of the nonlinear polarization. The task of coherent Raman spectroscopy is to relate the observed variations in the signal amplitude to the underlying quantum mechanics.

The third-order nonlinear susceptibility often seems a mysterious and complex quantity with contributions from a great many processes. In general, four or more quantized levels are required for a complete calculation of $\chi^{(3)}$. The form for $\chi^{(3)}$ is derived in Sect.7.2.10, but more physical insight can be gained from the simple model of Sect.7.2.1 which is appropriate when all of the incident frequencies — as well as the signal frequency — are far from allowed transitions.

It has become customary in recent years to describe Raman scattering in quantum electrodynamic terms using creation and annihilation operators. It is more convenient — and completely rigorous — to describe the coherent Raman effects using a semiclassical approach in which the optical field amplitudes are complex numbers and only the matter is quantized. Such a treatment was first employed by PLACZEK to explain spontaneous Raman scattering [7.2]. We shall employ a somewhat modernized notation in the Schrödinger representation. The identification

$$\underline{E}_j = \frac{i}{n_j}\left(\frac{\hbar\omega_j}{V}\right)^{\frac{1}{2}} \tilde{a}_j\hat{e}_j \qquad \underline{E}^*_j = \frac{-i}{n_j}\left(\frac{\hbar\omega_j}{V}\right)^{\frac{1}{2}} \tilde{a}^+_j\hat{e}_j \tag{7.3}$$

allows the results derived here to be cast in terms of quantized fields [7.46].

7.2.1 Extended Two-Level Model for Coherent Raman Spectroscopy

Consider a quantum mechanical system diagrammed in Fig.7.2. States |g> and |r> have the same parity, and thus the matrix element of the local field corrected dipole moment operator

$$\tilde{\mu} = [\varepsilon(\omega)+2]e\underline{r}/3 \tag{7.4}$$

$\hbar = h/2\pi$ (normalized Planck's constant).

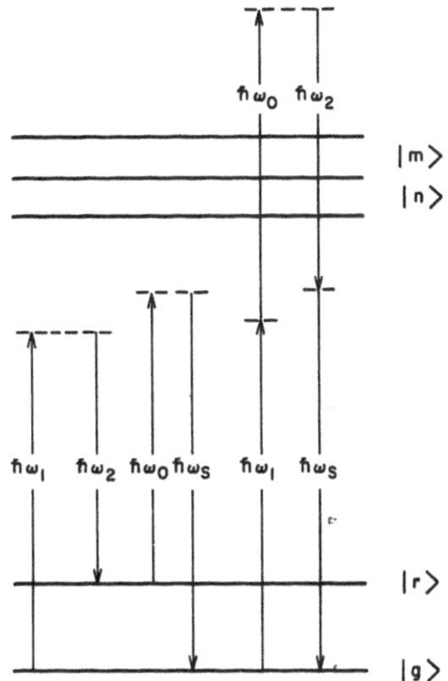

Fig. 7.2. Energy level diagram for third-order frequency mixing. States |g>, |r>, and |t> have the same parity while |m> and |n> have opposite parity. The diagram shows the relative magnitudes of the level splittings and the photon energies, and should not be interpreted as describing a time-ordered process

between them vanishes: $\langle g|\tilde{\mu}_\alpha|r\rangle = 0$ [7.27]. States of the opposite parity are labeled |n> and |m>, and lie well above states |g> and |r>. An electromagnetic wave with several Fourier components

$$\underline{E}(\underline{r},t) = \sum_{j=0}^{2} \frac{1}{2}\left(\underline{E}_j\, e^{-i(\omega_j t-\underline{k}_j\cdot\underline{r})} + \underline{E}_j^*\, e^{i(\omega_j t-\underline{k}_j\cdot\underline{r})}\right) \tag{7.5}$$

is incident upon the system. The frequencies ω_j are far from resonance, and fulfill the condition $|\Omega_{rn}-\omega_j| \gg \Omega_{rg}$, where $\Omega_{uv} = (E_u-E_v)/\hbar$. The interaction of such waves with this system is correctly described by the following "adiabatic" model [7.47,48].

The effect of the incident waves is to "transform" the states |r> and |g> by mixing in some amplitude for each of the states |n>. The transformed states are

$$|r'\rangle = U_{rn}^{-1}|n\rangle \quad \text{and} \quad |g'\rangle = U_{gn}^{-1}|n\rangle \tag{7.6}$$

where the sum is over all states and the transformation operator is

$$\tilde{U} = e^{-i\tilde{S}} = 1 - i\tilde{S} - \frac{1}{2}\tilde{S}\tilde{S} + \frac{i}{6}\tilde{S}\tilde{S}\tilde{S}\ \ldots \tag{7.7}$$

with the matrix elements of $\tilde{S}$ given by

$$S_{nu} = -\frac{i}{2\hbar}\sum_{j=0}^{2}\left(\frac{\underline{\mu}_{nu}\cdot\underline{E}_j\, e^{-i(\omega_j t-\underline{k}_j\cdot\underline{r})}}{\Omega_{nu}-\omega_j} + \frac{\underline{\mu}_{nu}\cdot\underline{E}_j^*\, e^{i(\omega_j t-\underline{k}_j\cdot\underline{r})}}{\Omega_{nu}+\omega_j}\right) . \tag{7.8}$$

The Hamiltonian for the transformed states $|r'\rangle$ and $|g'\rangle$ becomes

$$\tilde{H}' = \tilde{U}(\tilde{H}_0 - \underline{\mu}\cdot\underline{E})\tilde{U}^{-1} + \hbar\frac{\partial\tilde{S}}{\partial t} - \frac{\hbar}{2i}\left[\frac{\partial\tilde{S}}{\partial t},\tilde{S}\right] \tag{7.9}$$

where the second and third terms result from time dependence of $\tilde{S}$, and higher order terms have been neglected. To derive the Raman resonance in the third-order nonlinear susceptibility, it is sufficient to solve the equation of motion for the density matrix for the two-level system $|r'\rangle|g'\rangle$. Defining the transformed density matrix by

$$\tilde{\rho}' = \tilde{U}\tilde{\rho}\tilde{U}^{-1} \tag{7.10}$$

the Liouville equation becomes

$$\frac{\partial\tilde{\rho}'}{\partial t} = \frac{i}{\hbar}[\tilde{\rho}',\tilde{H}'] \quad . \tag{7.11}$$

The matrix elements relevant to the $|g'\rangle|r'\rangle$ level system are

$$H'_{g'g'} = \hbar\Omega_g + \Delta E_g \tag{7.12a}$$

$$H'_{r'r'} = \hbar\Omega_r + \Delta E_r \tag{7.12b}$$

$$H'_{r'g'} = H'^{*}_{g'r'} \tag{7.12c}$$

(to second order in S) where the off-diagonal matrix element is

$$H'_{r'g'} = \sum_{j=0}^{2}\sum_{k\neq j} H'(\omega_j-\omega_k)\, e^{-i[(\omega_j-\omega_k)t-(\underline{k}_j-\underline{k}_k)\cdot\underline{r}]} \tag{7.13}$$

(plus terms oscillating at $2\omega_0, \omega_0+\omega_1$, etc.) where

$$H'(\omega_j-\omega_k) = (\Omega_{ng}+\Omega_{nr})/8\hbar\left(\frac{(\underline{\mu}_{gn}\cdot\underline{E}_k^*)(\underline{\mu}_{nr}\cdot\underline{E}_j)}{(\Omega_{ng}-\omega_j)(\Omega_{nr}+\omega_j)} + \frac{(\underline{\mu}_{gn}\cdot\underline{E}_j)(\underline{\mu}_{nr}\cdot\underline{E}_k^*)}{(\Omega_{ng}+\omega_k)(\Omega_{nr}-\omega_k)}\right) \quad . \tag{7.14}$$

This Hamiltonian can be expressed in terms of the Raman susceptibility tensor elements employed by MAKER and TERHUNE [4.27] and others [7.42,49]

$$H'(\omega_j-\omega_k) = \frac{1}{8}\left[\alpha^R_{\alpha\beta}(-\omega_j,+\omega_k)+\alpha^R_{\alpha\beta}(\omega_k,-\omega_j)\right] \times E_{j\alpha}E^*_{k\beta} \tag{7.15}$$

where

$$\alpha^R_{\alpha\beta}(-\omega_j,+\omega_k) = \frac{1}{\hbar}\left(\frac{\langle g|\tilde{\mu}_\alpha|n\rangle\langle n|\tilde{\mu}_\beta|r\rangle}{\Omega_{ng}-\omega_j} + \frac{\langle g|\tilde{\mu}_\beta|n\rangle\langle n|\tilde{\mu}_\alpha|r\rangle}{\Omega_{ng}+\omega_k}\right) \tag{7.16}$$

and the relationship $|\Omega_{rg}-(\omega_j-\omega_k)| \ll \Omega_{ng}-\omega_j$ has been used to simplify the notation.

The optical Stark shifts are

$$\Delta E_g = -\frac{\Omega_{ng}}{2\hbar}\left(\frac{|\underline{\mu}_{ng}\cdot\underline{E}_0|^2}{\Omega_{ng}^2-\omega_0^2}+\frac{|\underline{\mu}_{ng}\cdot\underline{E}_1|^2}{\Omega_{ng}^2-\omega_1^2}+\frac{|\underline{\mu}_{ng}\cdot\underline{E}_2|^2}{\Omega_{ng}^2-\omega_2^2}\right) \tag{7.17a}$$

$$\Delta E_r = -\frac{\Omega_{nr}}{2\hbar}\left(\frac{|\underline{\mu}_{nr}\cdot\underline{E}_0|^2}{\Omega_{nr}^2-\omega_0^2}+\frac{|\underline{\mu}_{nr}\cdot\underline{E}_1|^2}{\Omega_{nr}^2-\omega_1^2}+\frac{|\underline{\mu}_{nr}\cdot\underline{E}_2|^2}{\Omega_{nr}^2-\omega_2^2}\right) \tag{7.17b}$$

$$\delta_{g'r'} = (\Delta E_g-\Delta E_r)/\hbar \tag{7.17c}$$

and summation over the repeated indices α, β and n is assumed. The equations of motion for the individual elements of the density matrix become

$$\dot{\rho}_{g'g'} = \frac{i}{\hbar}(\rho_{g'r'}H'_{r'g'}-H'_{g'r'}\rho_{r'g'}) - \Gamma_{g'g'}(\rho_{g'g'}-\rho^e_{gg}) \tag{7.18a}$$

$$\dot{\rho}_{r'r'} = \frac{i}{\hbar}(\rho_{r'g'}H'_{g'r'}-H'_{r'g'}\rho_{g'r'}) - \Gamma_{r'r'}(\rho_{r'r'}-\rho^e_{rr}) \tag{7.18b}$$

$$\dot{\rho}_{r'g'} = -i(\Omega_{rg}+\delta_{r'g'})\rho_{r'g'} + \frac{i}{\hbar}(\rho_{r'r'}-\rho_{g'g'})H'_{r'g'} - \Gamma_{r'g'}\rho_{r'g'} \tag{7.18c}$$

where $\rho_{rg}=\rho^*_{gr}$ and longitudinal and transverse relaxation has been handled in a comparatively general way; ρ^e_{gg} and ρ^e_{rr} are the populations of levels |g> and |r> at thermal equilibrium. These equations are isomorphic to the familiar equations describing allowed transitions in two level atoms or transitions in spin 1/2 systems [7.50,51]. When only two frequencies (i.e., ω_1 and ω_2) are incident, the usual practice would be to make a rotating wave approximation with rotation circular frequency $\omega_1-\omega_2$ and solve explicitly.

In coherent Raman spectroscopy the coupling Hamiltonian is generally weak; it is therefore justifiable to assume $\rho_{gg}-\rho_{rr}=\rho^e_{gg}-\rho^e_{rr}=\rho_d=$ constant. The Raman coordinate can be defined in terms of the off-diagonal elements of the density matrix as

$$Q = \rho_{r'g'} + \rho_{g'r'} \tag{7.19}$$

which obeys the equation of motion

$$\ddot{Q} + 2\Gamma\dot{Q} + \Omega_Q^2 Q = \frac{\rho_d\Omega_{rg}}{\hbar}\left[\left(1+\frac{i}{\Omega_{rg}}\frac{\partial}{\partial t}\right)H'_{rg}+\left(1-\frac{i}{\Omega_{rg}}\frac{\partial}{\partial t}\right)H'_{gr}\right]$$

$$\xrightarrow[\Omega_{rg}\ll\Omega_{ng}-\omega_j]{} \frac{\rho_d\Omega_{rg}}{2\hbar}\,\alpha^R_{\gamma\delta}\sum_{j,k\neq j} E_{j\gamma}E^*_{k\delta}\,e^{-i[(\omega_j-\omega_k)t-(\underline{k}_j-\underline{k}_k)\cdot\underline{r}]} \tag{7.20}$$

where $\Gamma=\Gamma_{g'r'}$ and

$$\Omega_Q^2 = (\Omega_{r'g'}-\delta_{r'g'})^2 - \Gamma^2_{r'g'} \quad ,$$

and the limiting case indicated by the arrow occurs in the Born-Oppenheimer approximation. Equation (7.20) describes a damped harmonic oscillator driven by a force bilinear in the incident fields. Such an equation has been used by previous authors as a starting point for an intuitive development of coherent Raman effects [7.33]. Such treatments are correct within the approximation employed to derive (7.20). In most experiments to date, H' varies so slowly that the density matrix is always essentially in the steady state. Expanding the steady-state solution of (7.20) in Fourier components

$$Q(t) = \sum_{\Delta\omega} Q(\Delta\omega)\, e^{-i(\Delta\omega t - \Delta\underline{K}\cdot\underline{r})} \tag{7.21}$$

where $\Delta\omega = \omega_0 - \omega_1$, $\omega_0 - \omega_2$ and $\omega_1 - \omega_2$ (and similarly for ΔK), and

$$Q(\Delta\omega) = \frac{\rho_d/\hbar\left\{\Omega_{rg}[H'(\Delta\omega)+H'(-\Delta\omega)]+\Delta\omega[H'(\Delta\omega)-H'(-\Delta\omega)]\right\}}{\Omega_Q^2-\Delta\omega^2-2i\Gamma\Delta\omega}\ . \tag{7.22}$$

Terms oscillating at $\omega_0 + \omega_1$, $2\omega_0$, etc., have been ignored.

7.2.2 The Nonlinear Polarization

The dielectric polarization that radiates the coherent waves detected in coherent Raman spectroscopy can be obtained using

$$\underline{P}(t) = N(\underline{\mu}_{g'r'}\rho_{r'g'}+\underline{\mu}_{r'g'}\rho_{g'r'}) \tag{7.23}$$

where N is the number density of quantum systems, (7.14), and

$$\langle g'|\tilde{\mu}_\alpha|r'\rangle = \langle g|U\tilde{\mu}_\alpha U^{-1}|r\rangle = \frac{1}{2}\sum_{j=0}^{2}\left(\alpha^R_{\alpha\beta}(-\omega_j-\Omega_{rg},\omega_j)E_{j\beta}\, e^{-i(\omega_j t-\underline{k}_j\cdot\underline{r})} + \alpha^R_{\alpha\beta}(\omega_j-\Omega_{rg},-\omega_j)E^*_{j\beta}\, e^{i(\omega_j t-\underline{k}_j\cdot\underline{r})}\right) \tag{7.24}$$

where $\langle r'|\mu_\alpha|g'\rangle = \langle g'|\mu_\alpha|r'\rangle^*$.

The formulae obtained in this fashion are equivalent to those that have been extensively quoted in the literature and will appear in a slightly different guise in Sect.7.2.10 [7.27,42,44,47,49,52-55]. If, however, $\Omega_{rg} \ll \omega_j \ll \Omega_{ng}$ for all of the optical frequencies and all intermediate states, the differences between the various energy denominators in (7.14) and (7.24) can be ignored. This approximation is equivalent to the KLEINMAN symmetry assumption [7.56] and the Born-Oppenheimer approximation [7.57]. The essence of the phenomena can then be described more simply, with an expected inaccuracy of under 10%. With this approximation, (7.22,24) become

$$Q(\omega_j-\omega_k) = \frac{\rho_d\Omega_{rg}\alpha^R_{\gamma\delta}E_{j\gamma}E^*_{k\delta}}{2\hbar\left[\Omega_Q^2-(\omega_j-\omega_k)^2-2i\Gamma(\omega_j-\omega_k)\right]} \tag{7.25}$$

$$<g'|\tilde{\mu}_\alpha(t)|r'> = \sum_{j=0}^{2} \frac{1}{2}\left[\alpha^R_{\alpha\beta}E_{j\beta}\, e^{-i(\omega_j t-\underline{k}_j\cdot\underline{r})}+cc\right] \quad (7.26)$$

where

$$\alpha^R_{\gamma\delta} = \alpha^R_{\gamma\delta}(0,0) = (\hbar\Omega_{ng})^{-1}\left\{<g|\tilde{\mu}_\gamma|n><n|\tilde{\mu}_\delta|r>+<g|\tilde{\mu}_\delta|n><n|\tilde{\mu}_\gamma|r>\right\} \quad (7.27)$$

is the Raman tensor quoted in Sect.7.2.6 and in (7.15) and $E_{j\gamma} = E_\gamma(\omega_j)$, etc. Summation over the repeated indices n, β, γ and δ is assumed and the Raman resonant nonlinear polarization is

$$P^Q_\alpha(t) = N<g'|\tilde{\mu}_\alpha(t)|r'>Q(t) = \sum_{\omega_s} \frac{1}{2} P^Q_\alpha(\omega_s)\, e^{-i(\omega_s t-\underline{k}_p\cdot\underline{r})} + cc \quad . \quad (7.28)$$

Each Fourier component of this polarization radiates a different coherent Raman signal, but all are related to one another.

Stimulated Raman Gain and Loss Spectroscopy, and the Raman Induced Kerr Effect

These techniques exploit signals at the same frequency as one of the two input waves. Setting $\underline{E}_0 = 0$ in (7.5,24,28), the relevant Fourier components are

$$P^Q_\alpha(\omega_1) = \frac{N\rho_d\Omega_{rg}\alpha^R_{\alpha\beta}\alpha^R_{\gamma\delta}}{2\hbar\left[\Omega_Q^2-(\omega_1-\omega_2)^2-2i\Gamma(\omega_1-\omega_2)\right]} E_{2\beta}E_{1\gamma}E^*_{2\delta} \quad (7.29)$$

$$P^Q_\alpha(\omega_2) = \frac{N\rho_d\Omega_{rg}\alpha^R_{\alpha\beta}\alpha^R_{\gamma\delta}}{2\hbar\left[\Omega_Q^2-(\omega_1-\omega_2)^2-2i\Gamma(\omega_2-\omega_1)\right]} E_{1\beta}E^*_{1\gamma}E_{2\delta} \quad (7.30)$$

which obviously transform into one another when ω_1 and ω_2 are exchanged. Since the wave vector of the polarization $\underline{k}_p$ is the same as that of the input beam at the same frequency (and polarization in an anisotropic medium), the coherent Raman beams are radiated collinearly with the input and coherently with it. Unlike other techniques, those based upon these polarizations automatically fulfill the wave vector matching conditions for any beam geometry. The phases of the polarizations insure that on resonance the higher frequency wave is attenuated by the nonlinear interaction while the lower frequency wave is amplified in a stimulated Raman experiment [7.9,39].

If an additional wave of frequency ω_0 is injected, an additional term appears in the polarization at ω_1 or ω_2 in which E_0 and ω_0 replace E_2 and ω_2 or E_1 and ω_1, respectively. Such techniques are useful in normalizing the Raman cross sections of modes with widely different frequencies [7.42,58].

Coherent Anti-Stokes and Coherent Stokes Raman Spectroscopy

In these most popular frequency mixing techniques the Raman induced nonlinear polarizations have the form

$$P_{\alpha}^{Q}(2\omega_1-\omega_2) = \frac{N\rho_d\Omega_{rg}\left(\alpha_{\alpha\beta}^{R}\alpha_{\gamma\delta}^{R}+\alpha_{\alpha\gamma}^{R}\alpha_{\beta\delta}^{R}\right)}{4\hbar\left[\Omega_Q^2-(\omega_1-\omega_2)^2-2i\Gamma(\omega_1-\omega_2)\right]} E_{1\beta}E_{1\gamma}E_{2\delta}^{*} \tag{7.31}$$

and the polarization at $2\omega_2-\omega_1$ is obtained by interchanging subscripts 1 and 2. Since the amplitude of the Fourier component at ω_1 appears twice in the CARS polarization, it has become conventional to explicitly symmetrize the formula for that polarization with respect to the indices β and γ [7.27,49].

The wave vector of the CARS polarization is $\underline{k}_p = 2\underline{k}_1 - \underline{k}_2$ and its magnitude does not necessarily equal that of a freely propagating wave at the output frequency. In a medium of normal dispersion, the angles between the input waves can readily be chosen such that

$$\Delta k = |\underline{k}_p| - \frac{n(\omega_s)\omega_s}{c} < \pi/\ell \tag{7.32}$$

where ℓ is the sample legnth. Phase matching then occurs in the direction of $\underline{k}_p$; a maximum CARS output results [7.55].

Four-Wave Mixing

These techniques to date have been used only for very special applications. The relevant Raman resonant nonlinear polarization is

$$P_{\alpha}^{Q}(\omega_0+\omega_1-\omega_2) = \frac{N\rho_d\Omega_{rg}}{4\hbar}\left(\frac{\alpha_{\alpha\beta}^{R}\alpha_{\gamma\delta}^{R}}{\Omega_Q^2-(\omega_1-\omega_2)^2-2i\Gamma(\omega_1-\omega_2)} + \frac{\alpha_{\alpha\gamma}^{R}\alpha_{\beta\delta}^{R}}{\Omega_Q^2-(\omega_0-\omega_2)^2-2i\Gamma(\omega_0-\omega_2)}\right) E_{0\beta}E_{1\gamma}E_{2\delta}^{*} \quad . \tag{7.33}$$

Again, the wave vector matching condition $\underline{k}_s = \underline{k}_0 + \underline{k}_1 - \underline{k}_2$ leads to geometrical constraints. If $\omega_0 = \omega_1$ but $\underline{k}_0 \neq \underline{k}_1$, the technique is termed "boxcars" [7.59]. If $\omega_0 \approx \omega_2$ and $\underline{k}_0 \approx \underline{k}_2$, the technique is analogous to RIKES and SRS in that the wave vector matching condition requires that the output beam be collinear with the input wave at ω_1 [7.58,60]. Phase matching is then less critical if the ω_0 and ω_2 beams propagate in a direction nearly opposite to the output and ω_1 beams.

Photoacoustic Raman Spectroscopy

Acoustooptic detection of coherent Raman resonances has recently been demonstrated by BARRET [7.45]. The detected quantity is the time integral of the rate of vibrational energy deposition in the medium. For a two-frequency experiment that rate can be expressed as

$$R = \frac{c}{4\pi} \iint_0^{\ell} \mathrm{Im}\left\{n_1\omega_1 \underline{E}(\omega_1)\cdot \underline{P}^Q(\omega_1) + n_2\omega_2 \underline{E}(\omega_2)\cdot \underline{P}^Q(\omega_2)\right\} dZ\, dA$$

$$\propto (n_1\omega_1 - n_2\omega_2)\ \mathrm{Im}\left\{\chi^{(3)}(\omega_2,-\omega_1,\omega_1,-\omega_2)\right\} |\underline{E}_1|^2 |\underline{E}_2|^2 \qquad (7.34)$$

where the integral is performed over the sample volume.

7.2.3 The Nonlinear Susceptibility Tensor

Nonlinear polarization densities such as those calculated in the previous section are most succinctly expressed in terms of the third-order nonlinear susceptibility tensor defined by

$$P_\alpha^{NL}(\omega_s) = D\chi^{(3)}_{\alpha\beta\gamma\delta}(-\omega_s,\omega_0,\omega_1,-\omega_2) E_{0\beta} E_{1\gamma} E^*_{2\delta} \quad . \qquad (7.35)$$

The factor D was introduced by MAKER and TERHUNE [7.27] to properly account for processes having different frequency degeneracies; it is equal to the number of distinguishable permutations of the incident frequency and polarization components. Explicit values for $D\chi^{(3)}_{eff}$ and for $P_\alpha^{NL}(\omega_s)$ appear in Tables 7.1, 2 and 3.

Only three of the four frequency arguments of $\chi^{(3)}$ are independent as $\omega_0 + \omega_1 - \omega_2 - \omega_s = 0$ must be fulfilled. In the Maker and Terhune notation, each frequency argument has an associated polarization subscript. The arguments and sub-

Table 7.1. The effective nonlinear susceptibilities and Raman tensors for five common coherent Stokes and anti-Stokes polarization conditions. The signal polarization is always vertical, and the numerical subscripts refer to the axes of an arbitrary coordinate system. In nonlinear ellipsometry polarizations 4 and 5 are used simultaneously

CARS/CSRS			$P^{NL}(2\omega_1-\omega_2) = D\chi^{(3)}_{eff} E_1^2 E_2^*$ [a)]	
	E_1	E_2	$D\chi^{(3)}_{eff}(-2\omega_1+\omega_2,\omega_1,\omega_1,-\omega_2)$	$\left(\alpha^R_{eff}\right)^2$
1	↑	↑	$3\chi^{(3)}_{1111}$	$\left(\alpha^R_{11}\right)^2$
2	→	↑	$3\chi^{(3)}_{1221}$	$\left(\alpha^R_{12}\right)^2$
3	↑ a	↑ a	$\frac{3}{2}\left(\chi^{(3)}_{1111}+2\chi^{(3)}_{1122}+\chi^{(3)}_{1221}\right)$	$\frac{1}{2}\left[\left(\alpha^R_{11}\right)^2+\alpha^R_{11}\alpha^R_{22}+2\left(\alpha^R_{12}\right)^2\right]$
4	↗	→	$3\chi^{(3)}_{1122}$	$\frac{1}{2}\left[\left(\alpha^R_{12}\right)^2+\alpha^R_{11}\alpha^R_{22}\right]$
5	↗	↑	$\frac{3}{2}\left(\chi^{(3)}_{1111}+\chi^{(3)}_{1221}\right)$	$\frac{1}{2}\left[\left(\alpha^R_{11}\right)^2+\left(\alpha^R_{12}\right)^2\right]$

[a] All polarizations at 45° to 1 axis in 1-2 plane

Table 7.2. The effective nonlinear susceptibilities and Raman tensors for the stimulated Raman and Raman induced Kerr effect polarization conditions. The signal polarization is always vertical and the numerical subscripts refer to the axes of an arbitrary Cartesian coordinate system

SRS/RIKES			$P^{NL}(-\omega_2) = D\chi^{(3)}_{eff}\lvert E_1\rvert^2 E_2^*$	
	E_1	E_2	$D\chi^{(3)}_{eff}(\omega_2,-\omega_1,\omega_1,-\omega_2)$	$\left(\alpha^R_{eff}\right)^2$
6	↑	↑	$6\chi^{(3)}_{1111}$	$\left(\alpha^R_{11}\right)^2$
7	→	↑	$6\chi^{(3)}_{1221}$	$\left(\alpha^R_{12}\right)^2$
8	↗	→	$3\left(\chi^{(3)}_{1122}+\chi^{(3)}_{1212}\right)$	$\frac{1}{2}\left[\alpha^R_{11}\alpha^R_{22}+\left(\alpha^R_{12}\right)^2\right]$
9	↻	→	$3i\left(\chi^{(3)}_{1122}-\chi^{(3)}_{1212}\right)$	$\frac{1}{2}\left[\alpha^R_{11}\alpha^R_{22}-\left(\alpha^R_{12}\right)^2\right]$

scripts can be permuted without altering the meaning of $\chi^{(3)}$ as long as the pairing is preserved. It has become conventional, however, to keep the frequency arguments in constant order, with the output frequency on the left and, in coherent Raman spectroscopy, a negative input frequency on the right. The definition in (7.35) differs from the naive time domain definition of nonlinear susceptibility

$$P^{NL}_\alpha(t) = (4)\chi^{(3)}_{\alpha\beta\gamma\delta}E_\beta(r,t)E_\gamma(r,t)E_\delta(r,t) \tag{7.36}$$

by the factor of 4 in parentheses [7.27].

The term in the third-order nonlinear susceptibility that corresponds to the Raman resonant CARS polarization of (7.31) is

$$\chi^Q_{\alpha\beta\gamma\delta}[-(2\omega_1-\omega_2),\omega_1,\omega_1,-\omega_2] = \frac{N\rho_d\Omega_{rg}\left(\alpha^R_{\alpha\beta}\alpha^R_{\gamma\delta}+\alpha^R_{\alpha\gamma}\alpha^R_{\beta\delta}\right)}{12\hbar\left[\Omega^2_Q-(\omega_1-\omega_2)^2-2i\Gamma(\omega_1-\omega_2)\right]} \tag{7.37}$$

whereas the term corresponding to stimulated Raman spectroscopy or RIKES is

$$\chi^Q_{\alpha\beta\gamma\delta}(\omega_2,-\omega_1,\omega_1,-\omega_2) = \frac{N\rho_d\Omega_{rg}\alpha^R_{\alpha\beta}\alpha^R_{\gamma\delta}}{12\hbar\left[\Omega^2_Q-(\omega_1-\omega_2)^2-2i\Gamma(\omega_1-\omega_2)\right]} \quad . \tag{7.38}$$

The Raman contribution to the nonlinear susceptibility is the sum of such terms

$$\chi^R_{\alpha\beta\gamma\delta} = \sum_Q \chi^Q_{\alpha\beta\gamma\delta} \quad . \tag{7.39}$$

There are other contributions to the nonlinear susceptibility tensor, the most interesting being those resonant at the sum of two incident frequencies. High lying

Table 7.3. The effective nonlinear susceptibilities and Raman tensors for the common four-wave mixing polarization conditions. The signal polarization is always vertical, and the numerical subscripts refer to the axes of an arbitrary Cartesian coordinate system. Background suppression can be obtained in conditions 14 ("helicopter"), 15 ("submarine") and 16 ("asterisk")

	4WM			$P^{NL}(\omega_0+\omega_1-\omega_2) = D\chi^{(3)}_{eff}E_0E_1E_2^*$	
	E_0	E_1	E_2	$D\chi^{(3)}_{eff}(-\omega_0-\omega_1+\omega_2,\omega_0,\omega_1-\omega_2)$	$\left(\alpha^R_{eff}\right)^2$
10	↑	↑	↑	$6\chi^{(3)}_{1111}$	$\left(\alpha^R_{11}\right)^2$
11	→	→	↑	$6\chi^{(3)}_{1221}$	$\left(\alpha^R_{12}\right)^2$
12	↑	→	→	$6\chi^{(3)}_{1122}$	$\alpha^R_{22}\alpha^R_{11}$ a), $\left(\alpha^R_{12}\right)^2$ b)
13	→	↑	→	$6\chi^{(3)}_{1212}$	$\left(\alpha^R_{12}\right)^2$ a), $\alpha^R_{11}\alpha^R_{22}$ b)
14	↺	↻	→	$3\left(\chi^{(3)}_{1122}-\chi^{(3)}_{1212}\right)$	$\pm\frac{1}{2}\left[\alpha^R_{11}\alpha^R_{22}-\left(\alpha^R_{12}\right)^2\right]$ c)
15	→	↻	↻	$3\left(\chi^{(3)}_{1212}-\chi^{(3)}_{1221}\right)$	0 a), $\frac{1}{2}\left[\alpha^R_{11}\alpha^R_{22}-\left(\alpha^R_{12}\right)^2\right]$ b)
16	↖ (at angle φ)	↗ (at angle φ)	→	$3\left(\chi^{(3)}_{1212}\sin\theta\cos\phi-\chi^{(3)}_{1122}\cos\theta\sin\phi\right)$	$\left[\left(\alpha^R_{12}\right)^2\sin\theta\cos\phi-\alpha^R_{11}\alpha^R_{22}\cos\theta\sin\phi\right]$ d)

P's always vertical

a) Resonant at $\omega_1-\omega_2=\Omega_Q$.

b) Resonant at $\omega_0-\omega_2=\Omega_Q$.

c) Plus sign for $\omega_1-\omega_2=\Omega_Q$, minus for $\omega_0-\omega_2=\Omega_Q$.

d) For $\omega_1-\omega_2=\Omega_Q$, for $\omega_0-\omega_2=\Omega_Q$ interchange θ and ϕ.

states with the same parity as the ground state |g> are labeled |t> in Fig.7.2. Their contribution to the four-wave mixing and CARS $\chi^{(3)}$ is

$$\chi^T_{\alpha\beta\gamma\delta}(-\omega_s,\omega_0,\omega_1,-\omega_2) = \sum_T \frac{N}{12\hbar}\,\frac{\alpha^T_{\beta\gamma}(\omega_0,\omega_1)\alpha^T_{\alpha\delta}(-\omega_s,-\omega_2)}{\Omega_T-(\omega_0+\omega_1)-i\Gamma_T} \tag{7.40}$$

and to the SRS-RIKES $\chi^{(3)}$ is

$$\chi^{T}_{\alpha\beta\gamma\delta}(\omega_2,-\omega_1,\omega_1,-\omega_2) = \sum_T \frac{N}{12\hbar} \frac{\alpha^{T}_{\alpha\gamma}(\omega_2,\omega_1)\alpha^{T}_{\beta\delta}(-\omega_1,-\omega_2)}{\Omega_T-(\omega_1+\omega_2)-i\Gamma_T} \tag{7.41}$$

where

$$\alpha^{T}_{\alpha\beta}(\omega_j,\omega_k) = \frac{\langle g|\tilde{\mu}_\alpha|n\rangle\langle n|\tilde{\mu}_\beta|t\rangle}{\Omega_{ng}+\omega_j} + \frac{\langle g|\tilde{\mu}_\beta|n\rangle\langle n|\tilde{\mu}_\alpha|t\rangle}{\Omega_{ng}+\omega_k} \tag{7.42}$$

and $\Omega_T = \hbar^{-1}(E_t-E_g)$ [7.27,49,61]. There are also potentially "zero frequency resonances" with Debye type line shapes due to orientational and collisional motions. These contributions are parametrized somewhat differently in isotropic materials [7.39,57,62]

$$\chi^{0}_{\alpha\beta\gamma\delta}(-\omega_s,\omega_0,\omega_1,-\omega_2) = \sum_M \left\{ \frac{2A_M\delta_{\alpha\beta}\delta_{\gamma\delta}+B_M\delta_{\beta\gamma}\delta_{\alpha\delta}+B_M\delta_{\beta\delta}\delta_{\alpha\gamma}}{24[\Gamma_M-i(\omega_1-\omega_2)]} \right.$$
$$\left. + \frac{B_M\delta_{\alpha\beta}\delta_{\gamma\delta}+B_M\delta_{\beta\gamma}\delta_{\alpha\delta}+2A_M\delta_{\beta\delta}\delta_{\alpha\gamma}}{24[\Gamma_M-i(\omega_0-\omega_2)]} + \frac{B_M\delta_{\alpha\beta}\delta_{\gamma\delta}+B_M\delta_{\beta\delta}\delta_{\alpha\gamma}+2A_M\delta_{\beta\gamma}\delta_{\alpha\delta}}{[24\ \Gamma_M-i(\omega_0+\omega_1)]} \right\} \ . \tag{7.43}$$

Anti-resonant terms and other nonresonant processes contribute a real background nonlinear susceptibility χ^B that varies weakly with the incident frequencies. The total nonlinear susceptibility then has the form

$$\chi^{(3)} = \chi^B + \chi^0 + \chi^T + \chi^R = \chi^E + \chi^0 + \chi^R = \chi^{NR} + \chi^R \ . \tag{7.44}$$

7.2.4 Doppler Broadening

The preceding treatment applies directly to condensed phases and to vapors at sufficiently high density that the Raman lines are collisionally broadened. At lower densities the thermal distribution of molecular velocities must be taken into account [7.34,63,64]. The chief effect – in vapors that do not absorb the input or output frequencies – is to alter the line shapes of the Raman and two-photon resonances.

Waves with frequency ω_1 and ω_2 in the laboratory frame appear to have the frequencies $\omega_1 - \underline{k}_1 \cdot \underline{v}$ and $\omega_2 - \underline{k}_2 \cdot \underline{v}$ in the rest frame of a molecule moving with velocity $\underline{v}$. The average nonlinear susceptibility for an ensemble of such molecules with a Maxwell-Boltzmann velocity distribution is

$$\left\langle \chi^{(3)}(-\omega_s,\omega_0,\omega_1,-\omega_2) \right\rangle = \iiint_{-\infty}^{\infty} \frac{\exp\left(-v_z^2/v_0^2\right)}{\left(\pi_1^3 v_0^3\right)^{1/2}} \chi^{(3)}(-\omega_s,\omega_0-\underline{k}_0\cdot\underline{v},\omega_1-\underline{k}_1\cdot\underline{v},-\omega_2+\underline{k}_2\cdot\underline{v})$$
$$\times\, dv_x dv_y dv_z \tag{7.45}$$

where $v^2 = v_x^2 + v_y^2 + v_z^2$, $v_0^2 = 2kT/m$ and $\omega_s = \omega_0 + \omega_1 - \omega_2 - (\underline{k}_0 + \underline{k}_1 - \underline{k}_2) \cdot \underline{v}$. For a gas with a single Raman mode and a nonresonant background, HENESIAN and BYER [7.65] have expressed this integral in terms of the complex plasma dispersion function

$$Z(\xi) = \frac{1}{\sqrt{\pi}} \int_{-\infty}^{\infty} \frac{e^{-\omega'^2}}{\omega' - \xi} d\omega' = 2i\, e^{-\xi^2} \int_{-\infty}^{i\xi} e^{-\omega'^2} d\omega' \tag{7.46}$$

where $\xi = [\omega_1 - \omega_2 - \Omega_Q + i\Gamma]/\omega_d$ and $\omega_d = |\underline{k}_1 - \underline{k}_2| \cdot \underline{v}_0$ is the Doppler width which reaches its minimum value $n(\omega_1 - \omega_2)v_0/c$ for collinear propagation in a medium with index of refraction n [7.65]. Evaluating (7.45) using (7.38) yields

$$\left\langle \chi^{(3)}_{\alpha\beta\gamma\delta} \right\rangle = \frac{N\rho_d \alpha^R_{\alpha\beta} \alpha^R_{\gamma\delta}}{24\hbar\omega_d} Z(\xi) \quad , \tag{7.47}$$

where the approximation $\omega_1 - \omega_2 \approx \Omega_Q$ and $\Gamma \ll \Omega_Q$ have been employed to simplify the integrals. $\mathrm{Im}\{Z(\xi)\}$ and $|Z(\xi)|^2$ are plotted versus $(\omega_1 - \omega_2) - \Omega_Q$ in Fig.7.3 for $\Gamma \approx \omega_d/3$.

These integrals also simplify when $\Gamma \ll \omega_d$ yielding the approximate formulae [7.65]

$$\mathrm{Im}\left\{\left\langle \chi^{(3)}_{\alpha\beta\gamma\delta}(\omega_2, -\omega_1, \omega_1, -\omega_2) \right\rangle\right\} = \frac{\sqrt{\pi}\, N\rho_d}{24} \alpha^R_{\alpha\beta} \alpha^R_{\gamma\delta} \omega_d^{-1} \exp\left(-[\Omega_Q - (\omega_1 - \omega_2)]^2/\omega_d^2\right) \tag{7.48a}$$

$$\mathrm{Re}\left\{\left\langle \chi^{(3)}_{\alpha\beta\gamma\delta} \right\rangle\right\} = \frac{N\rho_d \alpha^R_{\alpha\beta} \alpha^R_{\gamma\delta}}{12\hbar\omega_d} \exp\left(-[\Omega_Q - (\omega_1 - \omega_2)]^2\right) \int_0^{\Omega_Q - (\omega_1 - \omega_2)} \exp(\omega'^2) d\omega' \quad . \tag{7.48b}$$

It is then evident that the width of the resonance (HWHM) in $\mathrm{Im}\{\chi^R\}$ is 0.83 ω_d whereas that in $|\chi^R|^2$ is 1.02 ω_d.

Similar results also apply to the two-photon absorption type resonance, but the Doppler width can then be reduced to zero in four-wave mixing experiments where two

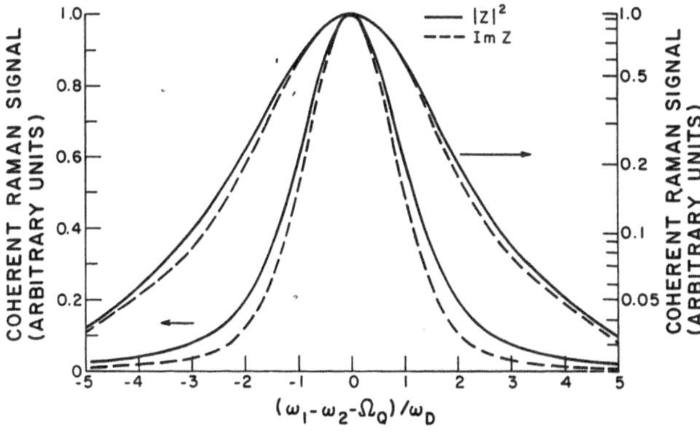

Fig. 7.3. The magnitude squared and imaginary part of the plasma dispersion function $Z(\xi)$ for $\Gamma = 1/3\omega_d$ plotted on logarithmic and linear scales. In vapors, the CRS line shapes can have this form with the blunt peak of a Gaussian and the extended wings of a Lorentzian. The spontaneous scattering line-shape function is Im{Z}

beams of equal frequency propagate in opposite directions through the nonlinear material [7.66]. Such complete cancellation is not possible in the Raman case.

7.2.5 Symmetry Considerations

Chi-three is a fourth-rank tensor with up to 81 nonvanishing elements. The symmetry of the nonlinear medium reduces the number of independent nonvanishing elements. In an isotropic material, there are four nonvanishing elements

$$\chi^{(3)}_{1111} \; , \quad \chi^{(3)}_{1122} \; , \quad \chi^{(3)}_{1212} \quad \text{and} \quad \chi^{(3)}_{1221}$$

which must fulfill the relationship

$$\chi^{(3)}_{1111} = \chi^{(3)}_{1122} + \chi^{(3)}_{1212} + \chi^{(3)}_{1221} \quad . \tag{7.49}$$

The numerical subscripts refer to the axes of an arbitrary Cartesian coordinate system. For the 32 point-group symmetries, the relationships among the nonvanishing elements of $\chi^{(3)}$ appear in Table 7.4, as tabulated by BUTCHER [7.67].

When the optical frequencies are much less than the frequencies of electronic transitions, the frequency arguments of the background (non-Raman resonant) nonlinear susceptibility may be permuted separately from the polarization subscripts. This condition is termed Kleinman symmetry and is inherent in the treatment of Sects. 7.2.1-3 [7.56]. For an isotropic medium, Kleinman symmetry implies

$$\chi^{(3)}_{1122} = \chi^{(3)}_{1212} = \chi^{(3)}_{1221} = \frac{1}{3}\chi^{(3)}_{1111} \quad . \tag{7.50}$$

Symmetry considerations also restrict the possible forms of the Raman tensors $\alpha^{R}_{\alpha\beta}$. In an isotropic material

$$\left(\alpha^{R}_{11}\right)^2 = \left(\alpha^{R}_{22}\right)^2 = \left(\alpha^{R}_{33}\right)^2 \tag{7.51a}$$

$$\alpha^{R}_{11}\alpha^{R}_{22} = \alpha^{R}_{22}\alpha^{R}_{33} = \alpha^{R}_{33}\alpha^{R}_{11} = (1-2\rho_Q)\left(\alpha^{R}_{11}\right)^2 \tag{7.51b}$$

and

$$\left(\alpha^{R}_{12}\right)^2 = \left(\alpha^{R}_{23}\right)^2 = \left(\alpha^{R}_{13}\right)^2 = \rho_Q\left(\alpha^{R}_{11}\right)^2 \tag{7.51c}$$

where ρ_Q is the usual Raman depolarization ratio. For "trace" modes $\rho_Q \ll 3/4$ while for "totally depolarized" modes $\rho_Q = 3/4$. Table 7.5 gives the possible forms for $\alpha^{R}_{\alpha\beta}$ for crystals and Raman modes of various symmetry [7.68].

HELLWARTH has developed an elegant spherical tensor representation of chi-three for isotropic materials in the Born-Oppenheimer approximation [7.57]. The electronic

Table 7.4. The form of the third-order nonlinear susceptibility tensor $\bar{\chi}^{(3)}_{\alpha\beta\gamma\delta}(-\omega_s,\omega_0,\omega_1,-\omega_2)$ for the thirty-two crystal classes and isotropic media. Each element is denoted by its subscripts in a Cartesian coordinate system with axes oriented along the directions of the principal crystalographic axes. A bar denotes the negative. [7.67]

Triclinic
For both classes, 1 and $\bar{1}$, there are 81 independent nonzero elements.

Monoclinic
For all three classes, 2, m, and 2/m, there are 41 independent nonzero elements, consisting of:

3 elements with suffixes all equal
18 elements with suffixes equal in pairs
12 elements with suffixes having two y's, one x, and one z
4 elements with suffixes having three x's and one z
4 elements with suffixes having three z's and one x.

Orthorhombic
For all three classes, 222, mm2, and mmm, there are 21 independent nonzero elements, consisting of:

3 elements with suffixes all equal
18 elements with suffixes equal in pairs.

Tetragonal
For the three classes 4, $\bar{4}$, and 4/m, there are 41 nonzero elements of which only 21 are independent. They are:

		xxxx = yyyy zzzz	
zzxx = zzyy	xyzz = $\overline{yxzz}$	xxyy = yyxx	xxxy = $\overline{yyyx}$
xxzz = yyzz	zzxy = $\overline{zzyx}$	xyxy = yxyx	xxyx = $\overline{yyxy}$
zxzx = zyzy	xzyz = $\overline{yzxz}$	xyyx = yxxy	xyxx = $\overline{yxyy}$
xzxz = yzyz	zxzy = $\overline{zyzx}$		yxxx = $\overline{xyyy}$
zxxz = zyyz	zxyz = $\overline{zyxz}$		
xzzx = yzzy	xzzy = $\overline{yzzx}$		

For the four classes 422, 4mm, 4/mmm, and $\bar{4}$2m, there are 21 nonzero elements of which only 11 are independent. They are:

	xxxx = yyyy	zzzz
yyzz = zzyy	zzxx = xxzz	xxyy = yyxx
yzyz = zyzy	zxzx = xzxz	xyxy = yxyx
yzzy = zyyz	zxxz = xzzx	xyyx = yxxy

Cubic
For the two classes 23 and m3, there are 21 nonzero elements of which only 7 are independent. They are:

xxxx = yyyy = zzzz
yyzz = zzxx = xxyy
zzyy = xxzz = yyxx
yzyz = zxzx = xyxy
zyzy = xzxz = yxyx
yzzy = zxxz = xyyx
zyyz = xzzx = yxxy

For the three classes 432, $\bar{4}$3m, and m3m, there are 21 nonzero elements of which only 4 are independent. They are:

xxxx = yyyy = zzzz
yyzz = zzyy = zzxx = xxzz = xxyy = yyxx
yzyz = zyzy = zxzx = xzxz = xyxy = yxyx
yzzy = zyyz = zxxz = xzzx = xyyx = yxxy

Table 7.4 (continued)

Trigonal

For the two classes 3 and $\bar{3}$, there are 73 nonzero elements of which only 27 are independent. They are:

$$
\begin{array}{llll}
\mathrm{zzzz} & & & (\mathrm{xxyy} = \mathrm{yyxx} \\
\mathrm{xxxx} = \mathrm{yyyy} = \mathrm{xxyy} + \mathrm{xyyx} + \mathrm{xyxy} & & & (\mathrm{xyyx} = \mathrm{yxxy} \\
 & & & (\mathrm{xyxy} = \mathrm{yxyx} \\
\mathrm{yyzz} = \mathrm{xxzz} & \mathrm{xyzz} = \overline{\mathrm{yxzz}} & & \\
\mathrm{zzyy} = \mathrm{zzxx} & \mathrm{zzxy} = \overline{\mathrm{zzyx}} & & \\
\mathrm{zyyz} = \mathrm{zxxz} & \mathrm{zxyz} = \overline{\mathrm{zyxz}} & & \\
\mathrm{yzzy} = \mathrm{xzzx} & \mathrm{xzzy} = \overline{\mathrm{yzzx}} & & \\
\mathrm{yzyz} = \mathrm{xzxz} & \mathrm{xzyz} = \overline{\mathrm{yzxz}} & & \\
\mathrm{zyzy} = \mathrm{zxzx} & \mathrm{zxzy} = \overline{\mathrm{zyzx}} & & \\
 & & & (\mathrm{yyxy} = \overline{\mathrm{xxyx}} \\
\mathrm{xxyy} = \overline{\mathrm{yyyx}} = \mathrm{yyxy} + \mathrm{yxyy} + \mathrm{xyyy} & & & (\mathrm{yxyy} = \overline{\mathrm{xyxx}} \\
 & & & (\mathrm{xyyy} = \overline{\mathrm{yxxx}} \\
\mathrm{yyyz} = \overline{\mathrm{yxxz}} = \overline{\mathrm{xyxz}} = \overline{\mathrm{xxyz}} & & & \\
\mathrm{yyzy} = \overline{\mathrm{yxzx}} = \overline{\mathrm{xyzx}} = \overline{\mathrm{xxzy}} & & & \\
\mathrm{yzyy} = \overline{\mathrm{yzxx}} = \overline{\mathrm{xzyx}} = \overline{\mathrm{xzxy}} & & & \\
\mathrm{zyyy} = \overline{\mathrm{zyxx}} = \overline{\mathrm{zxyx}} = \overline{\mathrm{zxxy}} & & & \\
\mathrm{xxxz} = \overline{\mathrm{xyyz}} = \overline{\mathrm{yxyz}} = \overline{\mathrm{yyxz}} & & & \\
\mathrm{xxzx} = \overline{\mathrm{xyzy}} = \overline{\mathrm{yxzy}} = \overline{\mathrm{yyxz}} & & & \\
\mathrm{xzxx} = \overline{\mathrm{xyzy}} = \overline{\mathrm{yxzy}} = \overline{\mathrm{yzyx}} & & & \\
\mathrm{zxxx} = \overline{\mathrm{zxyy}} = \overline{\mathrm{zyxy}} = \overline{\mathrm{zyyx}} & & &
\end{array}
$$

For the three classes 3m, $\bar{3}$m, and 32 there are 37 nonzero elements of which only 14 are independent. They are:

$$
\begin{array}{lll}
\mathrm{zzzz} & & (\mathrm{xxyy} = \mathrm{yyxx} \\
\mathrm{xxxx} = \mathrm{yyyy} = \mathrm{xxyy} + \mathrm{xyyx} + \mathrm{xyxy} & & (\mathrm{xyyx} = \mathrm{yxxy} \\
 & & (\mathrm{xyxy} = \mathrm{yxyx} \\
\mathrm{yyzz} = \mathrm{xxzz} & \mathrm{yyyz} = \overline{\mathrm{yxxz}} = \overline{\mathrm{xyxz}} = \overline{\mathrm{xxyz}} & \\
\mathrm{zzyy} = \mathrm{zzxx} & \mathrm{yyzy} = \overline{\mathrm{yxzx}} = \overline{\mathrm{xyzx}} = \overline{\mathrm{xxzy}} & \\
\mathrm{zyyz} = \mathrm{zxxz} & \mathrm{yzyy} = \overline{\mathrm{yzxx}} = \overline{\mathrm{xzyx}} = \overline{\mathrm{xzxy}} & \\
\mathrm{yzzy} = \mathrm{xzzx} & \mathrm{zyyy} = \overline{\mathrm{zyxx}} = \overline{\mathrm{zxyx}} = \overline{\mathrm{zxxy}} & \\
\mathrm{yzyz} = \mathrm{xzxz} & & \\
\mathrm{zyzy} = \mathrm{zxzx} & &
\end{array}
$$

Hexagonal

For the three classes 6, $\bar{6}$, and 6/m, there are 41 nonzero elements of which only 19 are independent. They are:

$$
\begin{array}{lll}
\mathrm{zzzz} & & (\mathrm{xxyy} = \mathrm{yyxx} \\
\mathrm{xxxx} = \mathrm{yyyy} = \mathrm{xxyy} + \mathrm{xyyx} + \mathrm{xyxy} & & (\mathrm{xyyx} = \mathrm{yxxy} \\
 & & (\mathrm{xyxy} = \mathrm{yxyx} \\
\mathrm{yyzz} = \mathrm{xxzz} & \mathrm{xyzz} = \overline{\mathrm{yxzz}} & \\
\mathrm{zzyy} = \mathrm{zzxx} & \mathrm{zzxy} = \overline{\mathrm{zzyx}} & \\
\mathrm{zyyz} = \mathrm{zxxz} & \mathrm{zxyz} = \overline{\mathrm{zyxz}} & \\
\mathrm{yzzy} = \mathrm{xzzx} & \mathrm{xzzy} = \overline{\mathrm{yzzx}} & \\
\mathrm{yzyz} = \mathrm{xzxz} & \mathrm{xzyz} = \overline{\mathrm{yzxz}} & \\
\mathrm{zyzy} = \mathrm{zxzx} & \mathrm{zxzy} = \mathrm{zyzx} & \\
 & & (\mathrm{yyxy} = \overline{\mathrm{xxyx}} \\
\mathrm{xxxy} = \overline{\mathrm{yyyx}} = \mathrm{yyxy} + \mathrm{yxyx} + \mathrm{xyyy} & & (\mathrm{yxyy} = \overline{\mathrm{xyxx}} \\
 & & (\mathrm{xyyy} = \overline{\mathrm{yxxx}}
\end{array}
$$

Table 7.4 (continued)

Hexagonal (cont.) For the four classes 622, 6mm, 6/mmm, and $\bar{6}$m2, there are 21 nonzero elements of which only 10 are independent. They are: zzzz (xxyy = yyxx xxxx = yyyy = xxyy + xyyx + xyxy (xyyx = yxxy (xyxy = yxyx yyzz = xxzz zzyy = zzxx zyyz = zxxz yzzy = xzzx yzyz = xzxz zyzy = zxzx
Isotropic Media There are 21 nonzero elements of which only 3 are independent. They are: xxxx = yyyy = zzzz yyzz = zzyy = zzxx = xxzz = xxyy = yyxx yzyz = zyzy = zxzx = xzxz = xyxy = yxyx yzzy = zyyz = zxxz = xzzx = xyyx = yxxy xxxx = xxyy + xyxy + xyyx

background and (off-resonant) two-photon terms are grouped together as a real electronic hyperpolarizability σ

$$\sigma = 24\left(\chi^{B}_{1122}+\chi^{T}_{1122}\right) = 24\left(\chi^{B}_{1212}+\chi^{T}_{1212}\right) = 24\left(\chi^{B}_{1221}+\chi^{T}_{1221}\right) \quad . \tag{7.52a}$$

The vibrational and orientational nuclear motions are parameterized in terms of two real, causal nuclear response functions a(t) and b(t). The Fourier transform of these functions, $A(\Delta\omega)$ and $B(\Delta\omega)$, contributes directly to chi-three

$$\begin{aligned}24\chi^{(3)}_{\alpha\beta\gamma\delta}(-\omega_s,\omega_0,\omega_1,-\omega_2) &= \delta_{\alpha\beta}\delta_{\gamma\delta}[\sigma+2A(\omega_1-\omega_2)+B(\omega_0+\omega_1)+B(\omega_0-\omega_2)]\\ &+ \delta_{\alpha\gamma}\delta_{\beta\delta}[\sigma+2A(\omega_0-\omega_2)+B(\omega_0+\omega_2)+B(\omega_1-\omega_2)]\\ &+ \delta_{\alpha\delta}\delta_{\beta\gamma}[\sigma+2A(\omega_0+\omega_1)+B(\omega_1-\omega_2)+B(\omega_0-\omega_2)] \quad .\end{aligned} \tag{7.52b}$$

In the following section, the imaginary parts of $A(\Delta\omega)$ and $B(\Delta\omega)$ are related to the cross sections for polarized and depolarized scattering. The parameters A(0) and B(0) fulfill

$$A(0) = \int\limits_0^\infty \frac{\mathrm{Im}\{A(\omega)\}}{\omega}\,d\omega \qquad B(0) = \int\limits_0^\infty \frac{\mathrm{Im}\{B(\omega)\}}{\omega}\,d\omega$$

and if all modes are depolarized $A(0) = -B(0)/3$.

Table 7.5. The form of the Raman tensors $\alpha^{R}_{\gamma\delta}$ and $\alpha'_{\gamma\delta}(\beta)$ for vibrational modes of various symmetries in the thirty-two crystal classes. The coordinate system is referenced to the principal crystalographic axes employed in Table 7.4. The direction of the dipole moment is given in parentheses for the polariton modes of acentric crystals. [7.68]

System	Class		Raman tensors			
Monoclinic			$\begin{pmatrix} a & & d \\ & b & \\ d & & c \end{pmatrix}$	$\begin{pmatrix} & e & \\ e & & f \\ & f & \end{pmatrix}$		
	2	C_2	A(y)	B(x,z)		
	m	C_{1h}	A'(x,z)	A"(y)		
	2/m	C_{2h}	A_g	B_g		
Orthorhombic			$\begin{pmatrix} a & & \\ & b & \\ & & c \end{pmatrix}$	$\begin{pmatrix} & d & \\ d & & \\ & & \end{pmatrix}$	$\begin{pmatrix} & & e \\ & & \\ e & & \end{pmatrix}$	$\begin{pmatrix} & & \\ & & f \\ & f & \end{pmatrix}$
	222	D_2	A	$B_1(z)$	$B_2(y)$	$B_3(x)$
	mm2	C_{2v}	$A_1(z)$	A_2	$B_1(x)$	$B_2(y)$
	mmm	D_{2h}	A_g	B_{1g}	B_{2g}	B_{3g}
Trigonal			$\begin{pmatrix} a & & \\ & a & \\ & & b \end{pmatrix}$	$\begin{pmatrix} c & d & e \\ d & -c & f \\ e & f & \end{pmatrix}$	$\begin{pmatrix} d & -c & -f \\ -c & -d & e \\ -f & e & \end{pmatrix}$	
	3	C_3	A(z)	E(x)	E(y)	
	$\bar{3}$	C_{3i}	A_g	E_g	E_g	
			$\begin{pmatrix} a & & \\ & a & \\ & & b \end{pmatrix}$	$\begin{pmatrix} c & & \\ & -c & d \\ & d & \end{pmatrix}$	$\begin{pmatrix} & -c & -d \\ -c & & \\ -d & & \end{pmatrix}$	
	32	D_3	A_1	E(x)	E(y)	
	3m	C_{3v}	$A_1(z)$	E(y)	E(-x)	
	$\bar{3}$m	D_{3d}	A_{1g}	E_g	E_g	
Tetragonal			$\begin{pmatrix} a & & \\ & a & \\ & & b \end{pmatrix}$	$\begin{pmatrix} c & d & \\ d & -c & \\ & & \end{pmatrix}$	$\begin{pmatrix} & & e \\ & & f \\ e & f & \end{pmatrix}$	$\begin{pmatrix} & & -f \\ & & e \\ -f & e & \end{pmatrix}$
	4	C_4	A(z)	B	E(x)	E(y)
	$\bar{4}$	S_4	A	B(z)	E(x)	E(-y)
	4/m	C_{4h}	A_g	B_g	E_g	E_g

Table 7.5 (continued)

System	Class		Raman tensors					
Tetragonal (cont.)			$\begin{pmatrix} a & & \\ & a & \\ & & b \end{pmatrix}$	$\begin{pmatrix} c & & \\ & -c & \\ & & \end{pmatrix}$	$\begin{pmatrix} & d & \\ d & & \\ & & \end{pmatrix}$	$\begin{pmatrix} & & e \\ & & \\ e & & \end{pmatrix}$	$\begin{pmatrix} & & \\ & & e \\ & e & \end{pmatrix}$	
	4mm	C_{4v}	$A_1(z)$	B_1	B_2	$E(x)$	$E(y)$	
	422	D_4	A_1	B_1	B_2	$E(-y)$	$E(x)$	
	$\bar{4}2m$	D_{2d}	A_1	B_1	$B_2(z)$	$E(y)$	$E(x)$	
	4/mmm	D_{4h}	A_{1g}	B_{1g}	B_{2g}	E_g	E_g	
Hexagonal			$\begin{pmatrix} a & & \\ & a & \\ & & b \end{pmatrix}$	$\begin{pmatrix} & & c \\ & & d \\ c & d & \end{pmatrix}$	$\begin{pmatrix} & & -d \\ & & c \\ -d & c & \end{pmatrix}$	$\begin{pmatrix} e & f & \\ f & -e & \\ & & \end{pmatrix}$	$\begin{pmatrix} f & -e & \\ -e & -f & \\ & & \end{pmatrix}$	
	6	C_6	$A(z)$	$E_1(x)$	$E_1(y)$	E_2	E_2	
	$\bar{6}$	C_{3h}	A'	E''	E''	$E'(x)$	$E'(y)$	
	6/m	C_{6h}	A_g	E_{1g}	E_{1g}	E_{2g}	E_{2g}	
			$\begin{pmatrix} a & & \\ & a & \\ & & b \end{pmatrix}$	$\begin{pmatrix} & & \\ & & c \\ & c & \end{pmatrix}$	$\begin{pmatrix} & & -c \\ & & \\ -c & & \end{pmatrix}$	$\begin{pmatrix} & d & \\ d & & \\ & & \end{pmatrix}$	$\begin{pmatrix} d & & \\ & -d & \\ & & \end{pmatrix}$	
	622	D_6	A_1	$E_1(x)$	$E_1(y)$	E_2	E_2	
	6mm	C_{6v}	$A_1(z)$	$E_1(y)$	$E_1(-x)$	E_2	E_2	
	$\bar{6}m2$	D_{3h}	A_1'	E''	E''	$E'(x)$	$E'(y)$	
	6/mmm	D_{6h}	A_{1g}	E_{1g}	E_{1g}	E_{2g}	E_{2g}	
Cubic			$\begin{pmatrix} a & & \\ & a & \\ & & a \end{pmatrix}$	$\begin{pmatrix} b & & \\ & b & \\ & & -2b \end{pmatrix}$	$\begin{pmatrix} -b\sqrt{3} & & \\ & b\sqrt{3} & \\ & & 0 \end{pmatrix}$	$\begin{pmatrix} & & \\ & & d \\ & d & \end{pmatrix}$	$\begin{pmatrix} & & d \\ & & \\ d & & \end{pmatrix}$	$\begin{pmatrix} & d & \\ d & & \\ & & \end{pmatrix}$
	23	T	A	E	E	F(x)	F(y)	F(z)
	m3	T_h	A_g	E_g	E_g	F_g	F_g	F_g
	432	O	A_1	E	E	F_2	F_2	F_2
	$\bar{4}3m$	T_d	A_1	E	E	$F_2(x)$	$F_2(y)$	$F_2(z)$
	m3m	O_h	A_{1g}	E_g	E_g	F_{2g}	F_{2g}	F_{2g}

7.2.6 Relationship Between χ^R and the Spontaneous Cross Section

The magnitude of the Raman contributions to chi-three can be related to the cross section for spontaneous Raman scattering. For the Lorentzian modes previously considered, the Raman tensors can be related to the total Stokes cross sections

$$\rho_{gg}|\alpha^R_{\alpha\beta}|^2 = (\lambda/2\pi)^4 \frac{d\sigma_{\alpha\beta}}{d\Omega} \tag{7.53}$$

where polarized scattering occurs for $\alpha = \beta$ and depolarized scattering for $\alpha \neq \beta$ [7.69]. For isotropic materials, the differential Raman cross sections can be related directly to the imaginary part of chi-three [7.39,57]

$$\begin{aligned}\mathrm{Im}\left\{\chi^{(3)}_{1111}(\omega_2,-\omega_1,\omega_1,-\omega_2)\right\} &= \frac{\pi c^4}{24\hbar\omega_1\omega_2^3}\,\frac{d^2\sigma_{11}}{d\Omega d(\omega_1-\omega_2)}\,(e^{-\hbar(\omega_1-\omega_2)/kT}-1)\\ &= \mathrm{Im}\{A(\omega_1-\omega_2)+B(\omega_1-\omega_2)\}/12\end{aligned} \tag{7.54}$$

$$\begin{aligned}\mathrm{Im}\left\{\chi^{(3)}_{1221}(\omega_2,-\omega_1,\omega_1,-\omega_2)\right\} &= \mathrm{Im}\left\{\chi^{(3)}_{1212}(\omega_2,-\omega_1,\omega_1,-\omega_2)\right\} = \frac{\pi c^4}{24\hbar\omega_1\omega_2^3}\\ &\times\frac{d^2\sigma_{12}}{d\Omega d(\omega_1-\omega_2)}\,(e^{-\hbar(\omega_1-\omega_2)/kT}-1) = \mathrm{Im}\{B(\omega_1-\omega_2)/24\}.\end{aligned} \tag{7.55}$$

7.2.7 The Coherent Raman Signal

While all of the Fourier components of the polarization radiate simultaneously, only one signal amplitude is detected. Since the amplitude of the signal wave varies little within a wavelength of light, the wave equation (7.2) can be cast in terms of the slowly varying signal amplitude

$$ik_s \cdot \frac{\partial \underline{E}_s}{\partial z} = -4\pi\,\frac{\omega_s^2}{c^2}\,\underline{P}^{NL}(\omega_s)\,e^{-i(k_s-k_p)z} \tag{7.56}$$

where

$$\underline{E}_s(\underline{r},t) = \frac{1}{2}\,\underline{E}_s(\underline{r})\,e^{-i(\omega_s t-\underline{k}_s\cdot\underline{r})} + cc$$

and the z axis has been chosen parallel to $\underline{k}_p = \underline{k}_0 + \underline{k}_1 - \underline{k}_2$ [7.25]. The total nonlinear polarization $\underline{P}^{NL}$ has contributions from all Raman modes and generally also a background term due to other types of transitions driven far from resonance

$$P^{NL}_\alpha(\omega_s) = D\chi^{(3)}_{\alpha\beta\gamma\delta}(\omega_p,\omega_0,\omega_1,-\omega_2) = D\Big(\sum_Q \chi^Q_{\alpha\beta\gamma\delta}+\chi^{NR}_{\alpha\beta\gamma\delta}\Big)E_{0\beta}E_{1\gamma}E^*_{2\delta} \quad . \tag{7.57}$$

Generally the background polarization varies little on the frequency scale of a Raman experiment, and can be treated as a real constant.

By definition, there is no coherent Raman signal amplitude entering the sample. If the incident beams can be treated as uniform plane waves, (7.56) can be integrated directly to give the output signal amplitude

$$\underline{E}_s(\ell) = -\frac{4\pi i}{n_s}\frac{\omega_s}{c}\,\ell\underline{P}^{NL}(\omega_s)\,\mathrm{sinc}[(k_s-k_p)\ell/2]\,e^{-i(k_s-k_p)z/2} \tag{7.58}$$

where sinc $x = \sin x/x$, ℓ is the sample legnth, and z is the position of a point outside the sample.

The quantity physically detected in coherent Raman spectroscopy is not an amplitude, but rather an optical intensity or a change in intensity. In CARS, CSRS, four-wave mixing and RIKES, the detected intensity is related to the signal amplitude and polarization by

$$I_s = \frac{n_s c}{8\pi}|\underline{E}_s|^2 = \frac{2\pi}{n_s}\frac{\omega_s^2}{c^2}\ell^2|P^{NL}(\omega_s)|^2\,\mathrm{sinc}^2[(k_s-k_p)\ell/2]$$

$$\propto \left|\chi^{NR}+\sum_Q \chi^Q(-\omega_s,\omega_0,\omega_1,-\omega_2)\right|^2 \mathrm{sinc}^2[(k_s-k_p)\ell/2] \quad . \tag{7.59}$$

The polarizations resulting from different processes and different modes interfere with one another to yield a somewhat complicated spectrum. An isolated Lorentzian Raman resonance produces the asymmetric line shape

$$I(\omega_s) = \frac{\left[N\rho_d\Omega_{rg}\left(\alpha_{eff}^R\right)^2/12\hbar\right]^2}{\left[\Omega_Q^2-(\omega_1-\omega_2)^2\right]^2+4\Gamma^2(\omega_1-\omega_2)^2}$$

$$+\frac{\left[\chi_{eff}^{NR}N\rho_d\Omega_{rg}\left(\alpha_{eff}^R\right)^2/6\hbar\right]\left[\Omega_Q^2-(\omega_1-\omega_2)^2\right]}{\left[\Omega_Q^2-(\omega_1-\omega_2)^2\right]^2+4\Gamma^2(\omega_1-\omega_2)^2}+\left(\chi_{eff}^{NR}\right)^2 \tag{7.60}$$

plotted in Fig.7.4 when the background term is nonzero.

In stimulated Raman spectroscopy both the incident laser and the nonlinear interaction contribute to the intensity at the detector. Since

$$I(\omega_2) = \frac{n_s c}{8\pi}|\underline{E}_s+\underline{E}_2|^2 \tag{7.61}$$

and $E_s \ll E_2$, the detected change in intensity due to the nonlinear interaction is

$$\Delta I = \frac{n_s c}{4\pi}\underline{E}_2\cdot\mathrm{Re}\{\underline{E}_s\} = -\frac{\omega_2\ell}{c}E_2\,\mathrm{Im}\left\{P^{NL}(\omega_2)\right\} \propto \sum_Q \mathrm{Im}\left\{\chi^Q(-\omega_2,-\omega_1,\omega_1,-\omega_2)\right\} \quad . \tag{7.62}$$

An isolated Lorentzian resonance gives the line shape in Fig.7.5a.

In optical heterodyne detected (OHD) Raman induced spectroscopy and in stimulated Raman interferometry, a local oscillator amplitude with controlled magnitude and phase is injected into the detector along with the nonlinearly generated signal amplitude [7.40]. The total intensity at the detector is then

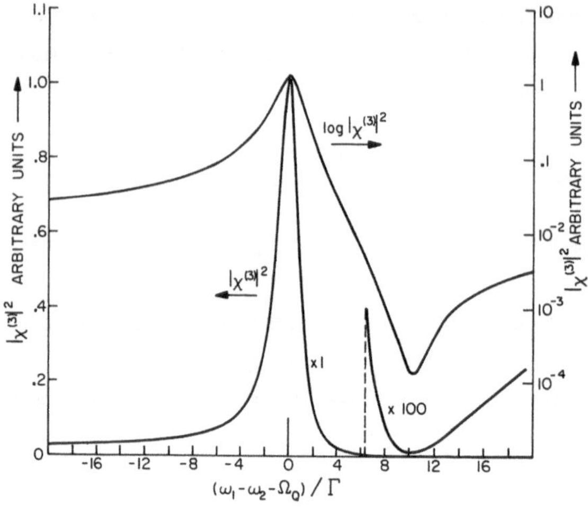

Fig. 7.4. The CARS line shape of (7.60) for $N(\alpha^R_{eff})^2/4\hbar D\chi^{NR}_{eff} = 10\Gamma$ on linear and logarithmic scales. This plot corresponds to the frequency dispersion of $|\chi^{(3)}|^2$. The deep minimum occurs when the nonresonant background is cancelled by the real part of the resonant Raman susceptibility

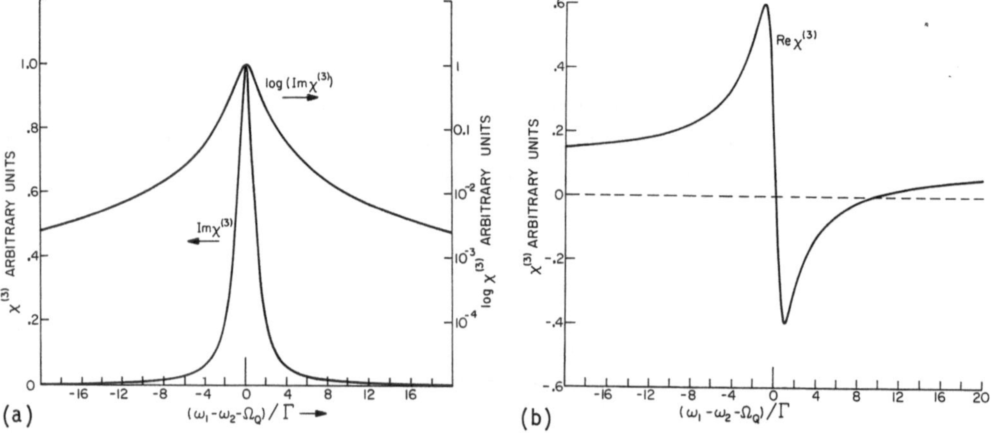

Fig. 7.5a,b. The real and imaginary parts of the Lorentzian resonance responsible for the CARS line shape in Fig.7.4. Part (a) shows $\mathrm{Im}\{\chi^{(3)}\}$ on linear and logarithmic scales while (b) gives $\mathrm{Re}\{\chi^{(3)}\}$

$$I = \frac{n_s c}{8\pi}|\underline{E}_{LO}+\underline{E}_s|^2 = I_{LO} + I_S + \frac{n_s c}{4\pi}\mathrm{Re}\left\{\underline{E}^*_{LO}\cdot\underline{E}_s\right\} \tag{7.63}$$

where I_{LO} is the local oscillator intensity and the signal intensity I_s can be obtained from (7.60). When $E_s \ll E_{LO}$, the detected change in intensity is

$$I_h = \frac{n_s c}{4\pi}\mathrm{Re}\left\{\underline{E}_{LO}\cdot\underline{E}_s\right\} = -\frac{\omega_s \ell}{c}\mathrm{Im}\left\{E_{LO}\cdot P^{NL}(\omega_s)\right\} \propto \sum_q \begin{Bmatrix}\mathrm{Re}\\ \mathrm{Im}\end{Bmatrix}\left\{\chi^Q(\omega_0)\right\} \quad . \tag{7.64}$$

Depending upon the local oscillator phase, an isolated Lorentzian resonance can give either the resonance or dispersion line shapes in Fig.7.5.

7.2.8 Focusing Considerations

If the incident laser beams can be treated as plane waves of finite area, the electric fields in the sample can be related to the incident intensity and power

$$I = \frac{P}{A} = \frac{n(\omega)c}{4\pi} |\underline{E}(\omega)|^2 \quad . \tag{7.65}$$

Substituting these formulae into (7.57-64) suggests that the maximum coherent Raman signal is obtained by focusing the beams as tightly as possible. A more detailed analysis of interacting collinearly TEM_{00} modes by BJORKLUND points out the limitations and benefits of this strategy [7.70].

For an incident electric field composed of TEM_{00} beams

$$\underline{E}(r,t) = \sum_{j=0}^{2} \frac{1}{2}\left[\underline{E}_j(\underline{r})\, e^{-i\omega_j \cdot t} + cc\right] \tag{7.66}$$

where

$$E_j(\underline{r}) = \underline{E}_j^0 \exp(ik_j z)(1+iE)^{-1} \exp\left[-k_j(x^2+y^2)/b(1+iE)\right] \tag{7.67}$$

as given by BOYD and GORDON with the confocal parameter defined as

$$b = k_j w_0^2 = n_j \frac{\omega_j}{c} w_0^2 \tag{7.68}$$

where w_0 is the beam waist radius [7.71]. The normalized coordinate is defined as

$$E = 2(z-f)/b \tag{7.69}$$

where f is the position of the beam waist along the z axis.

Equation (7.57) yields an $\underline{r}$ dependent nonlinear driving polarization when the $\underline{r}$ dependent amplitudes of (7.67) are substituted. Solving the wave equation and integrating the intensity of the radiated wave, one obtains the following expression for the total power radiated by the coherent Raman process:

$$P_s = (1.755\times10^{-5}) \frac{\omega_s^4 k_0 k_1 k_2}{c^4 k_s^2 k'} D^2 |\chi_{eff}^{(3)}|^2 P_0 P_1 P_2 F_2\left(b\Delta K, \frac{b}{\ell}, \frac{f}{\ell}, \frac{k''}{k'}\right) \quad . \tag{7.70}$$

In (7.70) all wave vectors are in units of cm^{-1}, b is in cm, as is the sample length ℓ, $\chi^{(3)}$ in esu/cm^3 and all powers are in watts. The wave vectors Δk, k', k'' are defined by

$$\Delta k = k_s - (k_0+k_1-k_2) \tag{7.71a}$$

$$k' = k_0 + k_1 - k_2 = k_p \tag{7.71b}$$

$$k'' = k_0 + k_1 + k_2 \tag{7.71c}$$

and the phase matching function is

$$F_2\left(b\Delta K, \frac{b}{\ell}, \frac{f}{\ell}, \frac{k''}{k'}\right) = \frac{2k'}{\pi b} \int_0^\infty 2\pi R dR \left| \int_{-\zeta}^{E} dE' \frac{\exp(-ib\Delta kE'/2)}{(1+iE')(k''-ik'E')H} \times \exp\left(\frac{-R^2}{bH}\right) \right|^2 \tag{7.72}$$

where

$$H = H(E,E') = \frac{(1+E')^2}{(k''-ik'E')} - i\left(\frac{E'-E}{k'}\right) \tag{7.73}$$

and

$$\zeta = 2f/b \quad . \tag{7.74}$$

In general, F_2 must be evaluated numerically; it is, however, always a maximum for $\Delta k = 0$. There are, in addition, two simple limiting cases. If $b \gg \ell$ and the cell is located in the beam waist,

$$F_2 = \frac{k'}{k''} \frac{4\ell^2}{b^2} \operatorname{sinc}^2(\Delta k\ell/2) \tag{7.75}$$

and the plane wave results of (7.58-64) are recovered. For the cases of (7.62,64) where $\Delta k \equiv 0$, $F_2 = (k'/k'')(4\ell^2/b^2)$.

In the tight focusing limit $b \ll \ell$ and the entire focal region is contained inside the cell; F_2 then becomes

$$F_2 = 4\pi^2 \frac{\exp[-(k''/k')|\Delta k|b]}{[1+(k''/k')]^2} \quad . \tag{7.76}$$

Again when $\Delta k \equiv 0$, $F_2 = 4\pi^2/[1+(k''/k')]^2$ which reproduces the usual tight focusing result.

IN OHD-RIKES, and SRS, the detected signal power is

$$P_H = 4.77 \times 10^8 \ \mathrm{Re}\left\{\int_0^\infty E^*_{LO}(R) E_s(R) 2\pi R dR\right\} \tag{7.77}$$

rather than the radiated power of (7.70). In (7.77) $E_{LO}(R)$ is the local oscillator or probe amplitude and the integral is over the detector surface. Since the power radiated as the result of the nonlinear interaction need not be in the same spatial mode as the probe, the quantities in (7.77,70) need not be simply related. EESLEY, however, has verified by direct calculation that

$$P_H = 2\eta(P_s \cdot P_{LO})^{1/2} \tag{7.78}$$

where the heterodyne efficiency $0.91 < \eta < 1.00$ when $\ell \geq b/2$ [7.72]. With Gaussian beams one can also expect a small component of the signal amplitude to be in quadrature with the phase defined in (7.58). Again, EESLEY has demonstrated that the amplitude of that component is more than five orders of magnitude smaller than the in-phase component.

It is apparent from the above that collinear, focused Gaussian laser beams can interact effectively only over a limited distance, and the advantages gained by tightly focusing are partly counteracted by the resulting decrease in the interaction length. As a rule of thumb, maximum signals are generated whenever $\ell/b \geq 10$.

The effective interaction length can be extended, however, by confining the interacting waves in a wave guide or optical fiber. STOLEN has demonstrated that wave guide dispersion can be used to phase match CARS processes in fibers where the index of refraction of the core exceeds that of the cladding [7.73]. The opposite condition on the indices is obtained in hollow wave guides as analyzed by MARCATILI and SCHMELTZER [7.74]. MILES et al. have demonstrated the advantages of doing CARS in such wave guides [7.75].

A Gaussian laser mode can be coupled to the low loss EH_{11} mode of a hollow wave guide with >98% efficiency if the beam is focused to a waist radius w_0 that is 73% of the guide radius [7.74]. Such wave guides do not significantly depolarize the radiation, and can be used effectively in all the coherent Raman techniques so long as

$$\Delta k \ell < \pi \tag{7.79}$$

where ℓ is the length of the guide. For such a guide the phase matching function becomes

$$F_2 \rightarrow \frac{1.28\times10^{22}}{\omega_1\omega_0}\frac{k'}{k''}\frac{\ell^2}{a^4} \tag{7.80}$$

which is easily 1000 times larger than the tight-focusing case.

The losses even in high quality hollow wave guides are not inconsiderable, typically 30 dB/meter. It will be shown later that the optimum length corresponds to a loss of ~5 dB which reduces the enhancement of wave guide geometry over tight focusing to a factor of 20 or so.

In CARS and four-wave mixing with collinear geometry Δkb is often greater than 3 in experiments on liquids and solids. The confocal parameter can be reduced by tighter focusing, but dielectric breakdown then limits the maximum incident power [7.11,76]. More rewarding is the strategy of crossing the input beams at an angle where

$$\Delta k = |\underline{k}_s-(\underline{k}_0+\underline{k}_1-\underline{k}_2)| = 0 \tag{7.81}$$

is fulfilled. One such geometry is shown in Fig.7.6. The angles are small, and can be calculated from

$$\omega_2 n(\omega_2)\cos\theta + \omega_s n(\omega_s)\cos\phi = \omega_0 n(\omega_0)\cos\theta' + \omega_1 n(\omega_1)\cos\phi' \tag{7.82a}$$

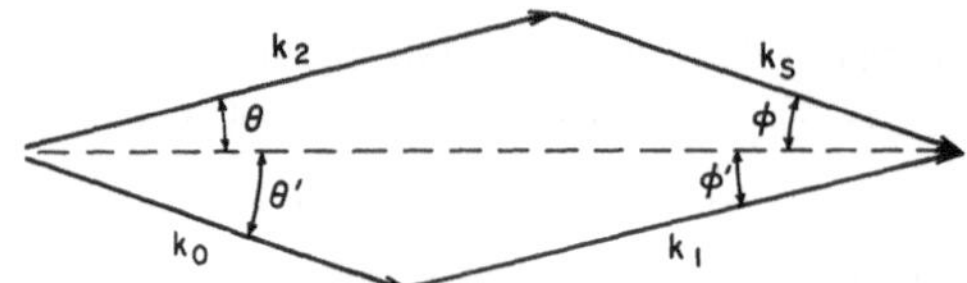

Fig. 7.6. The general wave vector matching condition in coherent Raman spectroscopy. In CARS $\theta' = \phi' = 0$

$$\omega_2 n(\omega_2) \sin\theta = \omega_s n(\omega_s) \sin\phi \quad ; \quad \omega_0 n(\omega_0) \sin\theta' = \omega_1 n(\omega_1) \sin\phi' \tag{7.82b}$$

where θ, ϕ, θ' and ϕ' are the angles inside the material. In the CARS case, where $\omega_0 = \omega_1$ and $\underline{k}_0 = \underline{k}_1$, the angle θ depends upon the frequency difference of the input beams. In an isotropic material, the index of refraction can be expanded as a power series in the frequency shift

$$n(\omega_1+\delta\omega) = n_1 + n_1'\delta\omega + \frac{1}{2} n_1''(\delta\omega)^2 \tag{7.83}$$

and substituting into (7.82) yields

$$\theta = \left[\left(\frac{n_1''}{n_1} + \frac{2n_1'}{n_1\omega_1}\right)\left(1 + \frac{\omega_1-\omega_2}{\omega_1}\right)\right]^{\frac{1}{2}}(\omega_1-\omega_2) \quad . \tag{7.84}$$

Analytic calculations have not been performed to date for focused Gaussian beams crossing at a finite angle. Such beams interact effectively over a length $\ell \approx 2w_0/\theta < 2b$ whenever this phase matching geometry is appropriate, and thus (7.75) becomes a reasonable approximation. The maximum CARS output then occurs at a crossing angle somewhat smaller than that indicated by (7.84).

7.2.9 Accentric Crystals and Polaritons

The nonlinear optics of materials without inversion symmetry is considerably more complicated than the previously discussed centrosymmetric case. The second-order nonlinear susceptibility $\chi^{(2)}_{\alpha\beta\gamma}$ can be nonzero in accentric materials and it acts to produce electric fields oscillating at the pairwise sums and differences of the incident laser frequencies [7.52]. The new difference frequency fields can then drive vibrational modes or mix directly with another input wave in a second chi-two process to give the CRS output frequency. The complete theory of these complex and interrelated interactions has been reviewed in [7.32,52,57]. Rather than develop it here in detail, we shall concern ourselves only with the processes leading to vibrational resonances in CRS signals.

In an accentric crystal, the optical phonons and photons couple together to form polariton modes with a mixed mechanical-electromagnetic character. The frequencies and wave vectors of the transverse modes fulfill the dispersion relation [7.77]

$$q^2 = \varepsilon(\omega_q,\Gamma)\omega_q^2/c^2 \tag{7.85}$$

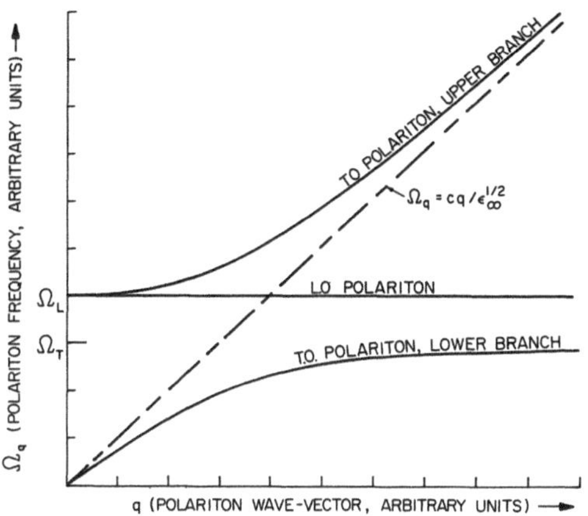

Fig. 7.7. Polariton dispersion curves

where

$$\varepsilon(\omega_q,\Gamma) = \varepsilon_\infty + \frac{4\pi Ne^{*2}}{\Omega_T^2-\omega_q^2-2i\omega_q\Gamma} \tag{7.86}$$

is the dielectric constant of the material, e^* is the effective charge for the mode, Ω_T is the transverse polariton frequency and ε_∞ is the asymptotic dielectric constant for frequencies above Ω_T. Figure 7.7 shows the dramatic variation of the resonant frequency with wave vector. Raman resonances occur where the line

$$q(\omega_1-\omega_2) = |\underline{k}_1(\omega_1)-\underline{k}_2(\omega_2)| \tag{7.87}$$

crosses one of the solid curves of Fig.7.7. The effective nonlinear susceptibility thus depends on both the frequency difference and the wave vector difference of the incident laser beams.

The development of the theory of the nonlinear polariton resonances begins with the equation of motion for the vibrational coordinate

$$\ddot{Q}_\beta + 2\Gamma\dot{Q}_\beta + \Omega_T^2 Q_\beta = e^* \cdot E_\beta + \frac{1}{2}\alpha'_{\gamma\delta}(\beta)E_\gamma E_\delta \tag{7.88}$$

where $\alpha'_{\gamma\delta}(\beta) = \partial\alpha_{\gamma\delta}/\partial Q_\beta = (\hbar/\Omega_{rg})^{-\frac{1}{2}}\alpha^R_{\gamma\delta}$ is the semiclassical Raman susceptibility used by WYNNE and FLYTZANIS [7.32,52]. Equation (7.88) is analogous to (7.20) for the centrosymmetric case [7.78]. The equation for the polariton resonant polarization similar to (7.28) is

$$P^P_\alpha = Ne^*Q_\alpha + N\alpha'_{\alpha\beta}(\gamma)Q_\gamma E_\beta + \chi^{(2)E}_{\alpha\beta\gamma}E_\beta E_\gamma \tag{7.89}$$

whereas the form of the wave equation appropriate for all of the possible frequency components of the electric field is

$$\nabla^2 E_\alpha - \nabla_\alpha(\underline{\nabla}\cdot\underline{E}) - \frac{\varepsilon_{\infty\alpha\alpha}}{c^2}\ddot{E}_\alpha = \frac{4\pi}{c^2}\ddot{P}^P_\alpha \tag{7.90}$$

and $\chi^{(2)E}$ above is the second-order electronic nonlinear polarizability.

Even in the absence of incident fields, there can be a purely longitudinal polariton solution to these equations where $\nabla^2 E_\alpha = \nabla_\alpha(\underline{\nabla}\cdot\underline{E})$. In this case (7.88) becomes

$$\ddot{Q}_L + 2\Gamma\dot{Q}_L + \Omega_L^2 Q_L = \left[\frac{1}{2}\alpha'_{\gamma\delta}(L) - \frac{4\pi e^*}{\varepsilon_{\infty L}}\chi^{(2)}_{L\gamma\delta}\right]E_\gamma E_\delta^* \tag{7.91}$$

where

$$\Omega_L^2 = \Omega_T^2 + \frac{4\pi N e^*}{\varepsilon_{\infty L}} \quad ,$$

$\varepsilon_{\infty L}$ is the asymptotic dielectric constant for fields polarized in the L direction and L specifies the axis of $\underline{k}_1 - \underline{k}_2$.

If $\omega_1 - \omega_2 \approx \Omega_L$ the resonant nonlinear polarization due to this longitudinal polariton is

$$P_\alpha(\omega_s) = D\chi^{LP}_{\alpha\beta\gamma\delta}(-\omega_s,\omega_0,\omega_1,-\omega_2)E_\beta(\omega_0)E_\gamma(\omega_1)E^*_\delta(\omega_2)\; e^{-i(\omega_s t - \underline{k}_P\cdot\underline{r})} \tag{7.92}$$

where

$$\chi^{LP}_{\alpha\beta\gamma\delta} = \frac{N}{4D}\,\frac{\left(\alpha'_{\alpha\beta}(L) - \frac{8\pi e^*}{\varepsilon_{\infty L}}\chi^{(2)}_{\alpha\beta L}\right)\left(\alpha'_{\gamma\delta}(L) - \frac{8\pi e^*}{\varepsilon_{\infty L}}\chi^{(2)}_{L\gamma\delta}\right)}{\Omega_L^2 - (\omega_1-\omega_2)^2 - 2i(\omega_1-\omega_2)\Gamma} \tag{7.93}$$

(plus similar terms with β, ω_0 exchanged with γ, ω_1). Equation (7.93) has exactly the same form as (7.37,38) except that the minus signs in the numerator substantially reduce the size of the resonance.

The case where $\nabla\cdot E = 0$ is more complex. Equations (7.88,90) can be solved for the amplitude oscillating at $\omega_1 - \omega_2 \approx \omega_R$ and with wave vector $\underline{k}_1 - \underline{k}_2$

$$Q(\omega_1-\omega_2) = \frac{\frac{1}{4}\alpha'_{\gamma\delta}(T) - 2\pi e^*\chi^{(2)}_{T\gamma\delta}/\left[\varepsilon_{\infty T} - 2c^2|\underline{k}_1-\underline{k}_2|^2(\omega_1-\omega_2)^{-2}\right]}{\left[\Omega_T^2 + \frac{4\pi N e^{*2}}{\varepsilon_{\infty T} - c^2|\underline{k}_1-\underline{k}_2|^2(\omega_1-\omega_2)^{-2}} - (\omega_1-\omega_2)^2 - 2i\Gamma(\omega_1-\omega_2)\right]}E_\gamma E_\delta \tag{7.94a}$$

$$E(\omega_1-\omega_2) =$$

$$-\frac{\left\{\pi N e^*\alpha'_{\gamma\delta} + \left[\Omega_T^2 - (\omega_1-\omega_2)^2 - 2i\Gamma(\omega_1-\omega_2)\right]2\pi\chi^{(2)}_{T\gamma\delta}\right\}E_\gamma E_\delta}{\left[\varepsilon_{\infty T} - c^2|\underline{k}_1-\underline{k}_2|^2(\omega_1-\omega_2)^{-2}\right]\left[\Omega_T^2 + \frac{4\pi N e^{*2}}{\varepsilon_{\infty T} - c^2|\underline{k}_1-\underline{k}_2|^2(\omega_1-\omega_2)^{-2}} - (\omega_1-\omega_2)^2 - 2i\Gamma(\omega_1-\omega_2)\right]} \; . \tag{7.94b}$$

The transverse polariton resonant contribution to the nonlinear polarization is

$$P^{TP}(\omega_s) = D\chi^{TP}_{\alpha\beta\gamma\delta}(-\omega_s,\omega_0,\omega_1,-\omega_2)E_\beta(\omega_0)E_\gamma(\omega_1)E_\delta(-\omega_2)\, e^{-i(\omega_s t - \underline{k}_p\cdot\underline{r})}$$

where

$$\chi^{TP}_{\alpha\beta\gamma\delta} = \frac{1}{4D}\left[\Omega_T^2 + \frac{4\pi N e^{*2}}{\varepsilon_{\infty T} - c^2|\underline{k}_1-\underline{k}_2|^2(\omega_1-\omega_2)^{-2}} - (\omega_1-\omega_2)^2 - 2i\Gamma(\omega_1-\omega_2)\right]^{-1}$$

$$\times\left[\frac{N\alpha'_{\alpha\beta}(T)\alpha'_{\gamma\delta}(T)}{2} - \frac{2\pi N e^*\left[\alpha'_{\alpha\beta}(T)\chi^{(2)}_{T\gamma\delta} + \alpha'_{\gamma\delta}(T)\chi^{(2)}_{\alpha\beta T}\right]}{\varepsilon_{\infty T} - c^2|\underline{k}_1-\underline{k}_2|^2(\omega_1-\omega_2)^{-2}}\right.$$

$$\left. + \frac{16\pi^2 N e^{*2}\chi^{(2)}_{\alpha\beta T}\chi^{(2)}_{T\gamma\delta}}{\left[\varepsilon_{\infty T} - c^2|\underline{k}_1-\underline{k}_2|^2(\omega_1-\omega_2)^{-2}\right]^2}\right] + \text{similar terms with } \beta,\ \omega_0 \text{ exchanged with } \gamma,\ \omega_1 \qquad (7.95)$$

which has appeared previously in slightly different guise [7.32,52,79]. The center of a resonance appears at different frequencies in each of the coherent Raman techniques, but if we define the center as the point where $\mathrm{Re}\{\chi^{TP}\} = 0$, the only accessible resonance condition is

$$|\underline{k}_1-\underline{k}_2|^2 = \varepsilon_T(\omega_1-\omega_2,0)(\omega_1-\omega_2)^2/c^2 \qquad (7.96)$$

which is the phase matching condition for generation of the polariton wave as in (7.85,87). In optically isotropic crystals (e.g., GaP) $\varepsilon_{\infty L} = \varepsilon_{\infty T}$ and geometries exist where both longitudinal and transverse resonances occur. The total polariton resonant susceptibility has the form

$$\chi^P = \chi^{LP}\cos^2\theta + \chi^{TP}\sin^2\theta \qquad (7.97)$$

where

$$\cos\theta = \frac{(\underline{k}_1-\underline{k}_2)\cdot\underline{E}(\omega_1-\omega_2)}{|\underline{k}_1-\underline{k}_2||\underline{E}(\omega_1-\omega_2)|}\ , \qquad (7.98)$$

in lower symmetry materials no such simple general result exists [7.52].

7.2.10 Resonant Effects and Absorbing Samples

As the input and signal frequencies approach the frequencies of allowed one-photon transitions, resonance effects similar to those in the resonant Raman effect enhance the third-order nonlinear susceptibility. The coherent Raman signals become stronger, facilitating the detection of species in dilute solution. The spectra can, however, be radically altered, with peaks becoming valleys, new non-Raman resonances appearing, and resonant background levels rising to obscure the entire spectrum [7.80,81]. Many of these effects can be qualitatively explained by noting that α^R and α^T which appear in (7.22,40) become complex under resonant conditions, but others must be ascribed to intrinsically new processes.

Various authors have developed techniques for calculating chi-three under particular resonant conditions, and the apparent disagreement among the various results continues to fuel controversy [7.53,54,67,81-85]. We have chosen to follow LYNCH, who obtained the steady-state solution of the set of equations [7.54,69]

$$\dot{\rho}_{uv} + [i(\Omega_u-\Omega_v)+\Gamma_{uv}]\rho_{uv} = -\frac{i}{\hbar}[H'_{vs}\rho_{sv}-\rho_{vs}H'_{sv}] + \Gamma_{ss}\delta_{uv}\delta_{vs}\rho^e_{ss} \tag{7.99}$$

which describe the evolution of the density matrix for a four-level system and u,v,s = g,r,n,m or t as in Fig.7.2. In (7.99) Γ_{uv} are the relaxation rates of the individual elements of the density matrix, with no assumptions made as to relationships among them. The diagonal terms of the density matrix relax towards their values at thermal equilibrium

$$\rho^e_{vv} = \exp(-\hbar\Omega_v/kT)/\sum_v \exp(-\hbar\Omega_v/kT) \tag{7.100}$$

at the population decay rates[1]. The off-diagonal elements relax towards zero at the transverse decay rates. In general

$$\Gamma_{uv} = \frac{1}{2}(\Gamma_{uu}+\Gamma_{vv}) + \gamma_{uv} \tag{7.101}$$

where γ_{uv} is the "pure dephasing" rate [7.86]. More general assumptions concerning relaxation processes are possible, but the calculations then become prohibitively complicated.

The coupling Hamiltonian is $H'_{uv} = -\underline{\mu}_{uv}\cdot\underline{E}(t)$ where the dipole matrix elements vanishes between states of the same symmetry $\langle g|\underline{\mu}|r\rangle = \langle m|\underline{\mu}|n\rangle = 0$ and $\underline{E}(t)$ appears in (7.5). The general result is so complex that special notation must be developed. We define first a nonlinear susceptibility tensor matrix element $\langle u|\tilde{\chi}^{(3)}_{\alpha\beta\gamma\delta}(-\omega_s,\omega_0,\omega_1,-\omega_2)|v\rangle$ which corresponds to the nonlinear susceptibility provided solely by transitions between state v and state u. The total third-order nonlinear susceptibility tensor can then be written as

$$\chi^{(3)}_{\alpha\beta\gamma\delta}(-\omega_s,\omega_0,\omega_1,-\omega_2) = N\sum_{u,v}\langle v|\tilde{\chi}^{(3)}_{\alpha\beta\gamma\delta}(-\omega_s,\omega_0,\omega_1,-\omega_2)|u\rangle\rho^e_{uu} \tag{7.102}$$

---

[1] Equation (7.100) applies strictly only to systems obeying Maxwell-Boltzmann statistics. For Bose-Einstein systems — such as the phonon modes of a perfect lattice — the correct result is obtained by setting $\rho_{vv} = \delta_{vg}\delta_{vg}$ where g is the state with no phonons excited. In this case $\rho_d \equiv 1$.

where ρ_{uu}^{e} is the density matrix of the system in thermal equilibrium, that is, with the incident lasers turned off. The tensor matrix elements can be written as

$$\left\langle v\left|\tilde{\chi}^{(3)}_{\alpha\beta\gamma\delta}(-\omega_s,\omega_0,\omega_1,-\omega_2)\right|u\right\rangle = \frac{1}{48\hbar^3}(\tilde{\mu}_{un}\tilde{\mu}_{nv}\tilde{\mu}_{vm}\tilde{\mu}_{mu})$$

$$\times\left\{\frac{1}{[\hat{\Omega}_{vu}-(\omega_1-\omega_2)](\hat{\Omega}_{nu}+\omega_2)}\left(\frac{\delta\gamma\beta\alpha}{\hat{\Omega}_{mu}-\omega_s}+\frac{\delta\gamma\alpha\beta[1+\kappa_1(\Omega_3,\Omega_2)]}{\hat{\Omega}^*_{mu}+\omega_0}\right)\right.$$

$$+\frac{1}{[\hat{\Omega}_{vu}-(\omega_1-\omega_2)](\hat{\Omega}_{nu}-\omega_1)}\left(\frac{\gamma\delta\beta\alpha}{\hat{\Omega}_{mu}-\omega_s}+\frac{\gamma\delta\alpha\beta[1+\kappa_1(\Omega_3,\Omega_1)]}{\hat{\Omega}^*_{mu}+\omega_0}\right)$$

$$+\frac{1}{[\hat{\Omega}_{vu}-(\omega_0-\omega_2)](\hat{\Omega}_{nu}+\omega_2)}\left(\frac{\delta\beta\gamma\alpha}{\hat{\Omega}_{mu}-\omega_s}+\frac{\delta\beta\alpha\gamma[1+\kappa_1(\Omega_1,\Omega_3)]}{\hat{\Omega}^*_{mu}+\omega_1}\right)$$

$$+\frac{1}{[\hat{\Omega}_{vu}-(\omega_0-\omega_2)](\hat{\Omega}_{nu}-\omega_0)}\left(\frac{\beta\delta\gamma\alpha}{\hat{\Omega}_{mu}-\omega_s}+\frac{\beta\delta\alpha\gamma[1+\kappa_1(\Omega_2,\Omega_1)]}{\hat{\Omega}^*_{mu}+\omega_1}\right)$$

$$+\frac{1}{[\hat{\Omega}_{vu}-(\omega_0+\omega_1)](\hat{\Omega}_{nu}-\omega_1)}\left(\frac{\gamma\beta\delta\alpha}{\hat{\Omega}_{mu}-\omega_s}+\frac{\gamma\beta\alpha\delta[1+\kappa_1(\Omega_1,\Omega_3)]}{\hat{\Omega}^*_{mu}-\omega_2}\right)$$

$$+\frac{1}{[\hat{\Omega}_{vu}-(\omega_0+\omega_1)](\hat{\Omega}_{nu}-\omega_0)}\left(\frac{\beta\gamma\delta\alpha}{\hat{\Omega}_{mu}-\omega_s}+\frac{\beta\gamma\alpha\delta[1+\kappa_1(\Omega_1,\Omega_2)]}{\hat{\Omega}^*_{mu}-\omega_2}\right)$$

$$+\frac{1}{(\hat{\Omega}^*_{vu}+\omega_1-\omega_2)(\hat{\Omega}^*_{mu}-\omega_2)}\left(\frac{\alpha\beta\gamma\delta}{\hat{\Omega}^*_{nu}+\omega_s}+\frac{\beta\alpha\gamma\delta[1+\kappa_2(\Omega_3,\Omega_2)]}{\hat{\Omega}_{nu}-\omega_0}\right)$$

$$+\frac{1}{(\hat{\Omega}^*_{vu}+\omega_1-\omega_2)(\hat{\Omega}^*_{mu}+\omega_1)}\left(\frac{\alpha\beta\delta\gamma}{\hat{\Omega}^*_{nu}+\omega_s}+\frac{\beta\alpha\delta\gamma[1+\kappa_2(\Omega_3,\Omega_1)]}{\hat{\Omega}_{nu}-\omega_0}\right)$$

$$+\frac{1}{(\hat{\Omega}^*_{vu}+\omega_0-\omega_2)(\hat{\Omega}_{mu}-\omega_2)}\left(\frac{\alpha\gamma\beta\delta}{\hat{\Omega}^*_{nu}+\omega_s}+\frac{\gamma\alpha\beta\delta[1+\kappa_2(\Omega_2,\Omega_3)]}{\hat{\Omega}_{nu}-\omega_1}\right)$$

$$+\frac{1}{(\hat{\Omega}^*_{vu}+\omega_0-\omega_2)(\hat{\Omega}_{mu}+\omega_1)}\left(\frac{\alpha\gamma\delta\beta}{\hat{\Omega}^*_{nu}+\omega_s}+\frac{\gamma\alpha\delta\beta[1+\kappa_2(\Omega_2,\Omega_1)]}{\hat{\Omega}_{nu}-\omega_1}\right)$$

$$+\frac{1}{(\hat{\Omega}^*_{vu}+\omega_0+\omega_1)(\hat{\Omega}^*_{mu}+\omega_1)}\left(\frac{\alpha\delta\beta\gamma}{\hat{\Omega}^*_{nu}+\omega_s}+\frac{\delta\alpha\beta\gamma[1+\kappa_2(\Omega_1,\Omega_3)]}{\hat{\Omega}_{nu}+\omega_2}\right)$$

$$\left.+\frac{1}{(\hat{\Omega}^*_{vu}+\omega_0+\omega_1)(\hat{\Omega}^*_{mu}+\omega_0)}\left(\frac{\alpha\delta\gamma\beta}{\hat{\Omega}^*_{nu}+\omega_s}+\frac{\delta\alpha\gamma\beta[1+\kappa_2(\Omega_1,\Omega_2)]}{\hat{\Omega}_{nu}+\omega_2}\right)\right\} \qquad (7.103)$$

where the complex frequencies $\hat{\Omega}_{uv} = \Omega_u - \Omega_v - i\Gamma_{uv}$ have been introduced to conserve space, and the symbol "$\alpha\beta\gamma\delta$" denotes the order of the Cartesian components of the product of the dipole matrix elements $\tilde{\mu}_{un}\tilde{\mu}_{nv}\tilde{\mu}_{vm}\tilde{\mu}_{mu}$. Two "correction factors" appear in this expression

$$\kappa_1(\Omega_j,\Omega_k) = \frac{i(\Gamma_{vn}-\Gamma_{vu}-\Gamma_{un})+i(\Gamma_{mn}-\Gamma_{mu}-\Gamma_{un})(\hat{\Omega}_{vu}-\Omega_j)/(\hat{\Omega}_{mn}-\Omega_k)}{\hat{\Omega}_{vn}-\omega_s} \tag{7.104a}$$

$$\kappa_2(\Omega_j,\Omega_k) = \frac{i(\Gamma_{vn}-\Gamma_{vu}-\Gamma_{un})+i(\Gamma_{mn}-\Gamma_{mu}-\Gamma_{un})(\hat{\Omega}^*_{vu}+\Omega_j)/(\hat{\Omega}_{mn}-\Omega_k)}{\hat{\Omega}^*_{vm}+\omega_s} \tag{7.104b}$$

with arguments $\Omega_1 = \omega_0 + \omega_1$, $\Omega_2 = \omega_0 - \omega_2$, and $\Omega_3 = \omega_1 - \omega_2$. In the damping approximation where pure dephasing is absent or the relaxation rates of the elements of the density matrix are the sums of the relaxation rates of the amplitudes of the corresponding wave functions, these "correction factors" vanish [7.53,54]. With these rather tenuous assumptions, the result of (7.102,103) becomes equivalent to that quoted previously in [7.42,53,67,81]. Feynman diagram techniques do not lead to expressions comparable to (7,103) because they cannot account for "collisional" effects which act differently on the wave function and its complex conjugate [7.87]. Double Feynman diagram techniques are more useful, but the correct form is obtained only after laboriously summing the effects of many diagrams [7.88,89].

To explore the meaning of (7.102,103) it is convenient to assume that only one ground state $|g\rangle$ and one Raman level $|r\rangle$ is populated at thermal equilibrium, and that all other states are unpopulated. Chi-three then becomes

$$\begin{aligned}\chi^{(3)}(-\omega_s,\omega_0,\omega_1,-\omega_2) \propto\ & \rho^e_{gg}\langle g|\tilde{\chi}^{(3)}|g\rangle + \rho^e_{rr}\langle r|\tilde{\chi}^{(3)}|r\rangle \\ & + \rho^e_{gg}\langle r|\tilde{\chi}^{(3)}|g\rangle + \rho^e_{rr}\langle g|\chi^{(3)}|r\rangle \\ & + \rho^e_{gg}\langle t|\tilde{\chi}^{(3)}|g\rangle + \rho^e_{rr}\langle t|\chi^{(3)}|r\rangle + \dots \quad .\end{aligned} \tag{7.105}$$

Comparing (7.102,103), it is evident that the first two terms in (7.105) contribute a new resonance peaked near $\omega_1 - \omega_2 = 0$ or $\omega_0 - \omega_2 = 0$ when the input frequencies lie within absorption bands. Such effects have been explored in CARS by YAJIMA [7.90] and in RIKES by SONG et al. [7.91], they are closely related to the grating dip effects predicted by SARGENT [7.92]. It is also evident from (7.104,105) that the third and fourth terms contribute the Raman resonant signals, but that the coefficients of ρ^e_{rr} and ρ^e_{gg} differ. Line strengths therefore are no longer proportional to population differences. Nor are the signals obtained with different frequency combinations simply related to one another. There are, in fact, relatively few substances with energy levels and relaxation processes well enough determined to allow even qualitative predictions to be made.

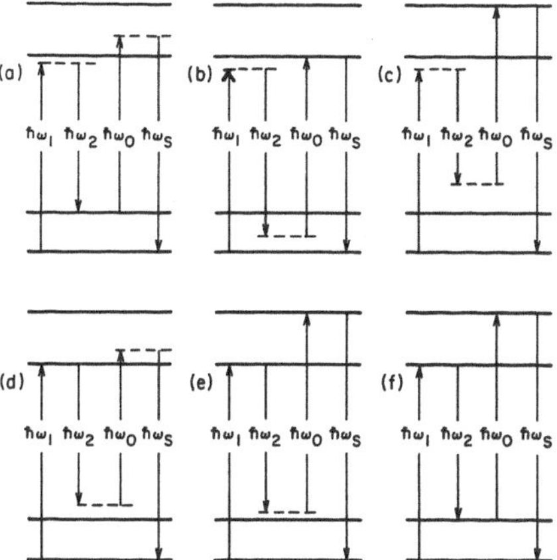

Fig. 7.8a-f. Some of the accessible resonance conditions in an absorbing sample. Part (a) shows the true Raman resonance; (b) shows an output absorption resonance accessed by tuning ω_2; (c) shows another output absorption resonance resulting from tuning ω_0 or ω_1, while (d) is an input absorption resonance. The doubly resonant condition in (e) involves Raman transitions among excited electronic states, and can be stronger than (a), while the triply resonant condition in (f) gives the strongest of all signals. These diagrams show the relative magnitudes of the photon energies and level splittings and should not be interpreted as describing time ordered processes

In CARS, CSRS and four-wave mixing, the Raman resonances that occur when $|\omega_1-\omega_2| = \Omega_{rg}$ or $|\omega_0-\omega_2| = \Omega_{rg}$ are accompanied by "output absorption" resonances occurring when $\omega_s = \Omega_{ng}$. Figure 7.8 diagrams these and other resonance conditions [7.53]. Previously weak two-photon resonances are also enhanced, becoming "excited state absorption" resonances and these effects are responsible for the remaining terms in (7.105). No effective two-level theory such as in [7.47] can account for these additional resonances; hence the need for this more detailed treatment.

In RIKES and SRS, $\omega_0 = -\omega_1$ and the situation is even worse. Terms appear in $\langle g|\chi^{(3)}|g\rangle$ and $\langle r|\chi^{(3)}|r\rangle$ that are independent of $\omega_1-\omega_2$, contributing a constant background signal whenever a one-photon resonance condition is fulfilled. These terms correspond to steady-state changes in the populations of the interacting levels, and are employed in saturation spectroscopy and polarization spectroscopy [7.91-96]. In SRS and RIKES, they contribute a large complex background susceptibility which limits the sensitivities of the techniques. Polarization selection often cannot suppress this background without also suppressing the Raman signals. In RIKES with circular polarization, for example, the Raman resonance vanishes for $\rho_Q = 1/3$, which is the most common case under resonance conditions. Excited state absorption resonances also make large contributions.

As complex as these formulae are, they do not encompass all the important expected or observed effects. In particular, to account for inhomogeneous broadening due to the Doppler effect in gases or random local strains in condensed phases, one must average (7.103) over the expected resonant frequencies. Also, at moderate to high intensities, the coherent Raman signal amplitude can cease to scale as $E(\omega_0)E(\omega_1)E^*(\omega_2)$ and saturate or even decrease [7.83,84]. To properly account for these phenomena, one must calculate at least the fifth-order nonlinear susceptibility [7.97].

7.3 Experimental Techniques

7.3.1 CARS in Liquids and Solids

Coherent anti-Stokes Raman spectroscopy (CARS) is the most widely practiced nonlinear mixing spectroscopy technique [7.36,43,44,55]. In CARS two beams at frequency ω_1 and ω_2 are mixed in the sample to generate a new frequency: $\omega_s = 2\omega_1 - \omega_2$. If ω_1 corresponds to the laser frequency in a spontaneous scattering experiment, and ω_2 to the Stokes-scattered photon, the output occurs at the corresponding anti-Stokes frequency. If $\omega_1 < \omega_2$, the analogous technique is called coherent Stokes-Raman spectroscopy (CSRS) [7.81,98].

The major experimental advantage of CARS — and of most coherent Raman techniques — is the large signal produced [7.99]. In a typical liquid or solid the effective third-order nonlinear susceptibility is approximately 10^{-13} esu, and in a typical CARS experiment, the laser powers are 10 kW or so and the interaction length is 0.1 cm. Under these circumstances, (7.70,75) imply a CARS output power of ~1 W, while conventional Raman scattering would give collected signal power of ~100 μW with the same lasers. Since the CARS output is directional, the collection angle can be five orders of magnitude smaller than that needed in spontaneous scattering. Taken together these two factors imply that CARS is nine orders of magnitude less sensitive to sample fluorescence than spontaneous scattering. The advantage is actually somewhat greater; since the output is at a higher frequency than the input, Stokes law of fluorescence implies that spectral filtering can further reduce the background fluorescence level.

The main disadvantages of CARS are 1) an unavoidable electronic background nonlinearity which alters the line shapes and can limit the detection sensitivity [7.49], 2) a signal which scales as the square of the spontaneous scattering signal (and as the cube of laser power), making the signals from weakly scattering samples difficult to detect [7.101], and 3) the need to fulfill the phase matching requirements of (7.81,84) [7.27]. While other techniques avoid these difficulties, CARS remains the most popular coherent Raman technique.

Many different kinds of lasers can be used as the sources of the two input beams. The main requirements are: good beam quality, adequate power and tunability. Figure 7.9 shows a system which uses two Hänsch-type dye lasers simultaneously pumped by a single nitrogen laser [7.98,99,101]. The peak power of these dye lasers can be in the range of 10-100 kW at about 10 Hz repetition rate with a linewidth of approximately 0.5 cm^{-1}. The beam divergence is typically twice that of a TEM_{00} mode. The second harmonic of a pulsed Q-switched Nd:YAG laser can be used as one of the CARS input beams and also as the pump for the tunable laser [7.99,102]. The power produced by such a system is much greater than that of a nitrogen laser based system ($P_1 \gtrsim 5$ MW $P_2 \gtrsim 0.1$ MW), but too much power is not entirely useful. Generally 100 kW peak power or less is sufficient for CARS in liquids and solids; greater power can lead to self-

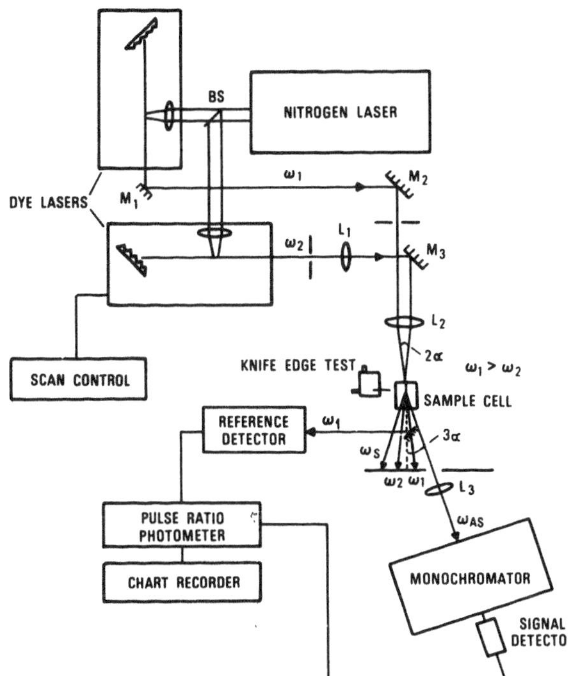

Fig. 7.9. Experimental apparatus for CARS. The two parallel dye laser beams are focused into a sample at the crossing angle 2α which equals θ in (7.84). The CARS output is directed into a monochromator, detected by a photodiode and averaged on a ratioing gated integrator. [7.98]

focusing, stimulated Raman oscillation, and damage to the sample [7.10,103]. Flashlamp pumped dye lasers and parametric oscillators have been used as laser sources for CARS, but have proven more difficult to scan than the laser pumped dye lasers which employ Littrow mount gratings as tuning elements [7.105-106]. To date only AKHMANOV has successfully employed CW lasers in a CARS experiment in liquids [7.107, 120]. The low power produced by such lasers leads to a prohibitively low CARS signal level.

A lens with a focal length between 10 and 20 cm focuses the two parallel incident beams into the sample. Care must be taken to ensure that the two beam waists lie at the same distance from the lens. The practice in our laboratory is to locate the beam waists by burning the emulsion off a piece of underexposed Polaroid film held in the focal region. The plane where the loudest popping sound is produced is the plane of the beam waist. CHABAY et al. have devised a knife edge test procedure to ensure that the beam waists coincide spatially [7.98]. Our experience is that the beams can be overlapped visually even though their diameters are only 50 microns or so.

The positions of mirrors M_2 and M_3 in Fig.7.9 must be adjusted so that the CARS wave vector matching condition is fulfilled [7.49]. If these mirrors are mounted on translation stages oriented so that translation of the mirrors does not affect the orientation of the plane defined by the crossing beams, the crossing angle can be adjusted without affecting the beam overlap. Spherical aberation will, however, cause the plane of maximum overlap to occur at slightly different distances from the lens. If the sample and the beam waists are more than 1 mm long, this translation of the overlap position is unimportant; it can be eliminated, however, by using a lens corrected for spherical abberation or by readjusting the angle of one of the mirrors. In CARREIRA's CARS microanalysis system, the mirror positions and angles are automatically optimized by computer [7.85].

The anti-Stokes output beam is generated at approximately the crossing angle of the input beams. A moveable knife edge or iris can block the laser beams while allowing the CARS signal into the detection system. A lens which images the interaction region in the sample onto the detector will collect the CARS signal beam radiated at any angle [7.49]. Alternatively, the entire detection system can be pivoted around the sample, with the collection angle optimized by computer [7.85].

Figure 7.9 shows a monochromator in the detection system, to spectrally filter the output and eliminate elastic scattering of the incident frequencies. If the Raman frequency is larger than about 200 cm^{-1}, a single 1/4 m monochromator will provide sufficient spectral filtering. For lower Raman frequencies, one should employ a tandem monochromator. Of course, a colored glass filter or an interference filter will suffice for the strongest and highest frequency modes.

The output beam can be detected by a p-i-n photodiode or an inexpensive photomultiplier. Because the CARS signal can span a dynamic range of eight decades, it is convenient to provide a calibrated variable attenuator to control the intensity at the detector surface (in some systems, the attenuators are adjusted by computer) [7.69]. The electrical pulses produced by the detector must be integrated and amplified, and then averaged using a gated integrator or computer system. Figure 7.9 shows a second reference detector which monitors and averages the intensity of the pump laser. Some of the noise resulting from fluctuations in the laser output can be eliminated by dividing the CARS signal by this monitor signal [7.98]. Other workers have monitored the CARS outputs from two different samples in order to reduce the noise due to laser fluctuations [7.49].

The CARS spectrum is actually the normalized and averaged signal intensity plotted as a function of difference frequency $\omega_1 - \omega_2$. In practice, ω_1 is often fixed while ω_2 is varied by tuning the dye laser grating [7.33]. It should be noted that the signal frequency ω_s varies with ω_2 and care should be taken that the signal frequency ω_s remains within the bandpass of the spectral filtering system. When a monochromator is used, its grating should be scanned synchronously with ω_2.

This synchronization is difficult to achieve mechanically, and early workers chose to scan the frequency ω_1 half as fast as ω_2 in order to keep ω_s fixed [7.49, 108]. CARS spectra of diamond taken in this way in various polarization conditions appear in Fig.7.10 [7.49]. For scans of a few tens of wave numbers, the monochromator bandpass can be made large enough to span the expected range of ω_s.

It remains difficult to take CARS spectra over an extended frequency range. Part of the difficulty results from the limited tuning range of individual laser dyes,

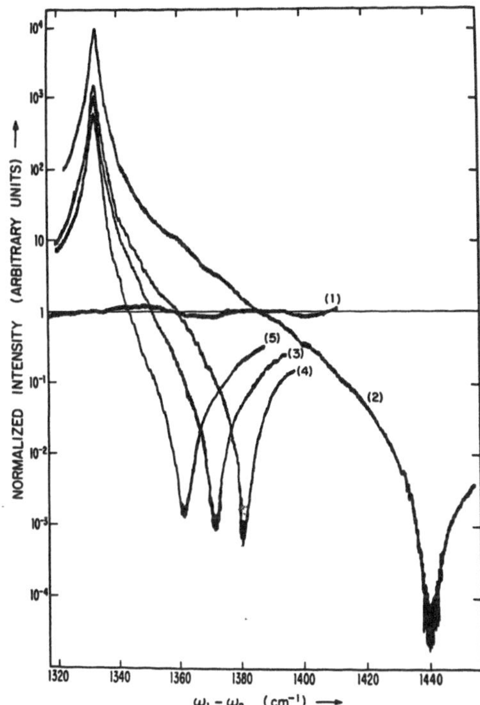

Fig. 7.10. CARS spectra of diamong. Each of the traces corresponds to one of the polarization conditions of Table 7.1. In unpolarized light, the deep minima would not appear. [7.49]

and part from the tendency of most dye laser cavities to misalign somewhat as the frequency selective element is rotated. In CARS there are additional difficulties related to the variation of the optimum crossing angles with frequency. To scan more than 100 cm^{-1}, it is necessary either to periodically readjust that angle by moving mirror M_2 or M_3 or to continuously vary the angle as the frequencies are changed [7.98]. If spherical aberration is significant, the angle of the translated mirror and the collection optics may also require adjustment. The intensity of the scanned laser will vary, altering the signal level. All of the factors can be managed in a system where the lasers, beam geometries, collection optics, and monochromator operate under computer control and in which the CARS signal and laser powers are monitored by the computer [7.85].

Another complication occurs in absorbing samples. Equations (7.56,57) must be modified to account for the attenuation of the incident and signal amplitudes which depend, in general, upon the frequencies and intensities. If the attenuation coefficients are independent of intensity, the amplitudes of the incident waves decrease as

$$E_j \propto e^{-\beta_j z} \quad .$$

Equation (7.56) implies that the sample length yielding the maximum CARS signal amplitude is

$$\ell_{OPT} = \left|\frac{\ln[\beta_s/(2\beta_1+\beta_2)]}{2\beta_1+\beta_2-\beta_s}\right| \quad . \tag{7.106}$$

If all of the attenuation coefficients are equal, (7.106) implies an optimum sample optical density of 0.48. Near this attenuation level, small changes in the absorption coefficient for one of the beams do not significantly affect the CARS spectrum.

The length of an absorbing sample should be less than the length of the region where the incident beams overlap. Standard cuvettes are available with sample thicknesses less than a millimeter, but CARREIRA has found that melting point capillaries can be more convenient [7.84]. Liquid samples can be flowed using such capillary cells, and photochemical decomposition thus avoided. In CARREIRA's setup, the capillary was in the plane of incidence and positioned such that the focused incident beams were not vertically deviated. This alignment condition assured that beams propagate parallel to the diameter of the capillary, and actually facilitated alignment.

7.3.2 CARS in Gases: Pulsed Laser Techniques

The nonlinear susceptibility of a gas at STP is 100 times smaller than that of a typical liquid, but the interaction length can be much longer since (7.84) implies a near-zero crossing angle for optimal phase matching. The coherence length for collinear beams is in fact

$$\ell_c = \frac{\pi}{\Delta k} = \frac{\pi c}{\Omega_Q^2}\,(2n_1'+\omega_1 n_1'') \tag{7.107}$$

which is generally longer than either the sample or the beam waist [7.109]. Since the optical breakdown threshold for gases is orders of magnitude higher than for liquids, all of the power produced by a Nd:YAG based CARS laser system can be profitably employed [7.99]. Thus the CARS signals can be nearly as large as in liquids.

The typical experimental setup for pulsed CARS in gases is very similar to that in Fig.7.9, the main differences being a longer sample cell and some provision for making the input beams collinear. High intensity beams can be combined using a dichroic beam splitter, a carefully aligned dispersing prism [7.102] or – if the planes of polarization are orthogonal – a glan-laser prism [7.103]. The signal beam is best separated from the incident beam using a prism or interference filter rather than an easily damaged monochromator.

7.3.3 Multiplex CARS

Multiplex CARS is an attractive alternative to the single frequency pulsed techniques for gas spectroscopy [7.107,112]. In multiplex CARS, the Stokes beam at ω_2 is produced by a broadband laser. Since the CARS output is linear in the Stokes input, the CARS signal due to each frequency component of the broadband laser can be separated by a spectrograph and detected with an optical multichannel analyzer or a photographic

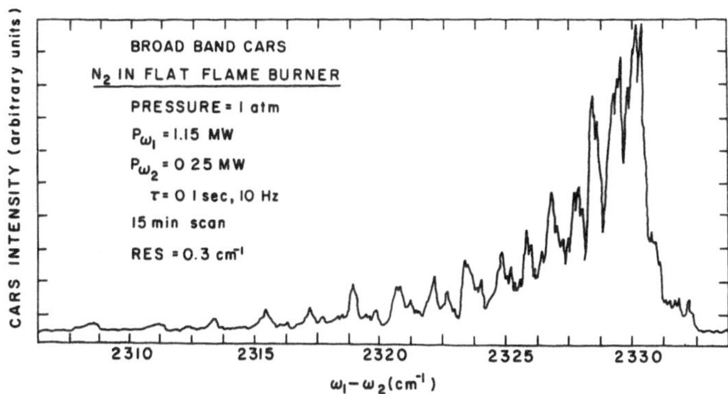

Fig. 7.11. Multiplex CARS spectrum of the Q branch of N_2 in a flame. The instantaneous concentration and temperature can be determined from such traces. [7.99]

plate. The resolution is obviously limited by the spectrograph employed, but much of the Raman spectrum can be obtained in a single laser shot [7.41]. That can be important in studies of the rapidly changing systems common in combustion. Averaging over many shots gives spectra of the same quality as single-channel CARS spectra if the lasers have comparable output powers [7.111].

Figure 7.11 shows a recent multiplex CARS spectrum taken from N_2 in an atmospheric pressure methane/air flame using an Nd:YAG laser as the pump source [7.99]. These researchers report that CARS spectra in laboratory flames (0.1 to 1.0 atm total pressure) can be recorded with minimum spectral slit widths. They also demonstrated that the single pulse experiments are feasible even in the sub-Torr pressure range by taking a CARS spectrum of CH_4 at 600 mT pressure.

It is also possible to take multiplex CARS spectra from condensed phases. In particular, when the sample length is very short as in one-photon resonant CARS experiments, the phase matching requirement can be relaxed and a broadband laser can be employed for the Stokes beam. LAU et al. obtained a single pulse CARS spectrum from several dye solutions (at 10^{-4} M concentration) employing a narrowband ruby laser and a broadband dye laser [7.110].

7.3.4 CW CARS

High resolution (3 MHz or 10^{-4} cm^{-1}) can be achieved in gas-phase CARS spectra with single mode CW lasers. The obvious disadvantage is the drastically reduced CARS signal level due to the low power levels of CW lasers. Thus the observation of CW CARS signals has been limited to relatively strong Raman transitions.

The first CW CARS experiment was reported by BARRETT and BEGLEY [7.113]. Their CARS signal level at the peak of the Q branch of the ν_1 mode of methane was actually lower than that obtained by spontaneous scattering under the same conditions. More recently BYER and co-workers have observed Raman lines in H_2, D_2 and CH_4 using the

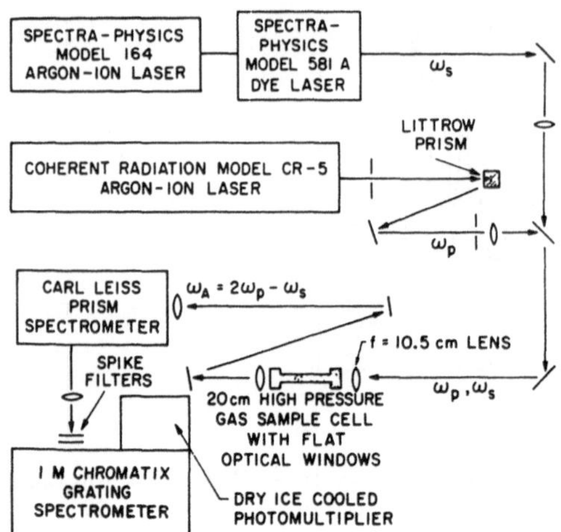

Fig. 7.12. CW CARS apparatus for high resolution spectra of gases. [7.114]

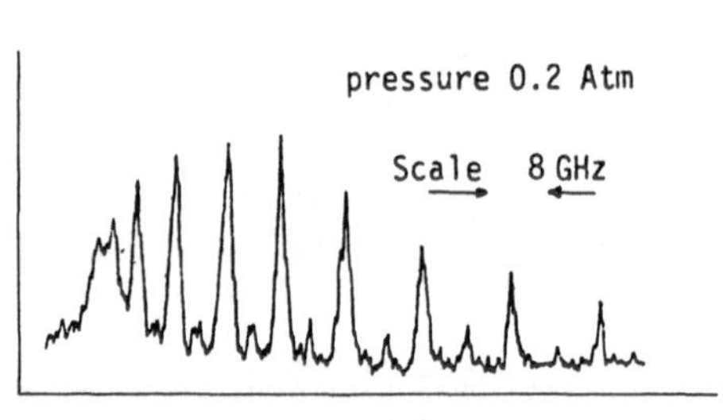

Fig. 7.13. CW CARS spectrum of the ν_2Q branch of C_2H_2 at 0.2 atm. The maximum signal corresponds to a few thousand photoelectron counts per second. [7.115]

CW CARS apparatus depicted in Fig.7.12 [7.65,114]. An argon-ion laser (5145 Å) and a CW dye laser (6054 Å) operating in single modes at power levels of 640 mW and 10 mW, respectively, were collinearly overlapped in the high pressure (21 atm) gas cell. The two beams were focused with a 10.5 cm lens to minimum radii of 6 μ and 20 μ for the argon and dye laser, respectively. The CARS signal due to the $Q(\nu_1)$ branch of CH_4 at 2416.7 cm^{-1} (or a Q-branch line of H_2 or D_2) was isolated from laser radiation and transmitted through a prism prefilter, spike filters and a 1 m grating spectrometer and then finally detected with a cooled photomultiplier tube operating in the photon counting mode.

By employing a similar apparatus to that shown in Fig.7.12, FABELINSKY et al. have obtained the high-resolution CARS spectrum of C_2H_2 and D_2 at various pressures between 0.06 and 40 atm [7.115]. Instead of a spectrometer they used a four-prism filter system which allowed a reduction of the pump laser light intensity by more than 17 orders of magnitude while transmitting more than 50% of the anti-Stokes signal beam. Figure 7.13 shows their CARS spectrum of C_2H_2 taken at 0.2 atm.

Finding these narrow Raman lines can be a significant problem in all such high-resolution experiments. It is necessary to search the region of the reported resonance systematically, resetting the tunable laser to a known frequency at the beginning of each scan. Precise frequency calibration techniques based upon the traveling-Michelson Lambda meter or upon fixed spacing Fabry-Perot or Fizeau interferometers are essential [7.116-118].

7.3.5 Nonlinear Ellipsometry

Another interesting CARS variation is the coherent Raman ellipsometry technique of AKHMANOV [7.119,120]. This technique is based on the measurement of the dispersion of the polarization parameters of the CARS signal upon tuning $\omega_1 - \omega_2$. If the input laser beams are linearly polarized at an angle to 45° to one another, the nonresonant signal field is also linearly polarized. As $\omega_1 - \omega_2$ approaches a Raman level, however, the signal field becomes elliptically polarized. The polarization state of the signal field can then be accurately analyzed by use of a Babinet-Soleil compensator. From the measurement of the ellipticity of the CARS signal and the direction of the major axis of the ellipse, it is possible to accurately determine the depolarization ratio ρ_Q of the Raman mode as well as the ratio of the nonresonant term χ^{NR} to the Raman susceptibility χ^R.

Since this method is based on measurement of "polarization dispersion" rather than "amplitude dispersion" as in conventional CARS, laser power fluctuations do not contribute to the uncertainties of the measurement and the parameters to be determined can be measured with relative accuracy up to 1 part in 10^4. Fine structure due to variation in the depolarization ratio (which does not appear in the spontaneous Raman data) can also be detected in the nonlinear ellipsometry spectrum.

7.3.6 Raman Induced Kerr Effect Spectroscopy (RIKES)

A strong pump wave propagating through a nonlinear medium induces an intensity dependent dichroism and birefringence which can alter the polarization of a weaker probe beam [7.121]. If the probe beam is initially linearly polarized, the change in polarization can be detected as an increase in the intensity transmitted through a crossed polarizer. The nonlinear susceptibility that produces this optical Kerr effect has contributions from the Raman modes of the material in addition to contributions from reorientation of molecules and electrons in the medium [7.27]. The Raman terms make their major contribution to the Kerr effect when the pump and probe frequencies differ by a Raman frequency, and the detection of these resonances is the basis of Raman induced Kerr effect spectroscopy (RIKES) [7.39,122].

The polarization conditions used in RIKES are numbered 7 and 8 in Table 7.2, and (7.57,58) give the corresponding RIKES signal amplitudes. The Raman resonant terms in the nonlinear susceptibilities in Table 7.2 are clearly comparable to the CARS resonances. The main advantages of RIKES are: 1) a phase-matching condition that is automatically fulfilled for every propagation direction and frequency combination; and 2) the suppression of the nonlinear background when the pump wave is circularly polarized as in polarization condition 8. The main disadvantage is a background signal due to stress induced birefringence in the sample and optics.

Figure 7.14 shows a typical RIKES setup [7.62,122]. The lasers used are essentially identical to those recommended for CARS in the previous sections, but in RIKES the probe is the laser that need not be tunable. A glan-laser prism ensures that the probe beam is linearly polarized and the beam is focused into a sample cell by a lens of roughly 15 cm focal length. After the sample cell, the probe beam is refocused into a Babinet-Soleil variable wave plate which partially compensates for

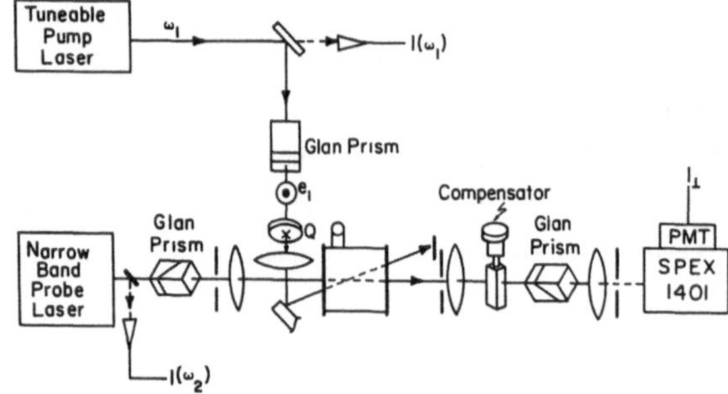

Fig. 7.14. Apparatus for Raman induced Kerr effect spectroscopy. The pump polarization is controlled by the quarter wave plate at Q. The Babinet-Soleil compensator cancels out the effects of uniform birefringence. The detected signal is proportional to the intensity in a plane perpendicular to the initial probe polarization

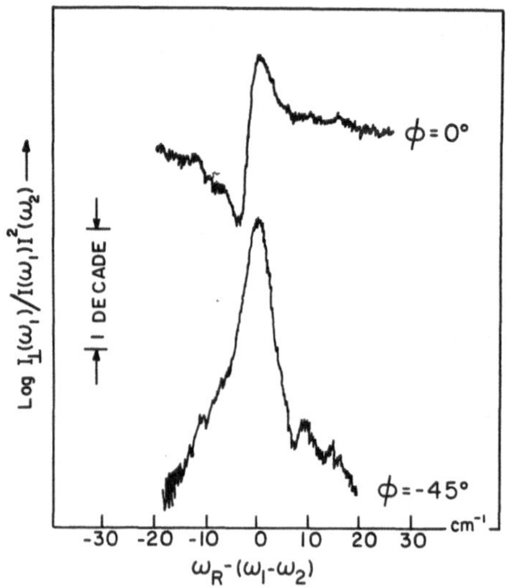

Fig. 7.15. RIKES traces of benzene near the 992 cm^{-1} mode. The lower trace shows the background suppression produced by a circularly polarized pump, while the upper trace shows the background level due to χ^{NR}

any depolarization due to the sample or optics. The probe beam is then blocked by a second carefully aligned glan prism and the RIKES signal beam is focused into a monochromator and a detector. Judiciously placed irises provide spatial filtering.

The polarization of the tunable pump laser is controlled by means of a linear polarizer and a rotatable quarter wave plate Q. The pump beam is focused into the cell and overlapped with the probe as in CARS, but because there is no wave vector matching condition to fulfill, the angle between the beams can be made very small and the length of the interaction region maximized. When the pump is linearly polarized, the nonresonant background susceptibility is sufficient to provide a RIKES signal whenever the beams overlap. If the quarter wave plate is adjusted to give circularly polarized light in the sample, that nonresonant background will disappear. Figure 7.15 shows RIKES spectra of Benzene taken by scanning the pump frequency in such a system with a linear and a circular polarized pump.

Stress induced birefringence in the sample and optics produces another troublesome background signal. The Babinet-Soleil compensator can eliminate the effects of uniform stress, but not the effects of variations in the stress. We have found that

Schott 57 glass has remarkably little stress birefringence, and have had good results with sample cell windows made of this material or of very thin microscope cover slides. Another promising material is [111] oriented calcium fluoride which theoretically should cause no depolarization of the probe if properly oriented [7.123].

To minimize the birefringence background signal, the compensator plate and polarization analyzer must be sequentially readjusted to reach a minimum. Our experience also indicates that it is also necessary to try various positions of the sample and focusing lenses in order to find the point of most uniform strain. With care, the birefringence background intensity can be reduced to less than 10^{-6} of the probe intensity, or two orders of magnitude less than a typical RIKES signal [7.122].

Multiplex RIKES is possible when the pump laser is narrowband and the probe broadband. A spectrograph separates the frequency components of the beam that is transmitted through the polarization analyzer and the RIKES spectrum can then be recorded on an OMA or photographic plate. Since the phase matching condition is automatically fulfilled, multiplex RIKES spectra can be taken over a wide frequency range in liquids, solids and gases [7.39].

7.3.7 Optical Heterodyne Detected RIKES

The sensitivity of Raman induced Kerr effect spectroscopy is radically improved and the importance of birefringence background signals is reduced when optical heterodyne techniques are employed to detect the signal amplitude [7.40]. The resulting technique — termed OHD-RIKES — has greater demonstrated sensitivity than any other coherent Raman technique, and yet is remarkably simple in operation. The signals obtained scale linearly with the concentration and Raman cross section, as does the spontaneous scattering intensity, and the line shapes are directly comparable to those in spontaneous scattering.

One form of OHD-RIKES apparatus is diagrammed in Fig.7.16, and the differences between this setup and that for ordinary RIKES are quite subtle. The pump laser is pulsed or modulated, and the probe laser is generally CW and stable. In Fig.7.16, the probe beam is shown to be slightly elliptically polarized. The horizontal component acts as a local oscillator field, which is out of phase by $\pi/2$ with the vertical component which corresponds to the linear "probe" beam of ordinary RIKES. In this configuration, the polarization analyzer transmits the horizontal component of the probe beam as well as the RIKES amplitude E_s. In Fig.7.16 the RIKES amplitude is shown as a pulse having the same shape as the pump laser pulse and lying on top of the local oscillator. Spatial and spectral filters separate these signal amplitudes from scattered pump light, and the corresponding intensity

$$I(t) = \frac{nc}{8\pi} |E_{LO}(t)+E_s(t)|^2 = I_{LO}(t) + I_S(t) + \frac{n_s c}{4\pi} \mathrm{Re}\{E^*_{LO}(t)E_s(t)\} \qquad (7.63)$$

is detected photoelectrically. An electrical bandpass filter separates the ac signals due to $E_S(t)$ from the near dc signals due to $I_{LO}(t)$. Because $E_{LO} \gg E_S$ the hetero-

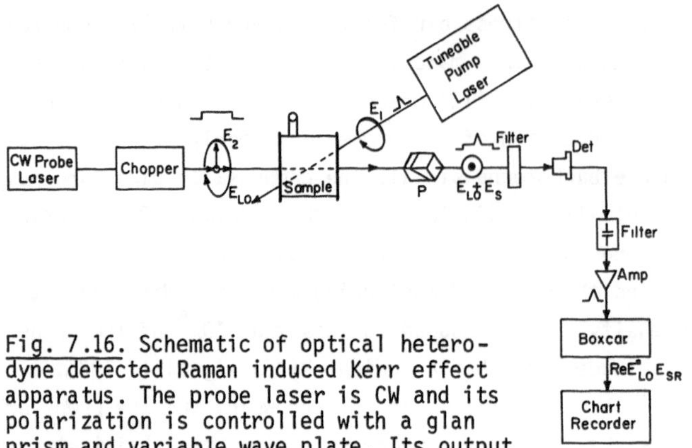

Fig. 7.16. Schematic of optical heterodyne detected Raman induced Kerr effect apparatus. The probe laser is CW and its polarization is controlled with a glan prism and variable wave plate. Its output is chopped to prevent excessive average power damaging the sample or detector. Electronic filtering separates the high-frequency OHD-RIKES signal from the near CW local oscillator signal

dyne term $(n_s c/4\pi)\mathrm{Re}\{E^*_{LO}(t)E_s(t)\}$ makes the dominant ac contribution to the electrical signal. The heterodyne term is then averaged and plotted as a function of the pump laser frequency.

Equations (7.53-58) imply that this heterodyne signal is proportional to the spontaneous scattering signal, but it is also sensitive to the phase of $\chi^{(3)}$. The detected phase depends upon the polarization of the pump and probe. There are essentially four cases yielding four different types of spectra. When the probe wave is elliptically polarized as in Fig.7.16 the local oscillator is in quadrature with the probe and the detected line shape corresponds to $\mathrm{Re}\left\{D\chi^{(3)}_{\mathrm{eff}}\right\}$ as given in Table 7.2. When the pump is circularly polarized, this choice of phases gives a resonance-type line shape similar to spontaneous scattering, except that polarized and depolarized peaks will have opposite signs. If the pump is linearly polarized, dispersion-type line shapes will appear. A local oscillator in phase with the probe can be obtained when the output of the probe laser is linearly polarized but the polarization analyzer P is rotated away from the angle of minimum probe transmission. In this case the spectrum reflects $\mathrm{Im}\left\{D\chi^{(3)}_{\mathrm{eff}}\right\}$ as in Table 7.2, and a linearly polarized pump gives the resonance line shapes while a circularly polarized pump gives dispersion-type line shapes.

The optimum strength of the local oscillator wave depends upon the characteristics of the lasers and detectors. The details of this optimization will be discussed in Sect.7.3.10. For the poorly stabilized 0.5 W single mode CW argon laser used as a probe in our laboratory, the optimum local oscillator power was 0.3 mW. Laser noise degraded the OHD-RIKES signals when the total oscillator level was larger, and shot noise degraded the spectra when the level was lower [7.72]. A 50 kW Hänsch-type dye laser was the pump laser in our experiments.

The detector used in OHD-RIKES must be able to withstand milliwatts of local oscillator power and yet not contribute excessive electrical noise. In our laboratory

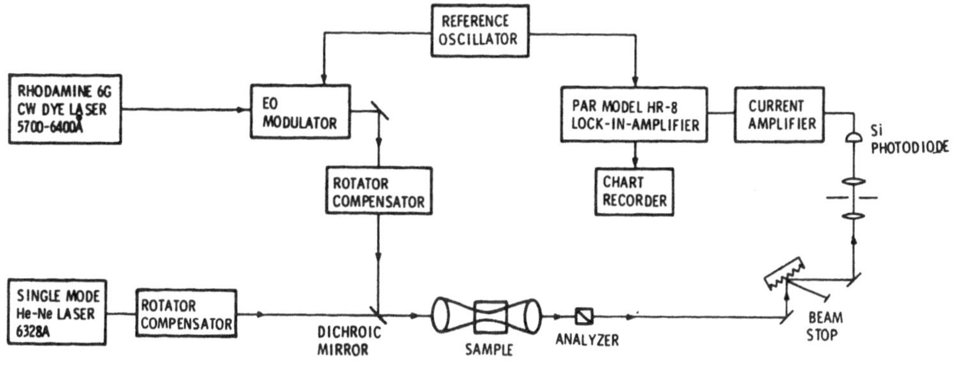

Fig. 7.17. CW OHD-RIKES apparatus for high-resolution spectroscopy of gases. [7.124]

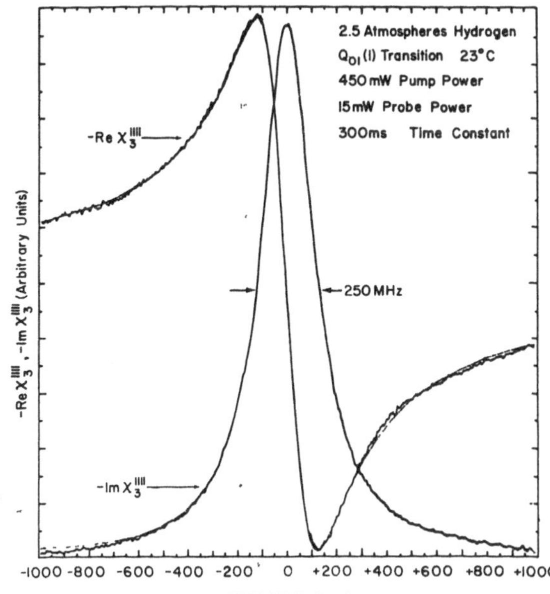

Fig. 7.18. OHD-RIKES traces of $Q(1)v=0\text{-}1$ line of H_2. [7.63]

we have found that photomultipliers can be used if the last several dynode stages are shorted to the anode, but that a silicon p-i-n photodiode integrally coupled to a transconductance amplifier is more reliable[2]. Additional electrical amplification is provided for the heterodyne signal following the bandpass filter. To suppress electrical noise, the detector, amplifiers and all necessary power supplies were housed in a solid copper enclosure [7.72].

OWYOUNG has taken OHD-RIKES spectra using CW lasers; one version of his apparatus is shown in Fig.7.17 [7.37]. The pump beam from a CW dye laser was amplitude modulated at 25 KHz by an electrooptic modulator which was driven by a reference oscillator. A well-stabilized HeNe laser operating on a single longitudinal mode at

[2] Meret Corp. MDA 7708 proved best.

6328 Å was the probe source. The pump and probe beams were combined on the dichroic mirror and focused collinearly through the sample. The probe beam was separated from the pump source by use of a spatial filter and two 1200 lines/mm gratings. Input beam polarizations were controlled by a pair of rotator-compensators which consisted of half and quarter wave retarders and the output polarization was selected with an analyzing polarizer. The output was detected using a silicon photodiode, amplified in a current amplifier and finally demodulated with a lock-in amplifier which was referenced to the pump modulator. The spectrum was scanned by tuning the pump frequency with respect to the probe.

Figure 7.18 illustrates the high-resolution capabilities of OHD-RIKES. This CW OHD-RIKES spectrum of the Q(1) $v=0\rightarrow 1$ line of hydrogen was taken by OWYOUNG using apparatus similar to Fig.7.17 except that a single-mode argon laser was used as a pump and the dye laser became the probe [7.63]. Note that both resonance and dispersion line shapes can be obtained. The slight asymmetry in the $\mathrm{Im}\{\chi^{(3)}\}$ line shape can be eliminated by modulating the pump polarization rather than the pump intensity. Since the OHD-RIKES signal scales as the product of two laser intensities, rather than three, it is better suited than CARS for low power CW experiments. At low densities, the CW OHD-RIKES signal level is orders of magnitude larger than that of CARS, and the sensitivity is correspondingly better.

The main problem with OHD-RIKES occurs in absorbing samples where the pump beam can alter the intensity of the local oscillator wave, thereby producing spurious signals. In such samples OHD-RIKES is also sensitive to the effects studied in "polarization spectroscopy," and it becomes difficult to separate features in the Raman spectrum from those in the saturated absorption or excited state absorption spectra [7.91,94].

7.3.8 Stimulated Raman Gain and Loss Spectroscopy

The modern techniques of stimulated Raman spectroscopy (SRS) employ stable CW probe lasers and detect the small ($\sim 10^{-5}$) changes in intensity due to Raman gain or loss induced by a pump laser [7.124]. SRS shares with OHD-RIKES the advantages of a signal linearly proportional to the spontaneous cross section (and to the product of two laser intensities), and an automatically fulfilled phase-matching condition. It has the additional advantage of being insensitive to depolarization. The main disadvantage is that SRS is much more sensitive to laser noise than OHD-RIKES.

Figure 7.19 shows the high-resolution SRS system developed by OWYOUNG et al. [7.125]. The pump source was a single mode Ar^+ laser operating at 5145 Å. Its output was modulated electrooptically and was combined with the probe on a dichroic mirror. The probe laser was a feedback stabilized CW dye laser that was capable of tuning over 1 cm^{-1} with a resolution of 1 MHz. An electrooptic demodulator suppressed the characteristic power fluctuations, and a reference detector allowed the remaining fluctuations to be subtracted from the SRS signal by a differential input lock-in amplifier.

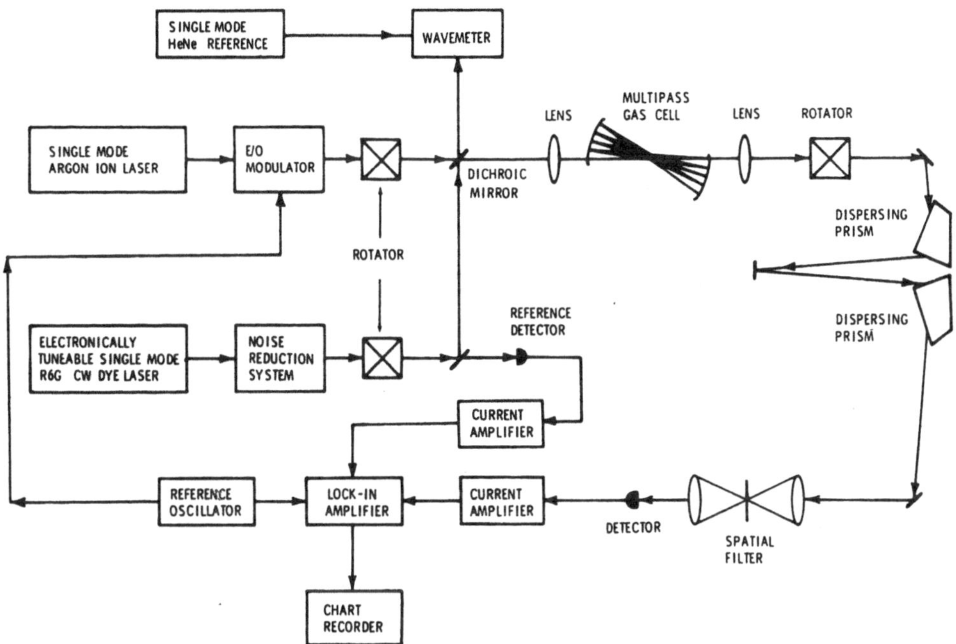

Fig. 7.19. High-resolution-high-sensitivity CW stimulated Raman spectrometer. The wavemeter at top is used to accurately determine the Raman frequency. Note the sophisticated signal enhancement and noise reduction features, such as the multipass cell and differential lock-in. [7.125]

The coincident beams were directed into a multipass cell. After 97 passes through the sample, the SRS signal had been enhanced by a factor of 50 over the single pass case, but the probe intensity had been reduced to 0.5 mW. The output beams were separated by dispersing prisms and a spatial filter, and the modulation of the probe due to the stimulated Raman effect was detected by a p-i-n diode and differential lock-in amplifier with quantum noise limited sensitivity.

The most striking demonstration of the resolution and sensitivity of this CW SRS system is shown in Fig.7.20. The spectrum displays the ν_1 fundamental mode of methane (near 2917 cm^{-1}) under the pressure of 35 torr [7.125]. This spectrum gives the first fully resolved Q-branch Raman spectrum of a polyatomic molecule.

The sensitivity of a "hybrid" SRS system employing a pulsed pump laser and a CW probe can be expected to be 1000 times greater than the all-CW system in Fig.7.19. Such systems may prove useful for coherent Raman spectroscopy in liquids and solids. OWYOUNG has already measured the stimulated Raman gain of benzene using such a system with a double Nd:YAG pump and a Jamin interferometer [7.126].

Sensitive multiplex SRS has been demonstrated by WERNCKE et al. using a technique where the Raman sample was within a broad band dye laser cavity [7.38,127]. The additional gain or loss due to the stimulated Raman effect altered the output spectrum of the laser, which was recorded photographically (see Fig.7.30). Other multiplex techniques await the development of suitable sources.

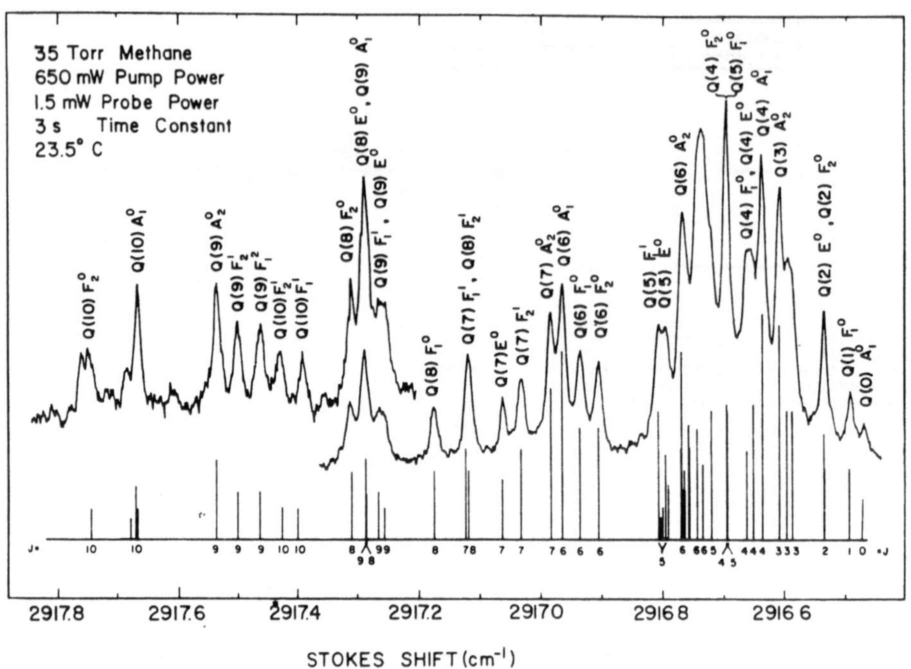

Fig. 7.20. High-resolution SRS trace of the Q branch of the ν_1 mode of methane at 35 Torr. The widths of these lines are due to the Doppler effect. [7.125]

7.3.9 Four-Wave Mixing

When three laser beams at three different frequencies, ω_0, ω_1 and ω_2 are mixed in the sample by the third-order nonlinear susceptibility, the coherently generated nonlinear signal can have the frequency $|\omega_s| = |\omega_0+\omega_1-\omega_2|$ [7.31,42,58]. Raman resonances occur when $|\omega_0-\omega_2| = \Omega_Q$ and when $|\omega_1-\omega_2| = \Omega_Q$. The process producing the output at $|\omega_s| = |\omega_0+\omega_1-\omega_2|$ reduces to the CARS process when $\omega_0 \rightarrow \omega_1$, and to the RIKES or SRS process when $\omega_0 \rightarrow -\omega_1$.

Four-wave mixing (4WM) techniques permit the suppression of troublesome background signals which limit the sensitivity of the corresponding three-wave mixing technique: the nonresonant electronic background signal in case of CARS and the strain-induced birefringence in case of RIKES [7.42,58]. Since the nonlinear susceptibility tensors that apply to 4WM have simultaneous resonances at more than one Raman frequency, it is possible to compare two different Raman cross sections by direct interference of the susceptibilities [7.42]. In some applications these advantages outweigh the difficulties inherent in an experiment employing three synchronized, overlapped and phase-matched laser beams. Less obvious are the difficulties connected with the low signal levels that result when the background signal is suppressed leaving a Raman signal that scales as the square of the Raman cross section and as a cubic product of laser intensities. CARS type processes produce outputs at $2\omega_1-\omega_2$ and $2\omega_0-\omega_2$ which are useful for aligning the apparatus but which must later be suppressed by poor

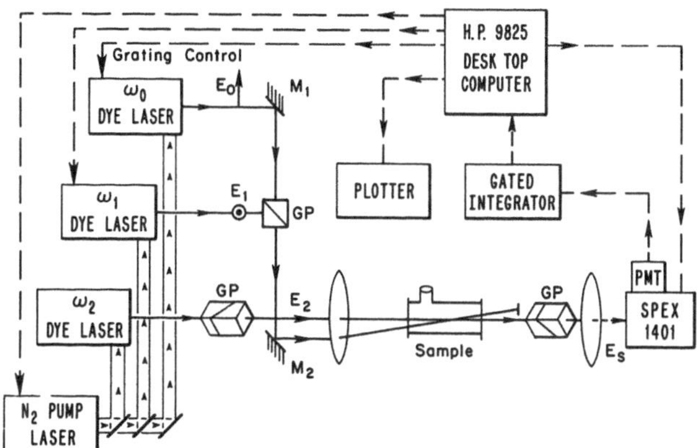

Fig. 7.21. A four-wave mixing spectrometer. The beams of two of the three Hänsch-type dye lasers are combined by a polarization beam splitter, labelled GP, and the combined beam is focused into the sample along with the output of the third laser

phase matching or separated from the 4WM output spatially or spectrally [7.60]. In scanning the four-wave CARS spectrum, it is convenient to keep $\omega_0-\omega_2$ constant while scanning the resonances as a function of $\omega_1-\omega_2$. That is, ω_0 and ω_2 are kept constant and ω_1 is tuned. Consequently, ω_s also changes and spectral filter frequency has to be varied [7.42].

Our four-wave mixing apparatus is diagrammed in Fig.7.21 [7.72]. This system employed three independent Hänsch-type dye lasers with two beams combined at a polarization beam splitter, but other authors have designed dye laser cavities capable of producing two distinct frequencies [7.128]. Our experience has been that such lasers are too noisy to be useful in spectroscopy [7.58]. The beams were focused into the sample by a lens with 15 cm focal length and the output frequency and polarization were selected by a glan prism and a tandem monochromator. The low signal levels necessitated the use of a photomultiplier tube as the detector.

Table 7.3 shows the effective nonlinear susceptibility and Raman tensor elements that are accessible in the possible 4WM polarization conditions. The most interesting are conditions (7.14-16) where some degree of nonlinear background suppression is feasible [7.58]. Of these, condition (7.16) – termed Asterisk – is the most useful. Whatever the nonresonant background susceptibility, its contribution to the 4WM signal will disappear when

$$\chi_{1122}^{NR} \cos\theta \sin\phi = \chi_{1212}^{NR} \cos\phi \sin\theta \quad . \tag{7.108}$$

Were Kleinman symmetry exact, this condition would be met at $\theta = \phi = 45^\circ$ [7.56]. If not, one must find the necessary angles experimentally. In practice it is convenient to set $\theta = \phi = 45^\circ$ to acquire the signal, tune off resonance, and adjust θ or the polarization analyzer angle until the background signal disappears.

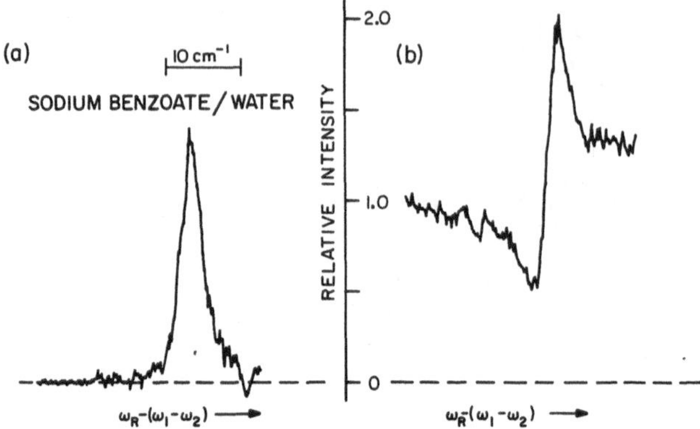

Fig. 7.22a,b. Sodium benzoate spectra taken in polarization conditions 16 (Trace a) and 12 (Trace b). The nonresonant background susceptibility is suppressed by a factor of 1000 in spectrum a taken using "asterisk"

Figure 7.22 shows four-wave RIKES traces of the 1005 cm^{-1} sodium benzoate mode in a 0.3 M aqueous solution. Figure 7.22a shows the background suppression possible in Asterisk, while trace (b) was taken in polarization condition 12. The frequency ω_0 was negative and $\omega_0+\omega_1$ was approximately 10 cm^{-1}; $|\omega_0|$ and $|\omega_1|$ were scanned in the same direction so that the output frequency $\omega_s = \omega_2 - (\omega_1+\omega_0)$ remained constant.

LOTEM et al. demonstrated a background suppression technique that works in all of the polarization conditions of Table 7.3 [7.42]. They set $\omega_0-\omega_2$ at the frequency corresponding to the minimum in the four-wave CARS signal where the real part of Raman susceptibility cancels the nonresonant background. By tuning ω_1 (and ω_s) they were able to scan $\omega_1-\omega_2$ over weak resonances and detect them with no background.

7.3.10 Signals, Noise and Sensitivity

The coherent Raman techniques produce immensely more intense optical signals than spontaneous scattering techniques, but the increased signal level does not immediately translate into better sensitivity or higher quality spectra. To see why not and to compare the capabilities of the various techniques, it is necessary to do a formal analysis of the signal-to-noise ratio [7.129].

The most convenient starting point for such an analysis is the nonlinear signal amplitude defined in (7.58). Only part of this amplitude results from the Raman process, the rest coming from background terms in the nonlinear susceptibility [7.130]. The true coherent Raman signal amplitude $\underline{E}_r$ is obtained from (7.58), when the proper coherent Raman polarization P^Q from (7.29-33) is substituted for P^{NL}.

At the photodetector surface, there are other amplitudes besides the Raman signal amplitude. The amplitudes that add coherently to the Raman amplitude can be added together to form the local oscillator amplitude $\underline{E}_{LO}$ [7.40,72,124]. In CARS, this amplitude results from the nonlinear background susceptibility, while in SRS it is the transmitted probe laser amplitude [7.37,100]. The intensity at the photodetector surface thus is

$$I(x,y) = \frac{nc}{8\pi}\,|\underline{E}_r(x,y)+\underline{E}_{LO}(x,y)|^2 + I_B(x,y)$$
$$= I_{LO}(x,y) + I_r(x,y) + I_B(x,y) + \frac{nc}{4\pi}\,\mathrm{Re}\left\{\underline{E}^*_{LO}(x,y)\cdot\underline{E}_r(x,y)\right\} \qquad (7.109)$$

where $I_{LO}(x,y)$, $I_r(x,y)$, and $I_B(x,y)$ are the local oscillator, coherent Raman, and incoherent background intensities, respectively. The total detector response is proportional to the integral of this intensity over the photodetector surface

$$\rho = \lambda \int I(x,y)dx\,dy \quad . \qquad (7.110)$$

The factor λ defines the dimensions of that detector response [7.130]. It is convenient to set λ equal to the quantum efficiency of the detector, and in this case the detector response has the dimensions of power. It is also convenient to separate the various contributions to the detector response

$$\rho = \rho_{LO} + \rho_r + \rho_B + \rho_h \qquad (7.111)$$

where each term on the right of (7.111) results from the corresponding term on the right of (7.109).

The total response due to the coherent Raman signal is $\rho_h+\rho_r$, and for the purpose of this argument both are assumed positive. The heterodyne response ρ_h and the Raman intensity response ρ_r can be related to the corresponding signal powers in (7.71,78, 79) by

$$\rho_r = \lambda P_r \qquad (7.112)$$

$$\rho_h = \lambda P_H = 2\lambda(P_r P_{LO})^{\frac{1}{2}}|\cos\Delta\phi| \approx 2\eta(\rho_r\rho_{LO})^{\frac{1}{2}} \qquad (7.113)$$

where again only the Raman term is included in the nonlinear susceptibility in (7.71). The phase angle between E_r and E^*_{LO} is $\Delta\phi$. The other terms in the detector response contribute no spectroscopically interesting information, but they do contribute noise.

Fluctuations in the laser power and mode structure produce noise on each of the terms in the detector response. The mean square noise fluctuation due to laser fluctuations is

$$<\delta\rho_L^2> = <\delta\rho_{LO}^2> + <\delta\rho_B^2> + <\delta\rho_r^2> + <\delta\rho_h^2> \qquad (7.114)$$

and

$$< \delta\rho_\sigma^2 > = \varepsilon_\sigma \rho_\sigma^2 \qquad (7.115)$$

where the contribution due to laser power fluctuations is

$$\varepsilon_\sigma = \frac{1}{\rho_\sigma^2} \sum_k \left(\frac{\partial\rho_\sigma}{\partial P_k}\right)^2 < \delta P_k^2 > \qquad (7.116)$$

and $< \delta P_k^2 >$ is the mean square fluctuation in the power of the kth laser, within the bandwidth of the detection system [7.72,130]. The noise levels due to fluctuations in the laser power can be reduced using detectors to monitor the laser powers and operational amplifiers to take the ratio or difference of the outputs of the signal and monitor detectors [7.33,37,122,124]. This strategy has the effect of reducing some of the ε_σ factors, but because of technical limitations, all of the ε_σ factors cannot be made simultaneously zero. Such detection schemes increase the thermal and quantum noise levels and have proved less practical for pulsed systems than for CW systems.

The shot noise that results from the quantized nature of light detection can be expressed in terms of a mean square fluctuation

$$< \delta\rho_q^2 > = 2\rho_1\rho = 2\rho_1(\rho_{LO}+\rho_r+\rho_B+\rho_h) \qquad (7.117)$$

where $\rho_1 = \hbar\omega_s \Delta\nu$ is the response produced by the detection of a single photon. The nonoptical noise sources — thermal noise, radio frequency interference, dark-current fluctuations, etc. — can be expressed in terms of the noise equivalent input power ζ

$$< \delta\rho_{NO}^2 > = \zeta^2 \Delta\nu \quad . \qquad (7.118)$$

All of these noise terms are defined for a given detection bandwidth $\Delta\nu$.

The signal-to-noise ratio can then be expressed in general as

$$\frac{S}{N} = \frac{2(\rho_r\rho_{LO})^{\frac{1}{2}} + \rho_r}{\left\{< \delta\rho_L^2 > + < \delta\rho_q^2 > + < \delta\rho_{NO}^2 >\right\}^{\frac{1}{2}}} \quad . \qquad (7.119)$$

To estimate the signal-to-noise level for a particular experiment, the parameters characterizing the apparatus used in that experiment must be inserted into (7.119) [7.40,72,124,130].

This formalism can be modified somewhat and applied to spontaneous scattering. In the spontaneous case, there is no local oscillator, so $\rho_{LO} \equiv 0$. The intensity response is proportional to the incident laser power, Raman cross section, concentration, sample length and collection efficiency λ_c

$$\rho_{rs} = \lambda\lambda_c N\ell \frac{d^2\sigma}{d\Delta\omega d\Omega} P_L \qquad (7.120)$$

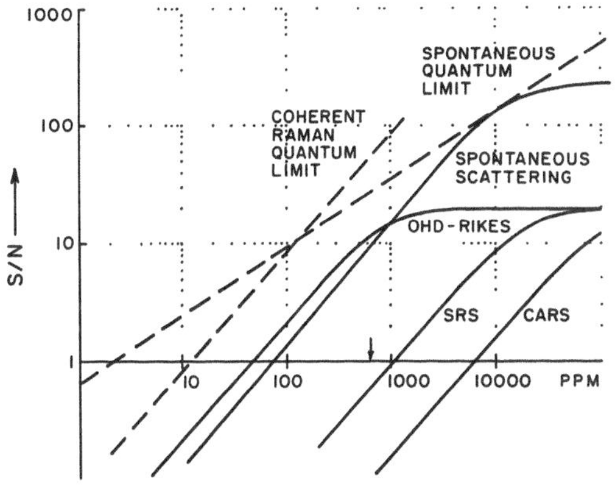

Fig. 7.23. Calculated signal-to-noise ratios for three forms of coherent Raman spectroscopy plotted as a function of sample concentration. Also plotted are the quantum-noise-limited signal-to-noise ratios for CRS and spontaneous scattering along with the proven practical limit for the latter. The CRS signal-to-noise ratios are calculated for a single laser shot with the system described in [7.72]: A 0.5 W CW argon probe laser and a 50 kW N_2 pumped dye pump laser. The spontaneous scattering calculations assume a 0.5 W CW laser and a 1 s integration time. The spontaneous scattering and OHD-RIKES curves were verified experimentally

while the corresponding intensity response in the coherent Raman case depends upon the *square* of the cross section and concentration. This difference in the scaling law is a fundamental disadvantage for coherent Raman spectroscopy for samples at low concentrations or with weak scattering cross sections [7.124].

Various authors have employed this sort of formalism to delineate the capabilities of various real and proposed coherent Raman systems [7.37,40,43,72,100,124,130]. Figure 7.23 shows a plot due to EESLEY of the signal-to-noise ratio expected for his laser system plotted as a function of the concentration of a Raman active species [7.72]. Also plotted are the quantum-noise-limited signal-to-noise ratios of the coherent Raman techniques and of spontaneous scattering. Because of photomultiplier dark current and residual sample luminescence, the practical sensitivity of spontaneous scattering falls significantly below the theoretical limit at low concentrations [7.124].

Even so, the sensitivity of spontaneous scattering exceeds that of all the coherent Raman techniques except OHD-RIKES. Increased laser power and greater stability would improve the performance of all of the coherent Raman techniques, with CARS benefiting more from increased power, and SRS from increased stability. Figure 7.23, however, does give a fair representation of the present state of the art. At high concentrations, the signal-to-noise levels are limited entirely by laser stability, while at low concentrations the limit is set for these CRS experiments by a combination of shot noise and laser stability.

If the nonoptical noise is negligible, (7.119) simplifies near the detection limit where I_r is the smallest intensity at the detector. TOLLES and TURNER [7.100] have shown that for CARS with pulsed lasers, the signal-to-noise ratio is limited to

$$\frac{S}{N} = \frac{2\rho_r}{(\varepsilon_{LO})^{\frac{1}{2}}\rho_{LO}} = \frac{2\chi^Q}{\chi^{NR}}\left(2<\delta P_1^2>/P_1^2+<\delta P_2^2>/P_2^2\right)^{-\frac{1}{2}} \tag{7.121}$$

by the laser power fluctuations. For CW CARS in gases, the "background-free" 4WM techniques, and SRS with a *very* well-stabilized CW probe, the signal-to-noise ratio is limited by the quantum noise

$$\frac{S}{N} = \left(\frac{\rho_r}{2\rho_1}\right)^{\frac{1}{2}} = \left(\frac{\lambda P_r}{2h\omega_s \Delta\nu}\right)^{\frac{1}{2}} \quad . \tag{7.122}$$

In RIKES, with a circularly polarized pump, the local oscillator signal must be completely eliminated, and the dominant noise source is usually fluctuations of the incoherent background signal [7.40]. Thus

$$\frac{S}{N} = \frac{\rho_r}{(\varepsilon_B)^{\frac{1}{2}}\rho_B} = \frac{P_r}{P_B}\,\frac{P_2}{(<\delta P_2^2>)^{\frac{1}{2}}} \tag{7.123}$$

where P_B is the power in the incoherent background, typically $10^{-6}\,P_2$.

In OHD-RIKES, the power in the local oscillator can be varied at will. The optimum sensitivity occurs for

$$\rho_{LO} = \rho_B\left(\frac{\varepsilon_B}{\varepsilon_{LO}} + \frac{2\rho_1}{\varepsilon_{LO}\rho_B} + \frac{\zeta^2\Delta\nu}{\varepsilon_{LO}\rho_B^2}\right)^{\frac{1}{2}} \tag{7.124}$$

and the signal-to-noise ratio is then

$$\frac{S}{N} = \left(\frac{2\rho_r}{\left[\rho_1 + \rho_B(\varepsilon_{LO}\varepsilon_B + 2\varepsilon_{LO}\rho_1/\rho_B)^{\frac{1}{2}}\right]}\right)^{\frac{1}{2}} \tag{7.125}$$

for $\zeta = 0$ [7.40].

In SRS the local oscillator power is constrained to equal the probe power, and fluctuations in the probe power are more important than in OHD-RIKES. If the probe noise dominates, the signal-to-noise ratio must be

$$\frac{S}{N} = 2(\rho_r/\varepsilon_{LO}\rho_{LO})^{\frac{1}{2}} = 2\left(P_r P_2/<\delta P_2^2>\right)^{\frac{1}{2}} \quad . \tag{7.126}$$

The foregoing analysis applies most simply when no one- or two-photon absorption occurs. When absorption does occur, the Raman and the nonlinear coherent background signals cannot be separated so clearly. In OHD-RIKES and SRS, new physical phenomena (e.g., thermal blooming) affect the propagation of the local oscillator wave and produce new coherent and incoherent background signals. In such cases, the noise on these new backgrounds often dominates the other noise sources. Also, at high intensities, self-focusing can alter the propagation of the beams in the sample, giving rise to spurious signals. Multiplex spectroscopy requires a somewhat more sophisticated treatment, but (7.119) still applies for the signal-to-noise ratio on a single channel [7.130].

7.3.11 Signal Enhancement with Interferometers, Intra-Cavity Techniques and Multipass Cells

The sensitivity of all of the coherent Raman spectroscopy techniques can be improved by the use of interferometry or multipass cells and such signal enhancement may be essential when CW lasers are used. A Jamin interferometer was employed in the earliest quasi CW stimulated Raman experiments of OWYOUNG [7.126]. The Jamin interferometer is a two-beam device, and in the application diagrammed in Fig.7.24a, the amplitude and phase of the beam undergoing stimulated Raman gain is compared with that of a reference beam. In our terminology, the local oscillator wave at the output of the interferometer has contributions from each arm and thus a controllable magnitude and phase [7.130]. By optimizing that local oscillator as suggested in Sect.7.3.9 the sensitivity of a stimulated Raman gain experiment can be enhanced. The Jamin interferometer is probably the most useful device of this sort for condensed samples where index of refraction inhomogeneities can degrade the performance of other interferometers. The magnitude of the coherent Raman signal amplitude E_s is not affected by this sort of interferometric heterodyne detection scheme, and thus this sort of interferometer is not advantageous in CARS, RIKES or 4WM.

The signal amplitude can be increased using the pump-resonant interferometer also diagrammed in Fig.7.24a. In this scheme the sample is contained in a near-spherical optical resonator tuned to the frequency of the pump wave. Inside the resonator, the pump power is enhanced by a factor of 60 or so leading to a similar enhancement of the signal amplitude [7.131]. In SRS and OHD-RIKES, the probe wave and signal waves need only pass through the sample once to undergo the maximum enhancement.

For such a scheme to work, it is necessary to lock the cavity frequency to the pump laser frequency. This can be readily accomplished using servo techniques, but

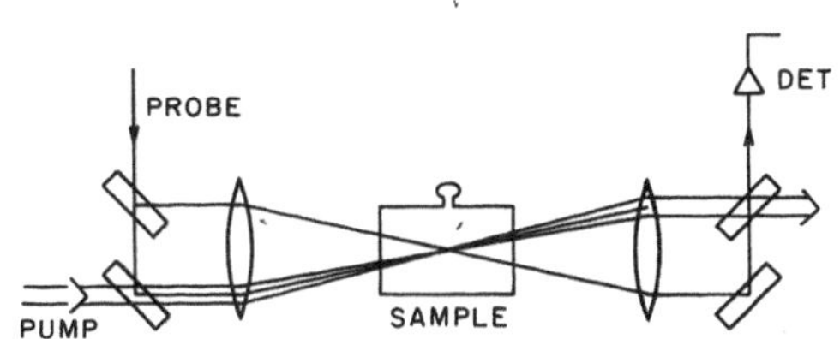

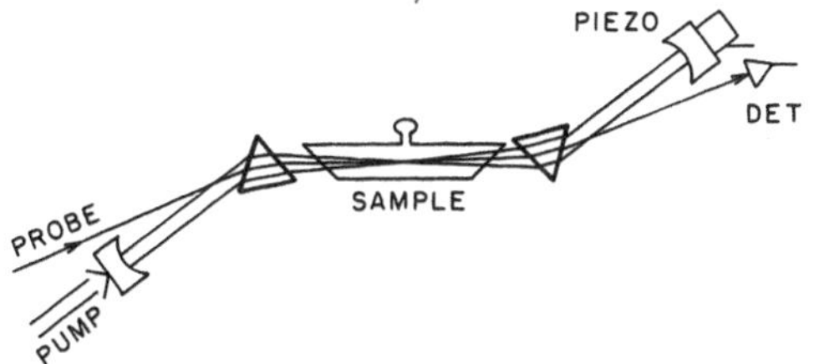

Fig. 7.24a. (*Top*) Nonlinear spectroscopy in a Jamin interferometer. The local oscillator amplitude has contributions from each of the two beam paths and thus can be controlled. (*Bottom*) Nonlinear spectroscopy inside a cavity resonant at the pump frequency. The piezoelectric translator on one mirror keeps the cavity frequency locked to the laser frequency

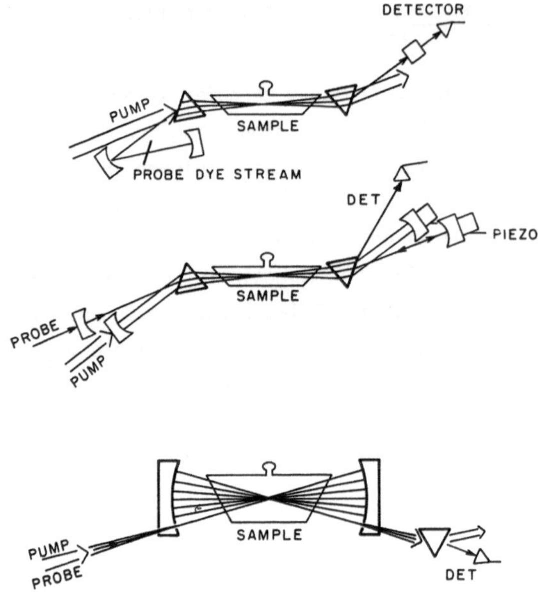

Fig. 7.24b. (*Top*) Nonlinear spectroscopy inside the probe laser cavity. Additional gain or loss due to the coherent Raman process alters the output spectrum of a broadband laser. (*Middle*) Doubly resonant interferometer system for nonlinear spectroscopy. Two intersecting near-spherical optical cavities are locked to the pump and probe frequencies. (*Bottom*) A multipass cell employing a misaligned spherical cavity. Input and output is through reflectionless spots on the cavity mirrors

it can be more convenient to put the sample cell inside the pump laser cavity itself. Clearly, this strategy is only appropriate for gases and other high optical quality samples that do not absorb significantly at the pump frequency.

Another sort of intra-cavity technique is diagrammed at the top of Fig.7.24b. Here the sample is placed within the probe laser cavity, and the additional gain or loss due to the stimulated Raman effect dramatically alters the output power. In principle, if the laser were just at threshold, an infinitesimal Raman gain or loss would produce finite changes in the output power. The difficulties with this technique result mostly from excessive sensitivity [7.38].

Doubly (or triply) resonant interferometer schemes are useful in CARS and 4WM. A typical doubly resonant apparatus appears in the middle of Fig.7.24b. Here both the pump and probe wave are resonated, with separate servos to lock independently to each frequency. The power in each wave is enhanced by a factor of 60 or so, implying an enhancement of the CARS output power of 2×10^5. Again it can be convenient to design laser cavities with a region where the beams overlap, and to place the sample cell in that region. Such was the technique of HIRTH and VOLRATH which gave CARS signal levels characteristic of a pulse experiment using CW lasers [7.132].

The frequency locking requirements of a resonant interferometer can be eliminated by substituting a multipass cell as diagrammed at the bottom of Fig.7.24b. In this case the nearly spherical optical cavity is purposedly misaligned so that the collinear pump and probe beams zigzag back and forth traversing an elliptical pattern on each mirror [7.125]. It is convenient to remove the reflective coating at one spot on each mirror to facilitate coupling in and out. Such a multipass cell enhances the interaction length, but not the power of the beams. The best demonstrated signal enhancement in an SRS experiment is presently a factor of 50. The same multipass cell would give a factor of 2500 enhancement in a CARS experiment if the coherence lengths were great enough.

7.4 Applications

Each of the coherent Raman spectroscopy techniques has proven useful as a tool for studying some physical phenomenon or some class of material system. It is not feasible even to list all of the experimental results obtained in a review of this length. Such a task has been performed by a number of admirable articles that have recently appeared [7.43,44,52,55,57,82,97,99,135,164,171]. What follows is an abbreviated discussion of proven applications with some typical examples cited for each.

7.4.1 Combustion Diagnostics: Concentration and Temperature Measurement

The coherent Raman signals can be easily separated from luminescent backgrounds, and this property of the coherent Raman techniques has made them especially useful in combustion diagnostics where luminescence often overwhelms the spontaneous scattering signal [7.133]. The first practical application of this sort was by REGNIER and TARAN who demonstrated that CARS could clearly detect hydrogen diluted in nitrogen down to 100 parts per million [7.35]. They did so using a Q-switched ruby laser to provide frequency ω_1 and an H_2 stimulated Raman oscillator to supply the ω_2 frequency The two beams were collinear. The anti-Stokes beam was separated from the incident laser beams with colored glass filters.

Although the pressure shift of the hydrogen vibration provided only slight tunability of the ω_2 frequency, TARAN was able to use this apparatus to map the distribution of H_2 formed by pyrolysis in a methane-air flame with a spatial resolution of a few millimeters. His initial results appear in Fig.7.25.

By imaging the sample onto a photographic plate, and using filters to block the laser frequencies and pass the anti-Stokes, TARAN was also able to take a CARS snapshot of a supersonic H_2 jet [7.134]. Detailed structures such as shock waves were clearly visible, thus demonstrating that the CARS phase-matching condition is not so restrictive as to reduce the resolution of such imaging techniques.

The spatial resolution of CARS can be enhanced in two ways: If the ω_1 and ω_2 beams cross at a small angle, the length of the active region where the beams overlap is greatly reduced. So too is the anti-Stokes signal which exits from the interaction region at an angle between the two incident waves, but the signals are often so large that this reduction is unimportant [7.135]. The "boxcars" technique of Fig.7.6 also

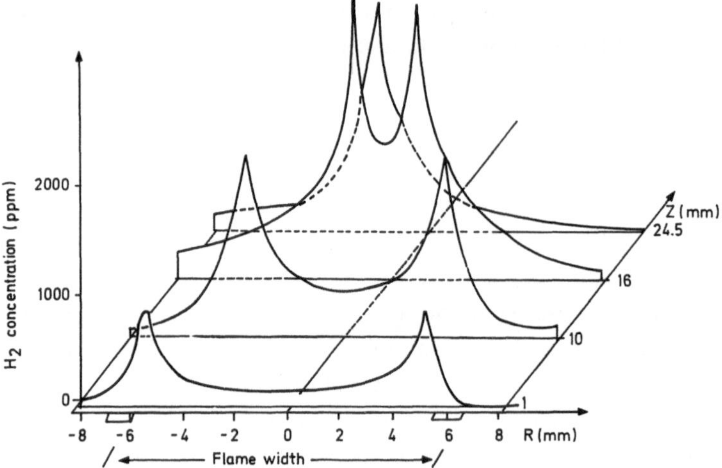

Fig. 7.25. Distribution of H_2 in an atmospheric pressure methane flame as determined by CARS. [7.35]

limits the length of the overlap region at the expense of signal intensity, but the angles involved can be larger than in the former scheme [7.59].

Equations (7.29,31,33) show that the Raman term in the nonlinear susceptibility depends upon the difference in populations of the two coupled levels. At thermal equilibrium

$$\rho_d = \rho_{gg} - \rho_{rr} = Z^{-1}(e^{-E_g/kT} - e^{-E_r/kT}) \tag{7.127}$$

where Z is the partition function for the sample species. By comparing the coherent Raman signal strengths for several transitions with known values of α^R, the temperature can be in principle determined. True thermal equilibrium, however, often does not exist in plasmas and flames, and a distinction must be made between rotational, vibrational and translational temperatures. When the fundamental and hot bands of a Raman transition are well resolved, the vibrational temperature can be readily assigned. The rotational temperature can also be unequivocally inferred if individual lines can be resolved.

TARAN's group also pioneered the measurement of rotational temperatures of the hydrogen formed in their methane flame, and obtained good agreement between the CARS measurement and a thermocouple measurement [7.136]. Subsequently they also inferred the rotational temperature of N_2 in a spherical diffusion flame by analyzing the slope of the unresolved Q branch [7.137]. Other workers have pointed out the limitations of the approximations employed and have advised comparing experimental and computer generated spectra in order to estimate rotational temperatures from poorly resolved bands [7.138]. HARVEY has obtained excellent agreement between experimental and calculated CARS traces of the resolved Q branch of N_2 in an oven at known temperatures [7.135].

NIBLER et al. employed CARS to measure the vibrational and rotational temperature in electric discharges. In D_2 at 48 Torr, the $V = 0 \to 1$ and $V = 1 \to 2$ Q branches were strong and well separated. The vibrational temperature inferred from their ratio was 1050 K even though the rotational temperature was only 400 K [7.103]. HARVEY et al. found much higher vibrational temperatures in a similar N_2 discharge. Their analysis technique assumed a Boltzmann distribution only among the lowest vibrational levels, and disclosed strong deviations from a Boltzmann distribution for levels above $V = 3$. The difference in population distributions explains why N_2 is a more effective energy transfer agent than D_2 in discharge lasers [7.139].

Similar data have been obtained by ROH et al. in multiplex CARS experiments, some performed upon the exhaust of a jet engine [7.41]. The multiplex techniques can obtain all the information necessary to assign a temperature in a single laser shot, and are thus more applicable to the study of turbulent or rapidly evolving systems.

Some of the difficulties in extracting concentration and temperature information from CARS spectra can be avoided in OHD-RIKES and SRS. In these techniques, the coherent spectrum can be made to reproduce the spontaneous scattering spectrum and previously proven techniques of analysis applied. Corrections for finite laser bandwidth and temperature-dependent linewidth can be made more readily, even for partially resolved bands, because of the absence of interference terms in the expression for the CRS signal. The signal is also linear in the concentration. Multiplex SRS and OHD-RIKES apparatus would also allow the simultaneous measurement of the intensity of enough Raman lines to assign a temperature using one laser shot.

7.4.2 Raman Cross Section and Nonlinear Susceptibility Measurements

Measurements of the total cross section for spontaneous scattering are among the most difficult experiments in Raman spectroscopy. Accurate corrections must be made for the collection of geometry, detector quantum efficiency, spectrometer transfer function, etc. It is thus not surprising that few such measurements exist with expected undertainty less than 10% [7.140,141].

In contrast, the precise measurement of the stimulated Raman gain or loss is relatively straightforward in a CW or quasi-CW experiment. For collinearly focused identical TEM_{00} pump and probe beams, the change in probe power due to Raman gain upon traversing a single focus is

$$\frac{P_H(\omega_2)}{P_1 P_2} = \frac{96\pi^2\omega_1\omega_2}{c^2} \operatorname{Im}\left\{\chi^{(3)}_{\alpha\beta\beta\alpha}(\omega_2,-\omega_1,\omega_1,-\omega_2)\right\} \tag{7.128}$$

which can be related to the total cross section using (7.54) [7.124]. To determine the gain and cross section directly, three powers must be measured accurately, one of them rather small. While satisfactory techniques exist for doing so, it is often simpler to compare the Raman gain for one sample with that of a well-known material such as benzene. Using an interferometric variation of SRS, OWYOUNG and PEERCY measured the peak gain of the benzene 992 cm^{-1} mode in terms of the well-known optical Kerr constant of CS_2 [7.126]. Their result

$$\frac{N(\alpha^R)^2}{12\hbar\Gamma} = 31.8 \pm 2.2 \times 10^{-14} \text{ esu} \tag{7.129}$$

agrees with the most accurate determinations by spontaneous scattering [7.141].

Coherent Raman techniques can also provide accurate spectroscopic measurements of the non-Raman contributions to the third-order nonlinear susceptibility χ^{NR}, and the need for such measurements provided much of the motivation for much of the early development of CARS [7.49,143]. In CARS, 4WM, and RIKES with a linearly polarized pump, the Raman line shape of an isolated mode has a maximum and a minimum as in Fig.7.4. The frequency difference between maximum and minimum depends upon the ratio of the background susceptibility to the Raman matrix element squared

$$\Delta\omega = \left\{4\Gamma^2 + \left[N\left(\alpha^R_{eff}\right)^2/4\hbar D\chi^{NR}_{eff}\right]^2\right\}^{\frac{1}{2}} \tag{7.130}$$

where α^R_{eff} and $D\chi_{eff}$ depend on the specific polarization and frequency condition. This ratio can thus be obtained without the need for potentially uncertain power or intensity measurements [7.49].

Tables 7.1, 2 and 3 show some of the combination of $\chi^{(3)}$ tensor elements that are accessible in different polarization and frequency conditions. Each combination of tensor elements gives a different frequency difference between maximum and minimum, as is shown in Fig.7.10. In a series of such experiments, the Raman and non-Raman contributions to each tensor element can thus be determined.

The non-Raman terms in $\chi^{(3)}$ have scientific and technological importance. For example, the term responsible for self-focusing is

$$n_2 = \frac{12\pi}{n}\chi^{(3)}_{1111}(-\omega,\omega,-\omega,\omega) \tag{7.131}$$

and self-focusing limits the performance of large solid-state laser systems [7.10, 143]. In liquids, glasses and cubic crystals $\chi^{(3)}_{1111}(-\omega,\omega,-\omega,\omega) = \chi^{NR} = \sigma/8 + [A(0)+B(0)]/6$ and thus a measurement of the background susceptibilities in CARS, RIKES, and interferometric SRS can completely characterize n_2 [7.57,62,126].

Once the ratio of the Raman and nonlinear background contributions to chi-three has been determined for one mode, the ratio of the cross sections for other modes can be obtained by repeating the experiment at different values of $\omega_1-\omega_2$. If the Raman frequencies are well separated, (7.60) can be applied directly, otherwise the observed line shape must be fitted using (7.59).

These techniques work well for materials with strong, well-characterized Raman lines and for mixtures when proper account is taken of local field effects [7.142].

The nonlinear susceptibilities of materials without such modes can be determined using a "sandwich sample" [7.108,144]. In this technique, a thin sample of Raman-active material, such as calcite, is sandwiched between two layers of the unknown sample. The entire assembly must be thinner than the region in which the laser beams completely overlap, and the angle between the beams may be adjusted for maximum destructive interference. The nonlinear susceptibilities of the unknown sample then add to that of the Raman sample, shifting the frequency of the minimum. From these shifts, various authors have estimated the nonlinear susceptibilities of technologically important organic and inorganic materials.

Two-photon absorption contributes additional resonances in $\chi^{(3)}$. When the two-photon absorption lines are narrow and Lorentzian, the resonance have the same general line shape as the Raman resonances, but with different frequency arguments. By comparing the dispersion of $\chi^{(3)}$ due to the two-photon resonances with that due to known Raman resonances, various authors have normalized two-photon absorption cross sections to Raman cross sections [7.142,145-147]. When the two-photon absorption lines are broad, they contribute a negative imaginary component in χ^{B}. In CARS, that imaginary component tends to fill in the distinct minimum seen in Figs.7.4 and 10. In CSRS, the effect is to deepen the minimum. By comparing CARS and CSRS line shapes using (7.40,44), LYNCH and LOTEM accurately estimated the small two-photon absorption cross section of CS_2 [7.61]. In RIKES, the interference between a narrow Raman resonance and a broad photon absorption resonance can actually invert a RIKES peak when $\omega_2 > \omega_1$.

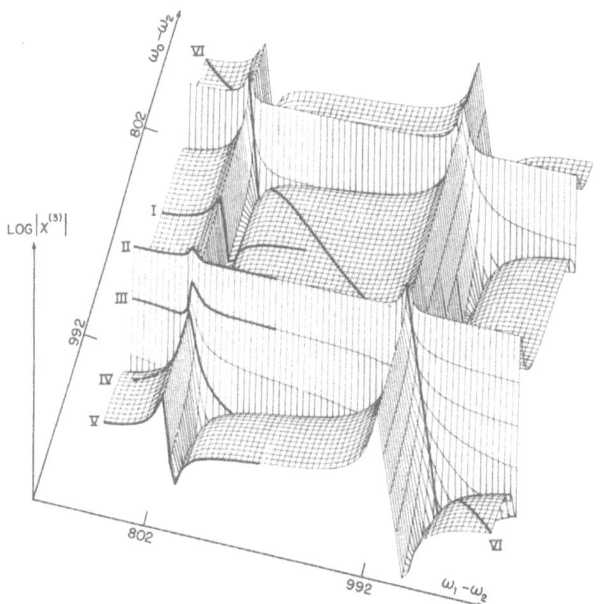

Fig. 7.26. Two-dimensional plot of the dispersion of $|\chi^{(3)}|^2$ vs $\omega_1-\omega_2$ and $\omega_0-\omega_2$ in a 1:1 mixture of benzene and cyclohexane. The heavy lines show the regions explored experimentally. [7.42]

In four-wave mixing, one can tune $\omega_1-\omega_2$ to one Raman frequency while scanning $\omega_0-\omega_2$ over another. Fitting the line shapes obtained in this way normalizes the cross sections of the two modes to one another. Figure 7.26 shows the dispersion of $|\chi^{(3)}|^2$ for a benzene-cyclohexane mixture as determined in this way by Lotem et al. [7.42].

7.4.3 High-Resolution Molecular Spectroscopy

Perhaps the most dramatic advantage of coherent Raman spectroscopy over conventional Raman scattering is in resolution and frequency precision. With great skill and patience, one can obtain spontaneous Raman scattering spectra with an instrumental resolution of 0.05 cm^{-1} or 1500 MHz [7.3]. Using known atomic transitions as reference frequencies, similar precision is possible. A considerable price must be paid, however, in intensity and integration time. In the practice of grating spectroscopy, high resolution means narrow slits or large f numbers or both, and little of the spontaneously scattered light can thus be collected.

In coherent Raman spectroscopy with CW sources, an instrumental resolution of 0.0001 cm^{-1} or 3 MHz can readily be obtained, and interferometric wavelength determination techniques with a frequency precision of ±10 MHz are also available [7.116, 148]. Since the resolution is built into the laser sources, the signal levels obtained are as large as for less well engineered CW sources. Recently, well-stabilized CW lasers have been used as oscillators and amplified to tens of kilowatts in laser pumped dye laser amplifiers. The resolution and frequency precision of coherent Raman spectrometers employing such lasers can be estimated as 20 MHz and ±50 MHz, respectively.

At these levels of resolution and precision, the Doppler effect gives the largest contribution to the linewidth and to the experimental uncertainty in the transition frequency. For typical vibrational frequencies, the Doppler width is of order 100 MHz; for rotational frequencies it will be less than 10 MHz. Even so, coherent Raman spectroscopy offers a potential improvement of two orders of magnitude in the precision of rotational and vibrational constants.

The earliest experiments of this sort were by DeMARTINI et al. who employed a pulsed, pressure tune H_2 stimulated Raman oscillator in a CARS experiment to study the Dicke narrowing and collisional broadening of the Q(1) V=0→1 line of H_2 [7.149]. More recently BYER et al. and KRYNETSKY et al. resolved the Q(2) line of D_2 and the Q(1) line of H_2 using CW CARS and measured the vibrational frequencies interferometrically [7.65,115,150,151]. FABELINSKY et al. also have obtained high-resolution (40 MHz) CW CARS spectra of these gases and of the Q branch of the ν_2 band in C_2H_2 in which they determined the rotational-vibrational coupling constant [7.115] (see Fig.7.13). OWYOUNG soon followed with a similar measurement of the D_2 Q(2) line using OHD-RIKES which revealed a vibrational frequency of 2987.293 cm^{-1}, a Doppler line width of 552 MHz, a motionally-narrowed minimum linewidth of 250 ± 5 MHz and a pressure broadening coefficient of 51.3 MHz/amagat [7.63].

Even more remarkable from a spectroscopic point of view has been OWYOUNG's recent work using stimulated Raman gain. Having reduced the amplitude fluctuations of his probe laser and developed a 97 pass gas cell, the Sandia group proceeded to resolve the Q branch of the ν_1 fundamental mode in methane near the band head at 2916.47 cm^{-1} The result along with a calculated spectrum appears in Fig.7.20 [7.125]. At 35 Torr, the lines are Doppler broadened, but an unexpected tetrahedral splitting of the rotational lines can clearly be resolved.

No spectra of similar quality have yet been produced by CARS or 4WM. ALIEV et al. and BOQUILLON et al. have, however, resolved the same methane branch by means of CARS, but at a higher gas density, and their results are similar to those of OWYOUNG at the same density [7.152,153]. In principle CARS, SRS, and OHD-RIKES should give the same resolution, but the CARS signal-to-noise ratio degrades more rapidly with decreasing density than the other two, and the multipass cell seems to work best in SRS. For the highest resolution work SRS may be the method of choice.

7.4.4 Raman Spectra of Fluorescent and Resonant Samples

By working with pulsed lasers, gated detection, judiciously prepared samples and carefully chosen (or oscillating) wavelengths, Raman spectroscopists have obtained spectra of fluorescent samples [7.154,155]. However, spontaneous scattering techniques must fail when the cross section for prompt fluorescence exceeds the cross section for spontaneous scattering by several orders of magnitude. In these cases, the coherent Raman techniques remain viable. Fluorescence can be suppressed by spatial and spectral filtering, and the coherent Raman signal enhanced by heterodyne detection. Particularly in CARS, fluorescence-free spectra can be easily obtained, since the signal frequency is at the anti-Stokes side of the input laser beams. In addition, CRS offers particular advantages for Raman studies of biologically important materials without the danger of sample degradation since ~1 mW of average power is sufficient to take CRS spectra under resonant conditions.

The fluorescence rejection capability of CARS was early demonstrated by HUDSON and co-workers by adding Rhodamine 6G, a brightly fluorescent laser dye, to a benzene sample [7.156]. They obtained fluorescence-free CARS traces even though the sample literally glowed. Similar demonstrations have been performed with RIKES, OHD-RIKES, and SRS [7.72].

More significant scientifically have been experiments in which the fluorescence was intrinsic to the sample. One example of this sort is the second-order Raman spectrum of diamond shown in Fig.7.27 as resolved by EESLEY [7.156]. This small second-order peak near twice the frequency of the first-order Raman peak has been the subject of some controversy [7.158]. In spontaneous scattering, this feature is 1000 times weaker than the first-order peak in Fig.7.10, and in the sample employed to obtain Fig.7.27, the peak was at least 20 times smaller than the fluorescent background due to impurities in the diamond sample. Nevertheless, EESLEY completely

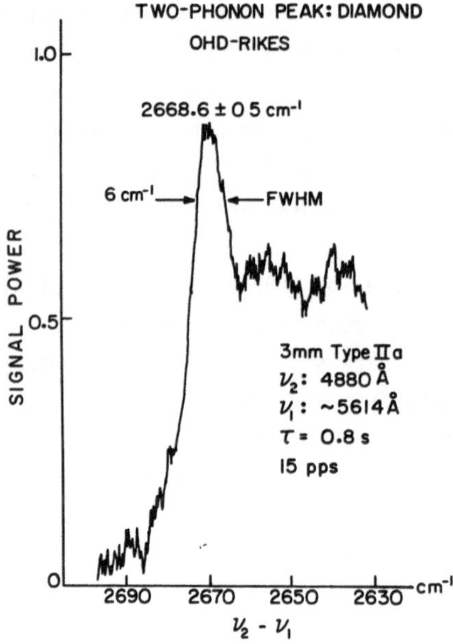

Fig. 7.27. A portion of the second-order Raman spectrum of diamond obtained by OHD-RIKES. This peak is 1000 times weaker than the first-order peak in Fig.7.10, 10 times weaker than the CARS background susceptibility and 20 times weaker than the background fluorescence in this sample. [7.157]

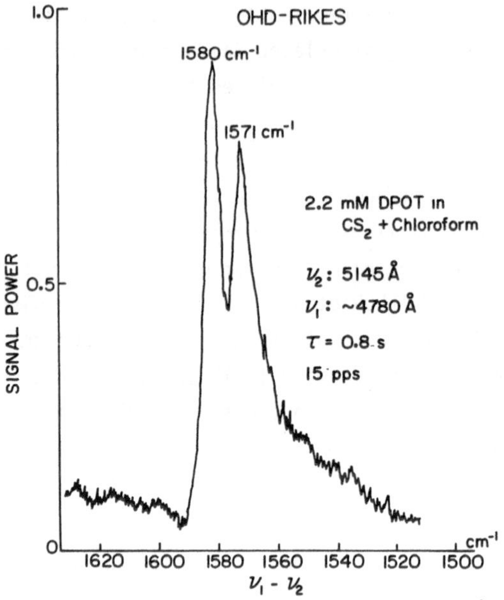

Fig. 7.28. Resonant OHD-RIKES spectrum of diphenyloctatetraene. Resonant enhancement makes these lines strong in pite of the low concentration. Similar results have been obtained in CARS. [7.72]

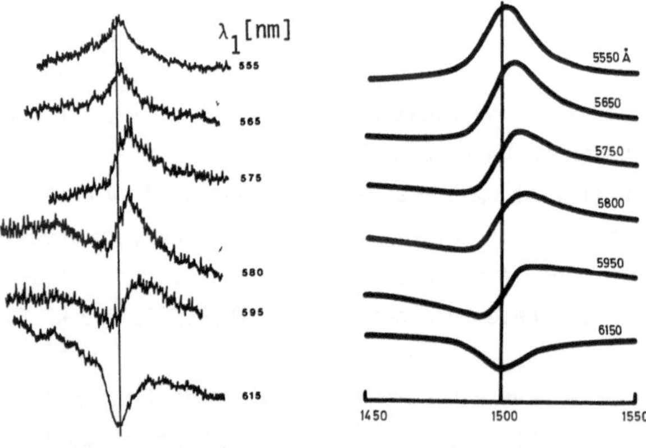

Fig. 7.29. Experimental and calculated CARS spectra of cyanoalbumin (vitamin B_{12}) for various pump wavelengths. The Raman frequency is 1500 cm^{-1}. Resonance effects alter the line shape to the extent of making maxima into minima. [7.80,81]

suppressed the fluorescence signal and nonlinear background in his OHD-RIKES experiment and accurately measured the frequency and linewidth.

Systematic application of resonantly enhanced CARS was first reported by HUDSON et al. who studied diphenyl octatatrane dissolved in benzene [7.159]. They observed enhancement of the CARS intensity without significant distortion of the line shapes.

A similar spectrum obtained in the same sample by EESLEY using OHD-RIKES is shown in Fig.7.28 [7.72].

Drastic line-shape changes were observed by NESTOR et al. from their CARS spectra of vitamin B_{12} in water [7.80]. Figure 7.29 shows a series of CARS traces taken at different laser wavelengths along with theoretical plots obtained by LYNCH et al. using (7.105) [7.54,69,81]. The Raman resonant structure no longer necessarily appears as a peak but can evolve into an "inverted peak" or dispersion shaped feature [7.80]. These line-shape distortions result from destructive interference between the resonant susceptibility of the solute species and the "background" susceptibility of the solvent. The nonlinear susceptibility of water is known to be quite low, with weak and broad Raman resonances [7.160]. The line-shape distortion observed by NESTOR would have been worse in almost any other solvent. LAU et al. also observed significant line-shape changes under resonant condition in the samples of 3,3'-diethyl thiacarbocyanine iodide (DTDC) in ethanol [7.110].

Detailed studies of CARS resonant excitation profiles were reported by CARREIRA et al. with several biological compounds [7.83-85,162]. By employing a computerized CARS microanalysis system, CARS detection of sample concentrations as low as $\sim 5\times 10^{-7}$ M was possible with β carotene in benzene [7.83]. The excitation profiles were compared to the curves obtained using a formula which is essentially the same as the one in (7.103). Reasonably good agreement was found between the theory and experiment. However, for incident power above 1 kW saturation effects reduced the CARS signal.

A different aspect of CARS in absorbing samples was investigated by AKHMANOV et al. [7.107]. They studied the change of $\chi^{(3)}$ in the vicinity of a solvent Raman mode in the mixture of toluene and rhodamine 6G. Different CARS line shapes were observed as the dye concentration was varied, due to the interference effects between the solvent and solute signals. Most interesting was the case where the nonresonant signals were suppressed at a certain concentration of the dye. Using this sort of background suppression AKHMANOV was able to improve the sensitivity of CARS and detect weak features of the Raman spectra of toluene.

Figure 7.30 shows the resonant inverse Raman spectrum of rhodamine 6G and rhodamine B as obtained by WERNCKE et al. [7.161]. These traces are similar to those obtained by CARS, but the CARS sensitivity seems somewhat better [7.162].

When the absorption lines are narrow, additional resonances appear in the CRS spectra due to processes similar to those diagrammed in Fig.7.8 [7.53]. In SRS and RIKES some of the resonances result from changes in the steady-state populations of excited electronic levels, and some can have linewidths narrower than ω_d. These points are illustrated in Fig.7.31. Figure 7.31a shows a near-resonant SRS spectrum of I_2 vapor taken with a CW argon laser pump and a CW dye laser probe. The cross sections of the two peaks are enhanced by the nearby P(13) and R(13) 43-0 X$\rightarrow$B absorption lines. The lines are Gaussian with width 0.83 ω_d (HWHM). When the pump is tuned onto resonance, the Raman peaks break up into narrower lines corresponding to individual hyperfine transitions with differing resonant velocity groups [7.163].

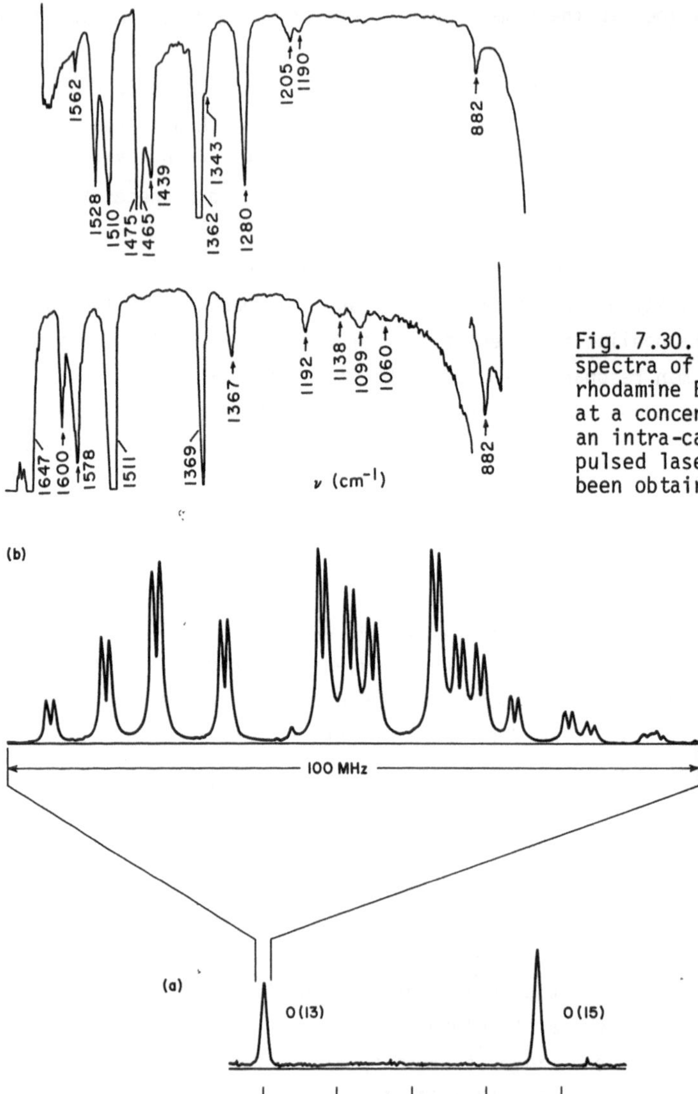

Fig. 7.30. Resonant stimulated Raman spectra of rhodamine 6G (top) and rhodamine B. These traces were taken at a concentration of 10^{-2} M using an intra-cavity technique with pulsed lasers. Similar results have been obtained using CARS. [7.161]

Fig. 7.31a,b. Resonant stimulated Raman spectra of iodine vapor. Trace (a) shows the O(13) and O(15) lines of the $V'' = 0 \rightarrow 9$ band (eigth overtone band) as they appear with the 5145 Å pump laser tuned 1 GHz from the P(13)R(15) 43-0 rovibronic absorption line of the $X \rightarrow B$ electronic transition. The two Raman lines are 46 MHz wide and 1.83 GHz apart. The expected Doppler width is 44 MHz. Trace (b) shows the additional structures that result when the pump laser is tuned into resonance with the P(13) 43-0 absorption line. Each pair of lines corresponds to amplification of the probe laser by a different inverted hyperfine transition. Related effects appear under resonant conditions in CARS and other CRS techniques. [7.163]

Somewhat analogous effects have also been seen in the resonant CARS spectrum of I_2 [7.53].

For resonant Raman work, CARS and its 4WM analogs have real advantages over SRS, OHD-RIKES. In the latter two techniques, thermal blooming and excited state absorption induced by the pump alter the local oscillator intensity at the detector, producing spurious signals and noise. Also, the depolarization ratio in the resonant case is often near $\rho_Q = 1/3$ resulting in a suppression of the Raman signal whenever RIKES or Asterisk-type background suppression techniques are employed. While resonant enhancement of the CARS background is possible, it has proved not to be as severe a problem as these other effects. All the coherent Raman techniques demonstrated to date are transmission techniques. For opaque and translucent samples, conventional spontaneous scattering continues to be the best available technique.

7.4.5 Polariton Dispersion: Spectroscopy in Momentum Space

In noncentrosymmetric media, the frequencies of the polariton modes depend dramatically upon the wave vector. In spontaneous scattering the wave vector of the polariton depends upon the difference of the wave vectors of the incident and scattered radiation: $\underline{q} = \underline{k}_i - \underline{k}_S$. Unfortunately, spontaneous scattering techniques must collect light over a finite range of wave vector $\Delta\underline{k}_S$ in order to have a finite signal level, and the uncertainty in the collected wave vector introduces an uncertainty in the wave vector of the polariton.

Since coherent Raman techniques employ diffraction limited beams propagating in definite directions, the wave vector of the excited polariton can be better defined than in spontaneous scattering, at least in principle. In fact, the usual practice of working with tightly focused beams in order to obtain the maximum signal from a finite length sample significantly reduces the precision with which $\underline{q}$ can be specified. Nevertheless, a number of workers have employed CARS and 4WM techniques to refine polariton dispersion curves.

COFFINET and DE MARTINI originated spectroscopy of this type using stimulated Raman oscillators [7.164]. Since their frequencies were fixed, they plotted the output signal as a function of the crossing angles of their beams. Today, their experiment would be considered an example of "boxcars," but they termed the technique "spectroscopy in k space" as the wave vector of the nonlinearly excited polariton, but not the frequency, varied as the crossing angle was scanned. Ultimately, using various pairs of input frequencies, DE MARTINI succeeded in plotting out the lower branch of the transverse polariton curve of GaP with high accuracy [7.165].

At the surfaces of noncentrosymmetric crystals, there exist surface polariton modes with frequencies between the LO and TO bulk polaritons. These modes have been studied by spontaneous scattering, but they can also be driven coherently by frequency mixing techniques [7.166]. DE MARTINI et al. have also plotted out the dispersion curve and linewidth for these modes in GaP using their k-space spectroscopy version of "boxcars" [7.167]. With due care, the driven surface polariton can also be coupled out of the sample and detected directly. The resonances observed in this case are more properly assigned to $\chi^{(2)}$ than to $\chi^{(3)}$ [7.168].

In a 4WM or CARS experiment in which the frequencies of the lasers vary, but the crossing angle is held fixed, the wave vector of the driven polariton varies along with the frequency. WYNNE first demonstrated the resonances in the CARS signal that result when the frequency and wave vectors correspond to a point on the dispersion curve [7.32]. In principle, similar experiments are possible using SRS and RIKES techniques.

7.4.6 Low Frequency Modes

Elastic scattering of the incident laser frequency often overwhelms spontaneous Raman scattering spectra at frequencies below 100 cm^{-1}. The most common technique for suppressing such scattering — the I_2 filter — imposes additional structure on the Raman spectrum.

Coherent Raman techniques can suppress the elastic peak with a combination of spectral and spatial filtering. There is an additional complication in that for $\hbar\Delta\omega \ll kT$ (7.54,55) predict that

$$\mathrm{Im}\{\chi^{(3)}\} \propto \frac{\hbar\Delta\omega}{kT}\frac{d^2\sigma}{d\Omega d\Delta\omega} \tag{7.132}$$

clearly altering the line shape of the resonance.

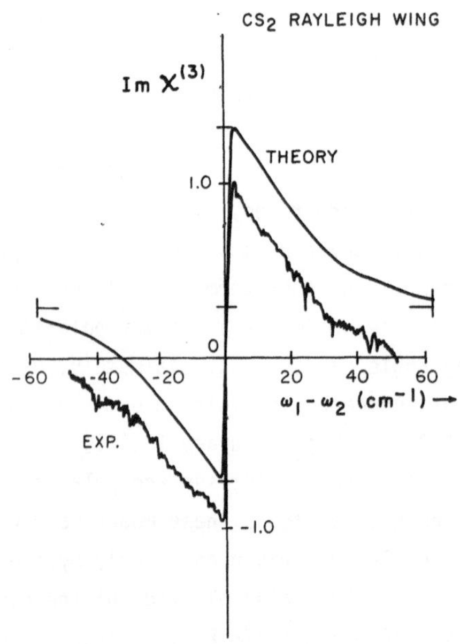

Fig. 7.32. Rayleigh wing spectrum of CS_2 taken by OHD-RIKES. This trace plots $\mathrm{Im}\{\chi^{(3)}\}$; the theoretical curve corresponding to (7.132,133) has been translated vertically for clarity. Note particularly the absence of any elastic scattering peak

Figure 7.32 shows an OHD-RIKES spectrum of $\mathrm{Im}\{\chi^{(3)}\}$ in CS_2. The line shape corresponds to a spontaneous scattering cross section of the form

$$\frac{d^2\sigma}{d\Omega d\Delta\omega} \propto \frac{\Gamma^2}{\Delta\omega^2+\Gamma^2} + \beta\, e^{-|\Delta\omega|/\Delta_0} \qquad (7.133)$$

with $\Gamma = 3\ \mathrm{cm}^{-1}$, $\Delta_0 = 20\ \mathrm{cm}^{-1}$ and $\beta = 0.1$, which is in agreement with the results of SHAPIRO [7.169]. In this figure there is no hint of an elastic peak. KOSTER et al. have explored the overdamped (soft) phonon modes of $BaTiO_3$ using SRS and found a line shape similar to Fig.7.32 [7.170]. An important advantage of this technique is the improved precision in assigning the wave vector of the driven vibration.

BLOOM has employed an extreme version of "boxcars" to explore the low frequency dispersion of chi-three in liquids [7.171]. In BLOOM's geometry, two pump waves at frequency ω_1 propagate in opposite directions through a very thin sample. The Stokes wave at ω_2 is incident at a small angle, and the anti-Stokes wave at $\omega_s = 2\omega_1 - \omega_2$ is radiated in the direction exactly opposite to the incident Stokes wave. This process phase matches only at $\omega_1 - \omega_2 = 0$ but for small frequency shifts the coherence length for the process is longer than the samples employed. As in other CARS techniques, the intensity of the detected signal is proportional to $|\chi^{(3)}|^2$, and the line shape in CS_2 is consistent with Fig.7.32 and (7.133).

7.4.7 Vibrational and Rotational Relaxation Measurements

When the optical fields used to drive a Raman mode are turned off, (7.18,20) imply that the coordinate for that mode continues to oscillate until dephasing and relaxation processes damp it out. The time scale involved ranges from picoseconds for most liquids and solids to many nanoseconds for low pressure gases. The dephasing and relaxation times can be measured using short pulsed lasers in the time domain, or by analyzing the CRS line shape in the frequency domain.

KAISER and his co-workers pioneered the time domain techniques using picosecond lasers and the stimulated Raman effect in the sample itself to drive the vibrational mode. A delayed picosecond probe laser sampled the coherent and incoherent part of the excitation at a later time [7.172]. Their results are reviewed extensively in Chap.6 of this volume.

LEE and RICARD have been studying vibrational relaxation using a transient version of CARS [7.173]. In this technique picosecond pulses from a doubled Nd: glass laser and a dye laser are overlapped in space and time to drive the vibrational mode of the sample, and a delayed pulse from the glass laser is used as a probe. Since each of the waves propagates in a slightly different direction, the phase-matching diagram in Fig.7.6 predicts a definite direction for the CARS pulse produced by the delayed probe. Vibrational relaxation can be studied in this way at intensity levels well below the threshold for stimulated Raman oscillation.

Similar studies have been performed by HERITAGE using a picosecond version of stimulated Raman gain spectroscopy [7.174]. Because of the advantages of heterodyne detection, HERITAGE could obtain vibrational relaxation data using synchronously pumped CW mode locked dye lasers.

7.5 Conclusions

We have shown that the technically demanding arts of coherent Raman spectroscopy can give better results than the conventional spontaneous scattering techniques in many applications. Even so, coherent Raman spectroscopy should be approached with caution. If an experiment can be done in a straightforward way using spontaneous scattering, it should be done that way. If not, the experimenter must decide upon the most feasible coherent Raman technique for his application. The diversity of such techniques is enormous. There are multiplex techniques to record spectra 1500 cm^{-1} wide in nanoseconds, high resolution CW techniques to determine Raman frequencies to seven figure accuracy, and picosecond techniques to probe the dynamics of the transitions themselves. The coherent Raman signal can be a new frequency or polarization component, a change in intensity of a probe beam, or a sound pulse emitted by the sample. The line shapes can be positive or negative peaks, dispersion curves or something in between, and so on.

One reason for this diversity is that each experimenter must assemble the coherent Raman system which best fits his needs. In some cases commercially available laser sources provide sufficient output power, repetition rate, stability, tunability, beam quality and reliability, but often the experimenter must design his own laser system as well. The proliferation of coherent Raman techniques resulted in part from attempts by various workers to invent techniques which best fit existing laser systems rather than designing systems to fit the best techniques.

The capabilities of the various techniques are now clear. For maximum sensitivity in a material without significant absorption, optical heterodyne detected Raman induced Kerr effect spectroscopy will give the best sensitivity. If the probe laser is well enough stabilized, stimulated Raman gain or loss techniques will give equivalent results. Stimulated Raman spectroscopy also gives the best precision in measurements of transition frequency or Raman cross section. CARS and CSRS should be used in resonant cases where saturation, excited state absorption or thermal blooming produce unacceptable backgrounds in OHD-RIKES or SRS. Nonlinear ellipsometry gives the best precision for the depolarization ratio. Most of the other techniques are curiosities with very limited applications.

No additional laboratory curiosities are needed. The time has come to apply these spectroscopic techniques to scientific issues outside the area of nonlinear optics.

In the years ahead, today's coherent Raman techniques will be increasingly accepted as standard laboratory tools by workers in physics, chemistry, biology, medicine and engineering, and the list of successful applications will continue to grow.

Acknowledgments. The authors would like to than Prof. N. Bloembergen, Prof. O. Schnepp and Prof. J.H. Marburger for critical reading of sections of the manuscript, and Prof. S Ezekiel for making Fig.7.31 available prior to publication. We would also like to express our appreciation to Mr. W.C. Wang for the preparation of certain figures and to Lorraine Volsky, Gwendy Romey and the staff of the JILA publication office for their diligent preparation of the manuscript. This research was partially supported by the National Science Foundation.

References

7.1 C.V. Raman: Indian J. Phys. *2*, 387 (1928)
see also G.S. Landsberg, L.I. Mandelstam: Naturwissenschaften *16*, 557 (1928)
7.2 G. Placzek: *Marx Handbuch der Radiologie*, 2nd ed., Vol.VI (Akademische Verlagsgesellschaft, Leipzig 1934) p.20-374
7.3 B.P. Stoicheff: *Methods in Experimental Physics*, Vol.3 (Academic Press, New York 1962) p.11
7.4 M.C. Tobin: *Laser Raman Spectroscopy* (Wiley, New York 1971)
7.5 T.R. Gilson, P.J. Hedra: *Laser Raman Spectroscopy* (Wiley, New York 1970)
7.6 J.A. Köningstein: Phys. Today *30* (No.12), 15 (Dec.1977)
7.7 G. Herzberg: *Molecular Spectra and Molecular Structure*, Vol.I: Spectra of Diatomic Molecules, Vol.II: Infrared and Raman Spectra of Polyatomic Molecules (Van Nostrand, Princeton, N.J. 1945)
7.8 E.J. Woodbury, W.E. Ng: Proc. IRE *50*, 2367 (1962)
7.9 G. Eckhardt, R.W. Hellwarth, F.J. McClung, J.E. Schwartz, D. Weiner, E.J. Woodbury: Phys. Rev. Lett. *9*, 455 (1962)
7.10 J.H. Marburger: In *Progress in Quantum Electronics*, Vol.4, ed. by J.H. Sanders, S. Stenholm (Pergamon Press, Oxford 1975) pp.35-110; also Y.R. Shen, *ibid.*, pp.1-34
7.11 M. Hercher: J. Opt. Soc. Am. *54*, 563A (1964)
7.12 R.W. Terhune: Bull. Am. Phys. Soc. *8*, 359 (1963)
7.13 N. Bloembergen: Am. J. Phys. *35*, 989-1023 (1967)
7.14 Y.R. Shen: In *Light Scattering in Solids*, ed. by M. Cardona, Topics in Applied Physics, Vol.8 (Springer, Berlin, Heidelberg, New York 1975) pp.275-328
7.15 N. Bloembergen, G. Bret, P. Lallemand, A. Pine, P. Simova: IEEE J. *QE-3*, 197 (1967)
7.16 P. Lallemand, P. Simova, G. Bret: Phys. Rev. Lett. *17*, 239 (1966)
7.17 W.T. Jones, B.P. Stoicheff: Phys. Rev. Lett. *13*, 657 (1964)
7.18 N. Bloembergen, Y.R. Shen: Phys. Rev. Lett. *12*, 504 (1964)
7.19 T. Yajima, M. Takatsuji: J. Phys. Soc. Jpn. *19*, 2343 (1964)
7.20 R.W. Terhune, P.D. Maker, C.M. Savage: Phys. Rev. Lett. *8*, 404 (1962)
7.21 R.W. Terhune: Solid State Design *4*, 38 (Nov.1963)
7.22 G. Mayer, F. Gires: Comp. Rend. *25*, 2039 (1964)
7.23 P.D. Maker, R.W. Terhune, C.M. Savage: Phys. Rev. Lett. *12*, 507 (1964)
7.24 W. Kaiser, C.G.B. Garrett: Phys. Rev. Lett. *7*, 229 (1961)
7.25 J.A. Armstrong, N. Bloembergen, J. Ducuing, P.S. Pershan: Phys. Rev. *127*, 1919 (1962)
N. Bloembergen: *Nonlinear Optics* (Benjamin, New York 1965)
7.26 N. Bloembergen: In *Quantum Electronics*, Proc. 3rd Quantum Electronics Conf., Paris, 1963, ed. by N. Bloembergen, P. Grivet (Dunod, Paris 1964) pp.1501-1512; also Proc. IEEE *51*, 124-131 (1963)

7.27 P.D. Maker, R.W. Terhune: Phys. Rev. *137*, A801 (1965)
7.28 J.A. Giordamaine, W. Kaiser: Phys. Rev. *144*, 676 (1966)
7.29 T. Yajima: J. Phys. Soc. Jpn. *21*, 1583 (1966)
7.30 E. Yablonovich, N. Bloembergen, J.J. Wynne: Phys. Rev. B *3*, 2060 (1971)
7.31 B.L. Stansfield, R. Nodwell, J. Meyer: Phys. Rev. Lett. *26*, 1219 (1971)
7.32 J.J. Wynne: Phys. Rev. Lett. *29*, 650 (1972)
7.33 M.D. Levenson, C. Flytzanis, N. Bloembergen: Phys. Rev. B *6*, 3962 (1972)
S. Akhmanov, V.G. Dmitriev, A.I. Kovrigin, N.I. Koroteev, V.G. Tunkin, A.I. Kholddnykh: JETP Lett. *15*, 600 (1972)
7.34 F. Demartini, F. Simoni, E. Santamato: Opt. Commun. *9*, 1976 (1973)
7.35 P.R. Regnier, J.P.E. Taran: Appl. Phys. Lett. *23*, 248 (1973)
7.36 R.F. Begley, A.B. Harvey, R.C. Byer: Appl. Phys. Lett. *25*, 387 (1974)
7.37 A. Owyoung: Opt. Commun. *16*, 266 (1976)
7.38 W. Werncke, J. Klein, A. Lau, K. Lenz, G. Hunsaltz: Opt. Commun. *11*, 159 (1974)
7.39 D. Heiman, R.W. Hellwarth, M.D. Levenson, G. Martin: Phys. Rev. Lett. *36*, 189 (1976)
7.40 G.L. Eesley, M.D. Levenson, W.M. Tolles: IEEE J. *QE-14*, 45 (1978)
7.41 Won B. Roh, P.W. Schreiber, J.P.E. Taran: Appl. Phys. Lett. *29*, 174 (1976)
7.42 H. Lotem, R.T. Lynch, Jr., N. Bloembergen: Phys. Rev. A *14*, 1748 (1976)
7.43 A.B. Harvey (ed.): *Chemical Applications of Nonlinear Spectroscopy* (Academic Press, New York 1979)
7.44 J.S. Druet, J.P.E. Taran: In *Chemical and Biochemical Applications of Lasers*, ed. by B. Moore (Academic Press, New York 1979)
7.45 J.J. Barrett: J. Opt. Soc. Am. *68*, 1433 (1978)
7.46 A.I. Akhiezer, V.B. Berestetskii: *Elements of Quantum Electrodynamics* (Old Bourne Press, London 1962) p.71
7.47 P. Grishkowsky, M.M.T. Loy, P.F. Liao: Phys. Rev. A *12*, 2514 (1975)
7.48 W. Heitler: *The Quantum Theory of Radiation*, 3rd ed. (Oxford University Press, London 1954) p.146
7.49 M.D. Levenson, N. Bloembergen: Phys. Rev. B *10*, 4447 (1974)
7.50 L. Allen, J.H. Eberly: *Optical Resonance and Two-Level Atoms* (Wiley, New York 1975)
7.51 R.P. Feynman, F.L. Vernon, R.W. Hellwarth: J. Appl. Phys. *28*, 49 (1957)
7.52 C. Flytzanis, M. Bloembergen: "Infrared Dispersion of Third Order Susceptibility in Dielectrics," Prog. Quant. Electr. *4*, 271 (1977)
C. Flytzanis: In *Quantum Electronics 1*, ed. by H. Rabin, C.L. Tang (Academic Press, New York 1975)
7.53 S.A.J. Druet, B. Attal, T.K. Gustafson, J.P. Taran: Phys. Rev. A *18*, 1529 (1978)
7.54 N. Bloembergen, H. Lotem, R.T. Lynch, Jr.: Ind. J. Pure Appl. Phys. *16*, 151 (1978)
7.55 J.W. Nibler, G.V. Knighten: In *Raman Spectroscopy of Gases and Liquids*, ed. by A. Weber, Topics in Current Physics, Vol.11 (Springer, Berlin, Heidelberg, New York 1978) p.253
7.56 D.A. Kleinman: Phys. Rev. *126*, 1977 (1962)
7.57 R.W. Hellwarth: "Third-Order Nonlinear Susceptibility of Liquids and Solids," in Prog. Quant. Electr. *5*, 1 (1977)
7.58 J.J. Song, G.L. Eesley, M.D. Levenson: Appl. Phys. Lett. *24*, 567 (1976)
7.59 A.C. Eckbreth: Appl. Phys. Lett. *32*, 421 (1978)
7.60 S. Chandra, A. Compaan, E. Wiener-Avnear: Appl. Phys. Lett. *33*, 867 (1978)
7.61 H. Lotem, R.T. Lynch, Jr.: Phys. Rev. Lett. *37*, 334 (1976)
7.62 J.J. Song, M.D. Levenson: J. Appl. Phys. *48*, 3496 (1977)
7.63 A. Owyoung: Opt. Lett. *2*, 91 (1978)
7.64 L. Galatry: Phys. Rev. *122*, 1218 (1961)
7.65 M.A. Henesian, R.L. Byer: Xth Int. Quant. Electron. Conf. (Atlanta, GA, May 1978) Paper G9
$Z(\xi)$ is defined in B.D. Fried, S.D. Conte: *The Plasma Dispersion Function* (Academic Press, New York 1961)
7.66 P.F. Liao, G.C. Bjorklund: Phys. Rev. A *15*, 2009-2018 (1977)
7.67 P.N. Butcher: "Nonlinear Optical Phenomena," Eng. Bulletin 200 (Ohio State University Press 1965)
7.68 R. Loudon: Proc. Roy. Soc. (London) *A275*, 218 (1963)
7.69 R.T. Lynch, Jr.: "Third-Order Nonlinear Spectroscopy of Liquids and Crystals"; Ph.D. Thesis, Harvard University (1977)

7.70 G.C. Bjorklund: IEEE J. *QE-11*, 287 (1975)
W.M. Shaub, A.B. Harvey, G.C. Bjorklund: J. Chem. Phys. *67*, 2547 (1977)
7.71 G.D. Boyd, J.P. Gordon: Bell Syst. Tech. J. *40*, 489 (1961)
7.72 G.L. Eesley: "Engineering Development of Coherent Raman Spectroscopy"; Ph.D. Thesis, University of Southern California, USCEE Rpt. 502 (1978). To be published in J. Quant. Spectrosc. Radiat. Transfer
7.73 R.H. Stolen: IEEE J. *QE-11*, 100 (1975)
7.74 E.A.J. Marcatili, R.A. Schmeltzer: Bell Syst. Tech. J. *43*, 1743 (1964)
7.75 R.B. Miles, G. Laufer, G.C. Bjorklund: Appl. Phys. Lett. *30*, 319 (1977)
7.76 A.J. Glass, A.H. Gunther: Appl. Opt. *16*, 1214 (1977)
7.77 R. Loudon: "Nonlinear Optics with Polaritons," in *Nonlinear Spectroscopy*, ed. by N. Bloembergen (North-Holland, Amsterdam 1977)
7.78 E. Yablonovich, C. Flytzanis, N. Bloembergen: Phys. Rev. Lett. *29*, 868 (1972)
7.79 S.D. Kramer, N. Bloembergen: Phys. Rev. B *14*, 4654 (1976)
7.80 J.R. Nestor, T.G. Spiro, G.K. Klauminzer: Proc. Nat. Acad. Sci. (Wash.) *73*, 3329 (1976)
7.81 R.T. Lynch, Jr., H. Lotem, N. Bloembergen: J. Chem. Phys. *66*, 4250 (1977)
7.82 R.S. Hudson, H.C. Andersen: *Molecular Spectroscopy*, Vol.5, A Specialist Periodic Report (Burlington House, London 1978) pp.142-201
7.83 L.A. Carreira, L.P. Goss, T.B. Malloy, Jr.: J. Chem. Phys. *69*, 855 (1978); *66*, 4360 (1977)
7.84 L.A. Carreira, T.C. Maguire, T.B. Malloy, Jr.: J. Chem. Phys. *66*, 2621 (1977); *68*, 280 (1978)
7.85 L.A. Carreira: In *Chemical Applications of Nonlinear Spectroscopy*, ed. by A.B. Harvey (Academic Press, New York 1979)
L.A. Carreira, L.P. Goss, T.B. Malloy, Jr.: J. Chem. Phys. *66*, 2762 (1977)
7.86 M. Sargent, M.O. Sculley, W.E. Lamb: *Laser Physics* (Addison Wesley, Reading, MA 1974) p.84
7.87 A. Yariv: IEEE J. *QE-13*, 943 (1977)
7.88 C.J. Bordé, J.L. Hall, C.V. Kunasz, D.G. Hummer: Phys. Rev. A *14*, 236 (1976)
C.J. Bordé: C. R. Acad. Sci. Paris *282B*, 341 (1976); in *Laser Spectroscopy III*, ed. by J.L. Hall, J.L. Carlsten, Springer Series in Optical Sciences, Vol.7 (Springer, Berlin, Heidelberg, New York 1977) pp.21-134
7.89 S.Y. Yee, T.K. Gustafson: Phys. Rev. A *18*, 1597 (1978)
A. Omont, E.W. Smith, J. Cooper: Astrophys. J. *175*, 185 (1972)
7.90 T. Yajima: Opt. Commun. *14*, 378 (1975)
T. Yajima, H. Souma, Y. Ishida: Phys. Rev. A *17*, 324 (1978)
7.91 J.J. Song, J.H. Lee, M.D. Levenson: Phys. Rev. A *17*, 1439 (1978)
7.92 M. Sargent, III: Appl. Phys. *9*, 127 (1976)
7.93 V.S. Letokhov, V.P. Chebotaev: *Nonlinear Laser Spectroscopy*, Vol.4, Springer Series in Optical Sciences (Springer, Berlin, Heidelberg, New York 1977)
7.94 C. Wieman, T.W. Hänsch: Phys. Rev. Lett. *34*, 1120 (1976)
R. Teets, R. Feinberg, T.W. Hänsch, A.L. Schawlow: Phys. Rev. Lett. *37*, 683 (1976)
7.95 T.W. Hänsch, P. Toschek: Z. Phys. *236*, 213 (1978)
7.96 Many of these effects are familiar from the atomic three-level case. In particular, see V.P. Chebotaev: In *High-Resolution Laser Spectroscopy*, ed. by K. Shimoda, Topics in Applied Physics, Vol.13 (Springer, Berlin, Heidelberg, New York 1976) pp.207-255
7.97 S.A. Akhmanov: In *Nonlinear Spectroscopy*, ed. by N. Bloembergen (North-Holland, Amsterdam 1977) pp.217-254
7.98 I. Chabay, G.K. Klauminzer, B.S. Hudson: Appl. Phys. Lett. *28*, 27 (1976)
7.99 A.B. Harvey, J.W. Nibler: Appl. Spect. Rev. *14*, 101 (1978)
7.100 W.M. Tolles, R.D. Turner: Appl. Spectrosc. *31*, 96 (1977)
7.101 T.W. Hänsch: Appl. Opt. *11*, 895 (1972)
7.102 N. Nitsch, W. Kiefer: Opt. Commun. *23*, 240 (1977)
7.103 J.W. Nibler, J.R. McDonald, A.B. Harvey: Opt. Commun. *18*, 371 (1976)
7.104 J.J. Black, T.R. Gilfand, D.A. Greenhalk, L.C. Laycock: Laser Focus, p.84 (March 1978)
7.105 R.L. Byer: "Parametric Oscillators", in *Tunable Lasers and Applications*, ed. by A. Mooradian, T. Jaeger, P. Stokseth, Springer Series in Optical Sciences, Vol.3 (Springer, Berlin, Heidelberg, New York 1976) p.378
S.A. Akhmanov, N.J. Koroteev: Sov. Phys. JETP *40*, 650 (1978)

7.106 M.G. Littman, J.H. Metcalf: Appl. Opt. *17*, 2224 (1978)
M.G. Littman: Opt. Lett. *3*, 138 (1978)
7.107 S.A. Akhmanov, A.F. Bunkin, S.G. Ivanov, N.I. Koroteev, A.I. Kovrigin, I.L. Shumay: In *Tunable Lasers and Applications*, ed. by A. Mooradian, T. Jaeger, P. Stoketh (Springer, Berlin, Heidelberg, New York 1976) p.389, and references therein
7.108 J.L. Oudar, D.S. Chemla, E. Batifol: J. Chem. Phys. *67*, 1626 (1977)
J.L. Oudar: J. Chem. Phys. *67*, 446 (1977)
7.109 P.R. Regnier, F. Moya, J.P.E. Taran: AIAA J. *12*, 826 (1974)
7.110 A. Lau, W. Werncke, J. Klein, M. Pfeiffer: Opt. Commun. *21*, 309 (1977)
7.111 There is some controversy over pressure dependence and linewidths, see W.M. Shaub, S. Lemont, A.B. Harvey: Appl. Spectrosc. (to be published)
W.B. Ron, P.W. Schreiber: Appl. Opt. *17*, 1418 (1978)
7.112 S.A. Akhmanov, N.I. Koroteev, A.I. Kholodnykh: J. Raman Spectrosc. *2*, 239 (1974)
7.113 J.J. Barrett, R.F. Begley: Appl. Phys. Lett. *27*, 129 (1975)
7.114 M.A. Henesian, L. Kulevskii, R.L. Byer: J. Chem. Phys. *65*, 5530 (1976)
7.115 V.I. Fabelinsky, B.B. Krynetsky, L.A. Kulevsky, V.A. Mishin, A.M. Prokhorov, A.D. Savel'ev, V.V. Smirnov: Opt. Commun. *20*, 389 (1977)
B.B. Krynetsky, L.A. Kulevsky, V.A. Mishin, A.M. Prokhorov, A.D. Savel'ev, V.V. Smirnov: Opt. Commun. *21*, 225 (1977)
7.116 J.L. Hall, S.A. Lee: Appl. Phys. Lett. *29*, 367 (1976)
7.117 J.J. Snyder: "Fizeau Wavelength Meter", in *Laser Spectroscopy III*, ed. by J.L. Hall, J.L. Carlsten, Springer Series in Optical Sciences, Vol.7 (Springer, Berlin, Heidelberg, New York 1977) p.419
7.118 R.L. Byer, S.J. Paul, M.D. Duncan: "A Wavelength Meter", in *Laser Spectroscopy III*, ed. by J.L. Hall, J.L. Carlsten, Springer Series in Optical Sciences, Vol.7 (Springer, Berlin, Heidelberg, New York 1977) p.414
7.119 S.A. Akhmanov, N.I. Koroteev: Sov. Phys. USP *20*, 899 (1977)
S.A. Akhmanov, A.F. Bunkin, S.G. Ivanov, N.I. Koroteev: JETP Lett. *25*, 416 (1977); JETP *25*, 444 (1977); JETP *74*, 1272 (1978)
A.F. Bunkin, S.G. Ivanov, N.I. Koroteev: Sov. Phys. Dokl. *22*, 146 (1977)
7.120 S.A. Akhmanov, F.N. Gadjiev, N.I. Koroteev, R.Yu. Orlov, I.L. Shumay: JETP Lett. *27*, 243 (1978)
7.121 M.D. Levenson: "Advanced Techniques of Coherent Raman Spectroscopy", in *Chemical Applications of Nonlinear Optics*, ed. by A.B. Harvey (Academic Press, New York 1979)
7.122 M.D. Levenson, J.J. Song: J. Opt. Soc. Am. *66*, 641 (1976)
7.123 R.E. Joiner, J. Marburger, W.H. Steier: Appl. Phys. Lett. *30*, 485 (1977)
7.124 A. Owyoung: "CW Stimulated Raman Spectroscopy", in *Chemical Applications of Nonlinear Spectroscopy*, ed. by A.B. Harvey (Academic Press, New York 1979)
A. Owyoung: IEEE J. *QE-14*, 192 (1978)
7.125 A. Owyoung, C.W. Patterson, R.S. McDowell: Chem. Phys. Lett. *59*, 156 (1978)
7.126 A. Owyoung, P.S. Peercy: J. Appl. Phys. *48*, 674 (1977)
7.127 A. Lau, W. Werncke, M. Pfeiffer, K. Lenz, H.J. Weigmann: Sov. J. Quant. Electron. *6*, 402 (1976)
7.128 H. Lotem, R.T. Lynch, Jr.: Appl. Phys. Lett. *27*, 344 (1975)
7.129 A. Yariv: *Introduction to Optical Electronics* (Holt, Rinehart and Winston, New York 1971) Chaps.10 and 11
7.130 M.D. Levenson, G.L. Eesley: Appl. Phys. *19*, 1 (1979)
7.131 J.L. Hall, O. Poulsen, S.A. Lee, J.C. Bergquist: Xth Int. Quant. Electron. Conf. (Atlanta, GA, May 29-June 1, 1978) Paper R-5
7.132 A. Hirth, K. Volrath: Opt. Commun. *18*, 213 (1976)
7.133 A.C. Eckbreth: J. Appl. Phys. *48*, 4473 (1977)
7.134 P.R. Regnier, J.P.E. Taran: In *Laser Raman Gas Diagnostics*, ed. by M. Lapp, C.M. Penney (Plenum Press, New York 1974) p.87
7.135 A.B. Harvey: Proc. Sixth Intern. Conf. Raman Spectrosc., Bangalore, India (1978)
7.136 F. Moya, S.A.J. Druet, J.P.E. Taran: Opt. Commun. *13*, 169 (1975)
7.137 F. Moya, S. Druet, M. Pealat, J.P.E. Taran: "Flame Investigation by Coherent Anti-Stokes Raman Scattering", AIAA Paper 76-29 AIAA 14th Aerospace Sciences Meeting, Washington, DC (1976)

7.138 J.P. McDonald, A.P. Baronavski, L. Pasternack, S. Lemont, A.B. Harvey: Proc. 10th Materials Research Symp. on Characterization of High Temperature Vapors and Gases (1978) National Bureau of Standards, Gaithersburg, MD
7.139 W.M. Shaub, J.W. Nibler, A.B. Harvey: J. Chem. Phys. *67*, 1883 (1977)
7.140 M.J. Colles, J.E. Griffiths: J. Chem. Phys. *56*, 3384 (1972)
7.141 Y. Kato, H. Takuma: J. Opt. Soc. Am. *61*, 347 (1971); J. Chem. Phys. *54*, 5398 (1971)
7.142 M.D. Levenson, N. Bloembergen: J. Chem. Phys. *60*, 1323 (1974)
E. Wiener-Avnear, S. Chandra, A. Compaan: Appl. Phys. Lett. *32*, 386 (1978)
7.143 C.C. Wang: Phys. Rev. *152*, 149 (1966)
7.144 M.D. Levenson: IEEE J. *QE-10*, 110 (1974)
7.145 S.D. Kramer, N. Bloembergen: Phys. Rev. B *14*, 4654 (1976)
7.146 G.R. Meredith, R.M. Hochstrasser, H.P. Trommsdorff: In *Advances in Laser Chemistry*, ed. by A.H. Zewail, Springer Series in Chemical Physics, Vol.3 (Springer, Berlin, Heidelberg, New York 1978)
7.147 R.J. Lynch, Jr., S.D. Kramer, H. Lotem, N. Bloembergen: Opt. Commun. *16*, 372 (1976)
7.148 S.A. Lee, J. Helmcke, J.L. Hall, B.P. Stoicheff: Opt. Lett. *3*, 141 (1978) also see J.L. Hall, S.A. Lee: Appl. Phys. Lett. *29*, 367 (1976)
7.149 F. De Martini, G.P. Giuliani, E. Santamato: Opt. Commun. *5*, 126 (1972)
F. De Martini, F. Simoni, E. Santamato: Opt. Commun. *9*, 176 (1973)
7.150 M.A. Henesian, L. Kulevskii, R.L. Byer, R.L. Herbst: Opt. Commun. *18*, 225 (1974)
7.151 M.A. Henesian, M.D. Duncan, R.L. Byer, P.D. May: Opt. Lett. *1*, 149 (1977)
7.152 M.R. Aliev, D.N. Kozlov, V.V. Smirnov: JETP Lett. *26*, 31 (1977)
7.153 J.P. Boquillon, J. Moret-Bailly, R. Chaj: C. R. Acad. Sci. (Paris) *284B*, 205 (1977)
7.154 P.P. Yaney: J. Raman Spectrosc. *5*, 219-242 (1976)
7.155 K.H. Levin, C.L. Tang: Appl. Phys. Lett. *33*, 817 (1978)
7.156 R.F. Begley, A.B. Harvey, R.L. Byer, B.S. Hudson: J. Chem. Phys. *61*, 2466 (1974); also Amer. Laboratory *6*, 11 (1974)
7.157 G.L. Eesley, M.D. Levenson: Opt. Lett. *3*, 178 (1978)
7.158 M.A. Washington, H.Z. Cummins: Phys. Rev. B *15*, 5840 (1972)
R. Tubino, J.L. Birman: Phys. Rev. B *15*, 5843 (1977) and references therein
7.159 B. Hudson, W. Hetherington III, S. Cramer, I. Chabay, G.K. Klauminzer: Proc. Natl. Acad. Sci. *73*, 3298 (1976)
7.160 I. Itzkan, D.A. Leonard: Appl. Phys. Lett. *28*, 106 (1975)
7.161 W. Werncke, A. Lau, M. Pfeiffer, J.H. Weigmann, G. Hansalz, K. Lenz: Opt. Commun. *16*, 128 (1976)
7.162 L.A. Carreira, L.P. Goss: In *Advances in Laser Chemistry* , ed. by A.H. Zewail, Springer Series in Chemical Physics, Vol.3 (Springer, Berlin, Heidelberg, New York 1978)
7.163 S. Ezekiel, R.P. Hackel: To be published
7.164 J.P. Coffinet, F. De Martini: Phys. Rev. Lett. *22*, 60, 752 (1969)
7.165 F. De Martini: In *Nonlinear Spectroscopy*, ed. by N. Bloembergen (North-Holland, Amsterdam 1977) pp.319-349
7.166 F. De Martini, Y.R. Shen: Phys. Rev. Lett. *36*, 216 (1976)
7.167 F. De Martini, G. Giuliani, P. Mataloni, E. Palange, Y.R. Shen: Phys. Rev. Lett. *37*, 440 (1976)
7.168 F. De Martini, M. Colocci, S.F. Kohn, Y.R. Shen: Phys. Rev. Lett. *38*, 1223 (1977)
7.169 S.L. Shapiro, M. McClintock, D.A. Jennings, R.L. Barger: IEEE *QE-2*, 89 (1966)
7.170 A. Koster, S. Biraud-Laval, R. Reinisch: J. Opt. Soc. *68*, 682 (1978)
R. Revinisch, S. Biraud-Laval, N. Paraire: J. Phys. (Paris) *31*, 227 (1976)
7.171 D. Bloom, G.C. Bjorklund, P.F. Liao: J. Opt. Soc. Am. *68*, 1367 (1978)
7.172 A. Laubereau, W. Kaiser: Rev. Mod. Phys. *50*, 607 (1978)
7.173 C.H. Lee, D. Ricard: Appl. Phys. Lett. *32*, 168 (1978)
7.174 J.P. Heritage: Appl. Phys. Lett. *34*, 470 (1979)
J.P. Heritage, C.P. Auschnitt, R.K. Jain: In *Picosecond Phenomena*, ed. by C.V. Shank, E.P. Ippen, S.L. Shapiro, Springer Series in Chemical Physics, Vol.4 (Springer, Berlin, Heidelberg, New York 1978) pp.8-11

Additional References with Titles

For recent references, see
Proceedings of the VIIth International Conference on Raman Spectroscopy, ed. by W.F. Murphys (North-Holland, New York 1980) Chap.VI, "CARS and Other High Order Processes," pp.639-705
J.J. Barrett, M.J. Berry: Photoacoustic Raman spectroscopy using cw laser sources. Appl. Phys. Lett. *34*, 144 (1979)
N. Bolembergen: "Recent Progress in Four-Wave Mixing Spectroscopy", in *Laser Spectroscopy*, ed. by H. Walther, K.W. Rothe, Springer Series in Optical Sciences, Vol.21 (Springer, Berlin, Heidelberg, New York 1979) pp.340-347
J.P. Boquillon, R. Bregier: High-resolution coherent stokes raman spectroscopy of ν_1 and ν_3 bands of methane. Appl. Phys. *18*, 195 (1979)
C.K. Chen, A.R.B. de Castro, Y.B. Shen, R. DeMartini: Surface Coherent anti-Stokes Raman Spectroscopy. Coherent anti-Stokes Raman scattering with counterpropagating laser beams. Phys. Rev. Lett. *43*, 946 (1979)
A. Compaan, S. Chandra: Opt. Lett. *4*, 170 (1979)
S. Druet, J.P. Taran, C.J. Borde: Lineshape and doppler broadening in resonant CARS and related nonlinear processes through a diagrammatic approach. J. de Phys. (Paris) *40*, 819 (1979)
M.D. Duncan, R.L. Byer: Very high resolution CARS spectroscopy in a molecular beam. IEEE J. *QE-15*, 63 (1979)
G.L. Eesley: Coherent raman spectroscopy. J. Quant. Spectrosc. Radiat. Transfer *22*, 507 (1979)
K.P. Goss, D.M. Guthals, J.W. Nibler: Electronic three wave mixing spectra of transient species produced by UV laser photolysis of benzene. J. Chem. Phys. *70*, 4673 (1979)
L.P. Goss, J.W. Fleming, A.B. Harvey: Pure rotational coherent anti-Stokes Raman scattering of simple gases. Opt. Lett. *5*, 345 (1980)
R.J. Hall, J.A. Shirely, A.C. Eckbreth: CARS: Spectra of water vapor in flames. Opt. Lett. *4*, 87 (1979)
A.G. Jacobson, Y.R. Shen: Coherent brillouin spectroscopy. Appl. Phys. Lett. *34*, 464 (1979)
F.M. Kamga, M.G. Sceats: Pulse-sequenced coherent anti-Stokes Raman scattering spectroscopy: a method for suppression of the nonresonant background. Opt. Lett. *5*, 126 (1980)
A. Lau, R. Konig, M. Pfeiffer: Lineshape of resonance CARS as an indicator for the scattering of molecules in ground or excited states. Opt. Commun. *32*, 75 (1980)
G. Laufer, R.B. Miles, D. Santavicca: Angularly resolved coherent Raman spectroscopy (ARCS) in gases. Opt. Comm. *31*, 242 (1979)
R.S. McDowell, C.W. Patterson, A. Owyoung: Quasi-cw inverse Raman spectroscopy of the ν_1 fundamental of $^{13}CH_4$. J. Chem. Phys. *72*, 1071 (1980)
D.V. Murphy, M.B. Long, R.K. Chang, A.C. Eckbreth: Spatially resolved coherent anti-Stokes Raman spectroscopy from a line across a CH_4 jet. Opt. Lett. *4*, 167 (1979)
J.L. Oudar, R.W. Smith, Y.R. Shen: Polarization-sensitive CARS. Appl. Phys. Lett. *34*, 464 (1979)
A. Owyoung: "High Resolution Coherent Raman Spectroscopy of Gases", in *Laser Spectroscopy*, ed. by H. Walther, K.W. Rothe, Springer Series in Optical Sciences, Vol.21 (Springer, Berlin, Heidelberg, New York 1979) pp.175-187
C.K.N. Patel, A.C. Tam: Optoacoustic Raman gain spectroscopy. Appl. Phys. Lett. *34*, 760 (1979)
L.A. Rahn, L.J. Zych, P.L. Mattern: Background-free CARS studies of carbon monoxide in a flame. Opt. Comm. *30*, 249 (1979)
M.A.F. Scarparo, J.J. Song, J.H. Lee, C. Crommer, M.D. Levenson: Improved geometry for polarization-sensitive nonlinear spectroscopy. Appl. Phys. Lett. *35*, 490 (1979)
J.C. Schaefer, I. Chabay: Generation of enchanced coherent anti-Stokes Raman spectroscopy signals in liquid-filled waveguides. Opt. Lett. *4*, 227 (1979)

J.A. Shirley, R.J. Hall, A.C. Eckbreth: Folded BOXCARS for rotational Raman studies. Opt. Lett. *5*, 380 (1980)
J. Tretzel, F.W. Schneider: Resonance CARS spectroscopy of bactereorhodopsin. Chem. Phys. Lett. *66*, 475 (1979)
W. Wernke, H.J. Weigmann, J. Patzold, A. Lau, K. Lenz, M. Pfeiffer: Rapid Raman spectroscopy of the excited electronic state of chrysene by coherent anti-Stokes Raman scattering. Chem. Phys. Lett. *61*, 105 (1979)
G.A. West, D.R. Siebert, J.J. Barrett: Gas phase photoacoustic Raman spectroscopy using pulsed laser excitation. J. Appl. Phys. *51*, 2823 (1980)

Subject Index

Picosecond Phenomena

Proceedings of the First International Conference on Picosecond Phenomena. Hilton Head, South Carolina, USA May 24–26, 1978

Editors: C. V. Shank, E. P. Ippen, S. C. Shapiro

1978. 222 figures, 10 tables. XII, 359 pages (Springer Series in Chemical Physics, Volume 4)

ISBN 3-540-09054-1

Contents:

Raman Spectroscopy

of Gases and Liquids

Editor: A. Weber

1979. 103 figures, 25 tables. XI, 318 pages (Topics in Current Physics, Volume 11)

ISBN 3-540-09036-3

Contents:

I. I. Sobelman, L. A. Vainshtein, E. A. Yukov

Excitation of Atoms and Broadening of Spectral Lines

1980. 34 figures, 40 tables. Approx. 370 pages (Springer Series in Chemical Physics, Volume 7)

ISBN 3-540-09890-9

Contents:

Ultrashort Light Pulses

Picosecond Techniques and Applications

Editor: S. L. Shapiro

1977. 173 figures. XI, 389 pages (Topics in Applied Physics, Volume 18)

ISBN 3-540-08103-8

Contents:

Springer-Verlag
Berlin
Heidelberg
New York